VOLUME THREE HUNDRED AND FIFTY NINE

INTERNATIONAL REVIEW OF CELL AND MOLECULAR BIOLOGY

Pancreatic β-Cell Biology in Health and Disease

INTERNATIONAL REVIEW OF CELL AND MOLECULAR BIOLOGY

VOLUME THREE HUNDRED AND FIFTY NINE

International Review of CELL AND MOLECULAR BIOLOGY

Pancreatic β-Cell Biology in Health and Disease

Edited by

IZORTZE SANTIN
University of the Basque Country, Leioa, Spain

LORENZO GALLUZZI
Weill Cornell Medical College, New York, NY, United States

ELSEVIER

ACADEMIC PRESS
An imprint of Elsevier

Academic Press is an imprint of Elsevier
50 Hampshire Street, 5th Floor, Cambridge, MA 02139, United States
525 B Street, Suite 1650, San Diego, CA 92101, United States
The Boulevard, Langford Lane, Kidlington, Oxford OX5 1GB, United Kingdom
125 London Wall, London, EC2Y 5AS, United Kingdom

First edition 2021

ISBN: 978-0-12-823755-7
ISSN: 1937-6448

For information on all Academic Press publications
visit our website at https://www.elsevier.com/books-and-journals

Publisher: Zoe Kruze
Acquisitions Editor: Ashlie M. Jackman
Developmental Editor: Tara Nadera
Production Project Manager: Denny Mansingh
Cover Designer: Alan Studholme

Typeset by SPi Global, India

Contents

Contributors

Maria Nicol Arroyo
ULB Center for Diabetes Research (UCDR), Université Libre de Bruxelles, Brussels, Belgium. http://www.ucdr.be/

Ignacio Babiloni-Chust
Instituto de Investigación, Desarrollo e Innovación en Biotecnología Sanitaria de Elche (IDiBE), Universitas Miguel Hernández, Elche; CIBER de Diabetes y Enfermedades Metabólicas Asociadas (CIBERDEM), Instituto de Salud Carlos III, Madrid, Spain

Asier Benito-Vicente
Department of Molecular Biophysics, Biofisika Institute (University of Basque Country and Consejo Superior de Investigaciones Científicas (UPV/EHU, CSIC)); Department of Biochemistry and Molecular Biology, University of the Basque Country (UPV/EHU), Leioa, Spain

Flora Brozzi
Department of Fundamental Neurosciences; Department of Biomedical Sciences, University of Lausanne, Lausanne, Switzerland

Reinaldo Sousa Dos Santos
Instituto de Investigación, Desarrollo e Innovación en Biotecnología Sanitaria de Elche (IDiBE), Universitas Miguel Hernández, Elche; CIBER de Diabetes y Enfermedades Metabólicas Asociadas (CIBERDEM), Instituto de Salud Carlos III, Madrid, Spain

Unai Galicia-Garcia
Department of Molecular Biophysics, Biofisika Institute (University of Basque Country and Consejo Superior de Investigaciones Científicas (UPV/EHU, CSIC)); Department of Molecular Biophysics, Fundación Biofísica Bizkaia, Leioa, Spain

Marta García-Arévalo
Brazilian Biosciences National Laboratory (LNBio), Brazilian Center for Research in Energy and Materials (CNPEM), Campinas, SP, Brazil

Minerva Gil-Rivera
Departamento de Fisiología, Genética y Microbiología, Universidad de Alicante, Alicante, Spain

Itziar González-Moro
Department of Biochemistry and Molecular biology, University of the Basque Country, Leioa; Biocruces Bizkaia Health Research Institute, Barakaldo, Spain

Vyron Gorgogietas
ULB Center for Diabetes Research (UCDR), Université Libre de Bruxelles, Brussels, Belgium. http://www.ucdr.be/

Mariana Igoillo-Esteve
ULB Center for Diabetes Research (UCDR), Université Libre de Bruxelles, Brussels, Belgium. http://www.ucdr.be/

Shifa Jebari-Benslaiman
Department of Molecular Biophysics, Biofisika Institute (University of Basque Country and Consejo Superior de Investigaciones Científicas (UPV/EHU, CSIC)); Department of Biochemistry and Molecular Biology, University of the Basque Country (UPV/EHU), Leioa, Spain

Asier Larrea-Sebal
Department of Molecular Biophysics, Biofisika Institute (University of Basque Country and Consejo Superior de Investigaciones Científicas (UPV/EHU, CSIC)); Department of Molecular Biophysics, Fundación Biofísica Bizkaia, Leioa, Spain

Laura Marroqui
Instituto de Investigación, Desarrollo e Innovación en Biotecnología Sanitaria de Elche (IDiBE), Universitas Miguel Hernández, Elche; CIBER de Diabetes y Enfermedades Metabólicas Asociadas (CIBERDEM), Instituto de Salud Carlos III, Madrid, Spain

Cesar Martin
Department of Molecular Biophysics, Biofisika Institute (University of Basque Country and Consejo Superior de Investigaciones Científicas (UPV/EHU, CSIC)); Department of Biochemistry and Molecular Biology, University of the Basque Country (UPV/EHU), Leioa, Spain

Juan Martínez-Pinna
Departamento de Fisiología, Genética y Microbiología, Universidad de Alicante, Alicante; Instituto de Investigación, Desarrollo e Innovación en Biotecnología Sanitaria de Elche (IDiBE), Universidad Miguel Hernández, Elche, Spain

Regla M. Medina-Gali
Centro de Investigación Biomédica en Red de Diabetes y Enfermedades Metabólicas Asociadas (CIBERDEM), Madrid; Instituto de Investigación, Desarrollo e Innovación en Biotecnología Sanitaria de Elche (IDiBE), Universidad Miguel Hernández, Elche, Spain

Beatriz Merino
Instituto de Biología y Genética Molecular (University of Valladolid-CSIC), Valladolid, Spain

Atenea Alexandra Perez-Serna
Instituto de Investigación, Desarrollo e Innovación en Biotecnología Sanitaria de Elche (IDiBE), Universitas Miguel Hernández, Elche; CIBER de Diabetes y Enfermedades Metabólicas Asociadas (CIBERDEM), Instituto de Salud Carlos III, Madrid, Spain

Laura Sanchez Caballero
ULB Center for Diabetes Research (UCDR), Université Libre de Bruxelles, Brussels, Belgium. http://www.ucdr.be/

Izortze Santin
Department of Biochemistry and Molecular biology, University of the Basque Country, Leioa; Biocruces Bizkaia Health Research Institute, Barakaldo; CIBER (Centro de Investigación Biomédica en Red) de Diabetes y Enfermedades Metabólicas Asociadas (CIBERDEM), Instituto de Salud Carlos III, Madrid, Spain

Sergi Soriano
Departamento de Fisiología, Genética y Microbiología, Universidad de Alicante, Alicante; Instituto de Investigación, Desarrollo e Innovación en Biotecnología Sanitaria de Elche (IDiBE), Universidad Miguel Hernández, Elche, Spain

Kepa B. Uribe
Center for Cooperative Research in Biomaterials (CIC biomaGUNE), Basque Research and Technology Alliance (BRTA), Donostia San Sebastián, Spain

CHAPTER ONE

Type I interferons as key players in pancreatic β-cell dysfunction in type 1 diabetes

Laura Marroqui[a,b,*], Atenea Alexandra Perez-Serna[a,b], Ignacio Babiloni-Chust[a,b], and Reinaldo Sousa Dos Santos[a,b,*]

[a]Instituto de Investigación, Desarrollo e Innovación en Biotecnología Sanitaria de Elche (IDiBE), Universitas Miguel Hernández, Elche, Spain
[b]CIBER de Diabetes y Enfermedades Metabólicas Asociadas (CIBERDEM), Instituto de Salud Carlos III, Madrid, Spain
*Corresponding authors: e-mail address: lmarroqui@umh.es; r.sousa@umh.es

Contents

International Review of Cell and Molecular Biology, Volume 359
ISSN 1937-6448
https://doi.org/10.1016/bs.ircmb.2021.02.011

Abstract

Type 1 diabetes (T1D) is a chronic autoimmune disease characterized by pancreatic islet inflammation (insulitis) and specific pancreatic β-cell destruction by an immune attack. Although the precise underlying mechanisms leading to the autoimmune assault remain poorly understood, it is well accepted that insulitis takes place in the context of a conflicting dialogue between pancreatic β-cells and the immune cells. Moreover, both host genetic background (i.e., candidate genes) and environmental factors (e.g., viral infections) contribute to this inadequate dialogue. Accumulating evidence indicates that type I interferons (IFNs), cytokines that are crucial for both innate and adaptive immune responses, act as key links between environmental and genetic risk factors in the development of T1D. This chapter summarizes some relevant pathways involved in β-cell dysfunction and death, and briefly reviews how enteroviral infections and genetic susceptibility can impact insulitis. Moreover, we present the current evidence showing that, in β-cells, type I IFN signaling pathway activation leads to several outcomes, such as long-lasting major histocompatibility complex (MHC) class I hyperexpression, endoplasmic reticulum (ER) stress, epigenetic changes, and induction of posttranscriptional as well as posttranslational modifications. MHC class I overexpression, when combined with ER stress and posttranscriptional/posttranslational modifications, might lead to sustained neoantigen presentation to immune system and β-cell apoptosis. This knowledge supports the concept that type I IFNs are implicated in the early stages of T1D pathogenesis. Finally, we highlight the promising therapeutic avenues for T1D treatment directed at type I IFN signaling pathway.

1. Introduction

Type 1 diabetes (T1D) is a chronic autoimmune disease characterized by pancreatic islet inflammation and specific destruction of pancreatic β-cells by the immune system. As a result, there is little to complete lack of endogenous insulin production. T1D affects millions of people worldwide, accounting for 5–10% of all cases of diabetes and its incidence has been alarmingly increasing in the past 30 years (Atkinson et al., 2014; Patterson et al., 2009).

The precise etiology and pathological mechanisms leading to the autoimmune attack are mostly unclear, even though significant advances to unveil such mechanisms have been made in the recent years. Up to day, the most accepted hypothesis is that an inadequate "dialogue" between the invading immune cells and pancreatic β-cells develops in the context of environmental signals acting upon a risk-enhancing genetic background, leading to islet inflammation (also known as insulitis) and progressive β-cell dysfunction and death (Eizirik et al., 2009, 2020; Mallone and Eizirik, 2020; Oram et al., 2019).

Although the triggers causing the autoimmune response that will culminate in T1D remain to be elucidated, it is known that the crosstalk between pancreatic β-cells and the immune system is determined by multiple combinations of the host genetic susceptibility (i.e., candidate genes) and environmental factors (e.g., viral infections and diet) (DiMeglio et al., 2018; Todd, 2010). Interestingly, the relative number of cases with a strong genetic component (e.g., high-risk HLA haplotypes) has decreased over the years, which suggest that changes in environment and/or lifestyle play a crucial role in the rise of T1D incidence (Ilonen et al., 2019; Rewers and Ludvigsson, 2016).

Type I interferons (IFNs), which are crucial for both innate and adaptive immune responses, may act as key connections between environmental and genetic risk factors in T1D. IFNα induced a coordinated antiviral response mediated by IFN-stimulated genes (ISGs) that protected human islets against coxsackievirus B (CVB) infection (Hultcrantz et al., 2007; Lind et al., 2013). Furthermore, several candidate genes for T1D expressed in β-cells, such *MDA5*, *TYK2*, and *PTPN2*, are implicated in the IFN response (Santin and Eizirik, 2013).

This chapter will overview the current knowledge of the relevant pathways involved in pancreatic β-cell dysfunction and demise, discussing how viral infections and genetic susceptibility impact islet inflammation. Notably, we will give particular focus on the mounting evidence supporting the notion that type I IFNs, especially IFNα, may be implicated in the early stages of T1D pathogenesis. Finally, we will highlight the promising therapeutic avenues for T1D treatment directed at type I IFN production and action.

2. Type I interferons

In 1957, Isaacs and Lindenmann described a soluble molecule with antiviral activity secreted in response to a viral infection (Isaacs and Lindenmann, 1957). This molecule, whose first name was *interfering factor*, is now known as interferon.

Three IFNs families have been described in humans: (i) type I IFN, which comprises at least 13 IFNα subtypes, IFNβ, IFNε, IFNω, and IFNκ; (ii) type II IFN, whose only member is IFNγ; and (iii) type III IFN, which is formed by four IFNλ subtypes (IFNλ1–4) (George et al., 2020).

Given that this chapter aims to discuss the evidence suggesting that type I IFNs may be a key trigger of β-cell dysfunction in T1D, the present section will focus on type I IFN biology, especially IFNα, and how these cytokines may be involved in different autoimmune diseases.

2.1 Induction of IFN synthesis

Nearly every nucleated cell type can produce type I IFNs. However, in the course of a viral infection, specialized immune cells known as plasmacytoid dendritic cells (pDCs) are responsible for the production of large quantities of IFNα (Liu, 2005; Siegal et al., 1999).

IFNs are rapidly produced in response to microbial products (e.g., viral nucleic acids or bacterial lipopolysaccharide) or endogenous ligands released from stressed/damaged cells (e.g., fragments of nucleic acids). These molecules are recognized by pattern-recognition receptors (PRRs), such as Toll like receptors (TLR), retinoic acid-inducible gene I (RIG-I), and melanoma differentiation-associated protein 5 (MDA5) (Jensen and Thomsen, 2012; Pichlmair and Reis e Sousa, 2007).

Upon nucleic acid detection, TLR3, TLR7, TLR8, and TLR9 induce type I IFN production through two different pathways. TLR7, TLR8 and TLR9 require the interaction with the myeloid differentiation primary response protein 88 (MYD88), which, in turn, can either directly induce type I IFN production, or interact with proteins from the interleukin-1 receptor-associated kinase (IRAK) family. The protein complex formed by MYD88 and IRAK1 recruit the IFN regulatory factor 7 (IRF7) (Tan et al., 2018). Once phosphorylated by IRAK1, IRF7 translocation into the nucleus promotes the expression of IFNα in plasmacytoid dendritic cells (Zhou et al., 2017). MYD88 activation also recruits TNF receptor-associated factor (TRAF) 6 (TRAF6) and transforming growth factor-β-activated kinase 1 (TAK1) proteins, thus inducing the NF-κB pathway and promoting the expression of proinflammatory cytokines (Uematsu and Akira, 2007).

TLR3 signaling is Toll/IL-1R domain-containing adaptor-inducing IFNβ (TRIF)-dependent. Once recruited, TRIF and TRAF3 activate

TANK-binding kinase 1 (TBK1) and IκB kinase ε (IKKε) signaling proteins, which promote type I IFN expression via IRF3 and IRF7. Like MYD88, TRIF activates NF-κB pathway, promoting the expression of proinflammatory cytokines (Tan et al., 2018).

RIG-I and MDA5, members of the RLR family, recognize double-stranded RNA (dsRNA), a key intermediary of single-stranded RNA viruses' life cycle, in the cytosol (Yong and Luo, 2018). Through their CARD domain, RIG-I and MDA5 interact with the mitochondrial antiviral signaling adaptor (MAVS), which is anchored to the outer mitochondrial membrane (Loo and Gale, 2012). This interaction leads to MAVS oligomerization, which is essential for the recruitment of TRAF2, 3, 5 and 6. Once recruited, TRAF proteins can induce type I IFNs expression either via TBK1 and IKKε activation, or through NF-κB pathway (Refolo et al., 2020).

2.2 Type I IFN signaling pathway

Type I IFNs bind to a specific cell surface, heterodimeric receptor, namely IFNα receptor (IFNAR), which is composed of two transmembrane subunits, IFNAR1 and IFNAR2. As IFNAR does not present intrinsic kinase activity, each subunit relies on the interaction with cytoplasmic tyrosine kinases belonging to the Janus kinase (JAK) family to phosphorylate downstream effectors. IFNAR1 constitutively pre-associates with tyrosine kinase 2 (TYK2), whereas IFNAR2 is associated with JAK1 (Platanias, 2005). Of note, the interaction between IFNAR1 and TYK2 is crucial to preserve IFNAR1 expression at the cell membrane (Ragimbeau et al., 2003).

In the classical signaling cascade, binding of type I IFNs to IFNAR results in the rearrangement and dimerization of IFNAR1 and IFNAR2 subunits, which is followed by autophosphorylation and transphosphorylation, and activation of the receptor-associated JAKs (Platanias, 2005). In turn, JAK1 and TYK2 activation promotes the phosphorylation and activation of the transcription factors signal transducer and activator of transcription 1 (STAT1), STAT2, STAT3, and STAT5 (Levy and Darnell, 2002). Once activated, phosphorylated STAT1 and STAT2 dimerize and associate with IRF9 to form a transcriptional trimolecular complex named IFN-stimulated gene factor 3 (ISGF3). ISGF3 is imported to the nucleus, where it binds to specific elements known as IFN-stimulated regulatory element (ISRE), which are in the promoter of certain IFN-stimulated genes (ISGs). Thus, ISGF3 binding to ISRE initiates the transcription of hundreds of ISGs (Schoggins, 2019).

More than 300 genes are induced by type I IFNs, many of which implicated in the antiviral response (de Veer et al., 2001; Schoggins et al., 2011). ISGs act at different stages of the viral life cycle. CH25H (Cholesterol-25-hidroxylase) and IFITM (IFN-induced transmembrane) families are involved in the inhibition of viral entry, while viral uncoating can be inhibited by MX1 (MX dynamin-like GTPase 1). Once the viral genetic material is released, ADAR1 (Adenosine deaminase RNA-specific 1) inhibits its transcription, while several ISGs, such as PKR (Protein kinase R), IFIT (IFN-induced proteins with tetratricopeptide repeats) family or ZAP (Zinc-finger antiviral protein) block the translation of viral genes. Finally, the assembly and release of the virion are also targeted by ISGs: ISG15 (Interferon stimulated gene 15) and Viperin block the viral assembly, while Viperin and Tetherin inhibit the viral release (Schneider et al., 2014; Schoggins, 2019; Schoggins et al., 2011) (Fig. 1).

Besides their role in the antiviral response, type I IFNs also play a myriad of immune system-related functions, participating in both innate and adaptive immune response (González-Navajas et al., 2012). For example, type I IFNs enhance antigen processing and presentation by major histocompatibility complex (MHC) class I and II molecules (Gessani et al., 2014; Keskinen et al., 1997), and are crucial for the differentiation of both $CD4^+$ and $CD8^+$ T cells (Huber and Farrar, 2011; Rogge et al., 1998; Strengell et al., 2004).

2.3 Negative regulation of type I IFN signaling

A fine-tuned, balanced activation of the type I IFN signaling cascade is crucial to avoid persistent IFN production and signaling, which could lead to pathological disorders known as interferonopathies (e.g., some autoimmune diseases). To keep this signaling under such tight control, several regulatory mechanisms regulate IFN production and response at multiple levels of the type I IFN pathway. Some of these mechanisms are cell intrinsic (e.g., receptor endocytosis and protein inhibitors of activated STAT), whereas others depend on the activity of ISGs (e.g., suppressor of cytokine signaling, or SOCS, and USP18) (Larner et al., 1986; Schneider et al., 2014). As this topic has been previously reviewed elsewhere (Arimoto et al., 2018; Coccia et al., 2006), here we will briefly discuss a cell intrinsic and an ISG-dependent mechanism, namely phosphatases and USP18, respectively.

2.3.1 Phosphatases

As the key point of the type I IFN signaling pathway is the tyrosine phosphorylation of downstream components upon IFN binding to its receptor, it

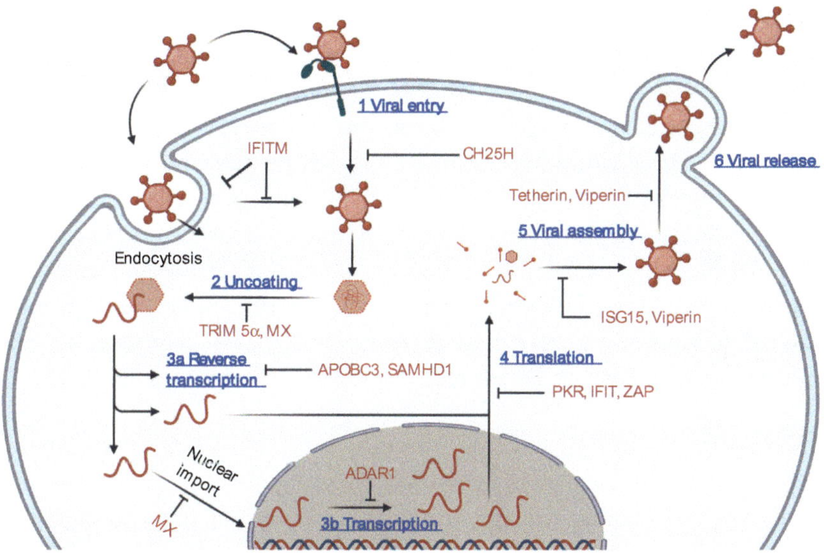

Fig. 1 Targets of type I IFN ISGs during a viral infection. Type I IFN ISGs can interfere with several steps of a viral infection, (1) viral entry is inhibited by CH25H (Cholesterol-25-hidroxylase) and IFITM (IFN-induced transmembrane) proteins; (2) viral uncoating is blocked by TRIM5α (Tripartite motif protein 5α) and MX (MX dynamin-like GTPase) proteins; (3) at the transcription level, two main pathways can be altered: (3a) reverse transcription is blocked by APOBEC3 (Apolipoprotein B mRNA editing enzyme, catalytic polypeptide-like 3) or SAMHD1 (SAM domain and HD domain-containing protein 1), while (3b) import to the nucleus is inhibited by MX, and the viral transcription is blocked by ADAR1 (Adenosine deaminase RNA-specific 1). Once the genetic material is transcribed, the following step is the protein translation (4), which can be prevented by several ISGs, such as PKR (Protein kinase R), IFIT (IFN-induced proteins with tetratricopeptide repeats) family or ZAP (Zinc-finger antiviral protein); (5) When the viral proteins are synthesized, they must be assembled, Interferon stimulated gene 15 (ISG15) and viperin are repressing this step. (6) Finally, the viral release is blocked by viperin and tetherin.

seems logical that this process is tightly regulated by phosphotyrosine phosphatases. JAK/STAT signaling can be regulated by 6 phosphotyrosine phosphatases, namely cluster of differentiation 45 (CD45), protein tyrosine phosphatase, receptor type (PTP-RT), protein tyrosine phosphatase 1B (PTP1B), T cell protein tyrosine phosphatase (TC-PTP), and Src homology region 2 domain-containing phosphatase 1 (SHP1) and 2 (SHP2) (Morris et al., 2018).

Protein tyrosine phosphatases (PTPs) are of particular interest for this chapter as genetic polymorphisms for two PTPs, non-receptor PTP 2 (PTPN2) and PTPN22, are associated with altered risk of T1D (see Section 5).

PTPN family is formed by several members involved in the regulation of ligand-mediated receptor signal or in cell cycle events. PTPN1, also called PTP1B, dephosphorylates TYK2 and JAK2, thus blocking downstream IFN signaling (Myers et al., 2001). PTP1B can also reduce type I IFN expression by decreasing TLR signaling (Xu et al., 2008). JAK proteins and STATs are also important targets of different PTPNs. PTPN2, also named TC-PTP, modulates IFN signal by dephosphorylating JAK1 and STATs in different cell types, such as β-cells (Moore et al., 2009), neurons (Loh et al., 2011), and hepatocytes (Gurzov et al., 2014). PTPN22 expression, on the other hand, is mostly restricted to immune cells, where it regulates signaling triggered by either T cell receptor (TCR) or B-cell receptor (BCR) (Fousteri et al., 2013).

2.3.2 USP18

USP18, previously known as UBP43, is a member of the ubiquitin-specific peptidases (USP) family strongly upregulated by type I IFNs (Basters et al., 2014). USP18 has two main mechanisms of action. In the first one, USP18 peptidase activity is crucial for correct removal of ISG15 from conjugated proteins that underwent ISGylation, an ubiquitination-like process (Malakhov et al., 2002). In the second mechanism, which is independent of its protease activity, USP18 decreases type I IFN signaling by directly interacting with the cytoplasmic region of IFNAR2 at its Box 1 and Box 2 domains (Villarroya-Beltri et al., 2017). As these domains constitute the JAK1 binding region, USP18 interaction establishes an inhibitory competition with JAK1 for IFNAR2 (Basters et al., 2018). It has also been suggested that USP18 inhibitory effects could be due to alterations in the stability of the receptor or receptor-ligand affinity, and that STAT2 participates as an adaptor molecule, promoting USP18 recruitment (Arimoto et al., 2017).

2.4 Type I IFNs and autoimmunity

As mentioned before, type I IFNs are one of the first mediators in the antiviral response, and their production and downstream pathways must be regulated in a very precise manner. Type I IFN pathway dysregulation can cause complex autoimmune diseases, whose etiology involves inflammation and loss of self-tolerance. This might be due to the implication of type I IFNs in antigen presentation, lymphocyte response, and chemokine secretion (Ivashkiv and Donlin, 2014).

Dysregulation of type I IFNs exposure can alter the functionality of several immune cells, promoting a pathological scenario. In B-cells physiology, type I IFNs induce the release of cytotoxic and neutralizing antibodies. However, a persistent exposure to type I IFNs induces B-cell switch and production of autoantibodies, a hallmark of many autoimmune diseases (Kiefer et al., 2012). Type I IFNs regulate biological processes in nearly all T cell populations: maturation of $CD4^+$ cells, TH17 differentiation, Th1 responses, and proliferation and expansion of $CD8^+$ cytotoxic T lymphocytes (Di Domizio and Cao, 2014). As for B-cells, a chronic or persistent exposure to type I IFN can lead to infiltration of autoreactive T lymphocytes in target tissues, organ damage, and loss of self-tolerance (i.e., autoimmunity) (Crow et al., 2019).

A profusely described autoimmune disease is systemic lupus erythematosus (SLE). SLE is a multiple organ pathology triggered by the loss of self-tolerance and subsequent autoantibodies production. Sensing of nucleic-acids--containing immune complexes induces pDCs activation and, consequently, an exacerbated increase in type I IFN production. Autoreactive monocytes and T cells, as well as a characteristic IFN signature, can be detected in the blood of SLE patients (Lee-Kirsch, 2017).

Despite the differences in pathogenesis among autoimmune diseases, all of them share an "IFN signature," referring to a characteristic dysregulation of IFN-regulated genes (Jean-Baptiste et al., 2017). The development of transcriptomic techniques made possible the study of the genetic profile. In 2003, Baechler et al. provided the first evidence of an IFN signature in an autoimmune disease (Baechler et al., 2003). Using global gene expression profiling to search for patterns of gene expression in peripheral blood mononuclear cells (PBMCs), the authors found that nearly half of the patients with severe SLE exhibited dysregulated expression of ISGs. Notably, this IFN signature could be used as a marker of disease severity (Baechler et al., 2003). Following this and other initial studies in SLE (Han et al., 2003; Peterson et al., 2004), many other reports have used a similar strategy to identify IFN signatures in other autoimmune diseases, such as rheumatoid arthritis and T1D (Ferreira et al., 2014; Seyhan et al., 2020).

Taken together, type I IFN signaling pathway is fundamental for the innate immune response (especially against viral infections) and its dysregulation is claimed to trigger autoimmunity. In fact, a growing body of evidence suggests that type I IFNs are driving factors in the development of T1D, which will be discussed along this chapter.

3. Pancreatic β-cell dysfunction in T1D

Endocrine pancreas is organized in secretory micro-organs that ensure glucose homeostasis through hormone secretion named islets. These structures, scattered throughout the exocrine pancreas, shelter five subsets of endocrine cells among which the most important are the insulin-producing β-cells (which make up to half of the cell content in human islets) and the glucagon-secreting α-cells (35–40% of cells in human islets) (Cabrera et al., 2006).

3.1 T1D etiology

Clinical diabetes onset occurs when hyperglycemia, or elevated glucose concentration in blood plasma, is detected. This key symptom of diabetes is also the final consequence of the loss of operational β-cells. In T1D, clinical onset manifests when 80% of β-cell mass has already been irreversibly destroyed (Klöppel et al., 1985). It is suggested that β-cell loss is a progressive and slow process that occurs over years of local islet inflammation (insulitis), rendered by an autoreactive immune system (Cnop et al., 2005).

The heterogeneity of islet infiltration by the immune system within the same pancreas (Campbell-Thompson et al., 2016), the β-cell necrosis scarcity in islets from T1D donors, and the existence of detectable autoantibodies in the serum of asymptomatic patients (Winter and Schatz, 2011) concord with the β-cell dysfunction preceding a slow immune infiltration.

Pancreatic β-cells are not a passive agent in their own fate. In fact, growing evidence shows the direct, active pathogenic role played by β-cells in T1D pathogenesis. For instance, genetic variants related to diabetes susceptibility have been associated to several loci, many of which are expressed in both immune cells and β-cells (Eizirik et al., 2012; Ramos-Rodríguez et al., 2019; Størling and Pociot, 2017). This discovery suggests a genetically modulated crosstalk between β-cells and immune system, which is impaired in T1D by gain or loss of function of these loci, supporting the active role of β-cells in their own demise (Santin and Eizirik, 2013; Størling and Pociot, 2017).

Mallone and Eizirik recently discussed three "weaknesses" in β-cell biology that may render β-cell susceptibility to an autoimmune assault: (1) "Making insulin and other granule proteins is a stressful job"; (2) "Islets are highly vascularised, which favours face-to-face encounters between immune cells and beta cells"; and (3) "Beta cells secrete their products into the bloodstream—The 'Wi-fi' dialogue with immune cells?" (Mallone and Eizirik, 2020).

Alterations in expression of genes related to endoplasmic reticulum (ER) stress relief have also been found both in β-cells and in islets from T1D patients (Engin et al., 2013; Marhfour et al., 2012) as well as in the quintessential murine model for T1D research, the nonobese diabetic (NOD) mouse (Engin et al., 2013; Song et al., 2008; Tersey et al., 2012). Another proof of β-cells implication in their own fate roots in their secretory behavior. β-Cell secretion of vascular endothelial growth factor-A (VEGF-A) contributes to the macrophage islet infiltration and proinflammatory cytokines production (Agudo et al., 2012). VEGF-A secretion is fundamental for ensuring islet vascularization (Brissova et al., 2006), which facilitates immune infiltration in pancreatic islets. Interestingly, an early abnormal microvascularity has been found in T1D patients, detectable even prior to clinical presentation (Khan et al., 2000). Moreover, during the progression of T1D, both in humans and in NOD mice, a subset of β-cells develops a senescence-associated secretory phenotype. This phenotype is characteristic of dying β-cells and is highly attractive to the immune system, which favors islet infiltration (Thompson et al., 2019).

Both clinical and preclinical studies suggest that declining β-cell function precedes T1D, following at least three metabolic stages of declining β-cell function, detectable 4–6 years prior to T1D onset (Sims et al., 2020). Failure of immune tolerance toward β-cells is a critical component of T1D pathogenesis, but it is not the initial feature nor is determinant for disease development. β-Cell autoantibodies are detectable years prior to clinical diabetes diagnosis (Winter and Schatz, 2011), while a portion of autoantibody-positive relatives of T1D individuals never progress to T1D clinical onset (Kulmala et al., 1998).

These findings suggest that β-cell dysfunction and stress may start years before insulitis, and that the canonical role of autoreactive T cells infiltration and consequent reduction in β-cell mass is a symptom more than a conductor of the T1D progression.

3.2 Pathways to β-cell dysfunction in T1D

3.2.1 Autoimmune-driven β-cell death

Loss of immune regulation is characterized by the appearance of autoantibodies directed against β-cell antigens, secreted by B lymphocytes (Fig. 2). The major β-cell autoantigens are (pro)insulin, the 65 kDa isoform of L-glutamic acid decarboxylase (GAD65), the zinc transporter 8 (ZnT8), the protein tyrosine phosphatase related islet antigen 2 molecule (IA2), and the islet-cell autoantigen 69 (ICA69). These proteins are presented at the β-cell surface through the MHC class I proteins and generate a strong immunogenic response (Arvan et al., 2012).

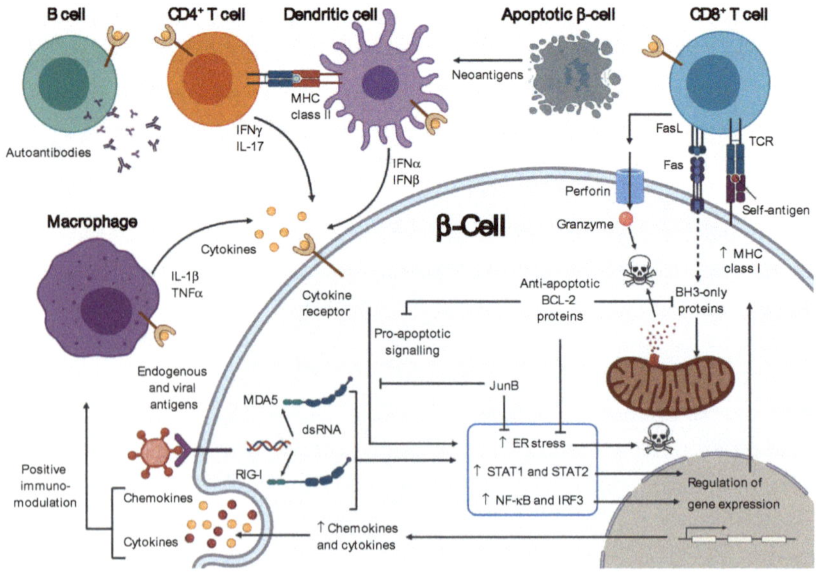

Fig. 2 Environmental triggers and a misguided dialogue between β-cells and the immune system lead to β-cell dysfunction and apoptosis in T1D. During insulitis, proinflammatory cytokines and chemokines are secreted and detected by both immune cells and β-cells. These molecules mediate the aberrant crosstalk between β-cells and the immune system during T1D pathogenesis, leading to positive immunomodulation and β-cell dysfunction and death. Activation of cytokine receptors in β-cells induces pro-apoptotic pathways that cause a deleterious increase in ER stress and activation of transcription factors (e.g., STAT1 and STAT2, NF-κB, and IRF3). These transcription factors upregulate β-cell expression and secretion of chemokines and cytokines, as well as the MHC class I upregulation. Enhanced self-antigen presentation to CD8$^+$ cytotoxic T cells causes β-cell apoptosis through Fas/FasL and granzyme/perforin pathways, as well as modulation of pro-apoptotic BH3-only proteins and caspase activation. Pro-survival proteins, such as JunB and anti-apoptotic BCL-2 proteins, mitigate cytokine-induced pro-apoptotic signals and ER stress, decreasing immune-mediated damage in β-cells. Once deceased, β-cells also contribute to insulitis by providing antigen-presenting cells with (neo)antigens and worsening the progressive loss of β-cell mass. *Adapted from Eizirik, D.L., Colli, M.L., 2020. Revisiting the role of inflammation in the loss of pancreatic β-cells in T1DM. Nat. Rev. Endocrinol. 16, 611–612. https://doi.org/10.1038/s41574-020-00409-6.*

3.2.1.1 Granzyme/perforin and Fas/FasL-induced β-cell death

CD8$^+$ cytotoxic T lymphocytes induce β-cell death through two main pathways. The Fas (CD95) pathway requires direct interaction between the Fas receptor at the β-cell surface and its ligand FasL (also known as CD178), located at the surface of T cells (Rouvier et al., 1993). Fas activation leads to its trimerization and interaction with the Fas-associated death domain

(FADD). FADD recruits pro-caspase 8, which is activated by autocleavage. Once activated, caspase 8 cleaves effector caspase 3 or recruits the intrinsic pathway of apoptosis by activating the pro-apoptotic BCL-2 protein BID.

The granzyme/perforin pathway consists of secretion of both molecules by $CD8^+$ T cells. Once perforin lands on the extracellular membrane of β-cells, it induces pore formation in a Ca^{2+}-dependent manner (Traore et al., 2013), enabling access of granzyme to the cell (Browne et al., 1999). This serin protease then cleaves and activates pro-apoptotic targets: the BH3-only pro-apoptotic protein BID and the effector caspase 3 (Estella et al., 2006; Yang et al., 1998).

3.2.1.2 PRR induction of type I IFN-driven pathways

Innate immunity starts when exogenous or endogenous ligands bind to the PRRs of antigen-presenting cells (APCs), such as pDCs and macrophages.

An increased number of pDCs has been detected in blood of subjects at-risk for T1D as well as recent-onset and established T1D patients (Allen et al., 2009; Xia et al., 2014). Notably, pDCs are responsible for almost all IFNα produced and secreted by PBMCs from T1D individuals, in response to a viral infection (Xia et al., 2014). In young NOD mice, Diana et al. demonstrated that pDCs infiltrate the pancreatic islets, where they produce IFNα, and that they are required for the initiation of the diabetogenic T cell response (Diana et al., 2013).

A crucial danger signal linked to T1D is dsRNA (Blum and Tse, 2020; Colli et al., 2010). Exposure to a mimic of dsRNA, namely polyinosinic-polycitidilic acid or poly(I:C), activates different PRRs in β-cells. Detection of poly(I:C) by TLR3 activates NF-κB pathway and STAT1-mediated responses in primary pancreatic β-cells (Rasschaert et al., 2005), while the IRF3 pathway is exclusively activated by intracellular dsRNA sensors, such as RIG-1 and MDA5 (Colli et al., 2010; Dogusan et al., 2008; García et al., 2009) (Fig. 2). As a result, pancreatic β-cells produce and secrete the immuno-attracting cytokines IFNα and IFNβ (Dogusan et al., 2008; Dos Santos et al., 2019; Marroqui et al., 2015a).

3.2.1.3 Dialogue between β-cells and immune system

It is increasingly recognized that a crosstalk between immune system and β-cells plays a key role in T1D pathogenesis. This dialogue occurs through chemokines, cytokines, and immunogenic peptides produced by apoptotic or stressed β-cells (Blachère et al., 2005; Eizirik and Colli, 2020). Briefly, upon recognition of danger signals, β-cells secrete numerous cytokines and chemokines that will recruit T lymphocytes and macrophages into the islet environment. In turn, activated macrophages, T lymphocytes and natural killer (NK) cells secrete the proinflammatory cytokines IL-1β, IFNγ and TNFα, which will bind to their respective receptors in β-cells and activate several intracellular signaling pathways (Fig. 2). Some of these cytokine-induced pathways will be discussed in more details below.

3.2.1.4 Cytokine-induced β-cell death

Proinflammatory cytokines, such as IL-1β, IFNγ, and TNFα, induce β-cell death through activation of diverse pathways, preferentially through the intrinsic pathway of apoptosis. Although these proinflammatory cytokines are unable to induce a deleterious effect in β-cells by themselves (Gysemans et al., 2008), their synergistic combination has a strong pro-apoptotic effect (Gurzov and Eizirik, 2011).

IL-1β, secreted by macrophages, NK cells, monocytes, and neutrophils, binds to interleukin-1 receptor and induces activation of NF-κB, the mitogen-activated protein kinases (MAPK) p38 and c-Jun N-terminal kinase (JNK), and protein kinase C (PKC) (Pirot et al., 2008). NF-κB induces the expression of the inducible nitric oxide synthase (iNOS), thus increasing nitric oxide production in β-cells (Flodström et al., 1996), which might contribute to exacerbated oxidative stress due to reactive nitrogen species production (Meares et al., 2013). It is noteworthy to mention that the effects of nitric oxide in β-cell demise are species-specific: nitric oxide plays a key role in cytokine-induced apoptosis in rat islets, but not in mouse or human islets (Brozzi et al., 2015; Chan et al., 2011; Delaney et al., 1997; Eizirik et al., 1994).

IFNγ is secreted by many immune cells, including cytotoxic T lymphocytes, T helper type 1 cells, macrophages, and NK cells. The main effects of IFNγ signaling are the activation of STAT1 and the induction of the extracellular signal-regulated kinase (ERK). Signal transduction through the IFNγ receptor (IFNGR) activates the JAK-STAT1 pathway, which regulates the expression of genes with IFN-gamma-activated sequence (GAS) elements. Most of these genes, such as *Nos2* (Darville and Eizirik, 1998; Yamada et al., 1996) and *Fas* (Yamada et al., 1996), participate in β-cell demise.

TNFα is mainly produced and secreted by activated macrophages, T cells and NK cells. Upon binding to its receptor (TNF receptor 1, TNF-R1), TNFα exerts its effect through activation of NF-κB, JNK, ERK and p38 (Sabio and Davis, 2014), and culminates with induction of cell death.

3.2.2 ER stress and pancreatic β-cells

Insulin represents half of the total protein production by β-cells (Schuit et al., 1988), and its secretion requires fast, fluctuating synthesis following a meal ingestion. Therefore, the role of β-cells in glucose homeostasis imposes a great burden to their ER due to the high requirements of proinsulin synthesis and folding. To cope with this constant situation and prevent ER stress, β-cells trigger an adaptive response known as unfolded protein response (UPR) (Eizirik et al., 2008).

The UPR response is mediated by three ER transmembrane sensors: Inositol-requiring transmembrane kinase/endoribonuclease 1α (IRE1α), activating transcription factor 6 (ATF6) and double-stranded RNA-activated kinase (PKR)-like ER kinase (PERK). The UPR aims to restore ER homeostasis and prevent cell death by, for instance, boosting the folding capacity, increasing the synthesis of ER chaperones (e.g., BIP), and inducing degradation of misfolded protein (Eizirik et al., 2008). However, under persistent or overwhelming ER stress, UPR fails to restore ER homeostasis and induces the intrinsic pathway of cell death (Eizirik et al., 2013). Extensive evidence supports that pancreatic β-cells undergo ER stress when exposed to different stimuli, such as proinflammatory cytokines (Brozzi et al., 2015), poly(I:C) (Colli et al., 2010), or a viral infection (Colli et al., 2019; Marroqui et al., 2015b).

In early stages of T1D, pancreatic β-cells recruit adaptive mechanisms to protect ER homeostasis in response to activation of cytokine-induced inflammatory pathways (Tersey et al., 2012). The unfolded protein response (UPR) reduces ER stress by slowing the general protein translation rate, increasing the production of ER chaperones, and engaging the ER associated degradation pathway (ERAD) to degrade newly synthetized proteins through ubiquitination and processing by the proteasome.

Persistent ER stress causes IRE1α to form a multiprotein complex with TRAF-2 and the apoptosis signal-regulating kinase-1 (ASK1) that then phosphorylates JNK, promoting β-cell apoptosis (Urano et al., 2000). Brozzi et al. propose that IRE1α-dependent JNK activation is fundamental for cytokine-induced β-cell death in rat and human (Brozzi et al., 2014, 2015). Simultaneously, the pro-survival physiological target of IRE1α, the spliced form of the transcription factor X-box binding protein 1 (XBP1s), gradually decreases its presence, favoring the pro-apoptotic program (Brozzi and Eizirik, 2016). Finally, prolonged activation of the PERK pathway induces the transcription of the C/EBP homologous protein (CHOP). CHOP inhibits the expression of the anti-apoptotic B-cell lymphoma 2 (BCL-2) proteins, thus favoring apoptosis (Brozzi and Eizirik, 2016).

3.2.3 BCL-2 family and the mitochondrial pathway of apoptosis

Pathways initiated by cytokines share an ending point in the intrinsic pathway of β-cell death: the inhibition of anti-apoptotic members of the BCL-2 family and consequent activation of the mitochondrial pathway of apoptosis (Gurzov and Eizirik, 2011).

Proteins belonging to the BCL-2 family play a crucial role in the intrinsic pathway of cell death. BCL-2 family consists of three groups of proteins, according to their function and domains: anti-apoptotic members are BCL-2, BCL-XL, MCL-1, BCL-W and A1. These proteins counteract pro-apoptotic members (BAX, BAK and BOK). Finally, BH3-only proteins can be subdivided in BH3-only sensitizers (DP5, BAD, BIK, BNIP3, BMF and NOXA) and BH3-only activators (PUMA, BIM and t-BID) (Youle and Strasser, 2008). In the presence of pro-apoptotic signals, the balance between pro-apoptotic and anti-apoptotic BCL-2 proteins determines whether the cell will survive or will continue toward the mitochondrial pathway of apoptosis. This outcome greatly depends on the duration of the pro-apoptotic signal (Grunnet et al., 2009).

BH3-only sensitizers detect pro-apoptotic signals through dephosphorylation (e.g., BAD) or transcriptionally (e.g., DP5, NOXA and BIK). Active BH3-only sensitizers rescue BH3-only activators (e.g., PUMA, BIM and t-BID) by competitively inhibiting anti-apoptotic proteins (BCL-2, BCL-XL, BCL-W, MCL-1 and A-1) in a highly selective protein-protein interaction fashion. Release of activators permits their interaction with pro-apoptotic BCL-2 proteins BAX, BAK and BOK. Pro-apoptotic proteins activation enables their homo-oligomerization, thus forming a pore in the outer mitochondrial membrane that leaks matrix proteins such as cytochrome *c*. Once in the cytosol, cytochrome *c* interacts with the apoptosis protease-activating factor-1 (APAF-1), generating the apoptosome, a protein complex able to recruit and cleave activator caspase 9, which, in turn, processes effector caspases 3 and 7, and initiates apoptosis (Gurzov and Eizirik, 2011; Kalkavan and Green, 2018).

Notably, β-cells express surprisingly low levels of the anti-apoptotic protein BCL-XL in comparison with pancreatic α-cells. This could shed some light over the mechanism by which α-cells survive the pro-apoptotic milieu of insulitis (Marroqui et al., 2015c).

In conclusion, β-cell death in T1D pathogenesis occurs because of many agents. While the role of the autoreactive immune system is still crucial for the β-cell demise, it has been proposed a protagonist role for pancreatic β-cells in their own death. In this disease, both the immune and β-cell responses against danger signals seem to be altered: cytokines and chemokines secreted by both the invading immune system and β-cells may contribute to the vicious cycle of β-cell stress and apoptosis that leads to aberrant antigen presentation by antigen-presenting cells that culminates with activation of autoreactive T lymphocytes.

4. Viruses as triggers of IFNα response in β-cells

Environmental cues are involved as triggers and potentiators of β-cell destruction during T1D pathogenesis. Studies in monozygotic twins determine that only 50% of the siblings with T1D (average value among several studies) develop the disease, and this percentage is age-dependent (Barnett et al., 1981; Jerram and Leslie, 2017; Redondo et al., 2001; Salvetti et al., 2000). Epidemiologic studies show great differences in T1D incidence following a geographic variation pattern, with almost a 20-fold difference in incidence. Among places whose population is mostly Caucasian, the highest incidences are in Finland (Harjutsalo et al., 2008) and the island of Sardinia, Italy (Songini et al., 2017), with approximately 45–60/100,000. Migration from countries with low-incidence to countries with high-incidence increases the probability to develop T1D, mainly in children younger than 10-years-old and offspring born in the high-incidence country (Ilonen et al., 2019). A sustained increase in the number of T1D cases has been reported over the last decades (Patterson et al., 2009), which cannot be explained by genetic differences alone. Therefore, changes in environment and lifestyle seem to be responsible for this raise in diabetes incidence.

Although still controversial, evidence indicated that dietary factors, toxins, vaccination, gut microbiota, or bacterial infection, contribute to T1D susceptibility (Ilonen et al., 2019; Knip and Simell, 2012). Among all these putative contributors, viral infections are the ones with the strongest evidence supporting their causal role. Even though many viruses, including cytomegalovirus, mumps, rotavirus, and rubella, have been linked to T1D development (Craig et al., 2013), here we will focus on enteroviruses because most of the available evidence implicate these viruses as triggers of T1D (Rodriguez-Calvo, 2019).

4.1 Enterovirus

Enterovirus is a genus from the family *picornaviridae*. This genus is divided into several species, being nine of enterovirus, including coxsackievirus and echovirus, and three rhinoviruses. Enteroviruses are causative agents for a wide variety of illnesses in humans and are among the most common infectious agents worldwide (Sinclair and Omar, 2020). All family members share the same characteristics: viruses with icosahedral capsids without lipid envelope and positive sense single-stranded RNA (+ssRNA). The cell machinery can directly translate the viral genetic material because it contains an internal ribosomal entry site (IRES) in 5′-UTR and is polyadenylated in 3′-UTR. This genetic material codifies one polyprotein that contains structural proteins (e.g., VP1) and non-structural proteins (polymerase and proteases) that are cleaved after translation (Rodriguez-Calvo, 2019; Sinclair and Omar, 2020).

Enteroviruses enter the cell by binding to two different receptors, namely decay accelerating factor (DAF) and coxsackie-adenovirus receptor (CAR). Of note, CAR, which is well-expressed in β-cells, is upregulated upon a viral infection (Hodik et al., 2016; Marroqui et al., 2015b; Rodriguez-Calvo, 2019; Sinclair and Omar, 2020). Once inside the cells, the virus is uncoated and its genetic material is recognized by PRRs, such as MDA5 and RIG-1, which triggers the ISG-mediated antiviral response (Fig. 2).

Enterovirus strains with β-cell tropism are candidates to perpetrate the diabetogenic outcomes, either via cytolytic effects on β-cells or via inducing a low-grade but persistent infection that trigger an autoimmune response that gradually destroys β-cells. Coxsackievirus B (CVB1–6) (Rodriguez-Calvo, 2019; Sinclair and Omar, 2020; Stone et al., 2018) and Echovirus (Sarmiento et al., 2017) are known for infecting and killing β-cells. However, not all coxsackieviruses are diabetogenic. While CVB1 was associated with increased risk of T1D (Oikarinen et al., 2014), CVB3 and CVB6 correlate with reduced risk of T1D; this might be because the antibodies generated after infection with CVB3 and CVB6 cross-react with diabetogenic viruses (Laitinen et al., 2014). Therefore, it seems feasible to create a coxsackievirus vaccine to prevent the initiation of the disease. In fact, this rationale has been approached by several groups and, in the last 5 years, more than 10 studies proposing different vaccine have been published. This point will be discussed later in Section 7.1.

4.2 Fulminant T1D

Fulminant diabetes is a rare variant of T1D where β-cells are infected in a massive way. Consequently, β-cell destruction is faster than classical T1D. Fulminant T1D is characterized by rapid hyperglycemia and ketoacidosis caused by an acute infection and an exacerbated innate and adaptive immune responses. All symptoms occur in a short period of time, usually less than 1 week (Morgan and Richardson, 2014; Tanaka et al., 2013), without the detection of autoantibodies, and with basal islet MHC class I expression (Imagawa et al., 2000). This rare disease has the highest prevalence in young adults of Asian population, counting for nearly 25% of T1D cases in Japan (Morgan and Richardson, 2014). One characteristic of this acute infection is that it affects both α- and β-cells, and the pancreatic islets are indiscriminately destroyed (Sayama et al., 2005). The most common enterovirus serotypes causing fulminant T1D are CVB4 and CVB1 (Akatsuka et al., 2009; Takeno et al., 2018). It is noteworthy to mention that other serotypes, such as CVA2 and CVA4–6, have been reported to cause fulminant T1D (Hanafusa et al., 2008; Ohara et al., 2016). These findings in fulminant T1D provide in vivo evidence of the enterovirus tropism for pancreatic β-cells.

4.3 Low-grade enteroviral infection in T1D

The prevalence of fulminant T1D worldwide is very low. In most cases what is observed is a low-grade, persistent infection of few islet cells (Dotta et al., 2007; Morgan and Richardson, 2014). In vivo findings show that only few islet cells are infected in a non-lytic way. These infected cells highly express virus markers (e.g., VP1) but represent only 0.005–0.01% of total pancreas section (Dotta et al., 2007). A low-grade, chronic CVB3 infection has been observed in cardiomyocytes (Kim et al., 2005) as well as in pancreas from NOD mouse (Tracy et al., 2015). This persistent infection was due to a defective virus that naturally lost part of the 5′-UTR after its replication in the first infected cells. These defective viruses are less replicative and non-cytopathic (Tracy et al., 2015). Although there is no direct evidence, it has been hypothesized that a less replicative, noncytopathic coxsackievirus might chronically infect β-cells, leading to cell responses that culminate in inflammation and autoimmunity (Morgan and Richardson, 2014; Op de Beeck and Eizirik, 2016). In vitro evidence also supports this persistent infection. Chehadeh et al. infected human islets with CVB3, and two CVB4 strains, being one diabetogenic (CVB4E2) and another nondiabetogenic (CVB4JBV). Detection of viral RNA and VP1 expression 30 days after infection without obvious cytolysis, accompanied by a chronic IFNa production, suggested a persistent CVB replication in human islets (Chehadeh et al., 2000a).

4.4 Enteroviruses as environmental triggers in T1D

As mentioned before, enteroviral infections rise as the main suspects of being the diabetogenic environmental triggers of β-cell autoimmunity. Enteroviruses are the most common human disease-causing viruses, usually related to mild or asymptomatic infections in children with characteristic annual infection patterns. Enterovirus peak of infection temporally coincides with the seasonal increase in seroconversion against β-cells (Gamble and Taylor, 1969; Kimpimäki et al., 2001; Watad et al., 2017). One of the first evidence of a virus-induced T1D was the higher incidence of autoantibodies against CVB4 in people with T1D as compared to control individuals (Gamble et al., 1969). Ten years later, CVB was isolated from the pancreas of a child with recent onset of diabetes (Yoon et al., 1979). Over the years, prospective studies aiming to assess the possible role of enterovirus infections in the initiation of islet autoimmunity have been performed (Cinek et al., 2014; Graves et al., 2003; Hiltunen et al., 1997; Honkanen et al., 2017; Hyoty et al., 1995; Laitinen et al., 2014; Oikarinen et al., 2011; Simonen-Tikka et al., 2011; Stene et al., 2010; Tapia et al., 2011). Analysis of pancreas samples from living patients with recent-onset T1D obtained from the Diabetes Virus Detection study (DiViD) cohort (for more information of this cohort, refer to Krogvold et al., 2014) detected VP1 in the islets of all individuals with T1D and enterovirus-specific RNA sequences in 67% of T1D patients. These findings indicate the possibility of a low-grade enteroviral infection in T1D patients (Krogvold et al., 2015). A study assessing young children enrolled in the Environmental Determinants of Diabetes in the Young (TEDDY) cohort (for more information of this cohort, refer to TEDDY Study Group, 2007) reported that prolonged enteroviral B infection may be involved in islet autoimmunity (Vehik et al., 2019). However, a direct correlation between enteroviral B infection and T1D was not observed. In the same line, Simonen-Tikka et al. examined children with genetic risk of islet autoimmunity from the BABYDIET cohort (for more information of this cohorts, refer to Hummel et al., 2011) and found no correlation between virus infection and development of islets autoantibodies (Simonen-Tikka et al., 2011). Finally, despite these controversial findings, a large meta-analysis including more than 4000 patients from 26 case-control studies (24 original papers and 2 abstracts) reported a clinically significant correlation between autoimmunity/T1D and enterovirus infection (Yeung et al., 2011).

In a study performed by Richardson et al. using a UK cohort, islets from 72 young recent-onset T1D donors were examined and compared with islets from 50 neonatal controls. It was observed that 60% of the T1D islets were positive for VP1; only insulin-containing β-cells presented VP staining. Interestingly, there was a strong correlation between VP1 positivity in islet cells and expression of PKR (dsRNA-dependent protein kinase) in insulin-containing islets (Richardson et al., 2009). These findings were further confirmed by the same group in samples from the Network for Pancreatic Organ donors with Diabetes (nPOD) cohort. Curiously, PKR upregulation in VP+ β-cells correlated with a decrease in MCL-1 expression (Richardson et al., 2013).

Currently, there have been coordinated efforts dedicated to strengthening the evidence implicating viruses as a causal role in T1D pathogenesis, such as the nPOD-virus group and the PEVNET (Persistent Virus Infection in Diabetes Network). The most recent findings from these studies were reviewed by Rodriguez-Calvo (2019).

One of the main issues faced by researchers investigating the possible presence or absence of enteroviruses in human samples is the limit of detection of the virus in such samples. One reason is the low grade of infection; only few islet cells are virus positive, accounting for nearly 0.01% of total mass of islet cells (Dotta et al., 2007; Richardson et al., 2014). Another important difficulty is the limitation in sample collection; virus detection is higher close to the disease onset, but access to these samples is very scarce (Rodriguez-Calvo, 2019). In recent years, new technologies have improved the sensitivity of virus detection in samples with a low-grade infection. The development of better antibodies, use of heat-induced epitope retrieval (Richardson et al., 2014), and development of new technologies, such as short fluorescently-labeled oligonucleotide probes (Busse et al., 2017), have allowed the detection of viral protein and RNA with higher sensitivity and higher specificity than classical methods.

4.5 Functional studies

Classical studies demonstrated that human islets could be infected by different strains of coxsackievirus, namely CVB3, CVB4, CVB5, and CVA9. In these studies, viral infection induced morphological changes and pyknosis in β-cells (Roivainen et al., 2000).

CVB3 infection augmented TRIM21 expression which, in turn, induced a MAVS-dependent, IRF-mediated increase in type I IFN signaling. In vivo, TRIM21 deficient mice showed a decreased IFNβ production and enhanced pancreatic CVB3 replication (Liu et al., 2018).

In human islets, CVB5 upregulated proinflammatory cytokines (e.g., IL-1β and TNFα), which are commonly described as mediators of β-cell dysfunction, while downregulating genes involved in glycolysis and insulin secretion (Ylipaasto et al., 2012). In line with these CVB5-induced changes, human islets infected with CVB1 have lower insulin gene expression (without changes in glucagon expression) compared with uninfected islets; this change was accompanied by an increase in the expression of chemokines (e.g., *CXCL10* and *CCL5*), cytokines (e.g., *IFNβ*), and ISGs (e.g., *MX1*, *IFIH1*, *OAS1*) (Anagandula et al., 2014). Immunodeficient mice lacking native β-cells but carrying human islet grafts were infected with CVB4. Nearly 50% of infected mice developed hyperglycemia. Human islet grafts derived from infected mice showed positivity for both viral RNA and viral protein, and lower insulin levels compared with grafts from uninfected mice. Moreover, grafts from infected mice exhibited a human-specific IFN-stimulated signature, where ISGs (e.g., *MX1*, *IFIH1*, *HLA-A*) and inflammatory genes (e.g., *CCL5* and *CXCL10*) were upregulated. Of note, CVB4 infection induced an increase in IFNα and IFNβ secretion in human islets (Gallagher et al., 2015).

PKR activation by the viral dsRNA promoted eIF2α phosphorylation, increased BIM expression and downregulated MCL-1. This change in the MCL-1/BIM balance led to activation of the mitochondrial intrinsic pathway of apoptosis (Colli et al., 2011). Recently, the same group showed that infection with two CVB strains, namely CVB4 and CVB5, induced a partial ER stress in rat and human β-cells (Colli et al., 2019). CVB5 infection triggered both IRE1α and PERK pathways. However, while IRE1α activation led to XBP1s upregulation and JNK activation, PERK activation did not change the expression of the downstream markers ATF3, ATF4, and CHOP. Activation of JNK1 was essential for viral replication and, ultimately, β-cell apoptosis, whereas JNK2 has anti-apoptotic effects (Colli et al., 2019).

As microRNAs (miRNAs) are essential for β-cell function and survival (Grieco et al., 2017; Poy et al., 2004), Kim et al. investigated whether CVB5 infection would impact the cellular expression of miRNAs in human islets (Kim et al., 2016). CVB5 infection dysregulated the expression of 33 miRNAs (out of 754 miRNAs studied). Notably, over 50 known T1D candidate genes, including *BACH2*, *DEXI*, *PTPN2*, *IFIH1*, were predicted to be targeted by these dysregulated miRNAs (Kim et al., 2016).

Aiming to uncover the CVB-triggered events in infected β-cells, Nyalwidhe et al. assessed transcriptional and proteomic profiles of human stem cell-derived β (SC-β) cells infected with CVB4 (Nyalwidhe et al., 2020). Interestingly, RNA sequencing (RNA-seq) and proteomic analyses showed that CVB4 activated two key pathways involved in the innate immune response and inflammation, namely type I IFN and NF-κB, and their downstream targets. Therefore, CVB4 upregulated numerous ISGs (e.g., *MX1* and *IFIT1*), chemokines (e.g., *CXCL1* and *CXCL8*), and genes involved in the MHC class I complex (e.g., *HLA-A*, *HLA-B*, and *B2M*). On the other hand, genes essential for β-cell identity, such as *INS*, *PDX1*, and *MAFB* were reduced following CVB4 infection. This downregulation was reflected on β-cell function, as CVB4-infected SC-β-cells exhibited impaired insulin release. Interestingly, α-cell-specific transcripts were not modulated by CVB4 infection (Nyalwidhe et al., 2020), which is in line with previous results showing that α-cells are more resistant to virus infection (Marroqui et al., 2015b; Op de Beeck and Eizirik, 2016).

4.6 Proposed hypothesis based on the available evidence

In 1987 Foulis and collaborators proposed a model with the sequence of events likely to happen in pancreatic islets over the course of T1D (Foulis et al., 1987b). Although this model was proposed more than 30 years ago, it is support by many empirical evidences gathered to this day (Richardson et al., 2011, 2014). This sequence of events is summarized in Fig. 3. In genetically predisposed individuals, an enteroviral infection (e.g., CVB) early in life could induce a low-grade but persistent infection in few pancreatic β-cells. This persistent infection is not efficient to produce new infective particles, but it can initiate the innate immune response, leading to increased production/secretion of cytokines (e.g., type I IFNs) and chemokines (e.g., CXCL10). Type I IFNs would then alert the rest of the islet cells to activate the self-defense mechanism, amplifying the production of IFNs and the expression of MHC class I in the islets. Concomitantly, the production of chemokines, which act as chemotactic attractors, guide immune cells toward the islet milieu. As a result of this sequence of events we have insulitis and subsequent β-cell destruction. This process occurs over several years, and, when around 80% of the β-cell mass is lost, the patient begins to have clinical symptoms and the debut in diabetes occurs.

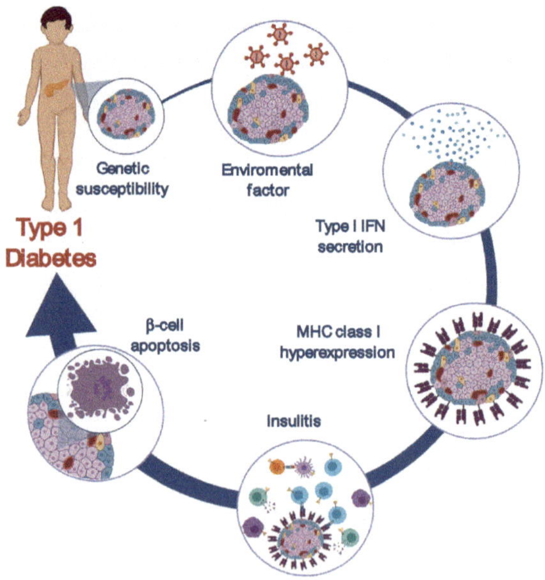

Fig. 3 Sequence of events proposed for T1D development. In genetically predisposed individuals, an enteroviral infection early in life could induce a low-grade but persistent infection in β-cells. This increases the production/secretion of type I IFNs, which would alert the rest of the islet cells to activate the self-defense mechanism, such as amplifying the production of IFNs, and the expression of MHC class I in the islets. The production of chemokines, which act as chemotactic attractors, guide immune cells toward the islet milieu. Eventually, these events culminate in insulitis, β-cell apoptosis, and finally T1D onset.

5. Candidate genes and type I IFN signaling pathway

T1D is a disease with a strong genetic component. Several genome-wide association studies (GWAS) have identified more than 50 non-HLA single-nucleotide polymorphisms (SNPs) associated with T1D risk (Størling and Pociot, 2017; Todd, 2010). Except for HLA gene polymorphisms that confer the highest risk (odds ratio (OR) >6.5) for disease, these alleles confer a moderate risk (OR <2) (Pociot et al., 2010).

During insulitis, human islets are exposed to proinflammatory cytokines, such as IL-1β and IFNγ, and undergo changes in expression in several of these causative genes (Eizirik et al., 2012; Ramos-Rodríguez et al., 2019). Cytokine-induced changes in gene expression may be explained by changes in chromatin accessibility. New evidence found by Ramos-Rodríguez et al. suggests that cytokine-driven changes in gene expression could be due to islet-specific transcription factors binding to genetic enhancers in open chromatin regions (Ramos-Rodríguez et al., 2019). These findings could explain cell type-specific responses to inflammation in T1D. In this context, genetic susceptibility variants are expressed in β-cells in basal conditions, but their function would be only evident in adverse conditions (e.g., inflammation) (Eizirik et al., 2020; Ramos-Rodríguez et al., 2019). Taken together, these findings suggest that some of these genetic loci modulate the dialogue between islets and immune cells during insulitis, and, therefore, their susceptibility-related polymorphisms induce an aberrant crosstalk that leads, at least partially, to the pathogenic mechanisms of T1D at the β-cell level.

As most of these candidate genes—around 80%—are expressed in β-cells (Eizirik et al., 2012; Marroqui et al., 2015a), a ingenuity pathway analysis of these β-cell-expressed gene variants revealed that the top three canonical pathways were related to interferon pathways: "Interferon signaling," "Role of JAK1, JAK2 and TYK2 in interferon signalling" and "Role of pattern recognition receptors in recognition of virus" (Marroqui et al., 2015a), suggesting a central role for IFN-regulated pathways in T1D pathogenesis. Functional studies performed at the β-cell level and/or in animal models have demonstrated the importance of these putative candidate genes for β-cells function and survival, indicating their role in the etiology of T1D. Here we will discuss the main T1D candidate genes involved in type I IFN signaling pathway and what is known about their function in the context of pancreatic β-cells and/or T1D.

5.1 TYK2

When we think about the type I IFN signaling pathway, *TYK2* emerges as the main T1D candidate gene regulating this pathway. TYK2, which is encoded by *TYK2*, is a tyrosine kinase crucial for the signaling of numerous cytokines that cause inflammation, including type I IFNs (Babon et al., 2014; Villarino et al., 2017). *TYK2* is located on chromosome region 19p13 and variants in this gene have been associated with several autoimmune diseases, such as SLE, multiple sclerosis, rheumatoid arthritis, and T1D (Tao et al., 2011; Wallace et al., 2010). Two protective SNPs are worth mentioning: rs34536443 and rs2304256. The SNP *TYK2* rs34536443: C > G (odds ratio for C vs G = 0.67) leads to the substitution of a highly conserved proline for an alanine at position 1104 in the TYK2 JH1 kinase domain, thus impairing TYK2 catalytic activity (Dendrou et al., 2016; Li et al., 2013). Primary human immune cells from rs34536443 minor allele homozygotes showed decreased type I IFN-induced STAT1 and STAT3 phosphorylation (Dendrou et al., 2016). The SNP *TYK2* rs2304256: C > A (odds ratio for A vs C = 0.86) has been also associated with protection against T1D. This SNP causes a missense mutation in *TYK2*, where the rare A allele causes a substitution of a valine for a phenylalanine at position 362 in the JAK-homology 4 region of the FERM domain (Tao et al., 2011; Wallace et al., 2010). Therefore, it has been proposed that the SNP *TYK2* rs2304256:C > A decreases TYK2 function, which would lead to a reduced susceptibility to IFN-related autoimmune diseases. Marroqui et al. reported that B lymphoblastoid cell lines from patients with genotype AA of the *TYK2* SNP rs2304256 showed lower IFNα-induced STAT1 phosphorylation when compared with patients with genotype CC (Marroqui et al., 2015a). Further functional studies in pancreatic β-cells demonstrated that TYK2 controls apoptotic and proinflammatory pathways via modulation of IFNα signaling, MHC class I expression, and regulation of cytokine and chemokine production. These processes are important for T cell recruitment to the islet milieu (Marroqui et al., 2015a).

Still in 2015, Izumi et al. noted that *Tyk2*-deficient mice as well as mice harboring a mutated *Tyk2* gene leading to reduced *Tyk2* expression were more susceptible to virus-induced diabetes than their wild-type controls (Izumi et al., 2015). Moreover, *Tyk2*-deficient mice presented higher pancreatic viral titer and augmented serum and pancreatic IFNα levels. Remarkably, β-cell-specific *Tyk2* recovery in the *Tyk2*-deficient animals was sufficient to prevent diabetes onset (Izumi et al., 2015).

Altogether, these findings place TYK2 as a critical regulator of type I IFN signaling pathway in pancreatic β-cells. As it will be discussed later in Section 7, TYK2 inhibition represents a promising therapeutic approach for T1D.

5.2 PTPN2

PTPN2 (Protein Tyrosine Phosphatase, Non-Receptor Type 2) is a tyrosine phosphatase acting as a negative regulator of different signaling pathways, including the type I IFN pathway, by dephosphorylating target proteins, such as JAKs and STATs (Gurzov et al., 2015). *PTPN2* is located on chromosome region 18p11 (Todd et al., 2007); several SNPs have been associated with T1D, all of them with a moderate increase in relative risk (Long et al., 2011; The Wellcome Trust Case Control Consortium, 2007; Todd et al., 2007).

For example, the SNP rs1893217 (A > G), whose minor allele correlates with decreased PTPN2 expression, predicts an increased relative risk of T1D (Long et al., 2011; Størling and Pociot, 2017; Todd et al., 2007).

In β-cells, siRNA-mediated silencing of *PTPN2* exacerbated inflammation and induced the intrinsic apoptotic pathway upon treatment with type I IFN (Santin et al., 2011), IFNγ (Moore et al., 2009), or poly(I:C) (Colli et al., 2010). As described by Santin et al., activation of JNK and BIM as well as BIM upregulation were involved in the mechanism whereby PTPN2 inhibition aggravated IFN-induced β-cell apoptosis (Santin et al., 2011). In MIN6 cells, PTPN2 knockdown alleviated palmitate- or tunicamycin-induced PERK/eIF2α ER stress signaling and conferred protection against ER stress-induced death (Bettaieb et al., 2011). Of note, PTPN2 expression is indirectly regulated by the transcription factor BACH2, which is another candidate gene for T1D (Marroqui et al., 2014).

Deletion of PTPN2 in pancreas affects glucose-stimulated insulin secretion in mice challenged with high fat diet and hyper activation of STAT1 and STAT3 (Xi et al., 2015). Moreover, NOD mice with PTPN2 deletion specifically in T cells have increase in the progression and the incidence of T1D (Wiede et al., 2019).

5.3 IFIH1

IFIH1 (Interferon induced with helicase C domain 1) is the gene encoding MDA5 (Melanoma differentiation-associated gene 5), a cytoplasmic viral sensor that can detect dsRNA resulting from a viral infection. *IFIH1* is located on a T1D-associated region, chromosome 2q24, that contains four genes, namely *IFIH1* itself, *FAD* (fibroblast activation protein), *GCA* (grancalcin), and *KCNH7* (potassium voltage-gated channel subfamily H member 7); only *IFIH1* is associated with risk of T1D in this region (Smyth et al., 2006). Several common and rare variants of *IFIH1* have been associated with T1D, including: rs1990760 (A946T; the allele TC is associated with protection while TT is associated with increased risk) and rs35667974 (I293V; rare variant) located in the conserved region of the gene; rs35337543 and 35,732,034 (respectively, IVS8 + 1 and IVS14 + 1, rare variants) located in a conserved splice site (Nejentsev et al., 2009; Smyth et al., 2006), and rs35744605 (Glu627X, rare variant) residing within exon 10. These are variants whose minor allele is associated with MDA5 loss-of-function or downregulation and conferred protection against T1D (Nejentsev et al., 2009; Smyth et al., 2006). Interestingly, the variant rs1990760 has been associated with seasonal manifestation of T1D (Jermendy et al., 2018), indicating that MDA5 can be a link between genetic susceptibility (candidate risk genes) and environmental factors (viral infection) in T1D.

MDA5 inhibition protected β-cells against poly(I:C)-induced activation of IFNβ and NF-κB, and the subsequent increase in expression and secretion of cytokines and chemokines. Curiously, MDA5 did not seem to be involved in the mechanism by which poly(I:C) induces β-cell apoptosis (Colli et al., 2010). These data indicate that MDA5 may contribute to β-cell death by increasing the local release of inflammatory mediators (i.e., cytokines and chemokines) in the course of a viral infection, thereby exacerbating insulitis.

NOD mice with homozygous *Ifih1* knockout ($MDA5^{-/-}$) were protected from spontaneous diabetes, whereas animals with heterozygous *Ifih1* knockout ($MDA5^{+/-}$) had a reduced incidence when compared to wild-type mice ($MDA5^{+/+}$). Additionally, $MDA5^{+/-}$ mice were completely protected from CVB4-induced diabetes (Lincez et al., 2015). Mice with a knock-in mutation encoding the risk variant $IFIH1^{T946}$ exhibited enhanced basal expression of type I IFNs and survived a lethal viral challenge but presented increased risk for autoimmunity (Gorman et al., 2017). Another crucial finding noted by Gorman et al. was the increased basal and ligand-triggered production of type I IFNs observed in human PBMCs and cell lines expressing the risk variant $IFIH1^{T946}$ (Gorman et al., 2017).

5.4 BACH2

BACH2 (BTB and CNC homology 1, basic leucine zipper transcription factor 2) is a transcription factor acting as a repressor that usually binds to the Maf recognition element (MARE) and the antioxidant-responsive element (ARE) (Muto et al., 1998; Tsukumo et al., 2013). BACH2 is usually recognized by its involvement in both innate and adaptive immune responses, where BACH2 is implicated in macrophage function, B-cell differentiation, and maintenance of the naive T cell state (Kometani et al., 2013; Muto et al., 2004; Roychoudhuri et al., 2013; Swaminathan et al., 2013; Zhou et al., 2016). *BACH2* is located on the chromosome 6, in the region 6q15 (365-kb), in which *BACH2* is the only gene. Numerous SNPs have been associated with risk of T1D (Cooper et al., 2008) as well as other autoimmune diseases, such as Crohn's disease, SLE, and rheumatoid arthritis (Franke et al., 2010; Ghodke-Puranik and Niewold, 2015; Kourilovitch et al., 2014). The minor allele A of the variant rs72928038, which is associated with an increased risk of T1D, decreases enhancer accessibility and BACH2 expression in T cells (Robertson et al., 2020). Furthermore, this variant is also associated with changes in distal genes, including some also in regions that have been associated with autoimmune diseases (Robertson et al., 2020).

Functional studies demonstrate that BACH2 is implicated in oxidative stress-induced apoptosis and virus-induced cell death in human cell lines (Hong et al., 2008; Muto et al., 2004). In β-cells, BACH2 regulates apoptosis induced by proinflammatory cytokines by modulation of the MKK7-JNK-BIM axis, as well as regulation of PTPN2 and anti-apoptotic members of the BCL-2 family (Marroqui et al., 2014).

5.5 CLEC16A and DEXI

Despite its name, *CLEC16A* (C-type lectin domain family 16 member A), previously known as KIAA0350, encodes a membrane-associated endolysosomal E3 ubiquitin ligase (Pearson et al., 2018; Soleimanpour et al., 2014). *CLEC16A* is located on the chromosome 16p13, in a region consisting of three other genes, namely *CIITA* (Class II major histocompatibility complex transactivator), *SOCS1*, and *DEXI* (dexamethasone-induced protein) (Gingerich et al., 2020). Several SNPs have been associated with different autoimmune diseases and rs12708716, localized in an intronic region, is the most associated with T1D (Gingerich et al., 2020; Todd et al., 2007).

Human islets from donors presenting one (i.e., AG) or two copies (i.e., GG) of the associated SNP rs12708716 exhibited lower CLEC16A expression than islets from donors possessing the genotype AA; importantly, reduced HOMA-β measurements and increased hemoglobin A1c were found in nondiabetic human islet donors homozygous GG for the rs12708716 SNP (Soleimanpour et al., 2014). Functional studies in pancreatic β-cells demonstrate that CLEC16A is involved in the control of mitophagy through the formation of a ubiquitin-dependent tripartite complex consisting of the E3 ubiquitin ligase NRDP1, USP18, and CLEC16A itself. By controlling mitophagy, CLEC16A regulates β-cell function, insulin secretion, and glucose homeostasis (Pearson et al., 2018; Soleimanpour et al., 2014, 2015). Moreover, mice with pancreas-specific *Clec16a* deletion displayed impaired glucose tolerance due to β-cell dysfunction (Soleimanpour et al., 2014). Of note, islets from these animals also exhibited mitochondrial dysfunction and age-dependent β-cell ER Stress (Soleimanpour et al., 2014). Finally, *Clec16a* knockdown in NOD mice protects against the diabetes progression due to changes in thymic epithelial cell autophagy and subsequent T cell hyporeactivity (Schuster et al., 2015).

CLEC16A was initially considered as the primary T1D candidate gene in its region, largely because most of the SNPs were within the gene sequence (Todd et al., 2007). However, recent evidence suggests that *DEXI* might be the main candidate risk gene in the 16p13 region (Davison et al., 2012). First, it was observed that polymorphisms within the *CLEC16A* gene regulate *DEXI* expression in B lymphoblastoid cell lines, which was associated with risk of T1D (Tomlinson et al., 2014). Later, using expression quantitative trait loci (eQTL) analysis and chromosome conformation capture, Davison et al. observed that a DNA enhancer loop formed between the intron 19 of *CLEC16A* and *DEXI* promoter region regulates *DEXI* expression (Davison et al., 2012).

Functional studies showed that DEXI knockdown partially decreased STAT1 activation and reduced chemokine production in β-cells exposed to poly(I:C) (Dos Santos et al., 2019). These effects were preceded by a reduction in IFNβ levels, which suggested that DEXI participates in the transcriptional activation of IFNβ. Of note, overexpression of DEXI exacerbated STAT1 activation and inflammation (Dos Santos et al., 2019). Even though this study by Dos Santos et al. provides evidence supporting *DEXI* as the etiological gene for T1D on chromosome 16p13, a recent study using a whole-body *Dexi* knockout NOD mouse showed that loss of *Dexi* did not affect the frequency of diabetes. Moreover, *Dexi* knockout did not impact the protection conferred by *Clec16a* knockdown (Nieves-Bonilla et al., 2020). Thus, as it is possible that differences in models (ex vivo/in vitro vs in vivo) and/or species (rat/human vs mouse) may account for the opposite conclusions, further studies are required to determine the causal gene within the chromosome 16p13.13 locus.

5.6 EBI2

GPR183 (G Protein-Coupled Receptor 183), also known as Epstein-Barr virus-induced receptor 2 (EBI2), is a receptor for oxysterols involved in the control of B-cell migration (Hannedouche et al., 2011; Liu et al., 2011; Pereira et al., 2009). *GPR183* is located on the chromosome 13, in the region 13q32, and the minor allele of SNP rs9585056 correlates with lower *EBI2* expression (Heinig et al., 2010; Wallace et al., 2012). Heinig et al. reported that GPR183 acts as a negative regulator of an IRF7-driven inflammatory network (IDIN) identified in rat and human monocytes; IDIN, which was enriched in genes involved in the innate antiviral response, was associated with the risk of T1D (Heinig et al., 2010). Studies in *Ebi2* knockout mouse showed that EBI2 negatively regulates type I IFN responses in pDCs and myeloid cells (Chiang et al., 2013). These findings indicate that EBI2 may be implicated in the pathogenesis of T1D via regulation of the innate immune response.

To our knowledge, there are no functional studies in pancreatic islets/β-cells, but it is known that *EBI2* is expressed in human islets (Eizirik et al., 2012). More studies are necessary to determine whether EBI2 plays a role in β-cells function and/or survival.

Visualizing the whole picture is evident that the type I IFN response plays a critical role in the progression of T1D and, in this context, the genetic background can modulate the predisposition and/or the progression of the disease. If we pay attention to the localization of the known T1D candidate genes related to type I IFN pathway (Fig. 4) we can see that these genes are

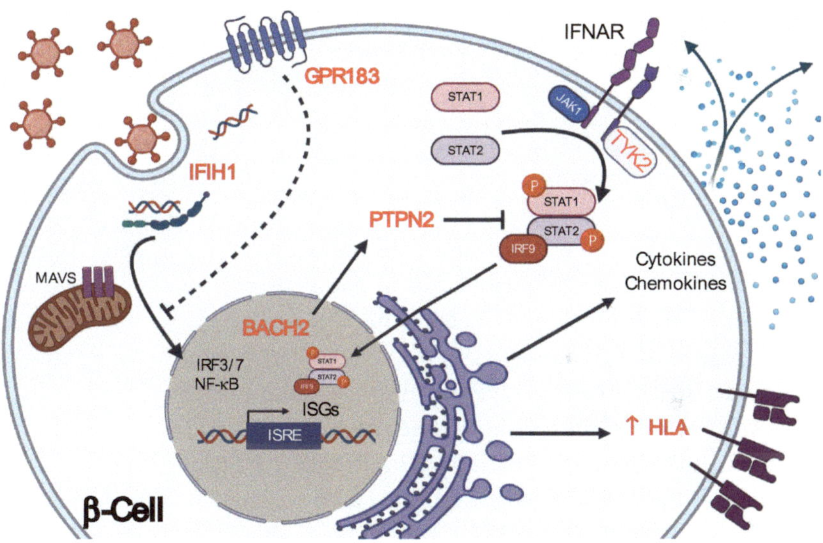

Fig. 4 Candidate genes related with interferon alpha in pancreatic β-cell. In red are highlighted the candidate genes for T1D expressed in β-cells and their putative role upon a virus infection. *IFIH1* detects dsRNA resulting from a viral infection and activates IRF3/7 and NF-κB pathways. *GPR183* acts as a negative regulator of an IRF7-driven inflammatory network. *TYK2* is a tyrosine kinase crucial for type I IFN signaling. *PTPN2* is a tyrosine phosphatase acting as a negative regulator of type I IFN signaling pathway, among others. *BACH2* is a transcription factor involved in both innate and adaptive immune responses, implicated in oxidative stress-induced apoptosis, virus-induced cell death, and involved in the regulation of *PTPN2* and BCL-2 proteins. *HLA* is responsible for antigen presentation. Of note, HLA gene polymorphisms confers the highest risk for T1D.

involved in all processes carried out in the β-cell; virus recognition, IFN signaling and amplification, and β-cell recognition by the immune system.

6. Type I IFNs in T1D pathogenesis

Type I IFNs are crucial for both innate and adaptive immune response due to their ability to directly induce the expression of genes involved in modulating such responses. Although type I IFNs play a crucial role in the early stages of human T1D pathogenesis, their effects on pancreatic β-cells have only recently begun to be studied. In this section we will discuss the evidence supporting a role for type I IFNs, especially IFNα, in the pathogenesis of human T1D as well as the known molecular mechanisms underlying type I IFN effects on pancreatic cells. We will focus mostly on pancreatic β-cells, as they seem to be the main target of immune cells during T1D development.

6.1 Evidence in humans

A series of studies published in the 1980s established the first associations between type I IFNs and the etiology of T1D. In 1985, aiming to identify the cells invading pancreas, Bottazo et al. studied islets collected from a 12-years-old donor with recent-onset diabetes (Bottazzo et al., 1985). Most of the inflammatory cells observed were T lymphocytes positive for HLA-DR, an indicative that these cells were cytotoxic T cells. Notably, they also noticed that endocrine cells in some affected islets had increased expression of HLA class I molecules. This was the first time that HLA class I hyperexpression was reported in insulitic islets. A couple of years later, Foulis et al. analyzed pancreases from T1D patients with different disease duration: 23 patients with recent-onset (died within a few weeks of onset) with residual β-cells, and 12 patients with longer duration (more than 1 year, maximum 9 years), eight of whom presented residual β-cells. In addition, normal and diseased pancreases from 95 patients were used as controls (Foulis et al., 1987a). Interestingly, all pancreases in which residual β-cells were present showed marked MHC class I hyperexpression in all endocrine cells (e.g., α- and δ-cells); this phenomenon was observed in 92% of insulin-containing islets, whereas only 1% of insulin-deficient islets exhibited such characteristic. All 95 control pancreases were negative to abnormal MHC class I expression, suggesting that this was a characteristic unique to T1D. In the light of these new findings, the authors proposed that hyperexpression of MHC class I molecules precedes insulitis and the aberrant expression of

MHC class II on some β-cells, which had also been previously reported (Bottazzo et al., 1985; Foulis and Farquharson, 1986), in the cascade of events leading to disease onset.

However, a link to explain why all endocrine cells in insulin-containing islets hyperexpressed MHC class I before the insulitis process was still missing. It was hypothesized that β-cells could be secreting a soluble substance with paracrine effects, which would, then, induce MHC class I hyperexpression on the adjacent endocrine cells. Based on prior data in human pancreatic islets in vitro (Pujol-Borrell et al., 1986), two candidates emerged as leading suspects: type I and type II IFNs. As type II IFNs are mainly produced by T lymphocytes, Foulis et al. focused on type I IFNs as prime suspects. To assess whether pancreatic β-cells could produce IFNα, Foulis et al. evaluated 80 control pancreases and 34 pancreases presenting residual β-cells from T1D patients (Foulis et al., 1987b). In pancreases from T1D, IFNα detection positively correlated with hyperexpression of MHC class I: IFNα was present in 93% of insulin-containing islets that also hyperexpressed MHC class I, whereas only 0.4% of IFNα-positive islets did not show MHC class I hyperexpression. Regarding the control patients, only 4 out of 80 pancreases expressed IFNα.

These seminal studies paved the way for subsequent studies implicating local IFNα production within the islet environment as trigger of insulitis, β-cell destruction by the immune system, and, consequently, progression toward clinical T1D. Since the first studies by Botazzo and Foulis, MHC class I hyperexpression has been confirmed in several other studies and is now extensively accepted as a defining feature of the disease (Benkahla et al., 2020; Coppieters et al., 2012; Krogvold et al., 2015; Richardson et al., 2009; Rodriguez-Calvo et al., 2015; Wyatt et al., 2019). Of particular interest is a recent study by Richardson et al. that analyzed pancreas samples from T1D individuals from three cohorts: the Network for Pancreatic Organ donors with Diabetes (nPOD), Diabetes Virus Detection study (DiViD), and an archival collection of postmortem samples from the United Kingdom (Richardson et al., 2016). In all three collections, insulin-containing islets of T1D patients hyperexpressed both "classical" (i.e. HLA-ABC) and a "non-classical" (HLA-F) HLA class I molecules at both RNA and protein levels. Moreover, STAT1 expression in β-cells, which was also elevated in T1D patients, strongly correlated with HLA class I hyperexpression. This study consolidated HLA class I hyperexpression as a hallmark in T1D pathogenesis. A recent study by Russell et al. has demonstrated that a subset of β-cells from T1D patients express HLA class II as well as its transcriptional regulator Class II MHC transactivator protein and other genes involved in the HLA class II pathway. These findings are particularly important because they leave open the possibility of a direct involvement of β-cells in autoantigen presentation to islet-infiltrating $CD4^+$ T cells during the T1D pathogenic process (Russell et al., 2019).

Several groups using different technical approaches have later confirmed the presence of IFNα in human islets from patients with T1D. Analysis of mRNA expression of several cytokines, such as type I and type II IFNs, and IL-1β, showed that IFNα expression was increased in partially purified pancreatic islets (Somoza et al., 1994) or pancreases (Huang et al., 1995) from individuals with recent-onset diabetes as compared to control pancreases. Dotta el al. performed immunohistochemical analysis of islets from 6 recent-onset T1D patients and 26 control donors (Dotta et al., 2007). IFNα-positive cells were detected in islets from three patients, all of whom showed reactivity to VP1 enteroviral peptide in pancreatic β-cells. Of note, no signs of IFNα or VP1 were observed in islets from control donors.

Complementing the studies in human islets and pancreases, some studies have shown that T1D patients exhibited higher circulating levels of IFNα (Chehadeh et al., 2000b; Mavragani et al., 2013; Xia et al., 2014). Chehadeh et al. detected IFNα mRNA in blood cells from 75% of T1D patients; remarkably, 50% of IFNα-positive patients were also positive for enterovirus RNA, while IFNα-negative patients did not exhibit enterovirus RNA (Chehadeh et al., 2000b). It is noteworthy to mention that the number of plasmacytoid dendritic cells, well known as the main type I IFN-producing cells (Siegal et al., 1999), were markedly elevated in the peripheral blood of both at-risk and T1D individuals; additionally, CVB4-induced IFNα production by PBMCs was higher in T1D patients than in control subjects (Allen et al., 2009).

As discussed in Section 2, type I IFNs stimulate the expression of hundreds of ISGs, which constitute a response commonly known as "IFN signature" or "type I IFN signature." This gene signature is characteristic of some autoimmune diseases (e.g., SLE) as well as type I interferonopathies (Jean-Baptiste et al., 2017; Rönnblom and Eloranta, 2013).

Besides HLA class I molecules, several other ISGs, such as IRF1, HLA-E, and PD-L1, had their expression elevated, at both RNA and protein levels, in islets from patients with T1D (Colli et al., 2018, 2020a; Lundberg et al., 2016; Richardson et al., 2016). In addition, recent evidence indicates that an IFN signature is present in human T1D, as previously reported in animal models. These IFN signatures have been identified in pancreases/islets as well as in blood from T1D patients. A whole genome transcriptomic profiling of pancreases and purified islets obtained from four patients with T1D not only revealed an IFN signature but also a global overexpression of genes related to the innate immune response (Planas et al., 2010). Despite its modest number of organ donors, this study by Planas et al. represents a valuable data set that established the ground for upcoming studies with larger cohorts and additional approaches. In 2016, assessment of 84 known ISGs in laser-captured islets from living patients with recent-onset T1D participating in the DiViD study showed that nearly 30% of these ISGs (e.g., *TLR3*, *OAS1*, and *STAT1*) were overexpressed in insulitic islets from T1D individuals as compared to nondiabetic, control donors. Curiously, ISGs were differentially overexpressed between islet core and peri-islet area containing the surrounding immune cells (Lundberg et al., 2016).

Due to the inherent difficulties to obtain pancreas samples from organ donors, exploring peripheral blood has emerged as a valuable tool for gene expression profiling in T1D patients (Cabrera et al., 2016). A global gene expression analysis comparing whole-blood samples from newly diagnosed children, their autoantibody-positive first-degree relatives (i.e., prediabetic individuals), and healthy controls revealed 107 distinct genes differently expressed among these groups. From these 107 genes, 12 were IFN response genes, including *ISG15*, *GBP1*, and *STAT1*, whose expression was upregulated in 30% of prediabetic children and characterized an IFN signature (Reynier et al., 2010).

Aiming to unveil the mechanisms underlying the autoimmune process leading to diabetes onset in genetically predisposed populations, two independent research groups studied participants enrolled in two longitudinal cohorts of children genetically at-risk for T1D, namely BABYDIET and the Finnish T1D Prediction and Prevention (DIPP) (for more information of DIPP cohort, refer to Kupila et al., 2001). Employing genome-wide transcriptomics (Kallionpaa et al., 2014) and targeted microarray (Ferreira et al., 2014), these studies analyzed gene expression in peripheral blood from children before and after detection of autoantibodies (seroconversion) through the clinical diagnosis of T1D. In susceptible children, an IFN signature was temporarily augmented prior to development of autoantibodies and throughout the development of T1D. More specifically, this transient signature was strongest just before seroconversion and began to decline after identification of the first clinical signs of diabetes (e.g., presence of autoantibodies). Ferreira et al. also noted a positive correlation between the IFN signature and episodes of upper respiratory tract infections. These studies demonstrated for the first time that type I IFN-mediated innate immune response is activated even prior to initiation of islet autoimmunity.

More recently, Meyer et al. reported that patients with autoimmune polyglandular syndrome type I (APS1, also known as APECED), a syndrome caused by mutations in the autoimmune regulator (*AIRE*) gene, harbored high-affinity, neutralizing autoantibodies, especially against specific cytokines (e.g., IFNα, IL-22, and IL-17). Furthermore, APS1/APECED patients harboring powerfully neutralizing anti-IFNα antibodies were protected against T1D development, whereas patients lacking IFNα-specific antibodies progressed to T1D (Meyer et al., 2016).

Finally, there have been several reports showing initiation of islet autoimmunity and T1D development upon chronic treatment with IFNα for different conditions, including chronic viral hepatitis (Fabris et al., 1992, 1998, 2003; Oka et al., 2011; Schreuder et al., 2007; Shiba et al., 1995; Zornitzki et al., 2015), autoimmune diseases (Uonaga et al., 2012), or hematologic malignancies (Guerci et al., 1994; Nakamura et al., 2011). It is worth noting that 50% of hepatitis C virus-positive patients were genetically susceptible and autoantibody-positive before treatment, which indicates that IFNα could induce or accelerate an ongoing diabetogenic process (Fabris et al., 2003).

6.2 Evidence in animal models

Numerous studies have used different rodent models (e.g., NOD mice and biobreeding rats) to study the role of type I IFNs in the pathogenesis of T1D (Huang et al., 1994; Li et al., 2008; Li and McDevitt, 2011; Sobel et al., 1994; Stewart et al., 1993). However, as this subject is outside the scope of the present review, and it has been thoroughly reviewed elsewhere, the interested reader is referred to some recent reviews on the section (Akhbari et al., 2020; Jean-Baptiste et al., 2017; Lombardi et al., 2018; Newby and Mathews, 2017; Qaisar et al., 2018). A summary of the most important findings from animal studies can be seen in Table 1.

6.3 Mechanisms underlying the role of type I IFN in T1D

Uncovering the different adaptive responses triggered by β-cells once they are exposed to type I IFNs may help us to understand the early changes occurring in β-cells over the course of T1D. Aiming to identify the "molecular footprints" left by the immune attack on β-cells, a recent study compared the transcriptomic signature observed in purified β-cells from

Table 1 Summary of the main findings observed in animal studies.

Animal models	Key findings	References
BB-DP rats	Poly(I:C) treatment induced IFNα production and accelerated the development of diabetes	Ewel et al. (1992)
	Poly(I:C) effect on IFNα production was dose-dependent; serum IFNα levels positively correlated with earlier diabetes onset	Sobel et al. (1994)
	Spontaneous expression of IFNα in pancreatic islets preceded lymphocytic infiltration and diabetes onset; poly(I:C) treatment accelerated the development of diabetes in BB-DP rats	Huang et al. (1994)
LEW.1WR1 rats	Deletion of the IFNAR1 subunit caused delayed onset and frequency of poly(I:C)- and Kilham rat virus-induced diabetes	Qaisar et al. (2017)
NOD mice	Transgenic NOD mice overexpressing human IFNβ in pancreatic β-cells developed accelerated autoimmune diabetes with a high incidence of the disease	Alba et al. (2004)
	NOD and NOD/BDC 2.5 T cell receptor transgenic mice presented increased expression of several ISGs in $CD4^+$ T cells; pancreatic draining lymph nodes showed increased number of IFNα-producing plasmacytoid dendritic cells (pDCs) and augmented IFNα production; IFNAR blockade with a neutralizing antibody delayed diabetes onset and decreased the incidence of the disease	Li et al. (2008)
	An extensive blockade of IFNAR1 with a specific monoclonal antibody almost completely prevented diabetes development; depletion of pDCs by a specific antibody delayed the onset and decreased diabetes incidence	Li and McDevitt (2011)
	Increased number of pDCs infiltrated in the pancreatic islets; augmented IFNα production and increased expression of ISGs in pancreatic islets	Diana et al. (2013)
	NOD islets presented an IFN signature that preceded islet infiltration by leukocytes, T cell activation, and diabetes onset	Carrero et al. (2013)
	Expression of several ISGs was detectable in NOD islets but not in islets from NOD mice deficient in IFNAR1 ($NOD.IFNAR1^{-/-}$); NOD controls and $NOD.IFNAR1^{-/-}$ mice developed insulitis and diabetes at a similar rate	Quah et al. (2014)

Continued

Table 1 Summary of the main findings observed in animal studies.—cont'd

Animal models	Key findings	References
SOCS1-tg mice	Transgenic NOD mice expressing SOCS1 under the control of the human insulin promoter presented defective β-cell response to IFNs, which rendered these mice more sensitive to coxsackievirus B4-induced β-cell destruction and diabetes development	Flodström et al. (2002)
Rip-LCMV mice	Immunization with lymphocytic choriomeningitis virus (LCMV) elevated serum IFNα and enhanced MHC class I expression in β-cells; treatment of immunized mice with IFNα or poly(I:C) resulted in a rapid and sometimes transient development of hyperglycemia; genetic ablation of *Ifnar* or *Myd88* prevented LCMV-induced diabetes	Lang et al. (2005)
	Mice challenged with LCMV presented an IFN signature that preceded CD8+ T cell infiltration into the islets, showed significant islet destruction, and developed diabetes; treatment with neutralizing antibodies targeting either the IFNAR1 or IFNα limited lymphocytic infiltration into the islets, reduced islet immunopathology, and prevented diabetes development	Marro et al. (2017)
Transgenic mice overexpressing type I IFNs	Transgenic mice overexpressing human IFNα in pancreatic β-cells developed a hypoinsulinemic diabetes; an anti-IFNα neutralizing antibody prevented diabetes development in these animals	Stewart et al. (1993)
	Transgenic mice overexpressing human IFNβ in pancreatic β-cells showed dysfunctional islets and presented a prediabetic state; treatment with a low dose of streptozotocin (STZ) caused lymphocytic islet infiltration and overt diabetes	Pelegrin et al. (1998)
	Transgenic mice overexpressing murine IFNκ in the β-cells developed diabetes and showed inflammatory infiltrates in the pancreatic islets with corresponding β-cell death	Vassileva et al. (2003)
	Transgenic nonobese-resistant mice overexpressing human IFNβ in pancreatic β-cells developed accelerated autoimmune diabetes with a high incidence of the disease	Alba et al. (2004)
STZ-treated mice	STZ promoted IFNα expression in pancreatic islets prior to STZ-induced diabetes; poly(I:C), which increases IFNα production, exacerbated STZ-induced insulitis; treatment with anti-IFNα antiserum abrogated the effects of STZ + poly(I:C)	Huang et al. (1994)
	STZ injection in NOD mice increased the expression of ISGs in the islets, which was dependent on the presence of pDCs	Diana et al. (2013)

Related to Section 6.2. *Abbreviations*: BB-DP, BioBreeding Diabetes Prone; Poly(I:C), polyinosinic-polycitidilic acid; IFNAR1, IFNα receptor subunit 1; ISG, IFN-stimulated genes; pDCs, plasmacytoid dendritic cells; *Rip-LCMV*, rat insulin promoter-lymphocytic choriomeningitis virus; Myd88, myeloid differentiation primary response protein 88; STZ, streptozot[illegible]

T1D patients with the signatures from human islets exposed in vitro to cytokines present at different phases of the disease, i.e., IFNα (early-stage) or the proinflammatory cytokines IL1β + IFNγ (later stages) (Colli et al., 2020b). The profile of transcripts induced by cytokines was similar to β-cells signatures observed in individuals with T1D. Enrichment analysis of the transcripts upregulated only in the intersection between the T1D and IFNα RNA-seq datasets identified biological processes (e.g., "Response to virus," "Type I interferon production," and "Response to type I interferon") and signaling pathways (e.g., "Interferon signaling" and "Class I MHC mediated antigen processing presentation"), reflecting the first responses initiated in β-cells during the early stages of diabetes. Overall, this integrated analysis allowed the identification of essential biological processes and signaling pathways activated over the course of T1D (Colli et al., 2020b).

We will discuss below some of the direct effects of IFNα on human β-cells.

6.3.1 HLA class I overexpression and upregulation of chemokines

The hyperexpression of MHC class I proteins observed in insulin-containing islets from T1D patients is of particular relevance to the pathogenesis of the disease, as β-cells expressing HLA class I antigens on the surface will be recognized by autoreactive $CD8^+$ T cells that will selectively attack and kill targeted β-cells (Coppieters et al., 2012; Eizirik et al., 2009). As discussed above, IFNα has emerged as the main responsible for the upregulation of HLA class I in the islet milieu, and here we will cover the recent evidence reinforcing this statement in human β-cells.

Based on previous evidence, Marroqui et al. hypothesized that IFNα might be a common mediator of three hallmarks of the pancreatic islets in early human T1D, namely HLA class I overexpression, ER stress and β-cell apoptosis (Marroqui et al., 2017). The authors demonstrated that IFNα-induced HLA class I overexpression as well as upregulation of the chemokine CXCL10 and the antiviral protein MX1 in human β-cells. Assessment of the signal transduction mediating these effects showed that TYK2, STAT2, and IRF9, which are part of the classical type I IFNs signaling pathway, played an indispensable role in IFNα-mediated effects. Moreover, inhibition of ubiquitin-specific peptidase 18 (USP18), an ISG15-specific protease that regulates IFN signaling in β-cells (Santin et al., 2012), exacerbated IFNα effects in β-cells.

A follow-up study investigated whether IFNα effects in β-cells were reversible upon cytokine withdrawal (Coomans de Brachène et al., 2018).

Interestingly, IFNα not only induced a robust increase in MHC class I expression but also seemed to stabilize the protein at the cell surface of human β-cells.

In 2020, an elegant study performed a multi-omics analysis comprised of RNA-seq, ATAC-seq (assay for transposase-accessible chromatin using sequencing), and proteomics to determine the global impact of this cytokine on human β-cells (Colli et al., 2020a). Among many interesting findings (some of them will be later discussed in the following sections), this integrative analysis identified IRF1, STAT1, and STAT2 as key transcription factors involved in the regulation of IFNα signaling. They found that IFNα-induced upregulation of the chemokines *CXCL1* and *CXCL10*, the HLA-I component beta-2-microglobulin (*B2M*), and *SOCS3* were prevented following siRNA-mediated silencing of these transcription factors (Colli et al., 2020a).

These findings indicate that IFNα has a central role in the early phases of diabetes as modulator of HLA class I overexpression and excessive inflammatory response. Thus, it seems that, along with the putative generation of β-cell neoantigens, a long-lasting HLA class I upregulation will likely amplify antigen presentation during the early stages of T1D.

6.3.2 Induction of ER stress and its putative role in β-cell function and apoptosis

Along with HLA class I hyperexpression (discussed above), ER stress and β-cell apoptosis comprise the three key hallmarks of human T1D (Marroqui et al., 2017).

Analysis of ER stress markers in insulin-containing islets from T1D patients showed that expression of CHOP and BIP was increased, whereas XBP-1 expression remained unchanged (Marhfour et al., 2012). Additionally, evaluation of UPR markers in pancreases from the nPOD cohort supported the presence of two ER stress markers, ATF6 and XPB1s, in insulin-positive β-cells of patients with T1D (Engin et al., 2013). Complementing human reports, mouse studies have also shown the importance of ER stress to diabetes pathogenesis (Austin et al., 2020; Engin et al., 2013; Song et al., 2008; Tersey et al., 2012).

Concurrently, two independent studies reported that IFNα upregulates several ER stress markers, including p-EIF2α, XBP1s, BIP, CHOP, ATF3, and ATF4, in the human insulin-producing cell line EndoC-βH1 and in human islets (Lombardi and Tomer, 2017; Marroqui et al., 2017). These findings have been confirmed in recent publications (Colli et al., 2020a;

Coomans de Brachène et al., 2018, 2020). ER stress induced by IFNα was prevented by either silencing of proteins of the TYK2-STAT2-IRF9 axis (Marroqui et al., 2017) or pre-treatment with the chemical chaperones tauroursodeoxycholic acid (TUDCA) and 4-phenyl butyric acid (PBA) (Lombardi and Tomer, 2017).

As ER stress may contribute to β-cell dysfunction and apoptosis (Eizirik et al., 2008, 2013; Tersey et al., 2012), IFNα effects on β-cell function and survival were investigated. In human β-cells, IFNα had no effects on the expression of *INS* or *PDX1* and did not affect glucose-stimulated insulin secretion. On the other hand, IFNα did increase proinsulin: insulin ratio and impaired proinsulin convertase 1/3 (*PCSK1*) and PC2 (*PCSK2*) expression (Coomans de Brachène et al., 2020; Lombardi and Tomer, 2017); once again, TUDCA and PBA prevented IFNα effects, which is indicative of an IFNα-induced ER folding/processing defect. The studies by Lombardi et al. and Coomans de Brachène et al. reported differences regarding IFNα effects on insulin content; the former noted a reduction in insulin content, while the latter did not observe any changes. These differences are likely due to differences in experimental design and methodology. These studies suggest that, if any, IFNα has only a mild/limited effect on human β-cell function.

Contrary to previous data in other cell types (Kotredes and Gamero, 2013; Thyrell et al., 2002; Yanase et al., 2000), IFNα alone did not induce apoptosis in human β-cells, indicating that IFNα-induced ER stress was not sufficient to induce β-cell death. Based on prior evidence that a mild ER stress sensitized rat β-cells to IL-1β-induced apoptosis (Miani et al., 2012), the authors tested the effects of the concomitant exposure to IFNα plus IL-1β on β-cell survival. This cytokine combination increased human β-cell death, which was accompanied by higher expression of ER stress markers when compared with IFNα or IL-1β alone. As further evidence that IFNα plus IL-1β-induced ER stress was, at least in part, leading to β-cell death, silencing of CHOP or co-treatment with TUDCA partially protected β-cells against cytokine-induced apoptosis (Marroqui et al., 2017).

The evidence above suggest that IFNα-induced ER stress may not be enough to cause β-cell dysfunction and apoptosis but may sensitize β-cells to the other cytokines (e.g., IL-1β) during early insulitis.

6.3.3 Putative involvement in the generation of neoantigens

As mentioned above, β-cell recognition by autoreactive T cells depends on the expression of specific autoantigens on the β-cell surface in the context of HLA class I. Although many of these β-cell peptides are recognized in their native state by $CD8^+$ T cells, stressful circumstances, such as exposure to proinflammatory cytokines or a viral infection, elicit several cellular processes that can lead to the generation of β-cell neoantigens (Gonzalez-Duque et al., 2018; Marré and Piganelli, 2017; Rondas et al., 2015). Once recognized by infiltrating $CD8^+$ T cells that may have escaped thymic deletion, these neoantigens can contribute to trigger and/or intensify the autoimmune response (James et al., 2018; Mallone and Eizirik, 2020). Among the processes inducing the production of neoantigens, ER stress-induced posttranslational modifications seem to be the most common (James et al., 2018; Marré et al., 2015). Additionally, novel mechanisms of neoantigens generation have been recently described, including, for instance, epigenetic modifications (Rui et al., 2016), translation of defective ribosomal initiation products (Kracht et al., 2017), and alternative splicing (Gonzalez-Duque et al., 2018).

The multi-omics analysis performed by Colli et al. showed that exposure of human β-cells to IFNα promoted major alterations in the alternative splicing landscape. Overall, IFNα improved the number of different splicing events; changed the usage of alternative transcription start sites, which increased the inventory of IFNα-regulated transcript; augmented intron removal and, consequently, protein expression; and lead to the generation of alternatively spliced proteins (Colli et al., 2020a). IFNα also induced pathways involved in posttranslational modifications (e.g., ubiquitination) and protein degradation (e.g., proteasome). Thus, it is possible that all these IFNα-induced posttranscriptional and posttranslational modifications, when associated to exacerbated ER stress, may lead to the generation of β-cell neoantigens.

6.3.4 Impact on epigenetic modifications

Epigenetic regulations, which are important molecular links between environment and genes, modulate pancreatic β-cell differentiation, function, and survival (Xu et al., 2020). Accumulating evidence implicate epigenetic modulations in the etiology of T1D (Jerram et al., 2017; Miao et al., 2008, 2012). For instance, DNA methylation and posttranscriptional modification of histones, two of the main epigenetic modifications, have been reported in whole blood, peripheral blood, and monocytes of T1D patients (Xie et al., 2020). A 2019 study demonstrated that exposure of human β-cells to proinflammatory cytokines caused deep changes in gene expression and protein production, which were associated with chromatin remodeling and plasticity of the β-cell regulatory landscape (Ramos-Rodríguez et al., 2019).

Adding to previous evidence suggesting that pathways related to type I IFNs might be involved in epigenetic changes in T1D (Bonifacio et al., 2011; Miao et al., 2013), recent omics analyses have contributed to unveil the impact that cytokine-induced epigenetic modifications have on β-cells and autoimmunity development.

Stefan-Lifshitz et al. exposed human islets to IFNα and analyzed the DNA methylation (DNAm) patterns and gene expression (Stefan-Lifshitz et al., 2019). Upon treatment with IFNα, around 70% of the CpG sites were hypomethylated; most of changes were in pathways related to protein phosphorylation, fatty acid oxidation, and apoptosis. Integration of RNA-seq with DEGseq analysis demonstrated that IFNα-upregulated genes were overrepresented in pathways related to the innate or adaptive immunity, such as antiviral response and antigen processing and presentation. Additionally, ingenuity pathway analysis identified IRFs and STATs as major regulators of the transcripts upregulated by IFNα exposure, with IRF7 playing a crucial role. It is interesting to note that IRF7 is the master regulator of type I IFN-dependent immune responses (Honda et al., 2005) and of an IRF7-driven inflammatory network whose regulatory locus have been implicated in the pathogenesis of T1D (Heinig et al., 2010). Mechanistically, IFNα-induced DNA hypomethylation by active DNA demethylation via the PNPT1-miR-26a-TET2 cascade (Stefan-Lifshitz et al., 2019). Briefly, the authors noted that IFNα upregulated the exoribonuclease PNPase old-35, also known as PNPT1, which, in turn, degraded a specific miRNA, miR-26a. Thus, as this miRNA is a negative regulator of Ten-eleven translocation (TET) enzymes, which catalyze DNA demethylation, IFNα indirectly promoted TET-dependent active DNA demethylation (Stefan-Lifshitz et al., 2019).

As discussed above, Colli et al. applied a multi-omics approach to study the different levels of gene regulation induced by IFNα, which allowed the assessment of early, intermediate, and late responses (Colli et al., 2020a). This approach indicated that IFNα induced early changes in chromatin accessibility associated with alterations in gene expression and protein abundance. More than 4000 regions of gained open chromatin regions (OCRs) were identified as early as 2h upon treatment IFNα, and this number declined to 1000 regions by 24h. Most of the gained OCRs were enriched for transcription factors binding motifs and might act as potential regulatory elements due to their localization in regions distally to gene transcription starting sites (Colli et al., 2020a). Interestingly, motifs of islet-specific ranscription factors were identified in these new regions of accessible chromatin, indicating that local responses during insulitis might be regulated by tissue-specific transcription factors, as previously described (Ramos-Rodríguez et al., 2019). For instance, islet-specific transcription factors could explain why HLA class I is preferentially expressed in pancreatic islet cells if compared to the neighboring exocrine pancreas.

Subsequent analysis showed that changes in chromatin remodeling were associated with variations in gene expression, where IFNα-induced genes associated preferentially with gained OCRs than with stable regions (Colli et al., 2020a). Similar to previous observations in IFNα-treated human islets (Stefan-Lifshitz et al., 2019), gene ontology analysis associated differentially IFNα-regulated abundant proteins to a number of biological processes involved in the pathogenesis of T1D, including antigen processing and presentation, defense responses to viruses, and apoptosis (Colli et al., 2020a). Furthermore, ER stress-related genes were also upregulated by IFNα at both the mRNA and protein levels, which agrees with previous findings (Lombardi and Tomer, 2017; Marroqui et al., 2017).

Altogether, these studies provide evidence to explain how IFNα could trigger autoimmunity by also impacting the β-cell epigenetic landscape.

6.3.5 Immune checkpoint proteins PD-L1 and HLA-E: An IFNα-mediated defense mechanism?

Notwithstanding IFNα effects on the induction of local innate immunity activation in early diabetes, not all responses induced by this cytokine are deleterious to pancreatic β-cells. New evidence demonstrates that IFNα upregulates β-cell expression of two immune checkpoint proteins, namely programmed death-ligand 1 (PD-L1) (Colli et al., 2018, 2020a; Osum et al., 2018) and HLA-E (Colli et al., 2020a) in an IRF1-dependent manner. PD-L1 is an immune regulatory protein involved in the regulation of antigen-specific $CD8^+$ T cell proliferation (Freeman et al., 2000; Karwacz et al., 2011), while HLA-E is a non-classical HLA-I molecule implicated in cell recognition by NK cells (Braud et al., 1998). As the expression of both proteins is increased in insulin-containing islets from people with T1D (Colli et al., 2018, 2020a; Lundberg et al., 2016; Planas et al., 2010), it has been speculated that upregulation of PD-L1 and HLA-E during insulitis may work as a defense mechanism aiming to attenuate or delay the autoimmune assault by blocking the actions of infiltrating immune cells

(Akhbari et al., 2020; Colli et al., 2020b). According to Colli et al. (2020b), this hypothesis is supported by the following evidence: (1) treatment of malignant cancers with antibodies targeting PD-1/PD-L1 increases the risk of developing T1D (Stamatouli et al., 2018); (2) PD-L1-expressing β-cells survive the immune attack in NOD mice (Rui et al., 2017); (3) incidence of diabetes was decreased in transgenic NOD mice overexpressing PD-L1 in β-cells (Wang et al., 2008); and (4) both PD-L1 and HLA-E are not expressed in insulin-deficient isles from T1D patients (Colli et al., 2018, 2020a).

In this context of insulitis, IRF1 arises as a key regulator of this defense response, as IRF1 upregulates checkpoint proteins (e.g., PD-L1 and HLA-E) (Colli et al., 2018, 2020a) and negative regulators of cytokine signaling (e.g., SOCS1 and SOCS3) (Colli et al., 2020a; Moore et al., 2011). Moreover, IRF1 provides a negative feedback on cytokine-induced, STAT1-driven chemokine production (Gysemans et al., 2009; Moore et al., 2011).

7. Targeting type I IFN production and/or signaling as potential therapeutic approaches

To date, there are no effective interventions to prevent T1D and the current therapies focused on stimulating β-cell regeneration or based on immunotherapy have limited clinical efficacy. Therefore, it is timely to pursue novel therapeutic strategies to prevent or treat T1D. Based on the evidence mentioned above, which places type I IFNs as key initiators of T1D development, we propose that targeting type I IFN signaling pathway in pre-symptomatic T1D subjects (i.e., positive for 2–3 autoantibodies but still normoglycemic) or in the very early stages of T1D onset might be an interesting adjuvant therapy for T1D. For this purpose, different approaches might be implemented as adjuvant therapies for T1D. According to Stewart, we could use three main strategies to target IFNα: (1) decrease IFNα extracellular levels; (2) block IFNα itself or its signal transduction via IFNAR; and (3) reduce the activity of the proteins mediating IFN effects (Stewart, 2003). In this section we will discuss some possible therapeutic interventions based on these strategies.

7.1 Targeting IFN production

As discussed in Section 3, β-cells and pDCs can locally produce IFNα in response to different stimuli, such as products of viral infections.

Although it seems reasonable to consider a therapy focused on blocking IFN release by these and other IFNα-producing cells, it may be difficult to find a way to reduce all agents responsible for the pathological production of IFNα. In any case, the removal of exogenous and endogenous "danger signals" to undermine signaling through host pattern-recognition receptors emerges as an interesting alternative to hamper IFN production and the subsequent IFN-induced inflammatory processes (Jean-Baptiste et al., 2017).

In vitro studies have shown that some compounds can protect human islets against the deleterious effects of enterovirus infection. The anti-enteroviral drug pleconaril reduced the replication of two CBV4 strains in human islets (Berg et al., 2007), while treatment with nicotinamide, a form of vitamin B3, nor only decreased viral replication but also reduced virus-induced islet destruction (Moëll et al., 2009).

Another potential therapeutic approach aiming viral infections is the development of vaccines against CVB strains. There has been a great increase in this area in the last 5 years, where more than 10 studies have reported different anti-CVB vaccines. We will discuss some examples below.

Flodström-Tullberg and co-workers have published preclinical studies designed to test the efficacy and safety of a formalin-inactivated CVB1 vaccine in experimental mouse models of T1D, namely NOD and *SOCS1*-tg mice (Hankaniemi et al., 2017; Larsson et al., 2015; Stone et al., 2018). Vaccinated mice showed higher titers of CVB1 neutralizing antibodies than mock-treated (i.e., non-vaccinated) animals. Vaccinated animals also presented lower levels of replicating virus in their blood and pancreas upon CVB1 infection. Notably, upon CVB1 infection, diabetes onset was accelerated in non-vaccinated prediabetic NOD mice, whereas vaccinated mice showed no accelerated disease manifestation (Larsson et al., 2015). Moreover, *SOCS1*-tg mice, which are highly susceptible to virus-induced diabetes due to the expression of *Socs1* in β-cells (Flodström et al., 2002), were protected from CVB1-induced diabetes, showing normal pancreas morphology and no signs of β-cell loss (Stone et al., 2018). Recently, the same group developed a novel hexavalent vaccine targeting the six known CVB serotypes (CVB1–6). Preclinical testing in murine models and nonhuman primates (rhesus macaques) showed that this novel CVB1–6 hexavalent vaccine was safe and induced a robust neutralizing antibody response to the six CVB serotypes, which conferred protection against acute CVB infections. Remarkably, immunization with the hexavalent vaccine prevented CVB1- and CVB3-induced diabetes in *SOCS1*-tg mice (Stone et al., 2020). Of note, 2 weeks after the last immunization, pancreas and spleen from non-vaccinated and vaccinated animals showed similar expression of genes related to innate antiviral immunity, such as *Ifnα*, *Ifnβ*, *Pkr*, *Oas1*, and *Nos2* (Stone et al., 2020).

Overall, these data encourage the development of antiviral strategies as valid alternatives to reduce the risk of T1D development.

7.2 Blocking IFN signaling pathway

Targeting the IFN binding to its receptor and activation of the downstream signaling pathway is possibly the most promising therapeutic approach to treat/prevent T1D. From antibodies against IFNα/IFNAR1 to inhibition of protein kinases involved in the IFN pathway, this has been the most studied therapeutic avenue so far. Despite its potential, one limitation of this approach is the increase in viral infections, as observed during the clinical trials with SLE patients treated with an anti-IFNα monoclonal antibody (Khamashta et al., 2016).

7.2.1 Monoclonal antibodies against IFNα and IFNAR1

Supporting a role for IFNα in the etiology of different autoimmune diseases, human monoclonal antibodies targeting IFNα (Sifalimumab; Khamashta et al., 2016; Petri et al., 2013) and its receptor IFNAR1 (Anifrolumab; Furie et al., 2017; Morand et al., 2020) have shown efficacy in clinical trials performed in patients with SLE. Sifalimumab binds to and specifically neutralizes multiple IFNα subtypes seen in the serum of lupus patients (Yao et al., 2009), whereas anifrolumab reduced cell surface IFNAR1 and blocked IFN-dependent signaling (Riggs et al., 2018). In addition, immunization with IFNα-kinoid, also known as IFN-K, a therapeutic vaccine composed of inactivated IFNα coupled to a carrier protein, induces polyclonal antibodies that neutralize all 13 human IFNα subtypes in a model of human IFNα transgenic mice (Lauwerys et al., 2013). A recent phase IIb trial showed that patients with moderate to severe SLE who received IFN-K presented higher levels of neutralizing anti-IFNα2b serum antibodies and decreased IFN gene signature (Houssiau et al., 2020).

In the context of T1D, preclinical data suggest that neutralizing antibodies against IFNAR1 prevents diabetes in different T1D mouse models. McDevitt and co-workers observed a delay in diabetes onset and reduced T1D incidence in 2- to 3-week-old NOD mice treated with an antibody specific to IFNAR1 (Li et al., 2008; Li and McDevitt, 2011).

Recently, Marro and co-workers used a viral mouse model of T1D that mimics several immunological and histopathological components of human T1D, to demonstrate that blockage of IFNα, but not IFNβ, signaling prevents T1D development (Marro et al., 2017). In this study, animals treated with neutralizing antibodies targeting either the IFNAR1 (anti-IFNAR) or six murine IFNα species (anti-IFNα) were normoglycemic and had limited trafficking of autoaggressive lymphocytes into the islets and reduced islet immunopathology (Marro et al., 2017).

Altogether, these findings indicate that targeting IFNα with monoclonal antibodies or vaccines represent a new therapeutic avenue that should be considered in the early stages of T1D, where they can help to preserve β-cell function and/or survival.

7.2.2 JAK inhibitors (JAKinibis)

Proteins belonging to the JAK family play critical roles as signal transducers downstream of IFNAR receptor activation and are implicated in several immune-mediated diseases. Small molecule JAK inhibitors, known as JAKinibs, have been successfully used to treat many autoimmune diseases (for thorough review, see Nogueira et al., 2020; Roskoski, 2016; Schwartz et al., 2017; Virtanen et al., 2019). For instance, phase II and phase III clinical studies conducted with first-generation JAKinibs ruxolitinib and baricitinib provided significant clinical benefits in patients with psoriasis (Ruxolitinib; Punwani et al., 2012), SLE (Baricitinib; Wallace et al., 2018), and rheumatoid arthritis (Baricitinib; Genovese et al., 2016; Taylor et al., 2017); a phase 2 trial with the selective TYK2 inhibitor BMS-986165 proved to be beneficial for patients with moderate-to-severe plaque psoriasis (Papp et al., 2018).

Preclinical studies suggest that JAK/TYK2 inhibitors could be repurposed for T1D treatment. Administration of AZD1480, a JAK1/JAK2 inhibitor, protected NOD mice against autoimmune diabetes and reversed diabetes in newly diagnosed NOD mice (Trivedi et al., 2017). Additionally, ruxolitinib and baricitinib prevented IFNα effects (e.g., MHC class I and CXCL10 upregulation) on human β-cells (Colli et al., 2020a; Coomans de Brachène et al., 2018, 2020). Finally, two novel TYK2 inhibitors protected human β-cells against the deleterious effects of IFNα without compromising β-cell function (i.e., glucose-stimulated insulin secretion) and susceptibility to potentially diabetogenic viruses (CVB1 and CVB5) (Coomans de Brachène et al., 2020). However, it is important to keep in mind that complete TYK2 inhibition/loss is not desired as it may lead to immunodeficiency and susceptibility to various microorganisms, such as mycobacteria, fungi, and virus (Minegishi et al., 2006). Conversely, partial TYK2 inhibition, as observed in individuals homozygous for the minor, protective allele of the TYK2 SNP rs34536443 (CC) (Dendrou et al., 2016), could provide maximal efficacy with minimized risk to develop secondary infections. Therefore, it is crucial to pursue/design drugs with the ability to only partially inhibit TYK2, especially for the treatment of early-stage disease.

A very recent case study reported a patient diagnosed with STAT1 gain-of-function disease, a monogenic syndrome associated with a heterozygous mutation in *STAT1* (c.1154C → T, pT385 M in this case), who was also diagnosed with T1D at 17 years of age (Chaimowitz et al., 2020). The patient began to receive ruxolitinib 9 months after being diagnosed with T1D. Twenty-one months after diabetes diagnosis, and 12 months upon initiation of treatment with ruxolitinib, exogenous insulin therapy was ceased and the patient remained euglycemic for at least 1 year after discontinuation of insulin therapy (Chaimowitz et al., 2020).

Therefore, the available preclinical evidence along with this single case study provide further evidence that JAKinibs might be a useful therapy to be used as early therapy of T1D or even to reverse diabetes in young adults.

7.3 Preventing IFNα effects on β-cells

As previously discussed, through the induction of several genes, IFNα elicits some deleterious effects on β-cells, including ER stress and apoptosis (when associated to IL-1β). Therefore, therapies directed at such molecular processes could represent new avenues for the prevention of β-cell demise and consequent onset of T1D. Aiming to identify drugs/compounds that could potentially revert the proinflammatory signatures present in β-cells during T1D, Colli et al. mined the RNA-sequencing of FACS-purified β-cells obtained from T1D individuals using the Connectivity Map database. Among the several classes of drugs/compounds identified were bile acids and bromodomain inhibitors, which will be briefly discussed below (Colli et al., 2020b).

7.3.1 Bile acids

It has been previously shown that several bile acids act as chemical chaperones and prevent ER stress in different tissues, such as liver and adipose tissue, thus preventing ER stress-related diseases (e.g., obesity and type 2 diabetes) (Engin and Hotamisligil, 2010; Kusaczuk, 2019; Özcan et al., 2006; Xie et al., 2002). One bile acid in particular, named TUDCA, restored a defective UPR in β-cells, which resulted in improved β-cell survival and less apoptosis, increased β-cell mass and number of β-cells per islet, and preserved insulin secretion with better control of glucose metabolism (Bronczek et al., 2019; Engin et al., 2013). Importantly, TUDCA administration at the prediabetic stage markedly reduced diabetes incidence in T1D mouse models (Engin et al., 2013). Studies using human β-cells in vitro have shown that pre-treatment with TUDCA prevented IFNα-induced ER stress (Lombardi and Tomer, 2017) and apoptosis (Marroqui et al., 2017). Of note, a phase 2 clinical trial aiming to assess the clinical efficacy of TUDCA as a mitigator of ER stress in patients with recent-onset T1D is currently ongoing (ClinicalTrials.gov Identifier: NCT02218619).

7.3.2 Bromodomain inhibitors

Bromodomain-containing proteins recognize histone modification marks that bind acetylated lysines in histone tails and, in association with other chromatin remodeling complexes, regulate gene transcription. As Bromodomain and Extraterminal domain (BET) proteins have been implicated in the pathogenesis of cancer and autoimmune diseases, BET inhibition showed promising therapeutic activity (Gilan et al., 2020; Pérez-Salvia and Esteller, 2017). In fact, treatment with a small-molecule BET inhibitor irreversibly suppressed T1D development in NOD mice (Fu et al., 2014). Furthermore, two BET protein inhibitors, namely JQ1 and I-BET-151, partially prevented the deleterious effects induced by IFNα on human β-cells (Colli et al., 2020a).

All findings discussed above point to type I IFN signaling pathway as an encouraging avenue for T1D treatment, especially considering IFNα direct effects on β-cells (Fig. 5). Novel insights into single-nucleotide polymorphisms in genes involved in type I IFN pathway will open new possibilities for the design and development of more specific monoclonal antibodies and drugs/compounds that target proteins either upstream or downstream of IFN signaling. Such new approaches would allow a more directed treatment strategy (Jean-Baptiste et al., 2017).

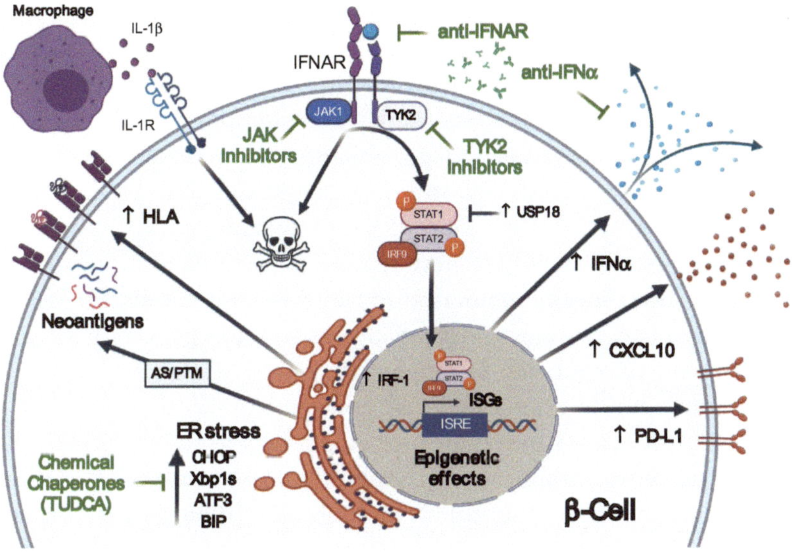

Fig. 5 Main type I IFN effects on pancreatic β-cells that may contribute to T1D pathogenesis. Upon activation of type I IFN signaling pathway, IFNα triggers several processes in β-cells: (1) increased production of chemokines and cytokines; (2) induction of posttranscriptional (e.g., alternative splicing) and posttranslational (e.g., ubiquitination) modifications; (3) induction of chromatin remodeling; and (4) long-lasting upregulation of antigen-presenting MHC class I proteins. All these IFNα effects contribute to β-cell dysfunction and demise. During this local islet inflammation, apoptotic β-cells function as additional "danger signals" and, together with presentation of (neo)antigens, lead to the activation of autoreactive $CD8^+$ T cells. Some defensive mechanisms are triggered to counteract the IFNα deleterious effects, such as upregulation of PD-L1 and HLA-E, two immune checkpoint proteins. As targeting type I IFN signaling pathway has arisen as a promising therapeutic approach, some strategies have been tested (in green), such as neutralizing antibodies against IFNα or its receptor IFNAR1, inhibitors of JAK proteins (JAK1 and TYK2), and chemical chaperones (e.g., TUDCA) aiming to attenuate IFNα-induced ER stress.

8. Concluding remarks

Significant progress has been made into unraveling the mechanisms whereby type I IFNs, especially IFNα, may bridge environmental triggers and genetic risk factors in early phases of T1D pathogenesis. As reviewed herein, locally produced type I IFNs, likely due to a viral infection, produce innate immune effects that may not only induce but also amplify an adaptive immune response against pancreatic β-cells. The genetic background also plays its role as T1D genetic risk variants involved in IFN signaling (e.g., *TYK2*) may build up this process.

Upon activation of its signaling pathway, IFNα regulates several cellular processes that put β-cells up against the wall: (1) increased production of chemokines and cytokines attracts monocytes, T lymphocytes, and NK cells to the islet milieu; (2) induction of posttranscriptional (e.g., alternative splicing) and posttranslational (e.g., ubiquitination) modifications; when associated to exacerbated ER stress, which could also be induced by IFNα, may lead to the generation of β-cell neoantigens; (3) induction of chromatin remodeling, where new regions of accessible chromatin were enriched for islets-specific transcription factor binding motifs, which could explain why HLA class I is preferentially expressed in pancreatic islet cells; (4) long-lasting upregulation of antigen-presenting MHC class I proteins; in conjunction with posttranscriptional/posttranslational modifications and ER stress, MHC class I overexpression might lead to sustained neoantigen presentation to the immune system. Amid this "IFNα storm", β-cells seem to have not much choice but to activate apoptotic pathways. Unfortunately, as once said Francis Bacon, sometimes "the remedy is worse than the disease": in this context of local islet inflammation, apoptotic β-cells function as additional "danger signals" and, together with (neo)antigen presentation by antigen-presenting cells, lead to the activation of autoreactive $CD8^+$ T cells. Trying to counteract the IFNα deleterious effects, β-cells upregulate PD-L1 and HLA-E, two immune checkpoint proteins, that may act as a defense response to attenuate or delay the autoimmune assault (Fig. 5).

As type I IFNs act as key initiators of T1D pathogenesis, targeting type I IFN signaling pathway has arisen as a promising therapeutic approach to treat either at-risk individuals or patients in the very early stages of T1D. Some strategies have been tested, such as neutralizing antibodies against IFNα or its receptor IFNAR1, inhibitors of JAK proteins (JAK1 and TYK2), and chemical chaperones aiming to attenuate IFNα-induced ER stress (Fig. 5). Moreover, Colli et al. have recently suggested that downregulation of HLA class I while keeping/increasing the expression of the checkpoint proteins PD-L1 and HLA-E would be an interesting approach to be pursued at the β-cell level (Colli et al., 2020b). More importantly, these pharmacological interventions should be safe and introduced as early as possible (pre-symptomatic T1D subjects or in the very early stages of T1D onset).

In conclusion, this chapter provides strong evidence suggesting that IFNα plays a pivotal role in the early phases of T1D pathogenesis by connecting innate and adaptive immune responses. However, we still have a long way to go, as there is still much to be learned regarding the underlying mechanisms regulating type I IFN production and function, especially in other islet-cell types, such as α- and δ-cells.

Acknowledgments

L.M. is supported by Generalitat Valenciana (SEJI/2018/023). CIBERDEM is an initiative of the Instituto de Salud Carlos III. The authors would like to recognize the work of many other colleagues that could not be cited herein due to space limitations. The authors acknowledge that all figures were created with BioRender.com.

Disclosure statement

The authors have nothing to disclose.

References

Agudo, J., Ayuso, E., Jimenez, V., Casellas, A., Mallol, C., Salavert, A., Tafuro, S., Obach, M., Ruzo, A., Moya, M., Pujol, A., Bosch, F., 2012. Vascular endothelial growth factor-mediated islet hypervascularization and inflammation contribute to progressive reduction of β-cell mass. Diabetes 61, 2851–2861. https://doi.org/10.2337/db12-0134.

Akatsuka, H., Yano, Y., Gabazza, E.C., Morser, J., Sasaki, R., Suzuki, T., Fujiwara, R., Katsuki, A., Takei, Y., Sumida, Y., 2009. A case of fulminant type 1 diabetes with coxsackie B4 virus infection diagnosed by elevated serum levels of neutralizing antibody. Diabetes Res. Clin. Pract. 84, e50–e52. https://doi.org/10.1016/j.diabres.2009.03.009.

Akhbari, P., Richardson, S.J., Morgan, N.G., 2020. Type 1 diabetes: interferons and the aftermath of pancreatic beta-cell enteroviral infection. Microorganisms 8, E1419. https://doi.org/10.3390/microorganisms8091419.

Alba, A., Puertas, M.C., Carrillo, J., Planas, R., Ampudia, R., Pastor, X., Bosch, F., Pujol-Borrell, R., Verdaguer, J., Vives-Pi, M., 2004. IFNβ accelerates autoimmune type 1 diabetes in nonobese diabetic mice and breaks the tolerance to β cells in nondiabetes-prone mice. J. Immunol. 173, 6667–6675. https://doi.org/10.4049/jimmunol.173.11.6667.

Allen, J.S., Pang, K., Skowera, A., Ellis, R., Rackham, C., Lozanoska-Ochser, B., Tree, T., Leslie, R.D.G., Tremble, J.M., Dayan, C.M., Peakman, M., 2009. Plasmacytoid dendritic cells are proportionally expanded at diagnosis of type 1 diabetes and enhance islet autoantigen presentation to T-cells through immune complex capture. Diabetes 58, 138–145. https://doi.org/10.2337/db08-0964.

Anagandula, M., Richardson, S.J., Oberste, M.S., Sioofy-Khojine, A.-B., Hyöty, H., Morgan, N.G., Korsgren, O., Frisk, G., 2014. Infection of human islets of langerhans with two strains of coxsackie B virus serotype 1: assessment of virus replication, degree of cell death and induction of genes involved in the innate immunity pathway. J. Med. Virol. 86, 1402–1411. https://doi.org/10.1002/jmv.23835.

Arimoto, K.I., Löchte, S., Stoner, S.A., Burkart, C., Zhang, Y., Miyauchi, S., Wilmes, S., Fan, J.B., Heinisch, J.J., Li, Z., Yan, M., Pellegrini, S., Colland, F., Piehler, J., Zhang, D.E., 2017. STAT2 is an essential adaptor in USP18-mediated suppression of type I interferon signaling. Nat. Struct. Mol. Biol. 24, 279–289. https://doi.org/10.1038/nsmb.3378.

Arimoto, K.-I., Miyauchi, S., Stoner, S.A., Fan, J.-B., Zhang, D.-E., 2018. Negative regulation of type I IFN signaling. J. Leukoc. Biol. 103, 1099–1116. https://doi.org/10.1002/JLB.2MIR0817-342R.

Arvan, P., Pietropaolo, M., Ostrov, D., Rhodes, C.J., 2012. Islet autoantigens: structure, function, localization, and regulation. Cold Spring Harb. Perspect. Med. 2, a007658. https://doi.org/10.1101/cshperspect.a007658.

Atkinson, M.A., Eisenbarth, G.S., Michels, A.W., 2014. Type 1 diabetes. Lancet 383, 69–82. https://doi.org/10.1016/S0140-6736(13)60591-7.

Austin, A.L.F., Daniels Gatward, L.F., Cnop, M., Santos, G., Andersson, D., Sharp, S., Gentry, C., Bevan, S., Jones, P., King, A.J.F., 2020. The KINGS *Ins2*$^{+/G32S}$ mouse: a novel model of beta cell endoplasmic reticulum stress and human diabetes. Diabetes 69, 2667–2677. https://doi.org/10.2337/db20-0570.

Babon, J.J., Lucet, I.S., Murphy, J.M., Nicola, N.A., Varghese, L.N., 2014. The molecular regulation of Janus kinase (JAK) activation. Biochem. J. 462, 1–13. https://doi.org/10.1042/BJ20140712.

Baechler, E.C., Batliwalla, F.M., Karypis, G., Gaffney, P.M., Ortmann, W.A., Espe, K.J., Shark, K.B., Grande, W.J., Hughes, K.M., Kapur, V., Gregersen, P.K., Behrens, T.W., 2003. Interferon-inducible gene expression signature in peripheral blood cells of patients with severe lupus. Proc. Natl. Acad. Sci. U. S. A. 100, 2610–2615. https://doi.org/10.1073/pnas.0337679100.

Barnett, A.H., Eff, C., Leslie, R.D.G., Pyke, D.A., 1981. Diabetes in identical twins—a study of 200 pairs. Diabetologia 20, 87–93. https://doi.org/10.1007/BF00262007.

Basters, A., Geurink, P.P., El Oualid, F., Ketscher, L., Casutt, M.S., Krause, E., Ovaa, H., Knobeloch, K.-P., Fritz, G., 2014. Molecular characterization of ubiquitin-specific protease 18 reveals substrate specificity for interferon-stimulated gene 15. FEBS J. 281, 1918–1928. https://doi.org/10.1111/febs.12754.

Basters, A., Knobeloch, K.P., Fritz, G., 2018. USP18—a multifunctional component in the interferon response. Biosci. Rep. 38, BSR20180250. https://doi.org/10.1042/BSR20180250.

Benkahla, M.A., Sabouri, S., Kiosses, W.B., Rajendran, S., Quesada-Masachs, E., von Herrath, M., 2020. Overexpression of HLA class I predominantly on alpha cell in at risk individuals for type 1 diabetes. bioRxiv. 2020.07.13.201079 https://doi.org/10.1101/2020.07.13.201079.

Berg, A.-K., Olsson, A., Korsgren, O., Frisk, G., 2007. Antiviral treatment of Coxsackie B virus infection in human pancreatic islets. Antiviral Res. 74, 65–71. https://doi.org/10.1016/j.antiviral.2006.12.001.

Bettaieb, A., Liu, S., Xi, Y., Nagata, N., Matsuo, K., Matsuo, I., Chahed, S., Bakke, J., Keilhack, H., Tiganis, T., Haj, F.G., 2011. Differential regulation of endoplasmic reticulum stress by protein tyrosine phosphatase 1B and T cell protein tyrosine phosphatase. J. Biol. Chem. 286, 9225–9235. https://doi.org/10.1074/jbc.M110.186148.

Blachère, N.E., Darnell, R.B., Albert, M.L., 2005. Apoptotic cells deliver processed antigen to dendritic cells for cross-presentation. PLoS Biol. 4, 223–231. https://doi.org/10.1371/journal.pbio.0030185.

Blum, S.I., Tse, H.M., 2020. Innate viral sensor MDA5 and coxsackievirus interplay in type 1 diabetes development. Microorganisms 8, 993. https://doi.org/10.3390/microorganisms8070993.

Bonifacio, E., Warncke, K., Winkler, C., Wallner, M., Ziegler, A.-G., 2011. Cesarean section and interferon-induced helicase gene polymorphisms combine to increase childhood type 1 diabetes risk. Diabetes 60, 3300–3306. https://doi.org/10.2337/db11-0729.

Bottazzo, G.F., Dean, B.M., McNally, J.M., MacKay, E.H., Swift, P.G.F., Gamble, D.R., 1985. In situ characterization of autoimmune phenomena and expression of HLA molecules in the pancreas in diabetic insulitis. N. Engl. J. Med. 313, 353–360. https://doi.org/10.1056/NEJM198508083130604.

Braud, V.M., Allan, D.S.J., O'Callaghan, C.A., Söderström, K., D'Andrea, A., Ogg, G.S., Lazetic, S., Young, N.T., Bell, J.I., Phillips, J.H., Lanier, L.L., McMichael, A.J., 1998. HLA-E binds to natural killer cell receptors CD94/NKG2A, B and C. Nature 391, 795–799. https://doi.org/10.1038/35869.

Brissova, M., Shostak, A., Shiota, M., Wiebe, P.O., Poffenberger, G., Kantz, J., Chen, Z., Carr, C., Jerome, W.G., Chen, J., Baldwin, H.S., Nicholson, W., Bader, D.M., Jetton, T., Gannon, M., Powers, A.C., 2006. Pancreatic islet production of vascular endothelial growth factor-A is essential for islet vascularization, revascularization, and function. Diabetes 55, 2974–2985. https://doi.org/10.2337/db06-0690.

Bronczek, G.A., Vettorazzi, J.F., Soares, G.M., Kurauti, M.A., Santos, C., Bonfim, M.F., Carneiro, E.M., Balbo, S.L., Boschero, A.C., Júnior, J.M.C., 2019. The bile acid TUDCA improves beta-cell mass and reduces insulin degradation in mice with early-stage of type-1 diabetes. Front. Physiol. 10, 561. https://doi.org/10.3389/fphys.2019.00561.

Browne, K.A., Blink, E., Sutton, V.R., Froelich, C.J., Jans, D.A., Trapani, J.A., 1999. Cytosolic delivery of granzyme B by bacterial toxins: evidence that endosomal disruption, in addition to transmembrane pore formation, is an important function of perforin. Mol. Cell. Biol. 19, 8604–8615. https://doi.org/10.1128/mcb.19.12.8604.

Brozzi, F., Eizirik, D.L., 2016. ER stress and the decline and fall of pancreatic beta cells in type 1 diabetes. Ups. J. Med. Sci. 121, 133–139. https://doi.org/10.3109/03009734.2015.1135217.

Brozzi, F., Gerlo, S., Grieco, F.A., Nardelli, T.R., Lievens, S., Gysemans, C., Marselli, L., Marchetti, P., Mathieu, C., Tavernier, J., Eizirik, D.L., 2014. A combined "omics" approach identifies N-Myc interactor as a novel cytokine-induced regulator of IRE1α protein and c-Jun N-terminal kinase in pancreatic beta cells. J. Biol. Chem. 289, 20677–20693. https://doi.org/10.1074/jbc.M114.568808.

Brozzi, F., Nardelli, T.R., Lopes, M., Millard, I., Barthson, J., Igoillo-Esteve, M., Grieco, F.A., Villate, O., Oliveira, J.M., Casimir, M., Bugliani, M., Engin, F., Hotamisligil, G.S., Marchetti, P., Eizirik, D.L., 2015. Cytokines induce endoplasmic reticulum stress in human, rat and mouse beta cells via different mechanisms. Diabetologia 58, 2307–2316. https://doi.org/10.1007/s00125-015-3669-6.

Busse, N., Paroni, F., Richardson, S.J., Laiho, J.E., Oikarinen, M., Frisk, G., Hyöty, H., de Koning, E., Morgan, N.G., Maedler, K., 2017. Detection and localization of viral infection in the pancreas of patients with type 1 diabetes using short fluorescently-labelled oligonucleotide probes. Oncotarget 8, 12620–12636. https://doi.org/10.18632/oncotarget.14896.

Cabrera, O., Berman, D.M., Kenyon, N.S., Ricordi, C., Berggren, P.O., Caicedo, A., 2006. The unique cytoarchitecture of human pancreatic islets has implications for islet cell function. Proc. Natl. Acad. Sci. U. S. A. 103, 2334–2339. https://doi.org/10.1073/pnas.0510790103.

Cabrera, S.M., Chen, Y.-G., Hagopian, W.A., Hessner, M.J., 2016. Blood-based signatures in type 1 diabetes. Diabetologia 59, 414–425. https://doi.org/10.1007/s00125-015-3843-x.

Campbell-Thompson, M., Fu, A., Kaddis, J.S., Wasserfall, C., Schatz, D.A., Pugliese, A., Atkinson, M.A., 2016. Insulitis and β-cell mass in the natural history of type 1 diabetes. Diabetes 65, 719–731. https://doi.org/10.2337/db15-0779.

Carrero, J.A., Calderon, B., Towfic, F., Artyomov, M.N., Unanue, E.R., 2013. Defining the transcriptional and cellular landscape of type 1 diabetes in the NOD mouse. PLoS One 8, e59701. https://doi.org/10.1371/journal.pone.0059701.

Chaimowitz, N.S., Ebenezer, S.J., Hanson, I.C., Anderson, M., Forbes, L.R., 2020. STAT1 gain of function, type 1 diabetes, and reversal with JAK inhibition. N. Engl. J. Med. 383, 1494–1496. https://doi.org/10.1056/NEJMc2022226.

Chan, J.Y., Cooney, G.J., Biden, T.J., Laybutt, D.R., 2011. Differential regulation of adaptive and apoptotic unfolded protein response signalling by cytokine-induced nitric oxide production in mouse pancreatic beta cells. Diabetologia 54, 1766–1776. https://doi.org/10.1007/s00125-011-2139-z.

Chehadeh, W., Kerr-Conte, J., Pattou, F., Alm, G., Lefebvre, J., Wattré, P., Hober, D., 2000a. Persistent infection of human pancreatic islets by coxsackievirus B is associated with alpha interferon synthesis in β cells. J. Virol. 74, 10153–10164. https://doi.org/10.1128/JVI.74.21.10153-10164.2000.

Chehadeh, W., Weill, J., Vantyghem, M., Alm, G., Lefèbvre, J., Wattré, P., Hober, D., 2000b. Increased level of interferon-α in blood of patients with insulin-dependent diabetes mellitus: relationship with coxsackievirus B infection. J Infect Dis 181, 1929–1939. https://doi.org/10.1086/315516.

Chiang, E.Y., Johnston, R.J., Grogan, J.L., 2013. EBI2 is a negative regulator of type I interferons in plasmacytoid and myeloid dendritic cells. PLoS One 8, e83457. https://doi.org/10.1371/journal.pone.0083457.

Cinek, O., Stene, L.C., Kramna, L., Tapia, G., Oikarinen, S., Witsø, E., Rasmussen, T., Torjesen, P.A., Hyöty, H., Rønningen, K.S., 2014. Enterovirus RNA in longitudinal blood samples and risk of islet autoimmunity in children with a high genetic risk of type 1 diabetes: the MIDIA study. Diabetologia 57, 2193–2200. https://doi.org/10.1007/s00125-014-3327-4.

Cnop, M., Welsh, N., Jonas, J.C., Jörns, A., Lenzen, S., Eizirik, D.L., 2005. Mechanisms of pancreatic β-cell death in type 1 and type 2 diabetes: many differences, few similarities. Diabetes 54, S97–107. https://doi.org/10.2337/diabetes.54.suppl_2.S97.

Coccia, E.M., Uzé, G., Pellegrini, S., 2006. Negative regulation of type I interferon signaling: facts and mechanisms. Cell. Mol. Biol. (Noisy-le-Grand) 52, 77–87.

Colli, M.L., Moore, F., Gurzov, E.N., Ortis, F., Eizirik, D.L., 2010. MDA5 and PTPN2, two candidate genes for type 1 diabetes, modify pancreatic β-cell responses to the viral by-product double-stranded RNA. Hum. Mol. Genet. 19, 135–146. https://doi.org/10.1093/hmg/ddp474.

Colli, M.L., Nogueira, T.C., Allagnat, F., Cunha, D.A., Gurzov, E.N., Cardozo, A.K., Roivainen, M., Op de Beeck, A., Eizirik, D.L., 2011. Exposure to the viral by-product dsrna or coxsackievirus b5 triggers pancreatic beta cell apoptosis via a bim/mcl-1 imbalance. PLoS Pathog. 7, e1002267. https://doi.org/10.1371/journal.ppat.1002267.

Colli, M.L., Hill, J.L.E., Marroquí, L., Chaffey, J., Dos Santos, R.S., Leete, P., Coomans de Brachène, A., Paula, F.M.M., Op de Beeck, A., Castela, A., Marselli, L., Krogvold, L., Dahl-Jorgensen, K., Marchetti, P., Morgan, N.G., Richardson, S.J., Eizirik, D.L., 2018. PDL1 is expressed in the islets of people with type 1 diabetes and is up-regulated by interferons-α and-γ via IRF1 induction. eBioMedicine 36, 367–375. https://doi.org/10.1016/j.ebiom.2018.09.040.

Colli, M.L., Paula, F.M., Marselli, L., Marchetti, P., Roivainen, M., Eizirik, D.L., Op de Beeck, A., 2019. Coxsackievirus B tailors the unfolded protein response to favour viral amplification in pancreatic β cells. J. Innate Immun. 11, 375–390. https://doi.org/10.1159/000496034.

Colli, M.L., Ramos-Rodríguez, M., Nakayasu, E.S., Alvelos, M.I., Lopes, M., Hill, J.L.E., Turatsinze, J.V., Coomans de Brachène, A., Russell, M.A., Raurell-Vila, H., Castela, A., Juan-Mateu, J., Webb-Robertson, B.J.M., Krogvold, L., Dahl-Jorgensen, K., Marselli, L., Marchetti, P., Richardson, S.J., Morgan, N.G., Metz, T.O., Pasquali, L.,

Eizirik, D.L., 2020a. An integrated multi-omics approach identifies the landscape of interferon-α-mediated responses of human pancreatic beta cells. Nat. Commun. 11, 2584. https://doi.org/10.1038/s41467-020-16327-0.

Colli, M.L., Szymczak, F., Eizirik, D.L., 2020b. Molecular footprints of the immune assault on pancreatic beta cells in type 1 diabetes. Front. Endocrinol. (Lausanne) 11, 666. https://doi.org/10.3389/fendo.2020.568446.

Coomans de Brachène, A., Dos Santos, R.S., Marroqui, L., Colli, M.L., Marselli, L., Mirmira, R.G., Marchetti, P., Eizirik, D.L., 2018. IFN-α induces a preferential long-lasting expression of MHC class I in human pancreatic beta cells. Diabetologia 61, 636–640. https://doi.org/10.1007/s00125-017-4536-4.

Coomans de Brachène, A., Castela, A., Op de Beeck, A., Mirmira, R.G., Marselli, L., Marchetti, P., Masse, C., Miao, W., Leit, S., Evans-Molina, C., Eizirik, D.L., 2020. Pre-clinical evaluation of TYK2 inhibitors for human beta cell protection in type 1 diabetes. Diabetes Obes. Metab. 22, 1827–1836. https://doi.org/10.1111/dom.14104.

Cooper, J.D., Smyth, D.J., Smiles, A.M., Plagnol, V., Walker, N.M., Allen, J.E., Downes, K., Barrett, J.C., Healy, B.C., Mychaleckyj, J.C., Warram, J.H., Todd, J.A., 2008. Meta-analysis of genome-wide association study data identifies additional type 1 diabetes risk loci. Nat. Genet. 40, 1399–1401. https://doi.org/10.1038/ng.249.

Coppieters, K.T., Dotta, F., Amirian, N., Campbell, P.D., Kay, T.W.H., Atkinson, M.A., Roep, B.O., von Herrath, M.G., 2012. Demonstration of islet-autoreactive CD8 T cells in insulitic lesions from recent onset and long-term type 1 diabetes patients. J. Exp. Med. 209, 51–60. https://doi.org/10.1084/jem.20111187.

Craig, M.E., Nair, S., Stein, H., Rawlinson, W.D., 2013. Viruses and type 1 diabetes: a new look at an old story. Pediatr. Diabetes 14, 149–158. https://doi.org/10.1111/pedi.12033.

Crow, M.K., Olferiev, M., Kirou, K.A., 2019. Type I interferons in autoimmune disease. Annu. Rev. Pathol. 14, 369–393. https://doi.org/10.1146/annurev-pathol-020117-043952.

Darville, M.I., Eizirik, D.L., 1998. Regulation by cytokines of the inducible nitric oxide synthase promoter in insulin-producing cells. Diabetologia 41, 1101–1108. https://doi.org/10.1007/s001250051036.

Davison, L.J., Wallace, C., Cooper, J.D., Cope, N.F., Wilson, N.K., Smyth, D.J., Howson, J.M.M., Saleh, N., Al-Jeffery, A., Angus, K.L., Stevens, H.E., Nutland, S., Duley, S., Coulson, R.M.R., Walker, N.M., Burren, O.S., Rice, C.M., Cambien, F., Zeller, T., Munzel, T., Lackner, K., Blankenberg, S., Fraser, P., Gottgens, B., Todd, J.A., 2012. Long-range DNA looping and gene expression analyses identify DEXI as an autoimmune disease candidate gene. Hum. Mol. Genet. 21, 322–333. https://doi.org/10.1093/hmg/ddr468.

de Veer, M.J., Holko, M., Frevel, M., Walker, E., Der, S., Paranjape, J.M., Silverman, R.H., Williams, B.R., 2001. Functional classification of interferon-stimulated genes identified using microarrays. J. Leukoc. Biol. 69, 912–920. https://doi.org/10.1189/jlb.1106655.

Delaney, C.A., Pavlovic, D., Hoorens, A., Pipeleers, D.G., Eizirik, D.L., 1997. Cytokines induce deoxyribonucleic acid strand breaks and apoptosis in human pancreatic islet cells. Endocrinology 138, 2610–2614. https://doi.org/10.1210/endo.138.6.5204.

Dendrou, C.A., Cortes, A., Shipman, L., Evans, H.G., Attfield, K.E., Jostins, L., Barber, T., Kaur, G., Kuttikkatte, S.B., Leach, O.A., Desel, C., Faergeman, S.L., Cheeseman, J., Neville, M.J., Sawcer, S., Compston, A., Johnson, A.R., Everett, C., Bell, J.I., Karpe, F., Ultsch, M., Eigenbrot, C., McVean, G., Fugger, L., 2016. Resolving TYK2 locus genotype-to-phenotype differences in autoimmunity. Sci. Transl. Med. 8, 363ra149. https://doi.org/10.1126/scitranslmed.aag1974.

Di Domizio, J., Cao, W., 2014. Fueling autoimmunity: type I interferon in autoimmune diseases. Expert Rev. Clin. Immunol. 9, 201–210. https://doi.org/10.1586/eci.12.106.

Diana, J., Simoni, Y., Furio, L., Beaudoin, L., Agerberth, B., Barrat, F., Lehuen, A., 2013. Crosstalk between neutrophils, B-1a cells and plasmacytoid dendritic cells initiates autoimmune diabetes. Nat. Med. 19, 65–73. https://doi.org/10.1038/nm.3042.

DiMeglio, L.A., Evans-Molina, C., Oram, R.A., 2018. Type 1 diabetes. Lancet 391, 2449–2462. https://doi.org/10.1016/S0140-6736(18)31320-5.

Dogusan, Z., Garcia, M., Flamez, D., Alexopoulou, L., Goldman, M., Gysemans, C., Mathieu, C., Libert, C., Eizirik, D.L., Rasschaert, J., 2008. Double-stranded RNA induces pancreatic β-cell apoptosis by activation of the toll-like receptor 3 and interferon regulatory factor 3 pathways. Diabetes 57, 1236–1245. https://doi.org/10.2337/db07-0844.

Dos Santos, R.S., Marroqui, L., Velayos, T., Olazagoitia-Garmendia, A., Jauregi-Miguel, A., Castellanos-Rubio, A., Eizirik, D.L., Castaño, L., Santin, I., 2019. DEXI, a candidate gene for type 1 diabetes, modulates rat and human pancreatic beta cell inflammation via regulation of the type I IFN/STAT signalling pathway. Diabetologia 62, 459–472. https://doi.org/10.1007/s00125-018-4782-0.

Dotta, F., Censini, S., Van Halteren, A.G.S., Marselli, L., Masini, M., Dionisi, S., Mosca, F., Boggi, U., Muda, A.O., Del Prato, S., Elliott, J.F., Covacci, A., Rappuoli, R., Roep, B.O., Marchetti, P., 2007. Coxsackie B4 virus infection of β cells and natural killer cell insulitis in recent-onset type 1 diabetic patients. Proc. Natl. Acad. Sci. U. S. A. 104, 5115–5120. https://doi.org/10.1073/pnas.0700442104.

Eizirik, D.L., Colli, M.L., 2020. Revisiting the role of inflammation in the loss of pancreatic β-cells in T1DM. Nat. Rev. Endocrinol. 16, 611–612. https://doi.org/10.1038/s41574-020-00409-6.

Eizirik, D.L., Sandler, S., Welsh, N., Cetkovic-Cvrlje, M., Nieman, A., Geller, D.A., Pipeleers, D.G., Bendtzen, K., Hellerström, C., 1994. Cytokines suppress human islet function irrespective of their effects on nitric oxide generation. J. Clin. Invest. 93, 1968–1974. https://doi.org/10.1172/JCI117188.

Eizirik, D.L., Cardozo, A.K., Cnop, M., 2008. The role for endoplasmic reticulum stress in diabetes mellitus. Endocr. Rev. 29, 42–61. https://doi.org/10.1210/er.2007-0015.

Eizirik, D.L., Colli, M.L., Ortis, F., 2009. The role of inflammation in insulitis and β-cell loss in type 1 diabetes. Nat. Rev. Endocrinol. 5, 219–226. https://doi.org/10.1038/nrendo.2009.21.

Eizirik, D.L., Sammeth, M., Bouckenooghe, T., Bottu, G., Sisino, G., Igoillo-Esteve, M., Ortis, F., Santin, I., Colli, M.L., Barthson, J., Bouwens, L., Hughes, L., Gregory, L., Lunter, G., Marselli, L., Marchetti, P., McCarthy, M.I., Cnop, M., 2012. The human pancreatic islet transcriptome: expression of candidate genes for type 1 diabetes and the impact of pro-inflammatory cytokines. PLoS Genet. 8, e1002552. https://doi.org/10.1371/journal.pgen.1002552.

Eizirik, D.L., Miani, M., Cardozo, A.K., 2013. Signalling danger: endoplasmic reticulum stress and the unfolded protein response in pancreatic islet inflammation. Diabetologia 56, 234–241. https://doi.org/10.1007/s00125-012-2762-3.

Eizirik, D.L., Pasquali, L., Cnop, M., 2020. Pancreatic β-cells in type 1 and type 2 diabetes mellitus: different pathways to failure. Nat. Rev. Endocrinol. 16, 349–362. https://doi.org/10.1038/s41574-020-0355-7.

Engin, F., Hotamisligil, G.S., 2010. Restoring endoplasmic reticulum function by chemical chaperones: an emerging therapeutic approach for metabolic diseases. Diabetes Obes. Metab. 12, 108–115. https://doi.org/10.1111/j.1463-1326.2010.01282.x.

Engin, F., Yermalovich, A., Ngyuen, T., Hummasti, S., Fu, W., Eizirik, D.L., Mathis, D., Hotamisligil, G.S., 2013. Restoration of the unfolded protein response in pancreatic β cells protects mice against type 1 diabetes. Sci. Transl. Med. 5, 211ra156. https://doi.org/10.1126/scitranslmed.3006534.

Estella, E., McKenzie, M.D., Catterall, T., Sutton, V.R., Bird, P.I., Trapani, J.A., Kay, T.W., Thomas, H.E., 2006. Granzyme B-mediated death of pancreatic β-cells requires the proapoptotic BH3-only molecule bid. Diabetes 55, 2212–2219. https://doi.org/10.2337/db06-0129.

Ewel, C.H., Sobel, D.O., Zeligs, B.J., Bellanti, J.A., 1992. Poly I:C accelerates development of diabetes mellitus in diabetes-prone BB rat. Diabetes 41, 1016–1021. https://doi.org/10.2337/diab.41.8.1016.

Fabris, P., Floreani, A., Lazzari, F.D., Betterle, C., Greggio, N.A., Naccarato, R., Chiaramonte, M., 1992. Development of type 1 diabetes mellitus during interferon alfa therapy for chronic HCV hepatitis. Lancet 340, 548. https://doi.org/10.1016/0140-6736(92)91744-S.

Fabris, P., Betterle, C., Greggio, N.A., Zanchetta, R., Bosi, E., Biasin, M.R., de Lalla, F., 1998. Insulin-dependent diabetes mellitus during alpha-interferon therapy for chronic viral hepatitis. J. Hepatol. 28, 514–517. https://doi.org/10.1016/S0168-8278(98)80328-0.

Fabris, P., Floreani, A., Tositti, G., Vergani, D., De Lalla, F., Betterle, C., 2003. Type 1 diabetes mellitus in patients with chronic hepatitis C before and after interferon therapy. Aliment. Pharmacol. Ther. 18, 549–558. https://doi.org/10.1046/j.1365-2036.2003.01681.x.

Ferreira, R.C., Guo, H., Coulson, R.M.R., Smyth, D.J., Pekalski, M.L., Burren, O.S., Cutler, A.J., Doecke, J.D., Flint, S., McKinney, E.F., Lyons, P.A., Smith, K.G.C., Achenbach, P., Beyerlein, A., Dunger, D.B., Clayton, D.G., Wicker, L.S., Todd, J.A., Bonifacio, E., Wallace, C., Ziegler, A.G., 2014. A type I interferon transcriptional signature precedes autoimmunity in children genetically at risk for type 1 diabetes. Diabetes 63, 2538–2550. https://doi.org/10.2337/db13-1777.

Flodström, M., Welsh, N., Eizirik, D.L., 1996. Cytokines activate the nuclear factor κB (NF-κB) and induce nitric oxide production in human pancreatic islets. FEBS Lett. 394, 263–267. https://doi.org/10.1016/0014-5793(96)00337-7.

Flodström, M., Maday, A., Balakrishna, D., Cleary, M.M., Yoshimura, A., Sarvetnick, N., 2002. Target cell defense prevents the development of diabetes after viral infection. Nat. Immunol. 3, 373–382. https://doi.org/10.1038/ni771.

Foulis, A.K., Farquharson, M.A., 1986. Aberrant expression of HLA-DR antigens by insulin-containing β-cells in recent-onset type I diabetes mellitus. Diabetes 35, 1215–1224. https://doi.org/10.2337/diab.35.11.1215.

Foulis, A.K., Farquharson, M.A., Hardman, R., 1987a. Aberrant expression of class II major histocompatibility complex molecules by B cells and hyperexpression of class I major histocompatibility complex molecules by insulin containing islets in type 1 (insulin-dependent) diabetes mellitus. Diabetologia 30, 333–343. https://doi.org/10.1007/BF00299027.

Foulis, A.K., Farquharson, M.A., Meager, A., 1987b. Immunoreactive α-interferon in insulin-secreting β cells in type 1 diabetes mellitus. Lancet 330, 1423–1427. https://doi.org/10.1016/S0140-6736(87)91128-7.

Fousteri, G., Liossis, S.N.C., Battaglia, M., 2013. Roles of the protein tyrosine phosphatase PTPN22 in immunity and autoimmunity. Clin. Immunol. 149, 556–565. https://doi.org/10.1016/j.clim.2013.10.006.

Franke, A., McGovern, D.P.B., Barrett, J.C., Wang, K., Radford-Smith, G.L., Ahmad, T., Lees, C.W., Balschun, T., Lee, J., Roberts, R., Anderson, C.A., Bis, J.C., Bumpstead, S., Ellinghaus, D., Festen, E.M., Georges, M., Green, T., Haritunians, T., Jostins, L., Latiano, A., Mathew, C.G., Montgomery, G.W., Prescott, N.J., Raychaudhuri, S., Rotter, J.I., Schumm, P., Sharma, Y., Simms, L.A., Taylor, K.D., Whiteman, D., Wijmenga, C., Baldassano, R.N., Barclay, M., Bayless, T.M., Brand, S., Büning, C., Cohen, A., Colombel, J.-F., Cottone, M., Stronati, L., Denson, T., De Vos, M.,

D'Inca, R., Dubinsky, M., Edwards, C., Florin, T., Franchimont, D., Gearry, R., Glas, J., Van Gossum, A., Guthery, S.L., Halfvarson, J., Verspaget, H.W., Hugot, J.-P., Karban, A., Laukens, D., Lawrance, I., Lemann, M., Levine, A., Libioulle, C., Louis, E., Mowat, C., Newman, W., Panés, J., Phillips, A., Proctor, D.D., Regueiro, M., Russell, R., Rutgeerts, P., Sanderson, J., Sans, M., Seibold, F., Steinhart, A.H., Stokkers, P.C.F., Torkvist, L., Kullak-Ublick, G., Wilson, D., Walters, T., Targan, S.R., Brant, S.R., Rioux, J.D., D'Amato, M., Weersma, R.K., Kugathasan, S., Griffiths, A.M., Mansfield, J.C., Vermeire, S., Duerr, R.H., Silverberg, M.S., Satsangi, J., Schreiber, S., Cho, J.H., Annese, V., Hakonarson, H., Daly, M.J., Parkes, M., 2010. Genome-wide meta-analysis increases to 71 the number of confirmed Crohn's disease susceptibility loci. Nat. Genet. 42, 1118–1125. https://doi.org/10.1038/ng.717.

Freeman, G.J., Long, A.J., Iwai, Y., Bourque, K., Chernova, T., Nishimura, H., Fitz, L.J., Malenkovich, N., Okazaki, T., Byrne, M.C., Horton, H.F., Fouser, L., Carter, L., Ling, V., Bowman, M.R., Carreno, B.M., Collins, M., Wood, C.R., Honjo, T., 2000. Engagement of the PD-1 immunoinhibitory receptor by a novel B7 family member leads to negative regulation of lymphocyte activation. J. Exp. Med. 192, 1027–1034. https://doi.org/10.1084/jem.192.7.1027.

Fu, W., Farache, J., Clardy, S.M., Hattori, K., Mander, P., Lee, K., Rioja, I., Weissleder, R., Prinjha, R.K., Benoist, C., Mathis, D., 2014. Epigenetic modulation of type 1 diabetes via a dual effect on pancreatic macrophages and β cells. eLife 3, e04631. https://doi.org/10.7554/eLife.04631.

Furie, R., Khamashta, M., Merrill, J.T., Werth, V.P., Kalunian, K., Brohawn, P., Illei, G.G., Drappa, J., Wang, L., Yoo, S., 2017. Anifrolumab, an anti-interferon-α receptor monoclonal antibody, in moderate-to-severe systemic lupus erythematosus. Arthritis Rheumatol. 69, 376–386. https://doi.org/10.1002/art.39962.

Gallagher, G.R., Brehm, M.A., Finberg, R.W., Barton, B.A., Shultz, L.D., Greiner, D.L., Bortell, R., Wang, J.P., 2015. Viral infection of engrafted human islets leads to diabetes. Diabetes 64, 1358–1369. https://doi.org/10.2337/db14-1020.

Gamble, D.R., Taylor, K.W., 1969. Seasonal incidence of diabetes mellitus. BMJ 3, 631–633. https://doi.org/10.1136/bmj.3.5671.631.

Gamble, D.R., Kinsley, M.L., FitzGerald, M.G., Bolton, R., Taylor, K.W., 1969. Viral antibodies in diabetes mellitus. BMJ 3, 627–630. https://doi.org/10.1136/bmj.3.5671.627.

García, M., Dogusan, Z., Moore, F., Sato, S., Hartmann, G., Eizirik, D.L., Rasschaert, J., 2009. Regulation and function of the cytosolic viral RNA sensor RIG-I in pancreatic beta cells. Biochim. Biophys. Acta, Mol. Cell Res. 1793, 1768–1775. https://doi.org/10.1016/j.bbamcr.2009.09.005.

Genovese, M.C., Kremer, J., Zamani, O., Ludivico, C., Krogulec, M., Xie, L., Beattie, S. D., Koch, A.E., Cardillo, T.E., Rooney, T.P., Macias, W.L., De Bono, S., Schlichting, D.E., Smolen, J.S., 2016. Baricitinib in patients with refractory rheumatoid arthritis. N. Engl. J. Med. 374, 1243–1252. https://doi.org/10.1056/NEJMoa1507247.

George, S., Deshpande, G.R., Sapkal, G.N., 2020. Interferons and their role in viral infection. In: Dynamics of immune activation in viral diseases. Springer, Singapore, pp. 61–80. https://doi.org/10.1007/978-981-15-1045-8_4.

Gessani, S., Conti, L., Del Cornò, M., Belardelli, F., 2014. Type I interferons as regulators of human antigen presenting cell functions. Toxins (Basel). 6, 1696–1723. https://doi.org/10.3390/toxins6061696.

Ghodke-Puranik, Y., Niewold, T.B., 2015. Immunogenetics of systemic lupus erythematosus: a comprehensive review. J. Autoimmun. 64, 125–136. https://doi.org/10.1016/j.jaut.2015.08.004.

Gilan, O., Rioja, I., Knezevic, K., Bell, M.J., Yeung, M.M., Harker, N.R., Lam, E.Y.N., Chung, C., Bamborough, P., Petretich, M., Urh, M., Atkinson, S.J., Bassil, A.K., Roberts, E.J., Vassiliadis, D., Burr, M.L., Preston, A.G.S., Wellaway, C., Werner, T., Gray, J.R., Michon, A.M., Gobbetti, T., Kumar, V., Soden, P.E., Haynes, A., Vappiani, J., Tough, D.F., Taylor, S., Dawson, S.J., Bantscheff, M., Lindon, M., Drewes, G., Demont, E.H., Daniels, D.L., Grandi, P., Prinjha, R.K., Dawson, M.A., 2020. Selective targeting of BD1 and BD2 of the BET proteins in cancer and immunoinflammation. Science 368, 387–394. https://doi.org/10.1126/science.aaz8455.

Gingerich, M.A., Sidarala, V., Soleimanpour, S.A., 2020. Clarifying the function of genes at the chromosome 16p13 locus in type 1 diabetes: CLEC16A and DEXI. Genes Immun. 21, 79–82. https://doi.org/10.1038/s41435-019-0087-7.

Gonzalez-Duque, S., Azoury, M.E., Colli, M.L., Afonso, G., Turatsinze, J.-V., Nigi, L., Lalanne, A.I., Sebastiani, G., Carré, A., Pinto, S., Culina, S., Corcos, N., Bugliani, M., Marchetti, P., Armanet, M., Diedisheim, M., Kyewski, B., Steinmetz, L.M., Buus, S., You, S., Dubois-Laforgue, D., Larger, E., Beressi, J.-P., Bruno, G., Dotta, F., Scharfmann, R., Eizirik, D.L., Verdier, Y., Vinh, J., Mallone, R., 2018. Conventional and neo-antigenic peptides presented by β cells are targeted by circulating naïve CD8+ T cells in type 1 diabetic and healthy donors. Cell Metab. 28, 946–960. https://doi.org/10.1016/j.cmet.2018.07.007.

González-Navajas, J.M., Lee, J., David, M., Raz, E., 2012. Immunomodulatory functions of type I interferons. Nat. Rev. Immunol. 12, 125–135. https://doi.org/10.1038/nri3133.

Gorman, J.A., Hundhausen, C., Errett, J.S., Stone, A.E., Allenspach, E.J., Ge, Y., Arkatkar, T., Clough, C., Dai, X., Khim, S., Pestal, K., Liggitt, D., Cerosaletti, K., Stetson, D.B., James, R.G., Oukka, M., Concannon, P., Gale, M., Buckner, J.H., Rawlings, D.J., 2017. The A946T variant of the RNA sensor IFIH1 mediates an interferon program that limits viral infection but increases the risk for autoimmunity. Nat. Immunol. 18, 744–752. https://doi.org/10.1038/ni.3766.

Graves, P.M., Rotbart, H.A., Nix, W.A., Pallansch, M.A., Erlich, H.A., Norris, J.M., Hoffman, M., Eisenbarth, G.S., Rewers, M., 2003. Prospective study of enteroviral infections and development of beta-cell autoimmunity. Diabetes Res. Clin. Pract. 59, 51–61. https://doi.org/10.1016/S0168-8227(02)00198-5.

Grieco, F.A., Sebastiani, G., Juan-Mateu, J., Villate, O., Marroqui, L., Ladrière, L., Tugay, K., Regazzi, R., Bugliani, M., Marchetti, P., Dotta, F., Eizirik, D.L., 2017. MicroRNAs miR-23a-3p, miR-23b-3p, and miR-149-5p regulate the expression of proapoptotic BH3-only proteins DP5 and PUMA in human pancreatic β-cells. Diabetes 66, 100–112. https://doi.org/10.2337/db16-0592.

Grunnet, L.G., Aikin, R., Tonnesen, M.F., Paraskevas, S., Blaabjerg, L., Størling, J., Rosenberg, L., Billestrup, N., Maysinger, D., Mandrup-Poulsen, T., 2009. Proinflammatory cytokines activate the intrinsic apoptotic pathway in β-cells. Diabetes 58, 1807–1815. https://doi.org/10.2337/db08-0178.

Guerci, A.-P., Guerci, B., Lévy-Marchal, C., Ongagna, J., Ziegler, O., Candiloros, H., Guerci, O., Drouin, P., 1994. Onset of insulin-dependent diabetes mellitus after interferon-alfa therapy for hairy cell leukaemia. Lancet 343, 1167–1168. https://doi.org/10.1016/S0140-6736(94)90276-3.

Gurzov, E.N., Eizirik, D.L., 2011. Bcl-2 proteins in diabetes: mitochondrial pathways of β-cell death and dysfunction. Trends Cell Biol. 21, 424–431. https://doi.org/10.1016/j.tcb.2011.03.001.

Gurzov, E.N., Tran, M., Fernandez-Rojo, M.A., Merry, T.L., Zhang, X., Xu, Y., Fukushima, A., Waters, M.J., Watt, M.J., Andrikopoulos, S., Neel, B.G., Tiganis, T., 2014. Hepatic oxidative stress promotes insulin-STAT-5 signaling and obesity by inactivating protein tyrosine phosphatase N2. Cell Metab. 20, 85–102. https://doi.org/10.1016/j.cmet.2014.05.011.

Gurzov, E.N., Stanley, W.J., Brodnicki, T.C., Thomas, H.E., 2015. Protein tyrosine phosphatases: molecular switches in metabolism and diabetes. Trends Endocrinol. Metab. 26, 30–39. https://doi.org/10.1016/j.tem.2014.10.004.

Gysemans, C., Callewaert, H., Overbergh, L., Mathieu, C., 2008. Cytokine signalling in the β-cell: a dual role for IFNγ. Biochem. Soc. Trans. 103, 737–742. https://doi.org/10.1042/BST0360328.

Gysemans, C., Callewaert, H., Moore, F., Nelson-Holte, M., Overbergh, L., Eizirik, D.L., Mathieu, C., 2009. Interferon regulatory factor-1 is a key transcription factor in murine beta cells under immune attack. Diabetologia 52, 2374–2384. https://doi.org/10.1007/s00125-009-1514-5.

Han, G.-M., Chen, S.-L., Shen, N., Ye, S., Bao, C.-D., Gu, Y.-Y., 2003. Analysis of gene expression profiles in human systemic lupus erythematosus using oligonucleotide microarray. Genes Immun. 4, 177–186. https://doi.org/10.1038/sj.gene.6363966.

Hanafusa, T., Imagawa, A., Iwahashi, H., Uchigata, Y., Kanatsuka, A., Kawasaki, E., Kobayashi, T., Shimada, A., Shimizu, I., Maruyama, T., Makino, H., 2008. Report of the Japan Diabetes Society's Committee on research on fulminant type 1 diabetes mellitus: analysis of antiviral antibodies at disease onset. J. Japan Diabetes Soc. 51, 531–536. https://doi.org/10.11213/tonyobyo.51.531.

Hankaniemi, M.M., Laitinen, O.H., Stone, V.M., Sioofy-Khojine, A., Määttä, J.A.E., Larsson, P.G., Marjomäki, V., Hyöty, H., Flodström-Tullberg, M., Hytönen, V.P., 2017. Optimized production and purification of Coxsackievirus B1 vaccine and its preclinical evaluation in a mouse model. Vaccine 35, 3718–3725. https://doi.org/10.1016/j.vaccine.2017.05.057.

Hannedouche, S., Zhang, J., Yi, T., Shen, W., Nguyen, D., Pereira, J.P., Guerini, D., Baumgarten, B.U., Roggo, S., Wen, B., Knochenmuss, R., Noël, S., Gessier, F., Kelly, L.M., Vanek, M., Laurent, S., Preuss, I., Miault, C., Christen, I., Karuna, R., Li, W., Koo, D.-I., Suply, T., Schmedt, C., Peters, E.C., Falchetto, R., Katopodis, A., Spanka, C., Roy, M.-O., Detheux, M., Chen, Y.A., Schultz, P.G., Cho, C.Y., Seuwen, K., Cyster, J.G., Sailer, A.W., 2011. Oxysterols direct immune cell migration via EBI2. Nature 475, 524–527. https://doi.org/10.1038/nature10280.

Harjutsalo, V., Sjöberg, L., Tuomilehto, J., 2008. Time trends in the incidence of type 1 diabetes in Finnish children: a cohort study. Lancet 371, 1777–1782. https://doi.org/10.1016/S0140-6736(08)60765-5.

Heinig, M., Petretto, E., Wallace, C., Bottolo, L., Rotival, M., Lu, H., Li, Y., Sarwar, R., Langley, S.R., Bauerfeind, A., Hummel, O., Lee, Y.A., Paskas, S., Rintisch, C., Saar, K., Cooper, J., Buchan, R., Gray, E.E., Cyster, J.G., Erdmann, J., Hengstenberg, C., Maouche, S., Ouwehand, W.H., Rice, C.M., Samani, N.J., Schunkert, H., Goodall, A.H., Schulz, H., Roider, H.G., Vingron, M., Blankenberg, S., Münzel, T., Zeller, T., Szymczak, S., Ziegler, A., Tiret, L., Smyth, D.J., Pravenec, M., Aitman, T.J., Cambien, F., Clayton, D., Todd, J.A., Hubner, N., Cook, S.A., Braund, P., Gracey, J., Krishnan, U., Moore, J.S., Nelson, C.P., Pollard, H., Attwood, T., Crisp-Hihn, A., Foad, N., Jolley, J., Lloyd-Jones, H., Muir, D., Murray, E., O'Leary, K., Rankin, A., Sambrook, J., Godfroy, T., Brocheton, J., Proust, C., Schmitz, G., Heimerl, S., Lugauer, I., Belz, S., Gulde, S., Linsel-Nitschke, P., Sager, H., Schroeder, L., Lundmark, P., Syvannen, A.C., Neudert, J., Scholz, M., Deloukas, P., Gray, E., Gwilliams, R., Niblett, D., 2010. A trans-acting locus regulates an anti-viral expression network and type 1 diabetes risk. Nature 467, 460–464. https://doi.org/10.1038/nature09386.

Hiltunen, M., Hyoty, H., Knip, M., Ilonen, J., Reijonen, H., Vahasalo, P., Roivainen, M., Lonnrot, M., Leinikki, P., Hovi, T., Akerblom, H.K., 1997. Islet cell antibody seroconversion in children is temporally associated with enterovirus infections. J Infect Dis 175, 554–560. https://doi.org/10.1093/infdis/175.3.554.

Hodik, M., Anagandula, M., Fuxe, J., Krogvold, L., Dahl-Jørgensen, K., Hyöty, H., Sarmiento, L., Frisk, G., 2016. Coxsackie–adenovirus receptor expression is enhanced in pancreas from patients with type 1 diabetes. BMJ Open Diabetes Res. Care 4, e000219. https://doi.org/10.1136/bmjdrc-2016-000219.
Honda, K., Yanai, H., Negishi, H., Asagiri, M., Sato, M., Mizutani, T., Shimada, N., Ohba, Y., Takaoka, A., Yoshida, N., Taniguchi, T., 2005. IRF-7 is the master regulator of type I interferon-dependent immune responses. Nature 434, 772–777. https://doi.org/10.1038/nature03464.
Hong, S.W., Kim, S., Lee, D.-K., 2008. The role of Bach2 in nucleic acid-triggered antiviral innate immune responses. Biochem. Biophys. Res. Commun. 365, 426–432. https://doi.org/10.1016/j.bbrc.2007.10.183.
Honkanen, H., Oikarinen, S., Nurminen, N., Laitinen, O.H., Huhtala, H., Lehtonen, J., Ruokoranta, T., Hankaniemi, M.M., Lecouturier, V., Almond, J.W., Tauriainen, S., Simell, O., Ilonen, J., Veijola, R., Viskari, H., Knip, M., Hyöty, H., 2017. Detection of enteroviruses in stools precedes islet autoimmunity by several months: possible evidence for slowly operating mechanisms in virus-induced autoimmunity. Diabetologia 60, 424–431. https://doi.org/10.1007/s00125-016-4177-z.
Houssiau, F.A., Thanou, A., Mazur, M., Ramiterre, E., Gomez Mora, D.A., Misterska-Skora, M., Perich-Campos, R.A., Smakotina, S.A., Cerpa Cruz, S., Louzir, B., Croughs, T., Tee, M.L., 2020. IFN-α kinoid in systemic lupus erythematosus: results from a phase IIb, randomised, placebo-controlled study. Ann. Rheum. Dis. 79, 347–355. https://doi.org/10.1136/annrheumdis-2019-216379.
Huang, X., Hultgren, B., Dybdal, N., Stewart, T.A., 1994. Islet expression of interferon-α precedes diabetes in both the BB rat and streptozotocin-treated mice. Immunity 1, 469–478. https://doi.org/10.1016/1074-7613(94)90089-2.
Huang, X., Yuan, J., Goddard, A., Foulis, A., James, R.F.L., Lernmark, Å., Pujol-Borrell, R., Rabinovitch, A., Somoza, N., Stewart, T.A., 1995. Interferon expression in the pancreases of patients with type I diabetes. Diabetes 44, 658–664. https://doi.org/10.2337/diab.44.6.658.
Huber, J.P., Farrar, J., 2011. Regulation of effector and memory T-cell functions by type I interferon. Immunology 132, 466–474. https://doi.org/10.1111/j.1365-2567.2011.03412.x.
Hultcrantz, M., Hühn, M.H., Wolf, M., Olsson, A., Jacobson, S., Williams, B.R., Korsgren, O., Flodström-Tullberg, M., 2007. Interferons induce an antiviral state in human pancreatic islet cells. Virology 367, 92–101. https://doi.org/10.1016/j.virol.2007.05.010.
Hummel, S., Pfluger, M., Hummel, M., Bonifacio, E., Ziegler, A.-G., 2011. Primary dietary intervention study to reduce the risk of islet autoimmunity in children at increased risk for type 1 diabetes: the BABYDIET study. Diabetes Care 34, 1301–1305. https://doi.org/10.2337/dc10-2456.
Hyoty, H., Hiltunen, M., Knip, M., Laakkonen, M., Vahasalo, P., Karjalainen, J., Koskela, P., Roivainen, M., Leinikki, P., Hovi, T., Akerblom, H.K., 1995. A prospective study of the role of coxsackie B and other enterovirus infections in the pathogenesis of IDDM. Diabetes 44, 652–657. https://doi.org/10.2337/diab.44.6.652.
Ilonen, J., Lempainen, J., Veijola, R., 2019. The heterogeneous pathogenesis of type 1 diabetes mellitus. Nat. Rev. Endocrinol. 15, 635–650. https://doi.org/10.1038/s41574-019-0254-y.
Imagawa, A., Hanafusa, T., Miyagawa, J.-I., Matsuzawa, Y., 2000. A proposal of three distinct subtypes of type 1 diabetes mellitus based on clinical and pathological evidence. Ann. Med. 32, 539–543. https://doi.org/10.3109/07853890008998833.
Isaacs, A., Lindenmann, J., 1957. Virus interference. I. The interferon. Proc. R. Soc. Lond. Ser. B Biol. Sci. 147, 258–267. https://doi.org/10.1098/rspb.1957.0048.

Ivashkiv, L.B., Donlin, L.T., 2014. Regulation of type I interferon responses. Nat. Rev. Immunol. 14, 36–49. https://doi.org/10.1038/nri3581.

Izumi, K., Mine, K., Inoue, Y., Teshima, M., Ogawa, S., Kai, Y., Kurafuji, T., Hirakawa, K., Miyakawa, D., Ikeda, H., Inada, A., Hara, M., Yamada, H., Akashi, K., Niho, Y., Ina, K., Kobayashi, T., Yoshikai, Y., Anzai, K., Yamashita, T., Minagawa, H., Fujimoto, S., Kurisaki, H., Shimoda, K., Katsuta, H., Nagafuchi, S., 2015. Reduced Tyk2 gene expression in β-cells due to natural mutation determines susceptibility to virus-induced diabetes. Nat. Commun. 6, 6748. https://doi.org/10.1038/ncomms7748.

James, E.A., Pietropaolo, M., Mamula, M.J., 2018. Immune recognition of β-cells: neoepitopes as key players in the loss of tolerance. Diabetes 67, 1035–1042. https://doi.org/10.2337/dbi17-0030.

Jean-Baptiste, V.S.E., Xia, C.Q., Clare-Salzler, M.J., Horwitz, M.S., 2017. Type 1 diabetes and type 1 interferonopathies: localization of a type 1 common thread of virus infection in the pancreas. eBioMedicine 22, 10–17. https://doi.org/10.1016/j.ebiom.2017.06.014.

Jensen, S., Thomsen, A.R., 2012. Sensing of RNA viruses: a review of innate immune receptors involved in recognizing RNA virus invasion. J. Virol. 86, 2900–2910. https://doi.org/10.1128/JVI.05738-11.

Jermendy, Á., Szatmári, I., Körner, A., Szabó, A.J., Tóth-Heyn, P., Hermann, R., 2018. Association between interferon-induced helicase (IFIH1) rs1990760 polymorphism and seasonal variation in the onset of type 1 diabetes mellitus. Pediatr. Diabetes 19, 300–304. https://doi.org/10.1111/pedi.12569.

Jerram, S., Leslie, R.D., 2017. The genetic architecture of type 1 diabetes. Genes (Basel) 8, 209. https://doi.org/10.3390/genes8080209.

Jerram, S.T., Dang, M.N., Leslie, R.D., 2017. The role of epigenetics in type 1 diabetes. Curr. Diab. Rep. 17, 89. https://doi.org/10.1007/s11892-017-0916-x.

Kalkavan, H., Green, D.R., 2018. MOMP, cell suicide as a BCL-2 family business. Cell Death Differ. 25, 46–55. https://doi.org/10.1038/cdd.2017.179.

Kallionpaa, H., Elo, L.L., Laajala, E., Mykkanen, J., Ricano-Ponce, I., Vaarma, M., Laajala, T.D., Hyoty, H., Ilonen, J., Veijola, R., Simell, T., Wijmenga, C., Knip, M., Lahdesmaki, H., Simell, O., Lahesmaa, R., 2014. Innate immune activity is detected prior to seroconversion in children with HLA-conferred type 1 diabetes susceptibility. Diabetes 63, 2402–2414. https://doi.org/10.2337/db13-1775.

Karwacz, K., Bricogne, C., MacDonald, D., Arce, F., Bennett, C.L., Collins, M., Escors, D., 2011. PD-L1 co-stimulation contributes to ligand-induced T cell receptor down-modulation on $CD8^+$ T cells. EMBO Mol. Med. 3, 581–592. https://doi.org/10.1002/emmm.201100165.

Keskinen, P., Ronni, T., Matikainen, S., Lehtonen, A., Julkunen, I., 1997. Regulation of HLA class I and II expression by interferons and influenza A virus in human peripheral blood mononuclear cells. Immunology 91, 421–429. https://doi.org/10.1046/j.1365-2567.1997.00258.x.

Khamashta, M., Merrill, J.T., Werth, V.P., Furie, R., Kalunian, K., Illei, G.G., Drappa, J., Wang, L., Greth, W., 2016. Sifalimumab, an anti-interferon-α monoclonal antibody, in moderate to severe systemic lupus erythematosus: a randomised, double-blind, placebo-controlled study. Ann. Rheum. Dis. 75, 1909–1916. https://doi.org/10.1136/annrheumdis-2015-208562.

Khan, F., Elhadd, T.A., Greene, S.A., Belch, J.J.F., 2000. Impaired skin microvascular function in children, adolescents, and young adults with type I diabetes. Diabetes Care 23, 215–220. https://doi.org/10.2337/diacare.23.2.215.

Kiefer, K., Oropallo, M.A., Cancro, M.P., Marshak-Rothstein, A., 2012. Role of type I interferons in the activation of autoreactive B cells. Immunol. Cell Biol. 90, 498–504. https://doi.org/10.1038/icb.2012.10.

Kim, K.-S., Tracy, S., Tapprich, W., Bailey, J., Lee, C.-K., Kim, K., Barry, W.H., Chapman, N.M., 2005. 5′-terminal deletions occur in coxsackievirus B3 during replication in murine hearts and cardiac myocyte cultures and correlate with encapsidation of negative-strand viral RNA. J. Virol. 79, 7024–7041. https://doi.org/10.1128/JVI.79.11.7024-7041.2005.

Kim, K.W., Ho, A., Alshabee-Akil, A., Hardikar, A.A., Kay, T.W.H., Rawlinson, W.D., Craig, M.E., 2016. Coxsackievirus B5 infection induces dysregulation of microRNAs predicted to target known type 1 diabetes risk genes in human pancreatic islets. Diabetes 65, 996–1003. https://doi.org/10.2337/db15-0956.

Kimpimäki, T., Kupila, A., Hämäläinen, A.M., Kukko, M., Kulmala, P., Savola, K., Simell, T., Keskinen, P., Ilonen, J., Simell, O., Knip, M., 2001. The first signs of β-cell autoimmunity appear in infancy in genetically susceptible children from the general population: the Finnish type 1 diabetes prediction and prevention study. J. Clin. Endocrinol. Metab. 86, 4782–4788. https://doi.org/10.1210/jc.86.10.4782.

Klöppel, G., Löhr, M., Habich, K., Oberholzer, M., Heitz, P.U., 1985. Islet pathology and the pathogenesis of type 1 and type 2 diabetes mellitus revisited. Surv. Synth. Pathol. Res. 4, 110–125. https://doi.org/10.1159/000156969.

Knip, M., Simell, O., 2012. Environmental triggers of type 1 diabetes. Cold Spring Harb. Perspect. Med. 2, a007690. https://doi.org/10.1101/cshperspect.a007690.

Kometani, K., Nakagawa, R., Shinnakasu, R., Kaji, T., Rybouchkin, A., Moriyama, S., Furukawa, K., Koseki, H., Takemori, T., Kurosaki, T., 2013. Repression of the transcription factor Bach2 contributes to predisposition of IgG1 memory B cells toward plasma cell differentiation. Immunity 39, 136–147. https://doi.org/10.1016/j.immuni.2013.06.011.

Kotredes, K.P., Gamero, A.M., 2013. Interferons as inducers of apoptosis in malignant cells. J. Interferon Cytokine Res. 33, 162–170. https://doi.org/10.1089/jir.2012.0110.

Kourilovitch, M., Galarza-Maldonado, C., Ortiz-Prado, E., 2014. Diagnosis and classification of rheumatoid arthritis. J. Autoimmun. 48–49, 26–30. https://doi.org/10.1016/j.jaut.2014.01.027.

Kracht, M.J.L., van Lummel, M., Nikolic, T., Joosten, A.M., Laban, S., van der Slik, A.R., van Veelen, P.A., Carlotti, F., de Koning, E.J.P., Hoeben, R.C., Zaldumbide, A., Roep, B.O., 2017. Autoimmunity against a defective ribosomal insulin gene product in type 1 diabetes. Nat. Med. 23, 501–507. https://doi.org/10.1038/nm.4289.

Krogvold, L., Edwin, B., Buanes, T., Ludvigsson, J., Korsgren, O., Hyöty, H., Frisk, G., Hanssen, K.F., Dahl-Jørgensen, K., 2014. Pancreatic biopsy by minimal tail resection in live adult patients at the onset of type 1 diabetes: experiences from the DiViD study. Diabetologia 57, 841–843. https://doi.org/10.1007/s00125-013-3155-y.

Krogvold, L., Edwin, B., Buanes, T., Frisk, G., Skog, O., Anagandula, M., Korsgren, O., Undlien, D., Eike, M.C., Richardson, S.J., Leete, P., Morgan, N.G., Oikarinen, S., Oikarinen, M., Laiho, J.E., Hyöty, H., Ludvigsson, J., Hanssen, K.F., Dahl-Jørgensen, K., 2015. Detection of a low-grade enteroviral infection in the islets of Langerhans of living patients newly diagnosed with type 1 diabetes. Diabetes 64, 1682–1687. https://doi.org/10.2337/db14-1370.

Kulmala, P., Savola, K., Petersen, J.S., Vähäsalo, P., Karjalainen, J., Löppönen, T., Dyrberg, T., Åkerblom, H.K., Knip, M., 1998. Prediction of insulin-dependent diabetes mellitus in siblings of children with diabetes: a population-based study. J. Clin. Invest. 101, 327–336. https://doi.org/10.1172/JCI119879.

Kupila, A., Muona, P., Simell, T., Arvilommi, P., Savolainen, H., Hämäläinen, A.-M., Korhonen, S., Kimpimäki, T., Sjöroos, M., Ilonen, J., Knip, M., Simell, O., 2001. Feasibility of genetic and immunological prediction of type I diabetes in a population-based birth cohort. Diabetologia 44, 290–297. https://doi.org/10.1007/s001250051616.

Kusaczuk, M., 2019. Tauroursodeoxycholate—bile acid with chaperoning activity: molecular and cellular effects and therapeutic perspectives. Cell 8, 1471. https://doi.org/10.3390/cells8121471.

Laitinen, O.H., Honkanen, H., Pakkanen, O., Oikarinen, S., Hankaniemi, M.M., Huhtala, H., Ruokoranta, T., Lecouturier, V., Andre, P., Harju, R., Virtanen, S.M., Lehtonen, J., Almond, J.W., Simell, T., Simell, O., Ilonen, J., Veijola, R., Knip, M., Hyoty, H., 2014. Coxsackievirus B1 is associated with induction of β-cell autoimmunity that portends type 1 diabetes. Diabetes 63, 446–455. https://doi.org/10.2337/db13-0619.

Lang, K.S., Recher, M., Junt, T., Navarini, A.A., Harris, N.L., Freigang, S., Odermatt, B., Conrad, C., Ittner, L.M., Bauer, S., Luther, S.A., Uematsu, S., Akira, S., Hengartner, H., Zinkernagel, R.M., 2005. Toll-like receptor engagement converts T-cell autoreactivity into overt autoimmune disease. Nat. Med. 11, 138–145. https://doi.org/10.1038/nm1176.

Larner, A.C., Chaudhuri, A., Darnell, J.E., 1986. Transcriptional induction by interferon. New protein(s) determine the extent and length of the induction. J. Biol. Chem. 261, 453–459.

Larsson, P.G., Lakshmikanth, T., Laitinen, O.H., Utorova, R., Jacobson, S., Oikarinen, M., Domsgen, E., Koivunen, M.R.L., Chaux, P., Devard, N., Lecouturier, V., Almond, J., Knip, M., Hyöty, H., Flodström-Tullberg, M., 2015. A preclinical study on the efficacy and safety of a new vaccine against coxsackievirus B1 reveals no risk for accelerated diabetes development in mouse models. Diabetologia 58, 346–354. https://doi.org/10.1007/s00125-014-3436-0.

Lauwerys, B.R., Hachulla, E., Spertini, F., Lazaro, E., Jorgensen, C., Mariette, X., Haelterman, E., Grouard-Vogel, G., Fanget, B., Dhellin, O., Vandepapelière, P., Houssiau, F.A., 2013. Down-regulation of interferon signature in systemic lupus erythematosus patients by active immunization with interferon α-kinoid. Arthritis Rheum. 65, 447–456. https://doi.org/10.1002/art.37785.

Lee-Kirsch, M.A., 2017. The type I interferonopathies. Annu. Rev. Med. 68, 297–315. https://doi.org/10.1146/annurev-med-050715-104506.

Levy, D.E., Darnell, J.E., 2002. STATs: transcriptional control and biological impact. Nat. Rev. Mol. Cell Biol. 3, 651–662. https://doi.org/10.1038/nrm909.

Li, Q., McDevitt, H.O., 2011. The role of interferon alpha in initiation of type I diabetes in the NOD mouse. Clin. Immunol. 140, 3–7. https://doi.org/10.1016/j.clim.2011.04.010.

Li, Q., Xu, B., Michie, S.A., Rubins, K.H., Schreriber, R.D., McDevitt, H.O., 2008. Interferon-α initiates type 1 diabetes in nonobese diabetic mice. Proc. Natl. Acad. Sci. U. S. A. 105, 12439–12444. https://doi.org/10.1073/pnas.0806439105.

Li, Z., Gakovic, M., Ragimbeau, J., Eloranta, M.-L., Rönnblom, L., Michel, F., Pellegrini, S., 2013. Two rare disease-associated Tyk2 variants are catalytically impaired but signaling competent. J. Immunol. 190, 2335–2344. https://doi.org/10.4049/jimmunol.1203118.

Lincez, P.J., Shanina, I., Horwitz, M.S., 2015. Reduced expression of the MDA5 gene IFIH1 prevents autoimmune diabetes. Diabetes 64, 2184–2193. https://doi.org/10.2337/db14-1223.

Lind, K., Richardson, S.J., Leete, P., Morgan, N.G., Korsgren, O., Flodstrom-Tullberg, M., 2013. Induction of an antiviral state and attenuated coxsackievirus replication in type III interferon-treated primary human pancreatic islets. J. Virol. 87, 7646–7654. https://doi.org/10.1128/JVI.03431-12.

Liu, Y.-J., 2005. IPC: professional type 1 interferon-producing cells and plasmacytoid dendritic cell precursors. Annu. Rev. Immunol. 23, 275–306. https://doi.org/10.1146/annurev.immunol.23.021704.115633.

Liu, C., Yang, X.V., Wu, J., Kuei, C., Mani, N.S., Zhang, L., Yu, J., Sutton, S.W., Qin, N., Banie, H., Karlsson, L., Sun, S., Lovenberg, T.W., 2011. Oxysterols direct B-cell migration through EBI2. Nature 475, 519–523. https://doi.org/10.1038/nature10226.

Liu, H., Li, M., Song, Y., Xu, W., 2018. TRIM21 restricts coxsackievirus B3 replication, cardiac and pancreatic injury via interacting with MAVS and positively regulating IRF3-mediated type I interferon production. Front. Immunol. 9, 2479. https://doi.org/10.3389/fimmu.2018.02479.

Loh, K., Fukushima, A., Zhang, X., Galic, S., Briggs, D., Enriori, P.J., Simonds, S., Wiede, F., Reichenbach, A., Hauser, C., Sims, N.A., Bence, K.K., Zhang, S., Zhang, Z.-Y., Kahn, B.B., Neel, B.G., Andrews, Z.B., Cowley, M.A., Tiganis, T., 2011. Elevated hypothalamic TCPTP in obesity contributes to cellular leptin resistance. Cell Metab. 14, 684–699. https://doi.org/10.1016/j.cmet.2011.09.011.

Lombardi, A., Tomer, Y., 2017. Interferon alpha impairs insulin production in human beta cells via endoplasmic reticulum stress. J. Autoimmun. 80, 48–55. https://doi.org/10.1016/j.jaut.2017.02.002.

Lombardi, A., Tsomos, E., Hammerstad, S.S., Tomer, Y., 2018. Interferon alpha: the key trigger of type 1 diabetes. J. Autoimmun. 94, 7–15. https://doi.org/10.1016/j.jaut.2018.08.003.

Long, S.A., Cerosaletti, K., Wan, J.Y., Ho, J.C., Tatum, M., Wei, S., Shilling, H.G., Buckner, J.H., 2011. An autoimmune-associated variant in PTPN2 reveals an impairment of IL-2R signaling in CD4+ T cells. Genes Immun. 12, 116–125. https://doi.org/10.1038/gene.2010.54.

Loo, Y.M., Gale, M., 2012. Immune signaling by RIG-I-like receptors. Immunity 34, 680–692. https://doi.org/10.1016/j.immuni.2011.05.003.

Lundberg, M., Krogvold, L., Kuric, E., Dahl-Jørgensen, K., Skog, O., 2016. Expression of interferon-stimulated genes in insulitic pancreatic islets of patients recently diagnosed with type 1 diabetes. Diabetes 65, 3104–3110. https://doi.org/10.2337/db16-0616.

Malakhov, M.P., Malakhova, O.A., Il Kim, K., Ritchie, K.J., Zhang, D.E., 2002. UBP43 (USP18) specifically removes ISG15 from conjugated proteins. J. Biol. Chem. 277, 9976–9981. https://doi.org/10.1074/jbc.M109078200.

Mallone, R., Eizirik, D.L., 2020. Presumption of innocence for beta cells: why are they vulnerable autoimmune targets in type 1 diabetes? Diabetologia 63, 1999–2006. https://doi.org/10.1007/s00125-020-05176-7.

Marhfour, I., Lopez, X.M., Lefkaditis, D., Salmon, I., Allagnat, F., Richardson, S.J., Morgan, N.G., Eizirik, D.L., 2012. Expression of endoplasmic reticulum stress markers in the islets of patients with type 1 diabetes. Diabetologia 55, 2417–2420. https://doi.org/10.1007/s00125-012-2604-3.

Marré, M.L., Piganelli, J.D., 2017. Environmental factors contribute to β cell endoplasmic reticulum stress and neo-antigen formation in type 1 diabetes. Front. Endocrinol. (Lausanne). 8, 262. https://doi.org/10.3389/fendo.2017.00262.

Marré, M.L., James, E.A., Piganelli, J.D., 2015. β cell ER stress and the implications for immunogenicity in type 1 diabetes. Front. Cell Dev. Biol. 3, 67. https://doi.org/10.3389/fcell.2015.00067.

Marro, B.S., Ware, B.C., Zak, J., De La Torre, J.C., Rosen, H., Oldstone, M.B.A., 2017. Progression of type 1 diabetes from the prediabetic stage is controlled by interferon-α signaling. Proc. Natl. Acad. Sci. U. S. A. 114, 3708–3713. https://doi.org/10.1073/pnas.1700878114.

Marroqui, L., Santin, I., Dos Santos, R.S., Marselli, L., Marchetti, P., Eizirik, D.L., 2014. BACH2, a candidate risk gene for type 1 diabetes, regulates apoptosis in pancreatic β-cells via JNK1 modulation and crosstalk with the candidate gene PTPN2. Diabetes 63, 2516–2527. https://doi.org/10.2337/db13-1443.

Marroqui, L., Dos Santos, R.S., Fløyel, T., Grieco, F.A., Santin, I., Op De Beeck, A., Marselli, L., Marchetti, P., Pociot, F., Eizirik, D.L., 2015a. TYK2, a candidate gene for type 1 diabetes, modulates apoptosis and the innate immune response in human pancreatic β-cells. Diabetes 64, 3808–3817. https://doi.org/10.2337/db15-0362.

Marroqui, L., Lopes, M., dos Santos, R.S., Grieco, F.A., Roivainen, M., Richardson, S.J., Morgan, N.G., de Beeck, A.O., Eizirik, D.L., 2015b. Differential cell autonomous responses determine the outcome of coxsackievirus infections in murine pancreatic α and β cells. Elife 4, 1–23. https://doi.org/10.7554/eLife.06990.

Marroqui, L., Masini, M., Merino, B., Grieco, F.A., Millard, I., Dubois, C., Quesada, I., Marchetti, P., Cnop, M., Eizirik, D.L., 2015c. Pancreatic α cells are resistant to metabolic stress-induced apoptosis in type 2 diabetes. eBioMedicine 2, 378–385. https://doi.org/10.1016/j.ebiom.2015.03.012.

Marroqui, L., Dos Santos, R.S., Op de Beeck, A., Coomans de Brachène, A., Marselli, L., Marchetti, P., Eizirik, D.L., 2017. Interferon-α mediates human beta cell HLA class I overexpression, endoplasmic reticulum stress and apoptosis, three hallmarks of early human type 1 diabetes. Diabetologia 60, 656–667. https://doi.org/10.1007/s00125-016-4201-3.

Mavragani, C.P., Niewold, T.B., Chatzigeorgiou, A., Danielides, S., Thomas, D., Kirou, K.A., Kamper, E., Kaltsas, G., Crow, M.K., 2013. Increased serum type I interferon activity in organ-specific autoimmune disorders: clinical, imaging, and serological associations. Front. Immunol. 4, 238. https://doi.org/10.3389/fimmu.2013.00238.

Meares, G.P., Fontanilla, D., Broniowska, K.A., Andreone, T., Lancaster, J.R., Corbett, J.A., 2013. Differential responses of pancreatic β-cells to ROS and RNS. Am. J. Physiol. Endocrinol. Metab. 304, E614–E622. https://doi.org/10.1152/ajpendo.00424.2012.

Meyer, S., Woodward, M., Hertel, C., Vlaicu, P., Haque, Y., Kärner, J., Macagno, A., Onuoha, S.C., Fishman, D., Peterson, H., Metsküla, K., Uibo, R., Jäntti, K., Hokynar, K., Wolff, A.S.B., Krohn, K., Ranki, A., Peterson, P., Kisand, K., Hayday, A., Meloni, A., Kluger, N., Husebye, E.S., Podkrajsek, K.T., Battelino, T., Bratanic, N., Peet, A., 2016. AIRE-deficient patients harbor unique high-affinity disease-ameliorating autoantibodies. Cell 166, 582–595. https://doi.org/10.1016/j.cell.2016.06.024.

Miani, M., Colli, M.L., Ladrière, L., Cnop, M., Eizirik, D.L., 2012. Mild endoplasmic reticulum stress augments the proinflammatory effect of IL-1β in pancreatic rat β-cells via the IRE1α/XBP1s pathway. Endocrinology 153, 3017–3028. https://doi.org/10.1210/en.2011-2090.

Miao, F., Smith, D.D., Zhang, L., Min, A., Feng, W., Natarajan, R., 2008. Lymphocytes from patients with type 1 diabetes display a distinct profile of chromatin histone H3 lysine 9 dimethylation. Diabetes 57, 3189–3198. https://doi.org/10.2337/db08-0645.

Miao, F., Chen, Z., Zhang, L., Liu, Z., Wu, X., Yuan, Y.-C., Natarajan, R., 2012. Profiles of epigenetic histone post-translational modifications at type 1 diabetes susceptible genes. J. Biol. Chem. 287, 16335–16345. https://doi.org/10.1074/jbc.M111.330373.

Miao, F., Chen, Z., Zhang, L., Wang, J., Gao, H., Wu, X., Natarajan, R., 2013. RNA-sequencing analysis of high glucose-treated monocytes reveals novel transcriptome signatures and associated epigenetic profiles. Physiol. Genomics 45, 287–299. https://doi.org/10.1152/physiolgenomics.00001.2013.

Minegishi, Y., Saito, M., Morio, T., Watanabe, K., Agematsu, K., Tsuchiya, S., Takada, H., Hara, T., Kawamura, N., Ariga, T., Kaneko, H., Kondo, N., Tsuge, I., Yachie, A., Sakiyama, Y., Iwata, T., Bessho, F., Ohishi, T., Joh, K., Imai, K., Kogawa, K., Shinohara, M., Fujieda, M., Wakiguchi, H., Pasic, S., Abinun, M., Ochs, H.D.,

Renner, E.D., Jansson, A., Belohradsky, B.H., Metin, A., Shimizu, N., Mizutani, S., Miyawaki, T., Nonoyama, S., Karasuyama, H., 2006. Human tyrosine kinase 2 deficiency reveals its requisite roles in multiple cytokine signals involved in innate and acquired immunity. Immunity 25, 745–755. https://doi.org/10.1016/j.immuni.2006.09.009.

Moëll, A., Skog, O., Åhlin, E., Korsgren, O., Frisk, G., 2009. Antiviral effect of nicotinamide on enterovirus-infected human islets in vitro: effect on virus replication and chemokine secretion. J. Med. Virol. 81, 1082–1087. https://doi.org/10.1002/jmv.21476.

Moore, F., Colli, M.L., Cnop, M., Esteve, M.I., Cardozo, A.K., Cunha, D.A., Bugliani, M., Marchetti, P., Eizirik, D.L., 2009. PTPN2, a candidate gene for type 1 diabetes, modulates interferon-γ- induced pancreatic β-cell apoptosis. Diabetes 58, 1283–1291. https://doi.org/10.2337/db08-1510.

Moore, F., Naamane, N., Colli, M.L., Bouckenooghe, T., Ortis, F., Gurzov, E.N., Igoillo-Esteve, M., Mathieu, C., Bontempi, G., Thykjaer, T., Ørntoft, T.F., Eizirik, D.L., 2011. STAT1 is a master regulator of pancreatic β-cell apoptosis and islet inflammation. J. Biol. Chem. 286, 929–941. https://doi.org/10.1074/jbc.M110.162131.

Morand, E.F., Furie, R., Tanaka, Y., Bruce, I.N., Askanase, A.D., Richez, C., Bae, S.C., Brohawn, P.Z., Pineda, L., Berglind, A., Tummala, R., 2020. Trial of anifrolumab in active systemic lupus erythematosus. N. Engl. J. Med. 382, 211–221. https://doi.org/10.1056/NEJMoa1912196.

Morgan, N.G., Richardson, S.J., 2014. Enteroviruses as causative agents in type 1 diabetes: loose ends or lost cause? Trends Endocrinol. Metab. 25, 611–619. https://doi.org/10.1016/j.tem.2014.08.002.

Morris, R., Kershaw, N.J., Babon, J.J., 2018. The molecular details of cytokine signaling via the JAK/STAT pathway. Protein Sci. 27, 1984–2009. https://doi.org/10.1002/pro.3519.

Muto, A., Hoshino, H., Madisen, L., Yanai, N., Obinata, M., Karasuyama, H., Hayashi, N., Nakauchi, H., Yamamoto, M., Groudine, M., Igarashi, K., 1998. Identification of Bach2 as a B-cell-specific partner for small Maf proteins that negatively regulate the immunoglobulin heavy chain gene 3′ enhancer. EMBO J. 17, 5734–5743. https://doi.org/10.1093/emboj/17.19.5734.

Muto, A., Tashiro, S., Nakajima, O., Hoshino, H., Takahashi, S., Sakoda, E., Ikebe, D., Yamamoto, M., Igarashi, K., 2004. The transcriptional programme of antibody class switching involves the repressor Bach2. Nature 429, 566–571. https://doi.org/10.1038/nature02596.

Myers, M.P., Andersen, J.N., Cheng, A., Tremblay, M.L., Horvath, C.M., Parisien, J.-P., Salmeen, A., Barford, D., Tonks, N.K., 2001. TYK2 and JAK2 are substrates of protein-tyrosine phosphatase 1B. J. Biol. Chem. 276, 47771–47774. https://doi.org/10.1074/jbc.C100583200.

Nakamura, K., Kawasaki, E., Imagawa, A., Awata, T., Ikegami, H., Uchigata, Y., Kobayashi, T., Shimada, A., Nakanishi, K., Makino, H., Maruyama, T., Hanafusa, T., 2011. Type 1 diabetes and interferon therapy. Diabetes Care 34, 2084–2089. https://doi.org/10.2337/dc10-2274.

Nejentsev, S., Walker, N., Riches, D., Egholm, M., Todd, J.A., 2009. Rare variants of IFIH1, a gene implicated in antiviral responses, protect against type 1 diabetes. Science 324, 387–389. https://doi.org/10.1126/science.1167728.

Newby, B.N., Mathews, C.E., 2017. Type I interferon is a catastrophic feature of the diabetic islet microenvironment. Front. Endocrinol. (Lausanne). 8, 232. https://doi.org/10.3389/fendo.2017.00232.

Nieves-Bonilla, J.M., Kiaf, B., Schuster, C., Kissler, S., 2020. The type 1 diabetes candidate gene Dexi does not affect disease risk in the nonobese diabetic mouse model. Genes Immun. 21, 71–77. https://doi.org/10.1038/s41435-019-0083-y.

Nogueira, M., Puig, L., Torres, T., 2020. JAK inhibitors for treatment of psoriasis: focus on selective TYK2 inhibitors. Drugs 80, 341–352. https://doi.org/10.1007/s40265-020-01261-8.

Nyalwidhe, J.O., Jurczyk, A., Satish, B., Redick, S., Qaisar, N., Trombly, M.I., Vangala, P., Racicot, R., Bortell, R., Harlan, D.M., Greiner, D.L., Brehm, M.A., Nadler, J.L., Wang, J.P., 2020. Proteomic and transcriptional profiles of human stem cell-derived β cells following enteroviral challenge. Microorganisms 8, 295. https://doi.org/10.3390/microorganisms8020295.

Ohara, N., Kaneko, M., Nishibori, T., Sato, K., Furukawa, T., Koike, T., Sone, H., Kaneko, K., Kamoi, K., 2016. Fulminant type 1 diabetes mellitus associated with coxsackie virus type A2 infection: a case report and literature review. Intern. Med. 55, 643–646. https://doi.org/10.2169/internalmedicine.55.5292.

Oikarinen, S., Martiskainen, M., Tauriainen, S., Huhtala, H., Ilonen, J., Veijola, R., Simell, O., Knip, M., Hyöty, H., 2011. Enterovirus RNA in blood is linked to the development of type 1 diabetes. Diabetes 60, 276–279. https://doi.org/10.2337/db10-0186.

Oikarinen, S., Tauriainen, S., Hober, D., Lucas, B., Vazeou, A., Sioofy-Khojine, A., Bozas, E., Muir, P., Honkanen, H., Ilonen, J., Knip, M., Keskinen, P., Saha, M.-T., Huhtala, H., Stanway, G., Bartsocas, C., Ludvigsson, J., Taylor, K., Hyoty, H., 2014. Virus antibody survey in different European populations indicates risk association between coxsackievirus B1 and type 1 diabetes. Diabetes 63, 655–662. https://doi.org/10.2337/db13-0620.

Oka, R., Hiroi, N., Shigemitsu, R., Sue, M., Oshima, Y., Yoshida-Hiroi, M., 2011. Type 1 diabetes mellitus associated with pegylated interferon-α plus ribavirin treatment for chronic hepatitis C: case report and literature review. Clin. Med. Insights Endocrinol. Diabetes 4, CMED.S7815. https://doi.org/10.4137/CMED.S7815.

Op de Beeck, A., Eizirik, D.L., 2016. Viral infections in type 1 diabetes mellitus-why the β cells? Nat. Rev. Endocrinol. 12, 263–273. https://doi.org/10.1038/nrendo.2016.30.

Oram, R.A., Sims, E.K., Evans-Molina, C., 2019. Beta cells in type 1 diabetes: mass and function; sleeping or dead? Diabetologia 62, 567–577. https://doi.org/10.1007/s00125-019-4822-4.

Osum, K.C., Burrack, A.L., Martinov, T., Sahli, N.L., Mitchell, J.S., Tucker, C.G., Pauken, K.E., Papas, K., Appakalai, B., Spanier, J.A., Fife, B.T., 2018. Interferon-gamma drives programmed death-ligand 1 expression on islet β cells to limit T cell function during autoimmune diabetes. Sci. Rep. 8, 8295. https://doi.org/10.1038/s41598-018-26471-9.

Özcan, U., Yilmaz, E., Özcan, L., Furuhashi, M., Vaillancourt, E., Smith, R.O., Görgün, C.Z., Hotamisligil, G.S., 2006. Chemical chaperones reduce ER stress and restore glucose homeostasis in a mouse model of type 2 diabetes. Science 313, 1137–1140. https://doi.org/10.1126/science.1128294.

Papp, K., Gordon, K., Thaçi, D., Morita, A., Gooderham, M., Foley, P., Girgis, I.G., Kundu, S., Banerjee, S., 2018. Phase 2 trial of selective tyrosine kinase 2 inhibition in psoriasis. N. Engl. J. Med. 379, 1313–1321. https://doi.org/10.1056/NEJMoa1806382.

Patterson, C.C., Dahlquist, G.G., Gyürüs, E., Green, A., Soltész, G., 2009. Incidence trends for childhood type 1 diabetes in Europe during 1989–2003 and predicted new cases 2005–20: a multicentre prospective registration study. Lancet 373, 2027–2033. https://doi.org/10.1016/S0140-6736(09)60568-7.

Pearson, G., Chai, B., Vozheiko, T., Liu, X., Kandarpa, M., Piper, R.C., Soleimanpour, S.A., 2018. Clec16a, Nrdp1, and USP8 form a ubiquitin-dependent tripartite complex that regulates β-cell mitophagy. Diabetes 67, 265–277. https://doi.org/10.2337/db17-0321.

Pelegrin, M., Devedjian, J.C., Costa, C., Visa, J., Solanes, G., Pujol, A., Asins, G., Valera, A., Bosch, F., 1998. Evidence from transgenic mice that interferon-β may be involved in the onset of diabetes mellitus. J. Biol. Chem. 273, 12332–12340. https://doi.org/10.1074/jbc.273.20.12332.

Pereira, J.P., Kelly, L.M., Xu, Y., Cyster, J.G., 2009. EBI2 mediates B cell segregation between the outer and centre follicle. Nature 460, 1122–1126. https://doi.org/10.1038/nature08226.

Pérez-Salvia, M., Esteller, M., 2017. Bromodomain inhibitors and cancer therapy: from structures to applications. Epigenetics 12, 323–339. https://doi.org/10.1080/15592294.2016.1265710.

Peterson, K.S., Huang, J.-F., Zhu, J., D'Agati, V., Liu, X., Miller, N., Erlander, M.G., Jackson, M.R., Winchester, R.J., 2004. Characterization of heterogeneity in the molecular pathogenesis of lupus nephritis from transcriptional profiles of laser-captured glomeruli. J. Clin. Invest. 113, 1722–1733. https://doi.org/10.1172/JCI19139.

Petri, M., Wallace, D.J., Spindler, A., Chindalore, V., Kalunian, K., Mysler, E., Neuwelt, C.M., Robbie, G., White, W.I., Higgs, B.W., Yao, Y., Wang, L., Greth, W., 2013. Sifalimumab, a human anti-interferon-α monoclonal antibody, in systemic lupus erythematosus: a phase I randomized, controlled, dose-escalation study. Arthritis Rheum. 65, 1011–1021. https://doi.org/10.1002/art.37824.

Pichlmair, A., Reis e Sousa, C., 2007. Innate recognition of viruses. Immunity 27, 370–383. https://doi.org/10.1016/j.immuni.2007.08.012.

Pirot, P., Cardozo, A.K., Eizirik, D.L., 2008. Mediators and mechanisms of pancreatic beta-cell death in type 1 diabetes. Arq. Bras. Endocrinol. Metabol. 52, 156–165. https://doi.org/10.1590/S0004-27302008000200003.

Planas, R., Carrillo, J., Sanchez, A., Ruiz de Villa, M.C., Nuñez, F., Verdaguer, J., James, R.F.L., Pujol-Borrell, R., Vives-Pi, M., 2010. Gene expression profiles for the human pancreas and purified islets in type 1 diabetes: new findings at clinical onset and in long-standing diabetes. Clin. Exp. Immunol. 159, 23–44. https://doi.org/10.1111/j.1365-2249.2009.04053.x.

Platanias, L.C., 2005. Mechanisms of type-I- and type-II-interferon-mediated signalling. Nat. Rev. Immunol. 5, 375–386. https://doi.org/10.1038/nri1604.

Pociot, F., Akolkar, B., Concannon, P., Erlich, H.A., Julier, C., Morahan, G., Nierras, C.R., Todd, J.A., Rich, S.S., Nerup, J., 2010. Genetics of type 1 diabetes: what's next? Diabetes 59, 1561–1571. https://doi.org/10.2337/db10-0076.

Poy, M.N., Eliasson, L., Krutzfeldt, J., Kuwajima, S., Ma, X., MacDonald, P.E., Pfeffer, S., Tuschl, T., Rajewsky, N., Rorsman, P., Stoffel, M., 2004. A pancreatic islet-specific microRNA regulates insulin secretion. Nature 432, 226–230. https://doi.org/10.1038/nature03076.

Pujol-Borrell, R., Todd, I., Doshi, M., Gray, D., Feldmann, M., Bottazzo, G.F., 1986. Differential expression and regulation of MHC products in the endocrine and exocrine cells of the human pancreas. Clin. Exp. Immunol. 65, 128–139.

Punwani, N., Scherle, P., Flores, R., Shi, J., Liang, J., Yeleswaram, S., Levy, R., Williams, W., Gottlieb, A., 2012. Preliminary clinical activity of a topical JAK1/2 inhibitor in the treatment of psoriasis. J. Am. Acad. Dermatol. 67, 658–664. https://doi.org/10.1016/j.jaad.2011.12.018.

Qaisar, N., Lin, S., Ryan, G., Yang, C., Oikemus, S.R., Brodsky, M.H., Bortel, R., Mordes, J.P., Wang, J.P., 2017. A critical role for the type I interferon receptor in virus-induced autoimmune diabetes in rats. Diabetes 66, 145–157. https://doi.org/10.2337/db16-0462.

Qaisar, N., Jurczyk, A., Wang, J.P., 2018. Potential role of type I interferon in the pathogenic process leading to type 1 diabetes. Curr. Opin. Endocrinol. Diabetes Obes. 25, 94–100. https://doi.org/10.1097/MED.0000000000000399.

Quah, H.S., Miranda-Hernandez, S., Khoo, A., Harding, A., Fynch, S., Elkerbout, L., Brodnicki, T.C., Baxter, A.G., Kay, T.W.H., Thomas, H.E., Graham, K.L., 2014. Deficiency in type I interferon signaling prevents the early interferon-induced gene signature in pancreatic islets but not type 1 diabetes in NOD mice. Diabetes 63, 1032–1040. https://doi.org/10.2337/db13-1210.

Ragimbeau, J., Dondi, E., Alcover, A., Eid, P., Uzé, G., Pellegrini, S., 2003. The tyrosine kinase Tyk2 controls IFNAR1 cell surface expression. EMBO J. 22, 537–547. https://doi.org/10.1093/emboj/cdg038.

Ramos-Rodríguez, M., Raurell-Vila, H., Colli, M.L., Alvelos, M.I., Subirana-Granés, M., Juan-Mateu, J., Norris, R., Turatsinze, J.-V., Nakayasu, E.S., Webb-Robertson, B.-J. M., Inshaw, J.R.J., Marchetti, P., Piemonti, L., Esteller, M., Todd, J.A., Metz, T.O., Eizirik, D.L., Pasquali, L., 2019. The impact of proinflammatory cytokines on the β-cell regulatory landscape provides insights into the genetics of type 1 diabetes. Nat. Genet. 51, 1588–1595. https://doi.org/10.1038/s41588-019-0524-6.

Rasschaert, J., Ladrière, L., Urbain, M., Dogusan, Z., Katabua, B., Sato, S., Akira, S., Gysemans, C., Mathieu, C., Eizirik, D.L., 2005. Toll-like receptor 3 and STAT-1 contribute to double-stranded RNA + interferon-γ-induced apoptosis in primary pancreatic β-cells. J. Biol. Chem. 280, 33984–33991. https://doi.org/10.1074/jbc.M502213200.

Redondo, M.J., Yu, L., Hawa, M., Mackenzie, T., Pyke, D.A., Eisenbarth, G.S., Leslie, R. D.G., 2001. Heterogeneity of type I diabetes: analysis of monozygotic twins in Great Britain and the United States. Diabetologia 44, 354–362. https://doi.org/10.1007/s001250051626.

Refolo, G., Vescovo, T., Piacentini, M., Fimia, G.M., Ciccosanti, F., 2020. Mitochondrial interactome: a focus on antiviral signaling pathways. Front. Cell Dev. Biol. 8, 8. https://doi.org/10.3389/fcell.2020.00008.

Rewers, M., Ludvigsson, J., 2016. Environmental risk factors for type 1 diabetes. Lancet 387, 2340–2348. https://doi.org/10.1016/S0140-6736(16)30507-4.

Reynier, F., Pachot, A., Paye, M., Xu, Q., Turrel-Davin, F., Petit, F., Hot, A., Auffray, C., Bendelac, N., Nicolino, M., Mougin, B., Thivolet, C., 2010. Specific gene expression signature associated with development of autoimmune type I diabetes using whole-blood microarray analysis. Genes Immun. 11, 269–278. https://doi.org/10.1038/gene.2009.112.

Richardson, S.J., Willcox, A., Bone, A.J., Foulis, A.K., Morgan, N.G., 2009. The prevalence of enteroviral capsid protein vp1 immunostaining in pancreatic islets in human type 1 diabetes. Diabetologia 52, 1143–1151. https://doi.org/10.1007/s00125-009-1276-0.

Richardson, S.J., Willcox, A., Bone, A.J., Morgan, N.G., Foulis, A.K., 2011. Immunopathology of the human pancreas in type I diabetes. Semin. Immunopathol. 33, 9–21. https://doi.org/10.1007/s00281-010-0205-0.

Richardson, S.J., Leete, P., Bone, A.J., Foulis, A.K., Morgan, N.G., 2013. Expression of the enteroviral capsid protein VP1 in the islet cells of patients with type 1 diabetes is associated with induction of protein kinase R and downregulation of Mcl-1. Diabetologia 56, 185–193. https://doi.org/10.1007/s00125-012-2745-4.

Richardson, S.J., Morgan, N.G., Foulis, A.K., 2014. Pancreatic pathology in type 1 diabetes mellitus. Endocr. Pathol. 25, 80–92. https://doi.org/10.1007/s12022-014-9297-8.

Richardson, S.J., Rodriguez-Calvo, T., Gerling, I.C., Mathews, C.E., Kaddis, J.S., Russell, M.A., Zeissler, M., Leete, P., Krogvold, L., Dahl-Jørgensen, K., von Herrath, M., Pugliese, A., Atkinson, M.A., Morgan, N.G., 2016. Islet cell hyperexpression of HLA class I antigens: a defining feature in type 1 diabetes. Diabetologia 59, 2448–2458. https://doi.org/10.1007/s00125-016-4067-4.

Riggs, J.M., Hanna, R.N., Rajan, B., Zerrouki, K., Karnell, J.L., Sagar, D., Vainshtein, I., Farmer, E., Rosenthal, K., Morehouse, C., De Los Reyes, M., Schifferli, K., Liang, M., Sanjuan, M.A., Sims, G.P., Kolbeck, R., 2018. Characterisation of anifrolumab, a fully human anti-interferon receptor antagonist antibody for the treatment of systemic lupus erythematosus. Lupus Sci. Med. 5, e000261. https://doi.org/10.1136/lupus-2018-000261.

Robertson, C.C., Inshaw, J.R.J., Onengut-Gumuscu, S., Chen, W.M., Flores Santa Cruz, D., Yang, H., Cutler, A.J., Crouch, D.J.M., Farber, E., Bridges, S.L., Edberg, J.C., Kimberly, R.P., Buckner, J.H., Deloukas, P., Divers, J., Dabelea, D., Lawrence, J.M., Marcovina, S., Shah, A.S., Greenbaum, C.J., Atkinson, M.A., Gregersen, P.K., Oksenberg, J.R., Pociot, F., Rewers, M.J., Steck, A.K., Dunger, D.B., Wicker, L.S., Concannon, P., Todd, J.A., Rich, S.S., 2020. Fine-mapping, trans-ancestral and genomic analyses identify causal variants, cells, genes and drug targets for type 1 diabetes. bioRxiv. 2020.06.19.158071 https://doi.org/10.1101/2020.06.19.158071.

Rodriguez-Calvo, T., 2019. Enterovirus infection and type 1 diabetes: unraveling the crime scene. Clin. Exp. Immunol. 195, 15–24. https://doi.org/10.1111/cei.13223.

Rodriguez-Calvo, T., Suwandi, J.S., Amirian, N., Zapardiel-Gonzalo, J., Anquetil, F., Sabouri, S., von Herrath, M.G., 2015. Heterogeneity and lobularity of pancreatic pathology in type 1 diabetes during the prediabetic phase. J. Histochem. Cytochem. 63, 626–636. https://doi.org/10.1369/0022155415576543.

Rogge, L., D'Ambrosio, D., Biffi, M., Penna, G., Minetti, L.J., Presky, D.H., Adorini, L., Sinigaglia, F., 1998. The role of Stat4 in species-specific regulation of Th cell development by type I IFNs. J. Immunol. 161, 6567–6574.

Roivainen, M., Rasilainen, S., Ylipaasto, P., Nissinen, R., Ustinov, J., Bouwens, L., Eizirik, D.L., Hovi, T., Otonkoski, T., 2000. Mechanisms of coxsackievirus-induced damage to human pancreatic β-cells. J. Clin. Endocrinol. Metab. 85, 432–440. https://doi.org/10.1210/jcem.85.1.6306.

Rondas, D., Crèvecoeur, I., D'Hertog, W., Ferreira, G.B., Staes, A., Garg, A.D., Eizirik, D.L., Agostinis, P., Gevaert, K., Overbergh, L., Mathieu, C., 2015. Citrullinated glucose-regulated protein 78 is an autoantigen in type 1 diabetes. Diabetes 64, 573–586. https://doi.org/10.2337/db14-0621.

Rönnblom, L., Eloranta, M.-L., 2013. The interferon signature in autoimmune diseases. Curr. Opin. Rheumatol. 25, 248–253. https://doi.org/10.1097/BOR.0b013e32835c7e32.

Roskoski, R., 2016. Janus kinase (JAK) inhibitors in the treatment of inflammatory and neoplastic diseases. Pharmacol. Res. 111, 784–803. https://doi.org/10.1016/j.phrs.2016.07.038.

Rouvier, E., Luciani, M.F., Golstein, P., 1993. Fas involvement in Ca(2+)-independent T cell-mediated cytotoxicity. J. Exp. Med. 177, 195–200. https://doi.org/10.1084/jem.177.1.195.

Roychoudhuri, R., Hirahara, K., Mousavi, K., Clever, D., Klebanoff, C.A., Bonelli, M., Sciumè, G., Zare, H., Vahedi, G., Dema, B., Yu, Z., Liu, H., Takahashi, H., Rao, M., Muranski, P., Crompton, J.G., Punkosdy, G., Bedognetti, D., Wang, E., Hoffmann, V., Rivera, J., Marincola, F.M., Nakamura, A., Sartorelli, V., Kanno, Y., Gattinoni, L., Muto, A., Igarashi, K., O'Shea, J.J., Restifo, N.P., 2013. BACH2 represses effector programs to stabilize T(reg)-mediated immune homeostasis. Nature 498, 506–510. https://doi.org/10.1038/nature12199.

Rui, J., Deng, S., Lebastchi, J., Clark, P.L., Usmani-Brown, S., Herold, K.C., 2016. Methylation of insulin DNA in response to proinflammatory cytokines during the progression of autoimmune diabetes in NOD mice. Diabetologia 59, 1021–1029. https://doi.org/10.1007/s00125-016-3897-4.

Rui, J., Deng, S., Arazi, A., Perdigoto, A.L., Liu, Z., Herold, K.C., 2017. β cells that resist immunological attack develop during progression of autoimmune diabetes in NOD mice. Cell Metab. 25, 727–738. https://doi.org/10.1016/j.cmet.2017.01.005.

Russell, M.A., Redick, S.D., Blodgett, D.M., Richardson, S.J., Leete, P., Krogvold, L., Dahl-Jørgensen, K., Bottino, R., Brissova, M., Spaeth, J.M., Babon, J.A.B., Haliyur, R., Powers, A.C., Yang, C., Kent, S.C., Derr, A.G., Kucukural, A., Garber, M.G., Morgan, N.G., Harlan, D.M., 2019. HLA class II antigen processing and presentation pathway components demonstrated by transcriptome and protein analyses of islet β-cells from donors with type 1 diabetes. Diabetes 68, 988–1001. https://doi.org/10.2337/db18-0686.

Sabio, G., Davis, R.J., 2014. TNF and MAP kinase signalling pathways. Semin. Immunol. 26, 237–245. https://doi.org/10.1016/j.smim.2014.02.009.

Salvetti, M., Ristori, G., Bomprezzi, R., Pozzilli, P., Leslie, R.D.G., 2000. Twins: mirrors of the immune system. Immunol. Today 21, 342–347. https://doi.org/10.1016/S0167-5699(00)01658-3.

Santin, I., Eizirik, D.L., 2013. Candidate genes for type 1 diabetes modulate pancreatic islet inflammation and β-cell apoptosis. Diabetes Obes. Metab. 15, 71–81. https://doi.org/10.1111/dom.12162.

Santin, I., Moore, F., Colli, M.L., Gurzov, E.N., Marselli, L., Marchetti, P., Eizirik, D.L., 2011. PTPN2, a candidate gene for type 1 diabetes, modulates pancreatic β-cell apoptosis via regulation of the BH3-only protein bim. Diabetes 60, 3279–3288. https://doi.org/10.2337/db11-0758.

Santin, I., Moore, F., Grieco, F.A., Marchetti, P., Brancolini, C., Eizirik, D.L., 2012. USP18 is a key regulator of the interferon-driven gene network modulating pancreatic beta cell inflammation and apoptosis. Cell Death Dis. 3, e49. https://doi.org/10.1038/cddis.2012.158.

Sarmiento, L., Frisk, G., Anagandula, M., Hodik, M., Barchetta, I., Netanyah, E., Cabrera-Rode, E., Cilio, C.M., 2017. Echovirus 6 infects human exocrine and endocrine pancreatic cells and induces pro-inflammatory innate immune response. Viruses 9, 1–25. https://doi.org/10.3390/v9020025.

Sayama, K., Imagawa, A., Okita, K., Uno, S., Moriwaki, M., Kozawa, J., Iwahashi, H., Yamagata, K., Tamura, S., Matsuzawa, Y., Hanafusa, T., Miyagawa, J., Shimomura, I., 2005. Pancreatic beta and alpha cells are both decreased in patients with fulminant type 1 diabetes: a morphometrical assessment. Diabetologia 48, 1560–1564. https://doi.org/10.1007/s00125-005-1829-9.

Schneider, W.M., Chevillotte, M.D., Rice, C.M., 2014. Interferon-stimulated genes: a complex web of host defenses. Annu. Rev. Immunol. 32, 513–545. https://doi.org/10.1146/annurev-immunol-032713-120231.

Schoggins, J.W., 2019. Interferon-stimulated genes: what do they all do? Annu. Rev. Virol. 6, 567–584. https://doi.org/10.1146/annurev-virology-092818-015756.

Schoggins, J.W., Wilson, S.J., Panis, M., Murphy, M.Y., Jones, C.T., Bieniasz, P., Rice, C.M., 2011. A diverse range of gene products are effectors of the type I interferon antiviral response. Nature 472, 481–485. https://doi.org/10.1038/nature09907.

Schreuder, T.C.M.A., Gelderblom, H.C., Weegink, C.J., Hamann, D., Reesink, H.W., DeVries, J.H., Hoekstra, J.B.L., Jansen, P.L.M., 2007. High incidence of type 1 diabetes mellitus during or shortly after treatment with pegylated interferon α for chronic hepatitis C virus infection. Liver Int. 28, 39–46. https://doi.org/10.1111/j.1478-3231.2007.01610.x.

Schuit, F.C., In't Veld, P.A., Pipeleers, D.G., 1988. Glucose stimulates proinsulin biosynthesis by a dose-dependent recruitment of pancreatic beta cells. Proc. Natl. Acad. Sci. U. S. A. 85, 3865–3869. https://doi.org/10.1073/pnas.85.11.3865.

Schuster, C., Gerold, K.D., Schober, K., Probst, L., Boerner, K., Kim, M.-J., Ruckdeschel, A., Serwold, T., Kissler, S., 2015. The autoimmunity-associated gene CLEC16A modulates thymic epithelial cell autophagy and alters T cell selection. Immunity 42, 942–952. https://doi.org/10.1016/j.immuni.2015.04.011.

Schwartz, D.M., Kanno, Y., Villarino, A., Ward, M., Gadina, M., O'Shea, J.J., 2017. JAK inhibition as a therapeutic strategy for immune and inflammatory diseases. Nat. Rev. Drug Discov. 16, 843–862. https://doi.org/10.1038/nrd.2017.201.
Seyhan, A.A., Gregory, B., Cribbs, A.P., Bhalara, S., Li, Y., Loreth, C., Zhang, Y., Guo, Y., Lin, L.L., Feldmann, M., Williams, L.M., Brennan, F.M., Taylor, P.C., 2020. Novel biomarkers of a peripheral blood interferon signature associated with drug-naïve early arthritis patients distinguish persistent from self-limiting disease course. Sci. Rep. 10, 8830. https://doi.org/10.1038/s41598-020-63757-3.
Shiba, T., Morino, Y., Tagawa, K., Fujino, H., Unuma, T., 1995. Onset of diabetes with high titer anti-GAD antibody after IFN therapy for chronic hepatitis. Diabetes Res. Clin. Pract. 30, 237–241. https://doi.org/10.1016/0168-8227(95)01188-9.
Siegal, F.P., Kadowaki, N., Shodell, M., Fitzgerald-Bocarsly, P.A., Shah, K., Ho, S., Antonenko, S., Liu, Y.-J., 1999. The nature of the principal type 1 interferon-producing cells in human blood. Science 284, 1835. https://doi.org/10.1126/science.284.5421.1835.
Simonen-Tikka, M.-L., Pflueger, M., Klemola, P., Savolainen-Kopra, C., Smura, T., Hummel, S., Kaijalainen, S., Nuutila, K., Natri, O., Roivainen, M., Ziegler, A.-G., 2011. Human enterovirus infections in children at increased risk for type 1 diabetes: the Babydiet study. Diabetologia 54, 2995–3002. https://doi.org/10.1007/s00125-011-2305-3.
Sims, E.K., Mirmira, R.G., Evans-Molina, C., 2020. The role of beta-cell dysfunction in early type 1 diabetes. Curr. Opin. Endocrinol. Diabetes Obes. 27, 215–224. https://doi.org/10.1097/MED.0000000000000548.
Sinclair, W., Omar, M., 2020. Enterovirus (enteroviral infections). StatPearls.
Smyth, D.J., Cooper, J.D., Bailey, R., Field, S., Burren, O., Smink, L.J., Guja, C., Ionescu-Tirgoviste, C., Widmer, B., Dunger, D.B., Savage, D.A., Walker, N.M., Clayton, D.G., Todd, J.A., 2006. A genome-wide association study of nonsynonymous SNPs identifies a type 1 diabetes locus in the interferon-induced helicase (IFIH1) region. Nat. Genet. 38, 617–619. https://doi.org/10.1038/ng1800.
Sobel, D.O., Ewel, C.H., Zeligs, B., Abbassi, V., Rossio, J., Bellanti, J.A., 1994. Poly I:C induction of alpha-interferon in the diabetes-prone BB and normal Wistar rats: dose-response relationships. Diabetes 43, 518–522. https://doi.org/10.2337/diab.43.4.518.
Soleimanpour, S.A., Gupta, A., Bakay, M., Ferrari, A.M., Groff, D.N., Fadista, J., Spruce, L.A., Kushner, J.A., Groop, L., Seeholzer, S.H., Kaufman, B.A., Hakonarson, H., Stoffers, D.A., 2014. The diabetes susceptibility gene Clec16a regulates mitophagy. Cell 157, 1577–1590. https://doi.org/10.1016/j.cell.2014.05.016.
Soleimanpour, S.A., Ferrari, A.M., Raum, J.C., Groff, D.N., Yang, J., Kaufman, B.A., Stoffers, D.A., 2015. Diabetes susceptibility genes Pdx1 and Clec16a function in a pathway regulating mitophagy in β-cells. Diabetes 64, 3475–3484. https://doi.org/10.2337/db15-0376.
Somoza, N., Vargas, F., Roura-Mir, C., Vives-Pi, M., Martí, M., Jaraquemada, D., Pujol-Borrell, R., Fernández-Figueras, M.T., Ariza, A., Gomis, R., Bragado, R., 1994. Pancreas in recent onset insulin-dependent diabetes mellitus: changes in HLA, adhesion molecules and autoantigens, restricted T cell receptor Vβ usage, and cytokine profile. J. Immunol. 153, 1360–1377.
Song, B., Scheuner, D., Ron, D., Pennathur, S., Kaufman, R.J., 2008. Chop deletion reduces oxidative stress, improves β cell function, and promotes cell survival in multiple mouse models of diabetes. J. Clin. Invest. 118, 3378–3389. https://doi.org/10.1172/JCI34587.
Songini, M., Mannu, C., Targhetta, C., Bruno, G., 2017. Type 1 diabetes in Sardinia: facts and hypotheses in the context of worldwide epidemiological data. Acta Diabetol. 54, 9–17. https://doi.org/10.1007/s00592-016-0909-2.

Stamatouli, A.M., Quandt, Z., Perdigoto, A.L., Clark, P.L., Kluger, H., Weiss, S.A., Gettinger, S., Sznol, M., Young, A., Rushakoff, R., Lee, J., Bluestone, J.A., Anderson, M., Herold, K.C., 2018. Collateral damage: insulin-dependent diabetes induced with checkpoint inhibitors. Diabetes 67, 1471–1480. https://doi.org/10.2337/dbi18-0002.

Stefan-Lifshitz, M., Karakose, E., Cui, L., Ettela, A., Yi, Z., Zhang, W., Tomer, Y., 2019. Epigenetic modulation of β cells by interferon-α via PNPT1/mir-26a/TET2 triggers autoimmune diabetes. JCI Insight 4, e126663. https://doi.org/10.1172/jci.insight.126663.

Stene, L.C., Oikarinen, S., Hyoty, H., Barriga, K.J., Norris, J.M., Klingensmith, G., Hutton, J.C., Erlich, H.A., Eisenbarth, G.S., Rewers, M., 2010. Enterovirus infection and progression from islet autoimmunity to type 1 diabetes: the diabetes and autoimmunity study in the young (DAISY). Diabetes 59, 3174–3180. https://doi.org/10.2337/db10-0866.

Stewart, T.A., 2003. Neutralizing interferon alpha as a therapeutic approach to autoimmune diseases. Cytokine Growth Factor Rev. 14, 139–154. https://doi.org/10.1016/S1359-6101(02)00088-6.

Stewart, T., Hultgren, B., Huang, X., Pitts-Meek, S., Hully, J., MacLachlan, N., 1993. Induction of type I diabetes by interferon-alpha in transgenic mice. Science 260, 1942–1946. https://doi.org/10.1126/science.8100367.

Stone, V.M., Hankaniemi, M.M., Svedin, E., Sioofy-Khojine, A., Oikarinen, S., Hyöty, H., Laitinen, O.H., Hytönen, V.P., Flodström-Tullberg, M., 2018. A coxsackievirus B vaccine protects against virus-induced diabetes in an experimental mouse model of type 1 diabetes. Diabetologia 61, 476–481. https://doi.org/10.1007/s00125-017-4492-z.

Stone, V.M., Hankaniemi, M.M., Laitinen, O.H., Sioofy-Khojine, A.B., Lin, A., Diaz Lozano, I.M., Mazur, M.A., Marjomäki, V., Loré, K., Hyöty, H., Hytönen, V.P., Flodström-Tullberg, M., 2020. A hexavalent coxsackievirus B vaccine is highly immunogenic and has a strong protective capacity in mice and nonhuman primates. Sci. Adv. 6, eaaz2433. https://doi.org/10.1126/sciadv.aaz2433.

Størling, J., Pociot, F., 2017. Type 1 diabetes candidate genes linked to pancreatic islet cell inflammation and beta-cell apoptosis. Genes (Basel) 8, 72. https://doi.org/10.3390/genes8020072.

Strengell, M., Julkunen, I., Matikainen, S., 2004. IFN-α regulates IL-21 and IL-21R expression in human NK and T cells. J. Leukoc. Biol. 76, 416–422. https://doi.org/10.1189/jlb.1003488.

Swaminathan, S., Huang, C., Geng, H., Chen, Z., Harvey, R., Kang, H., Ng, C., Titz, B., Hurtz, C., Sadiyah, M.F., Nowak, D., Thoennissen, G.B., Rand, V., Graeber, T.G., Koeffler, H.P., Carroll, W.L., Willman, C.L., Hall, A.G., Igarashi, K., Melnick, A., Müschen, M., 2013. BACH2 mediates negative selection and p53-dependent tumor suppression at the pre-B cell receptor checkpoint. Nat. Med. 19, 1014–1022. https://doi.org/10.1038/nm.3247.

Takeno, A., Kanazawa, I., Morita, M., Takedani, K., Miyake, H., Yamamoto, M., Nogami, K., Kaneko, S., Sugimoto, T., 2018. A case report of fulminant type 1 diabetes mellitus associated with drug-induced hypersensitivity syndrome in an elderly patient with coxsackie B4 virus infection and human leukocyte antigen-A24 haplotype. Endocr. J. 65, 129–132. https://doi.org/10.1507/endocrj.EJ17-0249.

Tan, X., Sun, L., Chen, J., Chen, Z.J., 2018. Detection of microbial infections through innate immune sensing of nucleic acids. Annu. Rev. Microbiol. 72, 447–478. https://doi.org/10.1146/annurev-micro-102215-095605.

Tanaka, S., Aida, K., Nishida, Y., Kobayashi, T., 2013. Pathophysiological mechanisms involving aggressive islet cell destruction in fulminant type 1 diabetes. Endocr. J. 60, 837–845. https://doi.org/10.1507/endocrj.EJ13-0222.

Tao, J.H., Zou, Y.F., Feng, X.L., Li, J., Wang, F., Pan, F.M., Ye, D.Q., 2011. Meta-analysis of TYK2 gene polymorphisms association with susceptibility to autoimmune and inflammatory diseases. Mol. Biol. Rep. 38, 4663–4672. https://doi.org/10.1007/s11033-010-0601-5.

Tapia, G., Cinek, O., Rasmussen, T., Witso, E., Grinde, B., Stene, L.C., Ronningen, K.S., 2011. Human enterovirus RNA in monthly fecal samples and islet autoimmunity in Norwegian children with high genetic risk for type 1 diabetes: the MIDIA study. Diabetes Care 34, 151–155. https://doi.org/10.2337/dc10-1413.

Taylor, P.C., Keystone, E.C., Van Der Heijde, D., Weinblatt, M.E., Del Carmen Morales, L., Gonzaga, J.R., Yakushin, S., Ishii, T., Emoto, K., Beattie, S., Arora, V., Gaich, C., Rooney, T., Schlichting, D., Macias, W.L., De Bono, S., Tanaka, Y., 2017. Baricitinib versus placebo or adalimumab in rheumatoid arthritis. N. Engl. J. Med. 376, 652–662. https://doi.org/10.1056/NEJMoa1608345.

TEDDY Study Group, 2007. The environmental determinants of diabetes in the young (TEDDY) study: study design. Pediatr. Diabetes 8, 286–298. https://doi.org/10.1111/j.1399-5448.2007.00269.x.

Tersey, S.A., Nishiki, Y., Templin, A.T., Cabrera, S.M., Stull, N.D., Colvin, S.C., Evans-Molina, C., Rickus, J.L., Maier, B., Mirmira, R.G., 2012. Islet β-cell endoplasmic reticulum stress precedes the onset of type 1 diabetes in the nonobese diabetic mouse model. Diabetes 61, 818–827. https://doi.org/10.2337/db11-1293.

The Wellcome Trust Case Control Consortium, 2007. Genome-wide association study of 14,000 cases of seven common diseases and 3,000 shared controls. Nature 447, 661–678. https://doi.org/10.1038/nature05911.

Thompson, P.J., Shah, A., Ntranos, V., Van Gool, F., Atkinson, M., Bhushan, A., 2019. Targeted elimination of senescent beta cells prevents type 1 diabetes. Cell Metab. 29, 1045–1060. https://doi.org/10.1016/j.cmet.2019.01.021.

Thyrell, L., Erickson, S., Zhivotovsky, B., Pokrovskaja, K., Sangfelt, O., Castro, J., Einhorn, S., Grandér, D., 2002. Mechanisms of interferon-alpha induced apoptosis in malignant cells. Oncogene 21, 1251–1262. https://doi.org/10.1038/sj.onc.1205179.

Todd, J.A., 2010. Etiology of type 1 diabetes. Immunity 32, 457–467. https://doi.org/10.1016/j.immuni.2010.04.001.

Todd, J.A., Walker, N.M., Cooper, J.D., Smyth, D.J., Downes, K., Plagnol, V., Bailey, R., Nejentsev, S., Field, S.F., Payne, F., Lowe, C.E., Szeszko, J.S., Hafler, J.P., Zeitels, L., Yang, J.H.M., Vella, A., Nutland, S., Stevens, H.E., Schuilenburg, H., Coleman, G., Maisuria, M., Meadows, W., Smink, L.J., Healy, B., Burren, O.S., Lam, A.A.C., Ovington, N.R., Allen, J., Adlem, E., Leung, H.-T., Wallace, C., Howson, J.M.M., Guja, C., Ionescu-Tîrgovişte, C., Simmonds, M.J., Heward, J.M., Gough, S.C.L., Dunger, D.B., Wicker, L.S., Clayton, D.G., 2007. Robust associations of four new chromosome regions from genome-wide analyses of type 1 diabetes. Nat. Genet. 39, 857–864. https://doi.org/10.1038/ng2068.

Tomlinson, M.J., Pitsillides, A., Pickin, R., Mika, M., Keene, K.L., Hou, X., Mychaleckyj, J., Chen, W.-M., Concannon, P., Onengut-Gumuscu, S., 2014. Fine mapping and functional studies of risk variants for type 1 diabetes at chromosome 16p13.13. Diabetes 63, 4360–4368. https://doi.org/10.2337/db13-1785.

Tracy, S., Smithee, S., Alhazmi, A., Chapman, N., 2015. Coxsackievirus can persist in murine pancreas by deletion of 5′ terminal genomic sequences. J. Med. Virol. 87, 240–247. https://doi.org/10.1002/jmv.24039.

Traore, D.A.K., Brennan, A.J., Law, R.H.P., Dogovski, C., Perugini, M.A., Lukoyanova, N., Leung, E.W.W., Norton, R.S., Lopez, J.A., Browne, K.A., Yagita, H., Lloyd, G.J., Ciccone, A., Verschoor, S., Trapani, J.A., Whisstock, J.C., Voskoboinik, I., 2013. Defining the interaction of perforin with calcium and the phospholipid membrane. Biochem. J. 456, 323–335. https://doi.org/10.1042/BJ20130999.

Trivedi, P.M., Graham, K.L., Scott, N.A., Jenkins, M.R., Majaw, S., Sutherland, R.M., Fynch, S., Lew, A.M., Burns, C.J., Krishnamurthy, B., Brodnicki, T.C., Mannering, S.I., Kay, T.W., Thomas, H.E., 2017. Repurposed JAK1/JAK2 inhibitor reverses established autoimmune insulitis in NOD mice. Diabetes 66, 1650–1660. https://doi.org/10.2337/db16-1250.

Tsukumo, S., Unno, M., Muto, A., Takeuchi, A., Kometani, K., Kurosaki, T., Igarashi, K., Saito, T., 2013. Bach2 maintains T cells in a naive state by suppressing effector memory-related genes. Proc. Natl. Acad. Sci. U. S. A. 110, 10735–10740. https://doi.org/10.1073/pnas.1306691110.

Uematsu, S., Akira, S., 2007. Toll-like receptors and type I interferons. J. Biol. Chem. 282, 15319–15323. https://doi.org/10.1074/jbc.R700009200.

Uonaga, T., Yoshida, K., Harada, T., Shimodahira, M., Nakamura, Y., 2012. Case of type 1 diabetes mellitus following interferon beta-1a treatment for multiple sclerosis. Intern. Med. 51, 1875–1877. https://doi.org/10.2169/internalmedicine.51.7609.

Urano, F., Wang, X.Z., Bertolotti, A., Zhang, Y., Chung, P., Harding, H.P., Ron, D., 2000. Coupling of stress in the ER to activation of JNK protein kinases by transmembrane protein kinase IRE1. Science 287, 664–666. https://doi.org/10.1126/science.287.5453.664.

Vassileva, G., Chen, S.-C., Zeng, M., Abbondanzo, S., Jensen, K., Gorman, D., Baroudy, B.M., Jiang, Y., Murgolo, N., Lira, S.A., 2003. Expression of a novel murine type I IFN in the pancreatic islets induces diabetes in mice. J. Immunol. 170, 5748–5755. https://doi.org/10.4049/jimmunol.170.11.5748.

Vehik, K., Lynch, K.F., Wong, M.C., Tian, X., Ross, M.C., Gibbs, R.A., Ajami, N.J., Petrosino, J.F., Rewers, M., Toppari, J., Ziegler, A.G., She, J.-X., Lernmark, A., Akolkar, B., Hagopian, W.A., Schatz, D.A., Krischer, J.P., Hyöty, H., Lloyd, R.E., 2019. Prospective virome analyses in young children at increased genetic risk for type 1 diabetes. Nat. Med. 25, 1865–1872. https://doi.org/10.1038/s41591-019-0667-0.

Villarino, A.V., Kanno, Y., O'Shea, J.J., 2017. Mechanisms and consequences of Jak-STAT signaling in the immune system. Nat. Immunol. 18, 374–384. https://doi.org/10.1038/ni.3691.

Villarroya-Beltri, C., Guerra, S., Sánchez-Madrid, F., 2017. ISGylation—a key to lock the cell gates for preventing the spread of threats. J. Cell Sci. 130, 2961–2969. https://doi.org/10.1242/jcs.205468.

Virtanen, A.T., Haikarainen, T., Raivola, J., Silvennoinen, O., 2019. Selective JAKinibs: prospects in inflammatory and autoimmune diseases. BioDrugs 33, 15–32. https://doi.org/10.1007/s40259-019-00333-w.

Wallace, C., Smyth, D.J., Maisuria-Armer, M., Walker, N.M., Todd, J.A., Clayton, D.G., 2010. The imprinted DLK1-MEG3 gene region on chromosome 14q32.2 alters susceptibility to type 1 diabetes. Nat. Genet. 42, 68–71. https://doi.org/10.1038/ng.493.

Wallace, C., Rotival, M., Cooper, J.D., Rice, C.M., Yang, J.H.M., McNeill, M., Smyth, D.J., Niblett, D., Cambien, F., Tiret, L., Todd, J.A., Clayton, D.G., Blankenberg, S., 2012. Statistical colocalization of monocyte gene expression and genetic risk variants for type 1 diabetes. Hum. Mol. Genet. 21, 2815–2824. https://doi.org/10.1093/hmg/dds098.

Wallace, D.J., Furie, R.A., Tanaka, Y., Kalunian, K.C., Mosca, M., Petri, M.A., Dörner, T., Cardiel, M.H., Bruce, I.N., Gomez, E., Carmack, T., DeLozier, A.M., Janes, J.M., Linnik, M.D., de Bono, S., Silk, M.E., Hoffman, R.W., 2018. Baricitinib for systemic lupus erythematosus: a double-blind, randomised, placebo-controlled, phase 2 trial. Lancet 392, 222–231. https://doi.org/10.1016/S0140-6736(18)31363-1.

Wang, C.-J., Chou, F.-C., Chu, C.-H., Wu, J.-C., Lin, S.-H., Chang, D.-M., Sytwu, H.-K., 2008. Protective role of programmed death 1 ligand 1 (PD-L1) in non-obese diabetic mice: the paradox in transgenic models. Diabetes 57, 1861–1869. https://doi.org/10.2337/db07-1260.

Watad, A., Azrielant, S., Bragazzi, N.L., Sharif, K., David, P., Katz, I., Aljadeff, G., Quaresma, M., Tanay, G., Adawi, M., Amital, H., Shoenfeld, Y., 2017. Seasonality and autoimmune diseases: the contribution of the four seasons to the mosaic of autoimmunity. J. Autoimmun. 82, 13–30. https://doi.org/10.1016/j.jaut.2017.06.001.

Wiede, F., Brodnicki, T.C., Goh, P.K., Leong, Y.A., Jones, G.W., Yu, D., Baxter, A.G., Jones, S.A., Kay, T.W.H., Tiganis, T., 2019. T-cell-specific PTPN2 deficiency in NOD mice accelerates the development of type 1 diabetes and autoimmune comorbidities. Diabetes 68, 1251–1266. https://doi.org/10.2337/db18-1362.

Winter, W.E., Schatz, D.A., 2011. Autoimmune markers in diabetes. Clin. Chem. 57, 168–175. https://doi.org/10.1373/clinchem.2010.148205.

Wyatt, R.C., Lanzoni, G., Russell, M.A., Gerling, I., Richardson, S.J., 2019. What the HLA-I!—Classical and non-classical HLA class I and their potential roles in type 1 diabetes. Curr. Diab. Rep. 19, 159. https://doi.org/10.1007/s11892-019-1245-z.

Xi, Y., Liu, S., Bettaieb, A., Matsuo, K., Matsuo, I., Hosein, E., Chahed, S., Wiede, F., Zhang, S., Zhang, Z.-Y., Kulkarni, R.N., Tiganis, T., Haj, F.G., 2015. Pancreatic T cell protein–tyrosine phosphatase deficiency affects beta cell function in mice. Diabetologia 58, 122–131. https://doi.org/10.1007/s00125-014-3413-7.

Xia, C.-Q., Peng, R., Chernatynskaya, A.V., Yuan, L., Carter, C., Valentine, J., Sobel, E., Atkinson, M.A., Clare-Salzler, M.J., 2014. Increased IFN-α-producing plasmacytoid dendritic cells (pDCs) in human Th1-mediated type 1 diabetes: pDCs augment Th1 responses through IFN-α production. J. Immunol. 193, 1024–1034. https://doi.org/10.4049/jimmunol.1303230.

Xie, Q., Khaoustov, V.I., Chung, C.C., Sohn, J., Krishnan, B., Lewis, D.E., Yoffe, B., 2002. Effect of tauroursodeoxycholic acid on endoplasmic reticulum stress-induced caspase-12 activation. Hepatology 36, 592–601. https://doi.org/10.1053/jhep.2002.35441.

Xie, Z., Chang, C., Huang, G., Zhou, Z., 2020. The role of epigenetics in type 1 diabetes. In: Chang, C., Lu, Q. (Eds.), Epigenetics in allergy and autoimmunity. Springer Singapore, Singapore, pp. 223–257. https://doi.org/10.1007/978-981-15-3449-2_9.

Xu, H., An, H., Hou, J., Han, C., Wang, P., Yu, Y., Cao, X., 2008. Phosphatase PTP1B negatively regulates MyD88- and TRIF-dependent proinflammatory cytokine and type I interferon production in TLR-triggered macrophages. Mol. Immunol. 45, 3545–3552. https://doi.org/10.1016/j.molimm.2008.05.006.

Xu, F., Liu, J., Na, L., Chen, L., 2020. Roles of epigenetic modifications in the differentiation and function of pancreatic β-cells. Front. Cell Dev. Biol. 8, 748. https://doi.org/10.3389/fcell.2020.00748.

Yamada, K., Takane-Gyotoku, N., Yuan, X., Ichikawa, F., Inada, C., Nonaka, K., 1996. Mouse islet cell lysis mediated by interleukin-1-induced Fas. Diabetologia 39, 1306–1312. https://doi.org/10.1007/s001250050574.

Yanase, N., Ohshima, K., Ikegami, H., Mizuguchi, J., 2000. Cytochrome c release, mitochondrial membrane depolarization, caspase-3 activation, and Bax-α cleavage during IFN-α-induced apoptosis in Daudi B lymphoma cells. J. Interferon Cytokine Res. 20, 1121–1129. https://doi.org/10.1089/107999000750053799.

Yang, X., Stennicke, H.R., Wang, B., Green, D.R., Jänicke, R.U., Srinivasan, A., Seth, P., Salvesen, G.S., Froelich, C.J., 1998. Granzyme B mimics apical caspases: description of a unified pathway for trans-activation of executioner caspase-3 and -7. J. Biol. Chem. 273, 34278–34283. https://doi.org/10.1074/jbc.273.51.34278.

Yao, Y., Richman, L., Higgs, B.W., Morehouse, C.A., De Los Reyes, M., Brohawn, P., Zhang, J., White, B., Coyle, A.J., Kiener, P.A., Jallal, B., 2009. Neutralization of interferon-α/β-inducible genes and downstream effect in a phase I trial of an anti-interferon-α monoclonal antibody in systemic lupus erythematosus. Arthritis Rheum. 60, 1785–1796. https://doi.org/10.1002/art.24557.

Yeung, W.-C.G., Rawlinson, W.D., Craig, M.E., 2011. Enterovirus infection and type 1 diabetes mellitus: systematic review and meta-analysis of observational molecular studies. BMJ 342, d35. https://doi.org/10.1136/bmj.d35.

Ylipaasto, P., Smura, T., Gopalacharyulu, P., Paananen, A., Seppänen-Laakso, T., Kaijalainen, S., Ahlfors, H., Korsgren, O., Lakey, J.R.T., Lahesmaa, R., Piemonti, L., Oresic, M., Galama, J., Roivainen, M., 2012. Enterovirus-induced gene expression profile is critical for human pancreatic islet destruction. Diabetologia 55, 3273–3283. https://doi.org/10.1007/s00125-012-2713-z.

Yong, H.Y., Luo, D., 2018. RIG-I-like receptors as novel targets for pan-antivirals and vaccine adjuvants against emerging and re-emerging viral infections. Front. Immunol. 9, 1379. https://doi.org/10.3389/fimmu.2018.01379.

Yoon, J.-W., Austin, M., Onodera, T., Notkins, A.L., 1979. Virus-induced diabetes mellitus. N. Engl. J. Med. 300, 1173–1179. https://doi.org/10.1056/NEJM197905243002102.

Youle, R.J., Strasser, A., 2008. The BCL-2 protein family: opposing activities that mediate cell death. Nat. Rev. Mol. Cell Biol. 9, 47–59. https://doi.org/10.1038/nrm2308.

Zhou, Y., Wu, H., Zhao, M., Chang, C., Lu, Q., 2016. The Bach family of transcription factors: a comprehensive review. Clin Rev Allergy Immunol 50, 345–356. https://doi.org/10.1007/s12016-016-8538-7.

Zhou, Y., He, C., Wang, L., Ge, B., 2017. Post-translational regulation of antiviral innate signaling. Eur. J. Immunol. 47, 1414–1426. https://doi.org/10.1002/eji.201746959.

Zornitzki, T., Malnick, S., Lysyy, L., Knobler, H., 2015. Interferon therapy in hepatitis C leading to chronic type 1 diabetes. World J. Gastroenterol. 21, 233–239. https://doi.org/10.3748/wjg.v21.i1.233.

CHAPTER TWO

Sexual hormones and diabetes: The impact of estradiol in pancreatic β cell

Beatriz Merino[a,†] and Marta García-Arévalo[b,*,†]
[a]Instituto de Biología y Genética Molecular (University of Valladolid-CSIC), Valladolid, Spain
[b]Brazilian Biosciences National Laboratory (LNBio), Brazilian Center for Research in Energy and Materials (CNPEM), Campinas, SP, Brazil
*Corresponding author: e-mail address: mgarciaarevalo@hotmail.com

Contents

Abstract

Diabetes is one of the most prevalent metabolic diseases and its incidence is increasing throughout the world. Data from World Health Organization (WHO) point-out that diabetes is a major cause of blindness, kidney failure, heart attacks, stroke and lower limb amputation and estimated 1.6 million deaths were directly caused by it in 2016. Population studies show that the incidence of this disease increases in women after

† Both authors contributed equally to this work.

International Review of Cell and Molecular Biology, Volume 359
ISSN 1937-6448
https://doi.org/10.1016/bs.ircmb.2021.02.004

menopause, when the production of estrogen is decreasing in them. Knowing the impact that estrogenic signaling has on insulin-secreting β cells is key to prevention and design of new therapeutic targets. This chapter explores the role of estrogen and their receptors in the regulation of insulin secretion and biosynthesis, proliferation, regeneration and survival in pancreatic β cells. In addition, delves into the genetic animal models developed and its application for the specific study of the different estrogen signaling pathways. Finally, discusses the impact of menopause and hormone replacement therapy on pancreatic β cell function.

1. Introduction

Sex hormones are important physiological regulators of glucose homeostasis and energy metabolism, so they have a role beyond sexual differentiation, puberty, and reproduction. And its dysregulation has been associated with different pathologies such as Diabetes. These hormones are present in women and men bloodstream in a different range of concentration and the main ones are estrogen, testosterone and progesterone. All of them are steroid hormones derived from cholesterol. In this chapter we will address the relationship between estrogen metabolism and the physiology and pathophysiology of pancreatic β cells, the principal endocrine cells involved in the control of glucose homeostasis and whose pathophysiology is closely linked to the development of diabetes mellitus.

The greatest evidence of the antidiabetic actions of the estrogen comes from the fact that prevalence of diabetes is lower in premenopausal women than in men. This trend changes after menopause, where it goes beyond the prevalence of men (Wild et al., 2004). Another interesting data which walks together with this antidiabetic estrogen actions, is the fact that common autoimmune diseases usually show a female predominance (Quintero et al., 2012), despite of this fact does not occur in type 1 diabetes mellitus (T1DM) (Beeson, 1994; Blohme et al., 1992). Under the 15s and in the early adult life there is an association between male sex and high frequency of T1DM in European populations (Gale and Gillespie, 2001). Other example of estrogen protection is in the ketosis-prone diabetes, also known as idiopathic diabetes, where the predominance in males is about 75% and women just present it under hypoestrogenic state, instead of women are protected (Louet et al., 2008; Mauvais-Jarvis et al., 2004).

In the case of type 2 diabetes mellitus (T2DM), the prevalence in adolescent females is higher than males, thus is associated with a greater insulin resistance in the former than latter during the puberty. In the middle age,

T2DM is more common in men than women. This change in trend may be occur by a different patterns of fat accumulation, where women have a majority subcutaneous fat presence and an improvement of insulin sensitivity (Sattar, 2013). The deficit of estrogen during the menopause increased also the risk of T2DM.

In 2004 with the women's health initiative study, where the study cohort include more than 15,000 women, the antidiabetic actions of estrogen were confirmed (Margolis et al., 2004). This study showed a decrease of 20% in the incidence of diabetes in the hormonal replace therapy (HRP) group at 5 year (Margolis et al., 2004). These results agree with those obtained in heart and Estrogen/progestin Replacement Study (Kanaya et al., 2003) (which had a less cohort of 3000 women), where the incidence of diabetes was reduced in 35% in the HRP group at 4 years. The number of studies where is observed an improvement on glucose stimulated insulin secretion after HRP in postmenopausal women are increasing (Godsland, 2005). Thus suggest that HRP could prevent diabetes improving β cell function or survival by direct or indirect mechanisms (Liu and Mauvais-Jarvis, 2010).

The fact stated above show that estrogen affects positively glucose homeostasis within a physiological window. Changes in this estrogen range, produced by menopause or by the use of contraceptives, are a risk factor for dysregulation of glucose homeostasis (Mauvais-Jarvis, 2018; Varlamov et al., 2014).

The evolution in health/diseases knowledge and the development of new drugs and therapeutic targets has as a result the increasing of life expectancy in men and women in developed countries. Thus, many women will spend approximately half of their adult lives in a state of estrogen deficiency. This fact could be related with the increasing number of cases in women of degenerative diseases of the cardiovascular, skeletal, and central nervous systems. Also estrogen deficiency enhances metabolic dysfunction predisposing to obesity, the metabolic syndrome, T2DM, and certain cancers (e.g., breast and colon, and hepatocellular carcinoma) (Carr, 2003; Mauvais-Jarvis, 2011; Mauvais-Jarvis et al., 2013).

1.1 Sex steroids hormones: Synthesis and physiological regulation

Males and females produce the same sex steroid hormones in a similar way. The synthesis of sex steroids is mainly produced by the gonads; in male in the testicular Leydig cells and in female in the ovarian granulosa cells. Nevertheless, other tissues as outer adrenal cortex, splanchnic organs, skin or adipose tissue, have the synthesizing machinery to produce them in mammals. Steroid hormones are formed by side-chain cleavage and oxidation of cholesterol. This process occurs in the mitochondria of steroidogenic tissues. The enzymes, involved in it, are the cytochrome P450 (CYP) members, specify the CYP P450 side-chain cleavage (CYP P450scc). The conversion by enzymatic cleavage of cholesterol to pregnenolone is a limiting step. The cholesterol is obtained from low-density lipoproteins in the extracellular fluid or from acetyl co-enzyme A in the cells. Pregnenolone has different intermediates until dihydroepiandrostenedione (DHEA) production. The pregnenolone is converted to progesterone by the 3-β-hydroxysteroid dehydrogenase (3β-HSD) or to DHEA by 17, 20-lyase. DHEA can be converted to androstenediol (or Adiol) by 17β-HSD or androstenedione (or Adione) by 3β-HSD. Both metabolites can be converted to testosterone (T) by 3β-HSD, in the first case, and 17β-HSD, in the second. T is then converted either to dihydrotestosterone (DHT), active biological metabolite, by 5α-reductase or to estradiol (E2) by CYP19-aromatase. All the transformation from pregnenolone to DHT or E2 happen in the smooth endoplasmic reticulum (Fig. 1). This aromatase has not only been found in the ovary and testis, but has also been found in extragonadal tissues as well as brain, prostate gland, breast, bone, muscle, liver and adipose tissue. The transformation of T to E2 occurs by the aromatization of the ring A in an aromatic ring by the addition of alternating double bonds. Besides, the estrone can be converted to E2 by the action of the 17β-HSD (Fig. 1) (Berne et al., 2009). Also there are evidences that islets of Langerhans have capacity of steroidogenesis, inasmuch as the necessary parts of the machinery to synthesize steroids are present in the pancreas. The Cytochrome P450scc was detected in dog pancreas mitochondria (Morales et al., 1999) and the 17β-HSD has been reported in canine, rat and human pancreas (Martel et al., 1992; Mendoza-Hernandez et al., 1988), especially the P450scc in islet β cells (Ogishima et al., 2008) and the type 12 of the 17β-HSD in whole islet (Sakurai et al., 2006). 3β-HSD is also find in the islet β cells. Thus, β cells, theoretically, can produce T, DTH and E2, which would act as local steroids in the islets (Mauvais-Jarvis, 2016). This local production could not have effects on distant tissues because the venous blood from the islets of Langerhans with the hormones, first passes through the liver via portal vein, where they are rapidly inactivate (Mauvais-Jarvis, 2016).

Both sexes have the capacity to produce the same steroid hormones by the process above explained. However, the physiological difference between sexes are produced by the concentration or activity of these hormones.

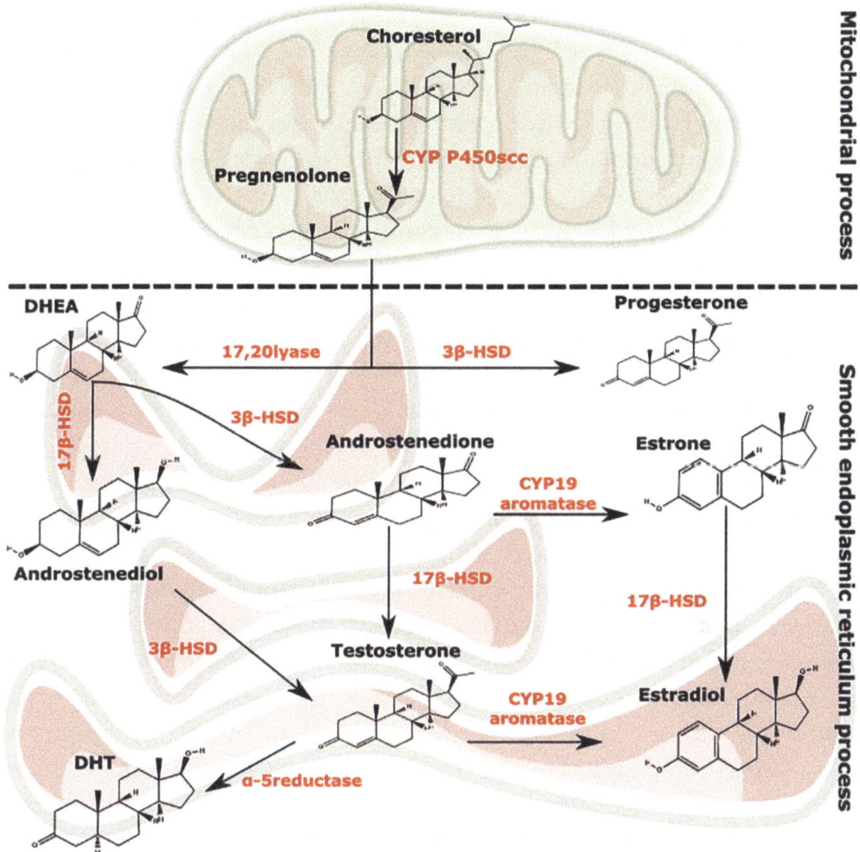

Fig. 1 Steroid hormones biosynthesis. Biological process that occurs inside of two different cell organelles the mitochondria and the smooth endoplasmic reticulum. A molecule of cholesterol is the precursor of the process and it is convert in the mitochondria to pregnenolone (a limiting step) and later in the smooth endoplasmic reticulum it is convert to testosterone or estradiol.

This balance is produced by two main process (1) how much androgen is made, and (2) what percentage of that quantity is converted into estrogen. In healthy people, the T and E2 is mainly synthesized in the gonads. In men, the testis makes approximately 7000 μg of T per day, and around 0.25% of it is converted in E2 (Gambineri and Pelusi, 2019). In the case of the women, the ovary makes 300 μg of T per day, approximately 20 times less than men, though half of it is convert to E2. Thus showed that the rate of T conversion to E2 in women is 200 times higher than men (Schiffer et al., 2017), facts that explain the difference between sexes steroids hormone concentration. On the other hand, not all E2 is in the active way. The estrogen sulfotransferase (EST) is a cytosolic enzyme that inactivates E2 conjugating a sulfonate group on it (estradiol sulfate [E2-S]). This means that the sulfonate group prevents binding the hormone to the estrogen receptors and enhances urinary hormone excretion (Strott, 1996) thereby, and it is inducing by T in both sexes. Consequently, in men, the higher T levels actives EST and increases in them the levels of E2-S, contributing to the physiological sexes differences (Mauvais-Jarvis et al., 2013) and controlling cellular estrogenic actions (Mauvais-Jarvis, 2012).

Other important difference between sexes is that, in healthy premenopausal women, E2 works as circulating hormone that acts on distant target tissues. However, in men and postmenopausal women, E2 acts as a paracrine and intracrine, and not as a circulating hormone, in extragonadal sites (where it is synthesized) as breast, brain, muscle, bone and adipose tissue (Simpson et al., 2005). When androgens and estrogens are transported in the blood, they are associated with two different types of proteins, the sex hormone-binding globulin (SHBG) and albumin. These proteins regulate the access of hormones to the target cells, and have an important role in sex-based difference by the regulation of free hormone levels (Gambineri and Pelusi, 2019). Albumin binds to both sex hormones, without any preference and with low affinity, however SHBG binds with high affinity to them. When comparing the affinity between T and E2, it is higher (twice) in the former than the latter. The affinity is also sex dependent, in females 77% of T is bound to SHBG, while in male only 53% is bound. Thus shows that change in the SHBG levels affects more to free T levels than E2 and have a higher impact on females than males (Vermeulen et al., 1999). In cross-sectional or prospective studies was predicted a risk of development of T2DM, in women, when was increased T blood levels and reduced SHBG concentration (Ding et al., 2009; Le et al., 2012; Lindstedt et al., 1991). Particularly, rodents do not express *shbg* in the liver postnatally and the trace amounts of SHBG detected in adult rats are produced by the gonads (Gunsalus et al., 1978; Joseph et al., 1987; Sullivan et al., 1991). The *shbg* expression in the gonads, specifically in Sertoli cells, is known as the androgen-binding protein (ABP) (Munell et al., 2002) and this protein has identical subunits and amino acid sequence homology with human SHBG (Joseph et al., 1987).

The E2 produces its physiological effects on the cells by binding to its specific receptor, the estrogen receptor (ER). ER was identified in the earlies 1970s with the use of E2 radioactive labeling in different animals models (Jensen and DeSombre, 1972). Twenty years after, in 1996 was identified a novel ER in rat prostate and ovary (Kuiper et al., 1996). To an easy identification was established that the first ER identified in 70s was called ER alpha (ERα) and the latter ER beta (ERβ), which are the two main forms of the receptor. Each of them has multiple isoforms and exhibit distinct tissue expression patterns and functions. These isoforms were derived from alternative splicing (Inoue et al., 1996; Zhang et al., 1993) and most of them lost the estrogen-dependent transactivation ability by the splicing action. Furthermore, some showed various effects on estrogen signaling, being able to have an inhibitory effect on the main receptor (the full-length receptor) (Inoue et al., 1996; Wang and Miksicek, 1991).

2. Receptors and modulation

2.1 Classic estrogen receptors ERα and ERβ

ER is a member of the nuclear receptor (NR) superfamily. Classically, it is associated with sex differentiation and reproductive function. However, it has an important role controlling glucose (Hevener et al., 2015) and energy homeostasis (Mauvais-Jarvis, 2011). ER shares a similar protein structure with the other NR members. It is composed by different domains. From the N-terminal portion to the C-terminal there are six domains. The domains A/B, as known as activation function 1 (AF-1), are responsible for transcription activity. The domain C, as known as DNA-binding domain (DBD), is the most conserved region among NRs (Berg, 1989). The DBD recognizes in the gene promotor the estrogen response element (ERE), formed by the palindromic nucleotide sequences GGTCAnnnTGACC (Heldring et al., 2007; Puzianowska-Kuznicka et al., 2013; Zhao et al., 2010). The domain D is the hinge region, a flexible domain that connects DBD with the ligand-binding domain (LBD), domains E/F, in the C-terminal. This last portion is an important region to dimerization and contains the ligand pocket and the activation function 2 (AF-2) (Pawlak et al., 2012).

There are two ER isoforms, ERα, in human described as the ESR1 gene (6q25.1), translated in the ERα protein with molecular mass of 66 kDa (Jensen and Jordan, 2003). ERβ is the other isoform; it is encoded by the ESR2 gene (14q23.2), translated in the ERβ protein with molecular mass of 58 kDa. It shares 47% amino acid sequence homology with ERα. This homology is even more evident in DBD and LBD with 97% and 60%, respectively (Enmark et al., 1997; Tremblay et al., 1997) (Fig. 2).

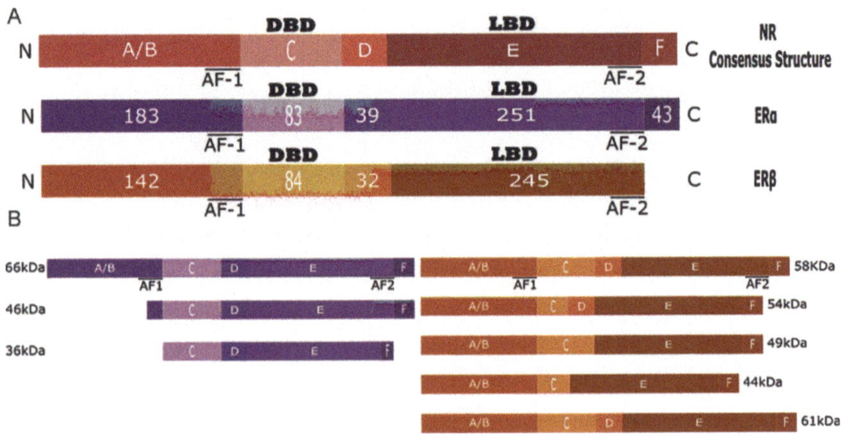

Fig. 2 Comparison of domain structure of estrogen receptors and their isoforms. In (A), the consensus structure of the nuclear receptor is form by six domains (A–F), scheme in the top of the picture in magenta shades. Estrogen receptor alpha (in blue shades) and beta (in orange tons) have these six domains and the number of amino acids and the function associated with the domains are expressed in each receptor. AF-1 and AF-2 refer to activation function. DBD and LBD are the DNA binding domain and ligand binding domain, respectively. In (B), there are represented the different human estrogen receptors isoforms in alpha (on the left) and beta (on the right), showing the differences between domains and the protein size.

Each isoform has different translational modification getting proteins with different sizes. In the case of ERα, there are two alternatively spliced truncated isoforms with 36 and 46kDa, which result from internal ATG transcription start codons. The 46kDa isoform lacks the AF-1 domain, and 36kDa isoform lacks both the AF-1 and AF-2 domains preventing efficient transcriptional activity but permitting heterodimerization with the full-length receptor (Figtree et al., 2003; Wang et al., 2005). ERβ has at least five different truncated isoforms that lack either the AF-1 or the AF-2 domain (Keselman and Heller, 2015). ERβ2, also knowns as ERβcx, has an alternative splicing after exon 7 and the addition of a small novel exon coding for a peptide of 26 amino acids. It is expressed in testis, ovary, thymus and prostate as well as in human cultured cell lines such as HEC-1, HOS-TE85 and Saos-2 cells. ERβ2 does not bind E2 or other tested ligands, and preferentially forms a heterodimer with ERα rather than that with ERβ, inhibiting DNA binding by ERα. Thus, ERβ2 was suggested to be a dominant-negative inhibitor of ERα (Ogawa et al., 1998). ERβ4 and ERβ5 isoforms are truncated transcripts containing only part of the common exon 7 and different exon 8 sequences and thus will not bind ligands. In vitro studies show that ERβ4 and β5 can heterodimerize with ERβ1 and enhance its transactivation in a ligand-dependent manner (Zhao et al., 2010). While ERβ1 is similar in mouse and human, the alternatively spliced variants are only expressed in primates (Gustafsson et al., 2019) (Fig. 2).

The ER, as other NR, can be found in the nucleus, however they exhibit a predominant extranuclear localization (Liu and Mauvais-Jarvis, 2009). The large number of studies on ER identified that these receptors have different mechanisms of action on the cells. These mechanisms mainly divided into nuclear and extranuclear-initiated signaling pathways, which will be explained below.

2.1.1 Nuclear-initiated signaling pathways

2.1.1.1 ERE-dependent signaling pathway (classical pathway)

The ER has a similar scenario for activation as other NFs. In general, when there are not any ligand, the ERs are sequestered by the heat shock proteins HSP90 preventing their activity. Ligands of the ER are lipophilic, which why they cross through the cell membrane by diffusion. The binding to the LBD of the receptor, produced the ER-HSP90 dissociation and the formation of ER dimers. The dimer could be form by two units of the same isoform, homodimer (ERα-ERα or ERβ-ERβ), or one unit of each isoform, heterodimer (ERα-ERβ). The dimer is translocated into the nucleus and modulated the expression of targets genes (Egea et al., 2000, Claessens and Gewirth, 2004, Green and Chambon, 1988). The recruitment of transcription factors, co-activators and co-repressors also occur in the LBD, acting as a platform, these proteins help the ER to exert their genomic actions (Darimont et al., 1998; Ham and Parker, 1989) (Fig. 3A).

2.1.1.2 ERE-independent signaling pathway (non-classical pathway)

ERs regulated around one third of the genes in humans that do not contain ERE-like sequences (O'Lone et al., 2004). Thus, the regulation happens without binding directly the ER to DNA, modulating the function of other transcription factors through protein–protein interactions in the nucleus (Gottlicher et al., 1998). The receptors are bound through protein–protein interactions to a transcription factor (TF) complex that contacts the DNA (Fig. 3B). The mechanism is commonly used by members of the NR superfamily and is known as transcriptional cross talk (Gottlicher et al., 1998). The interaction of ERs with the activator protein 1 (AP-1) TF complex is a typical example of such ERE-independent genomic actions. Several genes are activated by E2 through the interaction of ERs with Fos and

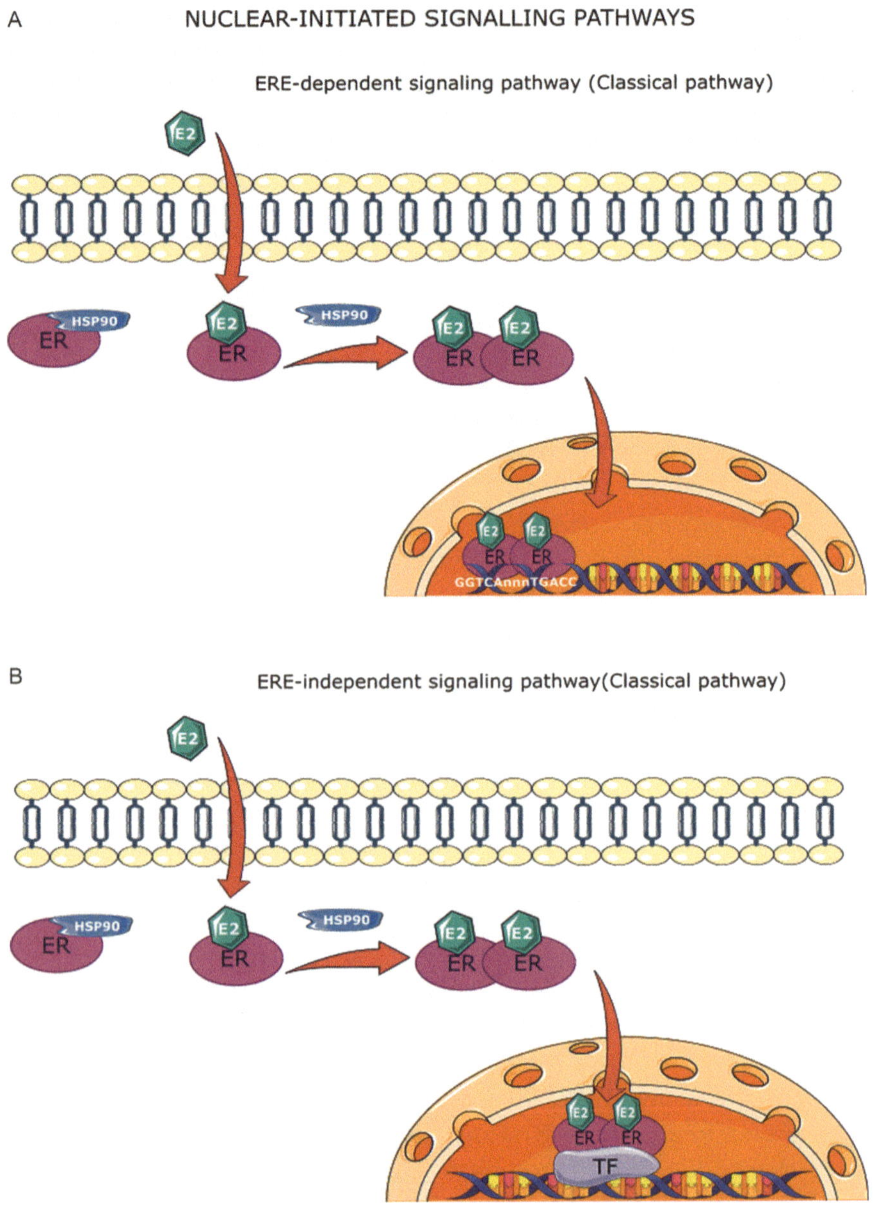

Fig. 3 Estrogen receptor genomic actions. (A) Classical pathway or ERE-dependent signaling pathway. After the bind of estradiol (E2) with the estrogen receptor (ER), the heat shock proteins HSP90 releases the ER and a dimer of it is translocate to the nucleus. Here the ER dimer binds directly to the DNA in the estrogen response element (ERE). (B) Non-classical pathway or ERE-independent signaling pathway. In this pathway after ER activation, it binds to another transcription factor (TF), as SP1 or AP1. The ER-DNA binding is indirect, through the FT.

Jun proteins at AP-1 binding sites, such as cyclin D1 (Liu et al., 2002; Sabbah et al., 1999). However, ERs also repress the activity of AP-1 TF (Schmitt et al., 1995).

ERs and AP-1 interact in an arduous way, being that the participation of the ER activation function (AF) domains depends both on the receptor subtype and on the type of ligand (Kushner et al., 2000). E2 induces the activation of ERα-AP-1 complexes and it requires the AF-2 domain of the receptor. AF-2 binds p160 coactivators and stabilizes, at the promoter, the formation of a multiprotein complex containing c-Jun, ERα, and transcriptional coactivators (Teyssier et al., 2001; Webb et al., 1999).

On the other hand, genes that contain GC-rich promoter sequences are regulated in a similar manner through the interaction between ERs and the Sp1 transcription factor (Porter et al., 1997). Examples of the modulation by E2 through ER-Sp1 complexes includes the LDL-R [low-density lipoprotein (LDL) receptor] (Li et al., 2001), c-fos (Duan et al., 1998) and cyclin D1 genes (Castro-Rivera et al., 2001). The actions of ERs on Sp1 binding sites, as is the case for the actions of ERs at AP-1 binding sites (Kushner et al., 2000), depend on the ligand, the cell type, and the receptor subtype (Safe, 2001).

The transcriptional cross talk, that involve the binding of ERs to other TF, does not require an interaction between the ER and DNA; however the DBD of the receptor is frequently involved, although intact DNA binding activity per se is not required (Bjornstrom et al., 2001; Faulds et al., 2001; Gaub et al., 1990; Jakacka et al., 2001; Sabbah et al., 1999; Schmitt et al., 1995; Stein and Yang, 1995; Umayahara et al., 1994; Webb et al., 1995). The generation of a Knock-In mouse model, where was introduced a mutation in the DBD (E207A/G208A; AA) of the ERα to inhibit the binding of the ER with the DNA, but maintaining the protein-protein interaction, is a clear show that the DBD may be required for proper protein-protein interactions or it may be involved in recruiting additional coregulator proteins to the promoter region (Bjornstrom and Sjoberg, 2005).

Besides, there are estrogen-responsive genes that contain ERE half-sites, or binding sites for the orphan nuclear hormone receptor SF-1 [SF-1 response elements (SFREs)] that serve as direct ER binding sites (O'Lone et al., 2004). However just ERα, but not ERβ, is able to bind to SFREs (Vanacker et al., 1999).

2.1.2 Extranuclear-initiated signaling pathways

A number of effects of estrogens are so rapid that they cannot depend on the activation of RNA and protein synthesis. These actions are known as non-genomic actions and are believed to be mediated mainly through membrane associated ERs. The actions are frequently associated with the activation of various protein-kinase cascades and are a common property of steroid hormones (Losel and Wehling, 2003). The non-genomic actions of estrogens may indirectly influence gene expression as an endpoint of the activation, because the signal transduction of the action could eventually act on TFs. This kind of action is important for gene expression regulation inasmuch as the functions of many TFs, including AP-1, are regulated through protein kinase-mediated phosphorylation. Therefore, a large number of TFs may be targets of non-genomic actions of estrogens. This signaling pathway can be referred to as "non-genomic-to-genomic" signaling. It offers for a mechanism by which ERs can modulate the functions of TFs, and thus regulate the expression of genes that do not contain EREs, distinct from non-classical genomic pathway (Bjornstrom and Sjoberg, 2005).

Varied non-genomic actions of E2 were reported (Fig. 4). There are include the mobilization of intracellular calcium (Improta-Brears et al., 1999), and

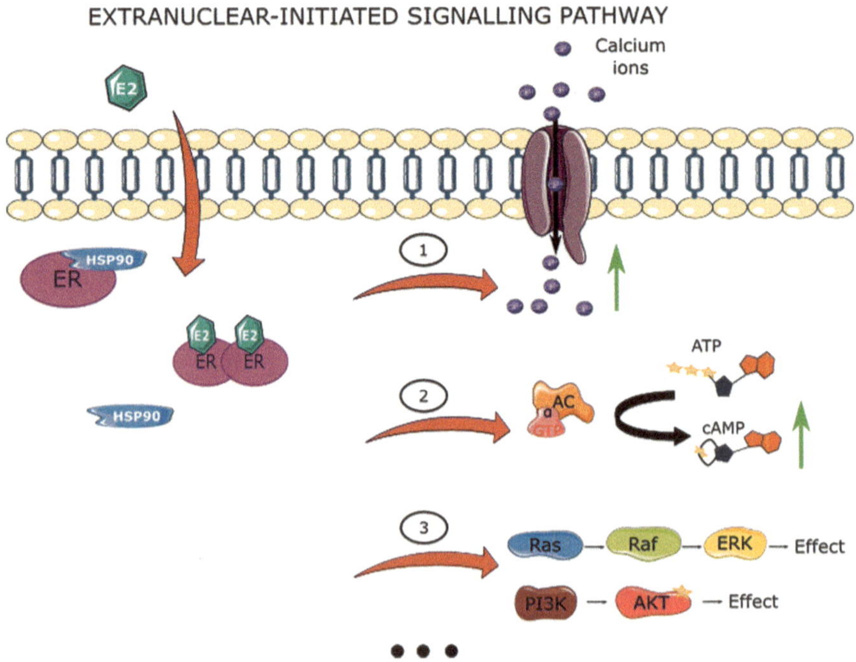

Fig. 4 Non-Genomic actions of the estrogen receptor. After the activation of the estrogen receptor (ER) by estrogen (E2), the receptor could activated many rapid actions as (1) mobilization of intracellular calcium (Ca^{2+}), (2) stimulation of adenylate cyclase (AC) activity and cAMP production, (3) activation of the MAPK or phosphoinositol 3-kinase (PI3K) signaling pathway.

the stimulation of adenylate cyclase (AC) activity and cAMP production (Aronica et al., 1994; Razandi et al., 1999). Activation of the MAPK signaling pathway by E2 has been extensively studied in several cell types, including breast cancer (Migliaccio et al., 1996), endothelial (Chen et al., 1999), bone (Endoh et al., 1997; Jessop et al., 2001), and neuroblastoma (Watters et al., 1997), pancreatic β (Alonso-Magdalena et al., 2008) cells. E2 also activates the phosphoinositol 3-kinase (PI3K) signaling pathway in endothelial (Chen et al., 1999), breast cancer (Castoria et al., 2001), liver (Marino et al., 2002) and skeletal muscle (García-Arévalo et al., 2019) cells. In pancreatic β cell, specifically, were observed non-genomic actions as mobilization of intracellular calcium (Soriano et al., 2009), activation of MAPK/ERK signaling pathway (Alonso-Magdalena et al., 2008), and increasing of the cGMP by the activation of the guanylate cyclase (Ropero et al., 1999; Soriano et al., 2009).

There is not a consensus of where are located the ERs when they trigger the non-genomic actions. Some authors have suggested that the non-genomic actions are mediated through a subpopulation of the classical ERs, ERα and ERβ, which is located at the plasma membrane (Pappas et al., 1995; Razandi et al., 1999). In the case of ERα represent the 5%–10% of the total pool depending on the cell type (Pedram et al., 2007). ERs do not contain a trans-membrane domain, and the association of ERα with the plasma membrane may be due to palmitoylation of the receptor at cysteine residues (Acconcia et al., 2004; Levin and Hammes, 2016; Li et al., 2003). However, plasma membrane ERs exist as functional dimers when activated by estrogens (Razandi et al., 2004). The LBD of the ERα targeted to the plasma membrane is sufficient for mediating non-genomic actions of estrogens (Kousteni et al., 2001). Nonetheless, other domains of the receptors might contribute to these actions by participating in protein–protein interactions. The interaction of ERs with various scaffold or signaling molecules facilitates the activity. ERs associate with the scaffold protein caveolin-1 (Chambliss et al., 2000; Razandi et al., 2002), and with a variety of proximal signaling molecules such as G proteins (Razandi et al., 1999; Wyckoff et al., 2001), SRC kinase, and Ras (Migliaccio et al., 1996, 1998), the p85α regulatory subunit and p110 catalytic subunit of PI3-kinase (García-Arévalo et al., 2019; Simoncini et al., 2000) and Shc (Song et al., 2002). These studies suggest that the molecular mechanisms underlying the non-genomic actions of ERs are specific for the cell type. The responses to estrogens may depend on the set of signal transduction molecules and downstream targets present in the target cell, and hence, the responses are likely to be diverse (Bjornstrom and Sjoberg, 2005). There are important events where ERα is membrane localized as example in males and females for reproduction (Nanjappa et al., 2016; Pedram et al., 2014), neuronal function (Meitzen et al., 2013), and vascular health (Adlanmerini et al., 2014). In the case of pancreatic islet, the activation of extranuclear ERα also promotes survival from apoptotic stress (Liu et al., 2009), prevents excess islet lipogenesis (Tiano et al., 2011; Tiano and Mauvais-Jarvis, 2012b), and stimulates insulin synthesis (Alonso-Magdalena et al., 2008; Wong et al., 2010).

This rapid and extranuclear effect of estrogen is potentially explained by the palmitoylation of both ERs, modification that allows their translocation to the membrane (Acconcia et al., 2004; Levin and Hammes, 2016; Marino and Ascenzi, 2006; Pedram et al., 2007). Palmitoylation is the covalent attachment of fatty acids, such as palmitic acid, to cysteine (S-palmitoylation). It has been shown that human ERα palmitoylation in Cys447 induced E2 rapid ERK phosphorylation (Marino et al., 2006). Furthermore, ERβ also has sites with cysteines susceptible to palmitoylation. This chemical modification is a reversible signal that allows the receptor to translocate from the nucleus to the plasma membrane, allowing rapid extranuclear-initiated E2 signaling (Marino and Ascenzi, 2006).

The non-genomic actions of estrogens may connect to genomic responses by the signal transduction pathways. This type of non-genomic-to-genomic signaling is a distinct mechanism by which ERs can regulate transcription at alternative response elements. This mechanism is different to the ERE-independent genomic actions described in non-classical pathway that involve protein-protein interactions in the nucleus.

On the other hand, non-genomic actions of estrogens may modulate the functions of ERs themselves, increasing the genomic actions of ER. Due to ERα and ERβ are also targets of phosphorylation by the MAPK signaling pathway (Kato et al., 1995; Tremblay et al., 1999). Furthermore, the transcriptional potency of ERs can be increased by the phosphorylation of the steroid receptor coactivator-3 (Src-3) (Wu et al., 2004). The functions of the ERs can be modulated through phosphorylation of certain residues in them in the absence of estrogens by IGF-1 and EGF signaling pathways (Driggers and Segars, 2002).

The regulation of gene expression by ERs is a multifactorial process, involving both genomic and non-genomic actions that often converge at certain response elements located in the promoters of target genes. The process depends on a number of conditions such as the combination of the nature of extracellular stimuli, the TFs bound to a specific gene promoter, the cellular localization of ERs, the expression levels of coregulatory proteins and signal transduction components. These variables are highly specific for the cell type. Therefore, the estrogen signaling depends on the cellular context, being able to use different mechanisms of regulation depending on the type of target gene.

2.2 G protein-coupled estrogen receptor

G protein-coupled estrogen receptor, GPER, (formerly knowns as G protein-couple receptor 30, GPR30), is a protein that in humans is encoded by the GPER gene and was first identified as an orphan member of the seven-transmembrane receptor family by multiple research groups in the late 1990's (Carmeci et al., 1997; O'Dowd et al., 1998; Owman et al., 1996; Takada et al., 1997) (Fig. 5A). GPER binds to E2 and this binding activates rapid non-genomic pathways in target cell (Prossnitz and Arterburn, 2015). Although there is controversy as to whether GPR30 is a "true" estrogen receptor (Levin, 2009; Otto et al., 2008), GPER plays a critical role in fast membrane coupling-mediated signaling of E2. This protein has been independently confirmed as genuine estrogen receptor by different authors before the evidence that E2 are able to bind to GPER activating specific estrogens signaling pathways (Prossnitz et al., 2008; Thomas et al., 2010). Although the main route of action of E2 is mediated by the action of classical nuclear receptors (ERα and ERβ), in cells that express both ERs and GPER there certainly exists the possibility of signaling interaction or cooperation (Albanito et al., 2007). It is widely described that the rapid cellular effects of estrogens such as cAMP production, intracellular calcium mobilization, kinases cascades and ion channels activation among other pathways, are mainly mediated by the ERα nuclear receptor (Edwards, 2005), however, it has also been shown that these effects may be mediated by membrane coupling through GPER (Prossnitz et al., 2008).

Filardo et al. described, for the first time in 2000, that there was an activation pathway of ERK1/2 carried out by E2 in a breast cancer cell line (SKBR3) but surprisingly this cell line does not express classic estrogen receptors, although it expresses GPER (Filardo et al., 2000). Subsequently, several groups published simultaneously proposed that GPER binds E2 and activates another pathways (Filardo and Thomas, 2005; Revankar et al., 2005; Thomas et al., 2005). Nowadays, GPR30 is officially classified as membrane

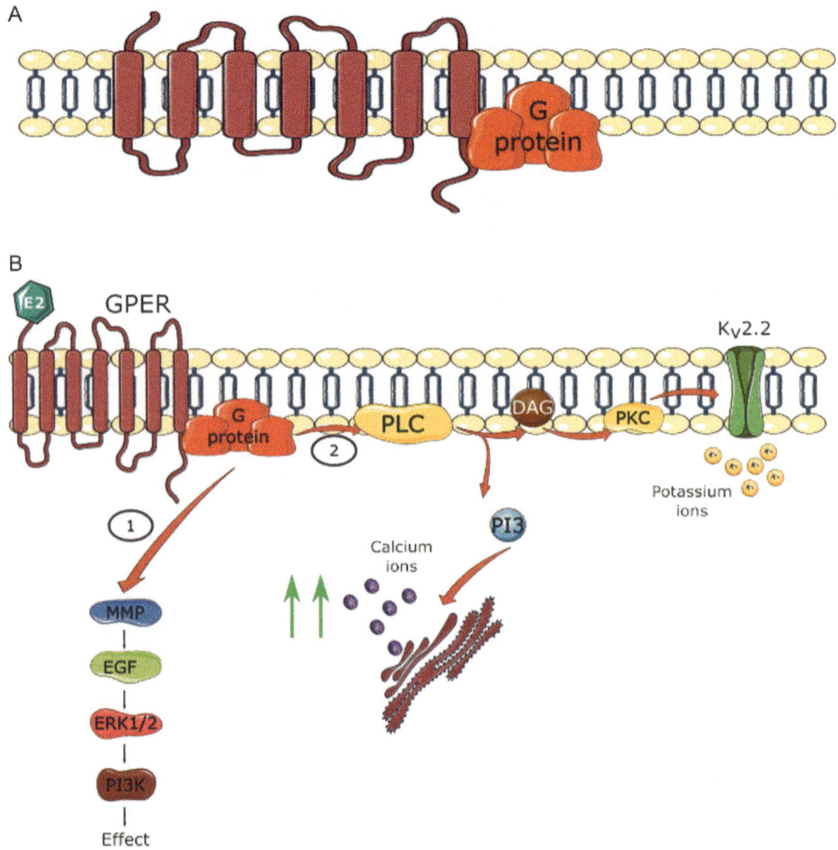

Fig. 5 G protein-coupled estrogen receptor (GPER). In (A) the GPER or GPR30 protein structure form by seven transmembrane α-helical regions, four extracellular and four cytosolic segments. It has a specific binding site for E2 in the extracellular portion, which affinity is lower than ERs. In the last cytosolic segment the G protein is anchored. In (B) GPER specific estrogens signaling pathways (1) the activation of metalloproteinases (MMP) releases the heparin-binding epidermal growth factor (EGF), and transactivation of the EGF receptor (EGFR), that produces the final activation of extracellular signal-regulated kinases 1/2 (ERK1/2) and phosphoinositide 3-kinase (PI3K). (2) GPER activation promotes the activity of the phospholipase C (PLC) that produces inositol triphosphate (IP3) and (DAG), thus promotes the intracellular calcium mobilization by the reticulum endoplasmic and voltage-gated potassium (Kv2.2) channel regulation.

estrogen receptor by the International Union of Pharmacology with the nomenclature GPER (Prossnitz and Arterburn, 2015).

Although the general location of GPCRs is practically exclusive to the plasma membrane, in the case of GPER there are studies that demonstrate its location in the plasma membrane (Filardo, 2002; Filardo et al., 2002) but also in intracellular membranes such as the endoplasmic reticulum (Revankar et al., 2005) and in the nucleus (Pupo et al., 2013; Smith et al., 2009). Recent studies have shown that the cellular localization of GPER depends on a complex trafficking in which clathrin vesicles are involved and new studies are necessary to clarify the differential signaling that may be regulating this trafficking (Cheng et al., 2014).

This receptor contains specific binding site for E2 and the affinity of this binding it has been reported in 3–6 mM range, this range is higher compare with ERα and ERβ that binds in the range of 0.1–0.5 nM (Revankar et al., 2005; Thomas et al., 2005). Therefore, GPER requires higher concentrations of estrogens to activate than the classical receptors. Thus suggest that classical receptors act under physiological E2 concentration and GPER is more involved in pathophysiological dysregulations. In addition to E2, it has been shown that many natural or synthetic estrogenic compounds can bind to GPER, activating its signaling in target cells. Therapeutic antiestrogens ICI182,780 (a selective ER antagonist) and 4-hydroxytamoxifen and raloxifene (selective ER modulator) acts as GPER agonists (Filardo et al., 2000; Petrie et al., 2013; Revankar et al., 2005), furthermore synthetic compounds act as endocrine disruptors with metabolic and reproductive effects such as Bisphenol-A, atrazine or daidzein among others (Yoon et al., 2014), as well as phytoestrogens like genistein, quercetin, resveratrol have also been reported to bind and/or activate GPER (Prossnitz and Arterburn, 2015).

In general, GPER is coupling in the plasma membrane to $G_{i/o}$ and G_s proteins (Filardo et al., 2000; Lindsey et al., 2014) that regulates downstream rapid signaling events. It has been described that the membrane activation of GPER by ligands binding leads to downstream activation of metalloproteinases (MMP), release of heparin-binding epidermal growth factor (EGF), transactivation of EGF receptor (EGFR) and finally activation of ERK1/2 and PI3K (Filardo et al., 2000). Consequently, GPER activation promotes intracellular calcium mobilization and potassium channels regulation (Dong et al., 2013; Meyer et al., 2012) (Fig. 5B). GPER activation can modulate, by this rapid pathway, the gene expression in different manner of ERα and ERβ (De Francesco et al., 2014).

3. Estrogen receptors in pancreatic β cells

Rodent and human β cells express both the full-length isoform (66 kDa) and a shorter 58 kDa or 52 kDa isoform of ERα (in mouse or human, respectively) (Le May et al., 2006) as well as ERβ (Liu et al., 2009). Between 2000 and 2002 was described for the first time the expression of ERα and ERβ in mouse and human islets (Contreras et al., 2002; Geisler et al., 2002; Nadal et al., 2000) and later in 2006 Mauvais-Jarvis and collaborators published the first physiological study on the impact of estrogens on the pancreatic β cell as a protector of cell apoptosis preventing diabetes onset (Le May et al., 2006). INS-1 cells is a well-establish model for the study of isolated pancreatic β cells in vitro. These cell lines come from a rat insulinoma and maintain the ability to secrete insulin in response to high glucose stimulus. INS-1 cells express also a short isoform of ERα (ERα36) (Tiano et al., 2011), this isoform does not contain the transcriptional activation domains and a functional DNA-binding domain (Kang et al., 2010). Also important are MIN6 cells as tool in β cell research, these cells from mouse insulinoma express the full-length isoform of ERα (Le May et al., 2006; Zhou et al., 2018) but not the sorter isoforms (Le May et al., 2006) and ERβ (Chuang et al., 2008). The predominant localization of classic ERs, in pancreatic β cells, is predominantly extranuclear (Alonso-Magdalena et al., 2008; Liu et al., 2009), but also can be found in the nucleus (Le May et al., 2006). This distribution suggests that estrogens can regulate pancreatic β cell physiology in a rapid manner via extranuclear receptors that produce cytoplasmic signaling that can end up modulating the expression of genes or can activate ERE on the promoter of target genes via nuclear ERs.

In addition to classic estrogens receptors, GPER is express in pancreatic β cells and also in INS-1 cells (Kooptiwut et al., 2014), MIN6 cells (Sharma and Prossnitz, 2011) and α, and δ cells (Balhuizen et al., 2010; Liu et al., 2009; Martensson et al., 2009) suggesting that probably is involved in the physiological regulation of glucagon, insulin and somatostatin release and therefore in glucose homeostasis. Previously, the group led by Nadal, described the existence of a membrane receptor capable of being activated by estrogens and with rapid signaling effects in pancreatic α and β cells (Alonso-Magdalena et al., 2005; Nadal et al., 2000, 2004; Ropero et al., 2002). This group demonstrated that this receptor is different from classical receptors ERα and ERβ and they named it non-classical membrane estrogen receptor (ncmER). With the knowledge that we currently have, there is a very high probability that ncmER receptor is GPER, since they present mimetic effects in pancreatic β cells.

3.1 Insulin secretion by pancreatic β cells

The main function of pancreatic β cells is produce and secrete insulin to maintain plasma glucose levels in normoglycemia. The central insulin secretagogue is glucose. Increasing levels of plasmatic glucose are the starting signal to glucose-stimulate insulin secretion (GSIS) by pancreatic β cells. GSIS "consensus mouse model" has four key metabolic events: Glucose uptake thought the specific high-glucose affinity transporter (GLUT2), step-limiting glucose metabolism in glucokinase (GK) action, ATP-sensitive potassium (K_{ATP}) channel closure and opening voltage dependent calcium channels (Ropero et al., 2002) (Fig. 6).

GLUT2 also known as solute carrier family 2 (SLC2A2) is a transmembrane carrier protein. The central role of GLUT2 in this process is well establish in rodents, but in humans GLUT1 and GLUT3 may play an important role also (van de Bunt and Gloyn, 2012). GLUT2 is express in the membrane of mouse pancreatic β cells and facilitates the entry of glucose to the insulin-secreting cell when glucose is in a satisfactory concentration for its opening, with low affinity but high capacity (high Km, ca. 15–20 mM) (McCulloch et al., 2011). After glucose uptake, GK catalyzes the first step of glycolysis with the increasing of ATP concentration, and consequently,

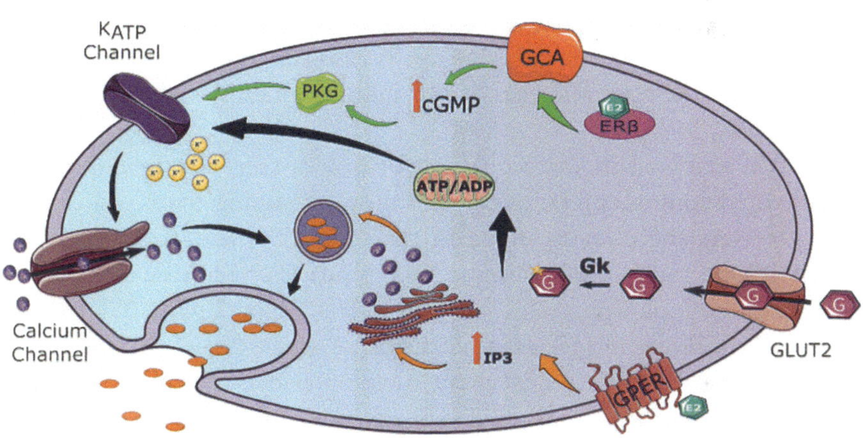

Fig. 6 Regulation of the insulin secretion in β pancreatic cells by the estrogen. The black arrows show the physiological process of glucose stimulate insulin secretion in β cells. The glucose is introduced into the cell by the GLUT2, inside it is phosphorylated by the glucokinase (GK) and metabolized in the mitochondria producing an increase of ATP by the Krebs cycle. The increasing of ATP produced a change in the ATP/ADP ration and the ATP-sensitive potassium (K_{ATP}) channel closure and opening voltage dependent calcium channels. This opening of calcium channel produced a Ca^{2+} influx that mobilized the insulin vesicles to the membrane and the insulin release. The green arrows show the potentiation effect of E2 by the estrogen receptor β (ERβ). It activates the guanylate cyclase A (GCA) activity and an increased the concentration of cyclic guanosine monophosphate (cGMP), which activates the PKG that produced the KATP channel closure and an increasing in the Ca^{2+} influx. The orange arrows show the effect of E2 by the GPER.

thus promotes an increased ATP/ADP ratio inside the cells (Boland et al., 2017). This increased in ATP/ADP ratio evokes the closure of K_{ATP} channels. With the closure of these channels, potassium ions will accumulate within the cell, causing a positive gradient that will lead to depolarization of the plasma membrane. The change in potential membrane will lead to the opening of voltage-gated calcium channels, allowing the flow of calcium into the cell. Finally, calcium influx will cause the movement of the insulin vesicles toward the membrane. Said cytoplasmic vesicles will fuse with the cell membrane, leading to insulin exocytosis (Ashcroft and Rorsman, 1990).

Although the greatest stimulus of insulin secretion is glucose, multiples factors can regulate also insulin secretion, such as other nutrients, ions, neurotransmitters, incretins, and many hormones, including sex steroids. E2 enhance, fast and directly, insulin release from pancreatic β cells to the bloodstream promoting the reduction of glucose blood levels due to physiological insulin effect in the liver. E2 can modulate insulin secretion via ERα, ERβ and GPER.

3.1.1 Estrogens regulation of insulin secretion

The mechanisms by which E2 acts on β cells function have been linked to membrane depolarization and Ca^{2+} influx to increase insulin secretion (Ripoll et al., 2008).

ERβ is involved in the regulation of insulin secretion in response to glucose mediating membrane depolarization by closing the K_{ATP} channel and increasing the calcium influx in the β cell (Soriano et al., 2009). In addition, E2 through the binding to ERβ produce an increase in the cyclic guanosine monophosphate (cGMP) by guanylate cyclase A (GC) in the membrane. This increase leads to the closure of the K_{ATP} channel by the action of PKG, and the consequent membrane depolarization and Ca^{2+} influx. Calcium entry into β cell and finally insulin is released to the bloodstream (Alonso-Magdalena et al., 2013; Soriano et al., 2009).

On the other hand, it has been shown that supraphysiological dose of E2 leads to activation of the plasma membrane receptor GPER also affects on insulin secretion. Mouse islets treated with G1, a specific GPER agonist, showed a similar GSIS pattern as with E2. However, the effect of GPER in GSIS is produced by an influence on the generation of both inositol triphosphate (IP3) and cAMP by activation of phospholipase C (PLC), in the former, and PKA, in the latter. This effect is independent of ERα and ERβ activation, because the broad antagonist ICI 182.780 (ICI) did not influence in the insulin secretion by G1; this effect is also observed in diabetic or nondiabetic human islets (Kumar et al., 2011).

As we will see in next sections, thanks to knock-out mouse models, many of the specific regulations carried out by E2 through to each type of ER have been elucidated. In the case of GPER it has been very relevant (Martensson et al., 2009). In part due to these studies, it is now known that GPER is capable of modulating insulin secretion in pancreatic β cells (Balhuizen et al., 2010; Martensson et al., 2009). E2 is able to enhanced insulin secretion via GPER under high glucose and supraphysiological glucose levels. Interestingly, as we have pointed out before, GPER is express in pancreatic α and δ cells. In addition to the action on pancreatic β cells, E2 regulates glucagon and somatostatin secretion in endocrine pancreas promoting the inhibition of both hormone release under stimulatory conditions (Balhuizen et al., 2010).

Interestingly, E2 is able to up-regulate the expression and content of glucokinase (GK) and GLUT2 in INS-1 insulin secreting cells and endocrine pancreas of diabetic rats and this up-regulation is associated with an up-regulation of GPER expression. This finding suggests that the improvement of insulin secretion by E2 can be promoted by GPER (Bian et al., 2019). E2 increased glucose uptake, GK activity and insulin secretion with up-regulation of Slc2a2, GK content, pAKT/AKT, p-mTOR/mTOR and GLUT2 protein under normal and high glucose conditions in vitro. On the other hand, G15 culture treatment, an antagonist of GPR30, show the opposite effect. These observations determine that the increase in insulin secretion by E2 GPR30-mediated directly regulates the AKT/mTOR pathway in pancreatic β cells, at least in rodent models (Bian et al., 2019). The estrogenic regulation of insulin secretion via GPER can have an important impact on physiological and pathophysiological regulation of these secretory cells. B cells failures in this signaling pathway could potentially trigger cellular dysfunction that lead to dysregulation of glucose homeostasis and therefore the development of diabetic phenotypes.

In a general way, the rapid insulinotropic effect of E2 in insulin secretion is mediated by ERβ and GPER, although the implication of ERα has not been formally excluded (Fig. 6).

3.2 Insulin biosynthesis in pancreatic β cells

Insulin is a hormone explicitly synthesized in pancreatic β cells. Before reaching insulin synthesis in mature form, a metabolic process of precursors (preproinsulin and proinsulin) processing occurs. Preproinsulin is modify in endoplasmic reticulum of the cells to give rise to proinsulin stabilized by three disulfide bonds. Proinsulin is correctly folded and transferred to the

Golgi apparatus in which is efficiently sorted in to secretory vesicles. Inside the vesicles the processing completes to yield insulin and C-peptide. These peptides and others such as amyloid polypeptide are stored in mature secretory granules waiting for the stimulus that leads to its secretion (Halban, 1991; Huang and Arvan, 1995; Steiner, 1982).

As with insulin secretion, there is evidence that E2 can regulate and modulate insulin biosynthesis in pancreatic β cells via ERα and GPER.

3.2.1 Estrogen regulation of insulin biosynthesis

E2 increase insulin biosynthesis through an ERE-independent signaling pathway (non-classical pathway), modifications in DNA to enhanced insulin synthesis, are not produced by direct binding of ER to DNA but by modulation of other transcription factors. E2 treatment of pancreatic islets leads to an increase in insulin biosynthesis and a consequent increase in insulin content. This increase is apparently mediated via ERα and not by ERβ. The specific pathway of E2 action has been proven with studies conducted in specific ERα and ERβ knock-out mouse models (see below in Section 4) and corroborated with specific agonist to ERα (PTT) and ERβ (DPN) (Alonso-Magdalena et al., 2008; Wong et al., 2010).

Additionally, studies carried out thanks to the generation of a mouse model characterized by a point mutation in the DBD of ERα (ERαAA/2 mice) resulting in an ERα deficient estrogenic response, demonstrated that E2-induced insulin biosynthesis was an effect independent of ERα binding at ERE sequences (Jakacka et al., 2002). In fact, ERα, in the cytosol, interacts with kinases such as SRC and ERK (Alonso-Magdalena et al., 2008; Wong et al., 2010). This extranuclear interaction between ERα and SRC regulates insulin biosynthesis through the binding of neurogenic differentiation factor 1 (NeuroD1) to insulin gene promoter, activating insulin gene transcription (Wong et al., 2010) (Fig. 7). It has also been reported that female mice lacking GPER have decreased pancreatic insulin content, suggesting that both ERα and GPER can affect insulin biosynthesis (Martensson et al., 2009). However, the specific mechanism that GPER is involved in the insulin content is not well understand.

Thus, insulin biosynthesis appear to be regulated mainly through ERα activation by extranuclear-initiated signaling pathway with genomic action, insulin release appears to be regulated mainly through extranuclear ERβ activation with rapid membrane response also; being that in both phenomena GPER is also involved. More studies are needed to understand the reasons why the different ERs may have different and selective effects on β cells.

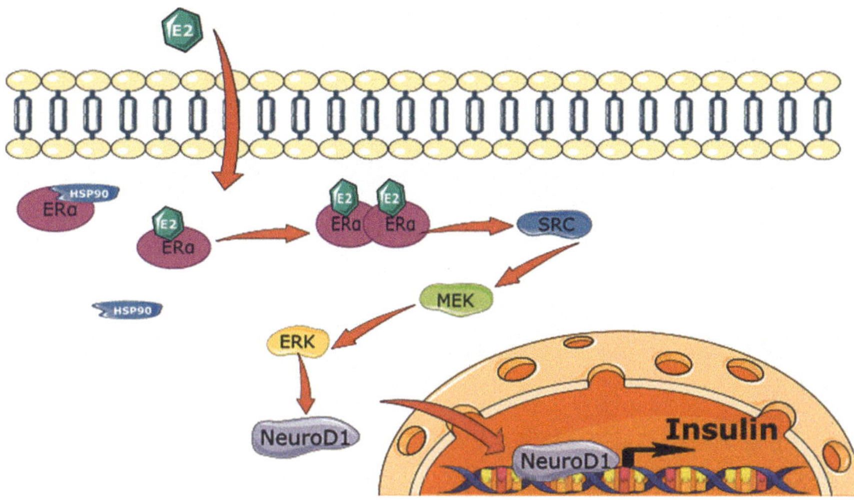

Fig. 7 Effect of estrogen receptor on the insulin biosynthesis. When ERα is activated by the E2 in the cytosol, it interacts with the kinase proto-oncogene tyrosine kinase (SRC) and an extracellular signal-regulated kinase (ERK). This extranuclear interaction between ERα and SRC activated the MERK/ERK pathway that regulates insulin biosynthesis through the binding of neurogenic differentiation factor 1 (NeuroD1) to the insulin gene promoter, activating insulin gene transcription.

3.3 β-Cell survival, proliferation, and regeneration

E2 is not only involved in insulin biosynthesis and GSIS, many studies are showing that ERs improve islet survival and hypertrophy (Alonso-Magdalena et al., 2006; Choi et al., 2005; Costrini and Kalkhoff, 1971; Geisler et al., 2002; Le May et al., 2006; Tiano et al., 2011). Males develop rapidly lethal diabetes in the most of animal models, whereas female are protected (Tiano and Mauvais-Jarvis, 2012a) (showed in Table 1). Thus, suggests a protective function of the E2 on the islet as happen in humans. When estradiol production is genetically suppressed by deletion of the aromatase gene, in both sexes, mice develop a dramatic vulnerability to streptozotocine (STZ)-induced insulin deficiency. Showing that circulating E2 is a protective hormone that prevents β cell apoptosis in vivo (Le May et al., 2006). Besides, E2 protects against oxidative stress and proinflammatory cytokines-induce apoptosis (Contreras et al., 2002; Eckhoff et al., 2003, 2004; Le May et al., 2006).

In human and mouse islets, E2 prevents islet apoptosis through ERE independent pathway. Both receptors promote survival, getting ERα a predominant effect (Liu et al., 2009). Through the use of an ERα knock-in mouse with a mutation of the DBD of ERα (that abolishes ERα binding

Table 1 Animal models and β cell failure.

Animal model	Species	Type of diabetes	Effect of E2 on β cell failure	Metabolic anomaly	Mechanism of β cell failure	References
Human islet amyloid polypeptide (hIAPP)	Mouse	2	Exogenous E2 protects males OVX females predisposed to amyloid formation	Selective β-cell failure	Overexpression of hIAPP	Janson et al. (1996), Geisler et al. (2002), van de Bunt and Gloyn (2012), Verchere et al. (1996)
db/db mouse	Mouse	2	OVX females predisposed on C3He3/FeJ background Exogenous E2 protects females on C57BL/6 background	Obesity, insulin resistance	Lepr mutation	Garris and Garris (2005), Leiter et al. (1989)
Akita	Mouse	2	OVX females not predisposed ERα-deficient females predisposed	Selective β-cell failure	Ins2 mutation	Hirosawa et al. (2008), Toyoshima et al. (2007), Wang et al. (1999), Yoshioka et al. (1997)
Irs2 −/−	Mouse	2	Without studies on the literature	Mild Obesity, insulin resistance	Targeted Irs2 knockout	Burks et al. (2000), Withers et al. (1998)
RIP-cHrasT	Mouse	2	OVX females predisposed Exogenous E2 protects males and OVX females	Selective β-cell failure	Expression of activated human H-ras in β cells	Efrat (1991), Efrat et al. (1990)
Zucker diabetic fatty (ZDF)	Rat	2	Exogenous E2 protects males	Obesity, insulin resistance	Lepr mutation (lipoapoptosis)	Lee et al. (1994)

Otsuka Long-Evans Tokushima Fatty (OLETF)	Rat	2	OVX females predisposed Hormone replacement restores incidence	Obesity, insulin resistance	Cckar mutation	Kawano et al. (1994), Shi et al. (1994), Shimizu et al. (1998)
Streptozotocin (STZ)	Mouse	1	Exogenous E2 protects males OVX females predisposed	Selective β-cell failure	Targeted β-cell apoptosis with chemical that induces DNA alkylation	Le May et al. (2006), Liu et al. (2009), Maclaren et al. (1980), Paik et al. (1982), Puah and Bailey (1985)
Alloxan	Mouse	1	Exogenous E2 protects males	Selective β-cell failure	Targeted β-cell destruction with chemical that induces oxidative stress	Goodman and Hazelwood (1974)

to the ERE) was showed that this process is independent of nuclear events, without de novo protein synthesis or gene transcription (Liu and Mauvais-Jarvis, 2009). Both ER are expressed in β cell, however they do not have antagonist effects on cell survival, because the pharmacological inhibition or genetic elimination of ERβ in islets do not enhance E2 anti-apoptotic action via ERα (Liu and Mauvais-Jarvis, 2009). On the other hand, the elimination of both receptors do not synergize to abolish E2 protection following exposure of islets to oxidative stress (Liu and Mauvais-Jarvis, 2009). Besides, the elimination of GPER in female mice predisposes to STZ-induced islet apoptosis (Le May et al., 2006). And its activation by G1 prevents oxidative stress-induction apoptosis in culture human and mouse islets (Bologa et al., 2006). In cultured GPER-deficient islets, the E2 protection from apoptosis was retained demonstrating that ERα can compensate for GPER deficiency. Suggesting that E2 protects islet survival by preventing apoptosis via ERα and ERβ and GPER pathways with a predominant ERα effect over the other two receptors (Liu and Mauvais-Jarvis, 2009). Nevertheless, in 2010 the study of Kang and collaborators, showed that GPER induces the expression of the ERα variant ERα36 (Kang et al., 2010), and experiments with siGPER transfected cells questioned that the G1 agonist is actually an ERα36 agonist and does not bind to GPER (Kang et al., 2010). ERα36 has been proposed with just membrane and cytosolic functions (Wang et al., 2005, 2006) and it is expressed in β cells (Tiano et al., 2011). Thus, probably the effect protective against STZ-induced by G1 in islet apoptosis is though ERα36 signaling. Furthermore, in brain-dead donors, were the proinflammatory cytokines were increased, the use of E2, in vivo and in vitro, induces cytoprotection in human islets and the recovery of functionality of brain-dead donors islets (Contreras et al., 2002; Eckhoff et al., 2004). These effects were produced by a decrease in NF-κB nuclear translocation, in c-Jun N-terminal kinase (JNK) activation, in mitochondrial cytochrome *c* release, and a lowered caspase-9 activation (Contreras et al., 2002; Eckhoff et al., 2003).

In conclusion, E2 is able to promote β cell survival via ERα, ERβ and GPER, but is suggest that ERα36 isoform in the main contributor to β cell protection. This promotion is possible due to the decrease in NF-ƘB nuclear translocation.

3.3.1 E2 in the regulation of β cells apoptosis

One of the main processes that produce β cell apoptosis is the reticulum stress. In T2DM, the high secretory burden of the islet, in the first stages, can compromise the efficient protein synthesis and folding, which are critical for β cell function and health. When the endoplasmic reticulum becomes overwhelmed by unfolded or misfolded proteins, the cell enters in endoplasmic reticulum stress and actives the unfolded protein response (UPR) signaling network. This signaling, depending upon the severity and duration of stress, initiates cytochrome *c* release from the mitochondria to trigger β cell apoptosis (Engin et al., 2014; Papa, 2012). Thus, there is a strong collaboration between endoplasmic reticulum and mitochondria in this type of cell. In ERα-deficient Min6 cells are imbalanced mitochondrial fission/fusion-mitophagy dynamics, increased expression of the mitochondrial stress gene *Oma1*, and the production of reactive oxygen species (ROS) is excessive. On the other hand, the absent of ERα produces an increase in CHOP expression, a reticulum stress marker, that is consistent with the increase in cleaved caspase levels in β cells and taken the cell to apoptosis (Zhou et al., 2018). However, the fact that ERα is an important regulator of autophagy maintenance by regulating the expression of key autophagy-related signaling factors, as *Atg3*, *-5*, *-7*, *-12* and *Beclin-1* genes (Zhou et al., 2018). And knowing that ERα is also a protective agent against mitochondrial-endoplasmic reticular stress through the control of *Chop* and *Oma1* expression; all of that indicates that this receptor has an important role in the maintenance of β cell mass.

In the case of transgenic mouse overexpressing human amyloid polypeptide (hIAPP), a model of T2DM, in pancreatic β cells, male mice are predisposed to develop islet amyloid accumulation and hyperglycemia (Verchere et al., 1996). The treatment with E2 prevents the amyloid formation and the β cell failure (Geisler et al., 2002), and in the case of ovariectomized mice, where the E2 production is suppressed, amyloid formation in islets was increased (Kahn et al., 2000). It knows that Chop inhibition is sufficient to normalize cleaved caspase-3, restore autophagy flux, reduce β cell apoptosis, delaying the onset of diabetes in this transgenic mice (Gurlo et al., 2016; Rivera et al., 2011).

Not only E2 has a role in the apoptosis process, selective estrogen receptor modulator (SERM) and conjugated equine estrogens (CE) can modulate this process. The combination of CE with the SERM bazedoxifene and bazedoxifene alone, reduced the severity of β cells destruction induced by STZ in female mice (Kim and Mauvais-Jarvis, 2016). The prevention of STZ-induced insulin-deficient diabetes in mice is a marker of estrogen agonist activity in preventing β cell apoptosis (Le May et al., 2006), thus bazedoxifene acts as an estrogen agonist in β cells. In the islet of male Akita mice (model animal described in Table 1), the accumulation of unfolded insulin promotes endoplasmic reticulum stress which induces the expression of the protein CHOP. The treatment with CE/bazedoxifene reduced islet CHOP expression, and prevented islet destruction and the development of insulin-deficient diabetes (Xu et al., 2018).

Both ER isoforms are involved in the apoptosis prevention via redundant pathways, inhibiting the activation of caspases in different kind of cells, however in β cell is no clear its role in apoptosis.

3.3.2 E2 in the promotion of β cell regeneration

The role of ERs in β cell proliferation remains in discussion, inasmuch as there are different studies showing different effects in it. In partially pancreatectomized rats treated with/without hormone replacement (E2, 50 μg/kg; T, 1 mg/kg) after ovariectomy, E2 promoted β cell hypertrophy but no expansion of the β cell population improving diabetes (Zhu et al., 1998). In another study, E2 treatment increased β cell proliferation in ovariectomized rats after 90% pancreatectomy and fed with high fat diet with corresponding administration of E2 (30 μg/kg body weight) (Choi et al., 2005). It is also known that in STZ-challenged, hyperglycemic mice, E2 treatment administrated by s.c. pellets (0.48 mg) did not induce β cell proliferation (Le May et al., 2006), and in cultured islets treated in a range of 0.1–100 nM E2, there is no effect of E2 on β cell proliferation (Alonso-Magdalena et al., 2008).

In pregnancy, when the metabolic demand is increased, the β cell population and function is augmented. The increase in E2 serum concentration during pregnancy may participate to the islet adaptation to satisfy this metabolic demand, in as much as exposure to physiological concentrations of E2 increased insulin gene expression, insulin content, and insulin release via an ERα though SRC and ERK kinases, as above mentioned (Alonso-Magdalena et al., 2008).

In the partial duct ligation (PDL) model, PDL-induced proliferation of β cells depended on aromatase activity, this event promotes Neurogenin3 (NGN^{3+}) cells generation near duct epithelium that can differentiate to β cells (Xu et al., 2008). The treatment with tamoxifen, an ERα antagonist, or genetic elimination of ERα in male mice decreased factor NGN^{3+} expression and β cell proliferation, suggesting that ERα is involved in this process (Yuchi et al., 2015). ERα has a cytosolic localization under normal condition, however in PDL, where E2 is increased in the ligated portion of the pancreas, ERα in β cells is nuclear localization. The inhibition of ERα by tamoxifen in the embryonic pancreas, or its deletion in ERα-deficient mouse, also decreased NGN^{3+} expression and NGN^{3+} progenitors at the end of gestation (Yuchi et al., 2015). These facts indicate that ERα is involved in the generation of NGN^{3+} cells and the subsequent β cell mass expansion in developing or injured mouse pancreas (Yuchi et al., 2015). On the other hand, the β cell proliferation, also, occurred concomitantly with an increased expression of pancreatic duodenal homeobox-1 (PDX-1). In orchidectomized (ORX) male transgenic (Tg) mice, that is, with bilateral removal of the testicles (castration), the treatment with E2 enhanced β cell proliferation by inducing the neogenesis of new β cells in the ducts and the replication of existing β cells in the islets (Inada et al., 2014). Male control Tg mice develops severe diabetes with β cell loss due to the overexpression of inducible cAMP early repressor (ICER)-Iγ, this repressor is a member of the cAMP-responsive element modulator (CREM) family that lacks an activation of DNA-binding domain (Inada et al., 2004). In this model, only male mice demonstrate severe hyperglycemia throughout life and develop diabetic nephropathy, whereas females exhibit mild hyperglycemia with age without nephropathy (Inada et al., 2007). This sexual dimorphism suggests the possibility that E2 are involved in protective effect on β cells or androgens have a deleterious effect in this endocrine cell type. In order to check this hypothesis, ORX or castration allow change the levels and pattern ratios of sexual hormones and E2-pellet implantation alone or in combination with ORX mimics the possible protective effect observed in females. E2 circulating levels and E2/T ratio levels affects β cell regeneration, when E2 is increase above a certain level, it achieves a full dramatic effect beyond the influence of T, which results in a higher increase in β cells induction in the pancreas, thereby rescuing male mice from severe diabetes. This enhance in β cell proliferation occurs with the activation of PDX-1 regulatory cascade (Inada et al., 2014).

In addition, it is has been described that GPER has been implicated in the expansion of functional β cell mass observed during pregnancy. In rodents, GPER expression is markedly upregulated during pregnancy, which is linked with decreased expression of the islet microRNA miR-338-3p (Liu and Mauvais-Jarvis, 2009). The downregulation of miR-338-3p promoted β cell proliferation, while increased miR-338-3p expression decreased β cell mass. In isolated rat islets, exposure to E2 or the GPER agonist G1 also decreased miR-338-3p to levels observed in gestation, this level is associated with increased β cell proliferation. This effect depends on cAMP and protein kinase A (PKA). The E2 exposure also reduced miR-338-3p level in cultured human islet cells (Liu and Mauvais-Jarvis, 2009), however E2 or silencing of miR-338-3p did not bring out the replication of human β cells in culture

(Choi et al., 2005). This no effect on proliferation also happen in human β cells transplanted in male mice that were treated with E2 or G1 (Shi et al., 1994). These data show that the effect of GPER activation and the resulting suppression of miR-338-3p observed in rodent β cell proliferation are not observed in human β cells, getting a main role to ERα.

Until now, the role of ERβ in β cell regeneration was showed in a study where male normal C57BL6 mice rendered mildly diabetic with STZ and diabetic db/db mice were treated with the selective ERβ agonist (WAY200070) (Alonso-Magdalena et al., 2013). This study shows that after 2 weeks treatment of WAY200070, diabetic mice regenerated pancreatic β cell mass. In consequences, they improved plasma insulin levels and glucose tolerance (Alonso-Magdalena et al., 2013).

As a summary (Table 2), the role of estrogen receptors in β cell can be reviewed as follows: ERα is involved in (1) β cell survival modulation Chop expression and inhibiting cell apoptosis, (2) insulin biosynthesis through the activation of SRC, ERK and NeuroD1, and (3) β cell proliferation/regeneration through the modulation of neurogenin3 expression and NGN^{3+} precursor cells. While ERβ enhances (1) GSIS due to the increase of intracellular cGMP and in consequence potentiating the Ca^{2+} intracellular influx and (2) increases β cell proliferation. On the other hand, the GPER is implicated in GSIS due to the increase PCL and PI3 concentration and in consequence, increasing the Ca^{2+} intracellular influx and β cell proliferation by miR-338-3p modulation. In a general point of view, in β cell ERs reside

Table 2 The role of estrogen receptors in β cell.

Impact on β cell	Receptor	Effect
Insulin secretion	ERβ	Increased of cGMP and Ca^{2+} influx
	GPER	Increased PI3 and Ca^{2+} influx
Insulin biosynthesis	ERα	Activation of SRC, ERK and NeuroD1
	GPER	KO mice have lower concentration
β cell proliferation	ERα	Involved in nerogenin3 expression and generation of $NGN3^+$ cells
	ERβ	Increased β cell mass
	GPER	Modulation of miR-338-3p
Apoptosis	ERα	Modulation of chop expression and mitochondria function

mainly in extranuclear locations, and they exert their effect via cytosolic interactions (extranuclear-initiated signaling pathways) with kinases such as SRC, ERK, and AMPK, or via transcription factors (genomic action by ERE-independent signaling pathway) that could finally lead to genomic effects.

4. Knock-out mice models: A useful tool in the study of estrogens impact in pancreatic β cells

The study of impact of E2 in β cell physiology date back 1940s, but it was not until the 90s when a significant advance was made in the knowledge of the effects of estrogens on the function of pancreatic β cells thanks to the development of animal models with genetic deficiency for steroid hormone receptors: ER knock out models.

4.1 Generation of ER knock-out models

Knock-out mouse models is a very useful tool to explore the absence of a gene product in a certain metabolic context. The metabolic actions of estrogens are mediated via the estrogen receptor ERα, ERβ, and the membrane-bound G-protein-coupled ER (GPER). In this sense, four different global knock-out models has been use as experimental approach in the study of estrogens in metabolic disorders: ERα (αERKO) (Lubahn et al., 1993), ERβ (βERKO) (Krege et al., 1998) individually, or that lack both ERα and ERβ genes (αβERKO) and GPR30 (GPRKO) (Couse et al., 2000; Dupont et al., 2000; Korach et al., 2003).

4.1.1 Global knock-outs

Focus on the effect in pancreatic β cells:

4.1.1.1 αERKO

One of the most widely used models for studying the impact of estrogens on glucose metabolism has been global knock-out of ERα (ERαKO), several studies demonstrates that this model develop insulin resistance and glucose intolerance both in males and females (Heine et al., 2000; Riant et al., 2009; Ribas et al., 2010; Zhu et al., 2013).

As we have previously described, ERα is express in pancreatic β cells with cytosolic and nucleus localization. A global null ERα has been used in many physiological studies to verify the role of ERα in metabolism. It should be noted that this mouse has increased fasting glucose and insulin circulatory levels (Bryzgalova et al., 2006) and develops obesity and insulin resistance (Heine et al., 2000), which may be indirect evidence of the impact that deletion of ERα has on the physiology of pancreatic β cells and glucose homeostasis. Focus on pancreatic β cells, this model has been used to evaluate the role and effect of estrogens via ERα on insulin secretion, insulin production and cell survival. In this sense, it has been shown that ERα plays an important role in the regulation of insulin biosynthesis, insulin secretion and β cell survival.

In order to evaluate the respective importance of nuclear ERα (NOER) and membrane ERα (MOER), Camille Allard et al. conducted an interesting study evaluating glucose homeostasis, insulin secretion and insulin sensitivity in male and female mice expressing either NOER or MOER pools (Allard et al., 2019). Both MOER genders developed hyperglycemia and glucose intolerance. Although both genders share this general phenotype, there are differences between the two in the evolution of the phenotype. Female MOER mice displayed hyperinsulinemia in association to impaired central insulin signaling due to constitutive hepatic gluconeogenesis and did not show alterations in insulin secretion from endocrine pancreas. On the other hand, male MOER mice islets displayed impaired GSIS associated to a decreased in brain glucose sensitivity and did not show general insulin resistance. In addition, female NOER mice develop glucose intolerance and hepatic insulin resistance. These results pointed out that nuclear ERα signaling is predominant in the maintaining of glucose homeostasis in both genders and lack of nuclear ERα promotes dysregulation of central insulin sensitivity in females and impairs insulin secretion from pancreatic β cells in mice (Allard et al., 2019).

Regarding ERα role in β cell survival, Le May et al. described for the first time, E2 as a protective hormone for β cell survival in mice using this ERα loss of function model. E2 protects pancreatic β cell from apoptosis induced by oxidative stress and this effect protects insulin production and diabetes development in mice of both genders. This protection is mediated, at least in part through ERα (Le May et al., 2006). Subsequent studies by Alonso-Magdalena et al. have shown that estrogens ERα-mediated regulate insulin content of pancreatic β cells, suggesting the up-regulation of mRNA insulin levels by E2 (Alonso-Magdalena et al., 2008).

Furthermore, recent studies carried out in this model have shown that signaling through ERα is involved in the control of β cell progenitor activation and β cell proliferation. Genetic ablation and competitive inhibition of ERα lead with a decreased number of Ngn^{3+} endocrine pancreas progenitors in mice embryos. Ngn^{3+} is the main transcription factor that is activated in the step of epithelial cells to endocrine cells in the development of the endocrine pancreas. With the activation of Ngn^{3+} expression starts the development events that give rise to the different endocrine lineages that conform mature islets of Langerhans. This could mean that E2 signals via ERα may be essential for the development of the endocrine pancreas and therefore for the physiological maturation of β cells. In addition, adult pancreas after partial duct ligation (PDL-treated mice) had increased E2 levels, ERα activity, Esr1 (which encodes ERα) transcript levels and nuclear localization of ERα in β cells, in concert with increased β cell proliferation. This important finding supports not only the regulatory role of estrogens via ERα during pancreas development but also during pancreas injure recovery (Yuchi et al., 2015).

4.1.1.2 βERKO

Another estrogen receptor, involved in the physiological signaling of E2, which is expressed on pancreatic β cells is ERβ. In order to know the function of this receptor on insulin-producing pancreatic cells, different studies have been carried out in recent years using the ERβ global knock-out mouse (βERKO) as a tool. Metabolic characterization of this model has shown that ERβ is involved in insulin signaling and glucose metabolism and regulates adipose function via PPARγ (Foryst-Ludwig et al., 2008). This impairment in insulin signaling and glucose regulation could be related, at least in part, with the direct regulation on pancreatic β cells. In order to elucidate this, Nadal et al. demonstrates that circulating hormone E2 and the xenoestrogens bisphenol-A (BPA) and Diethylstilbestrol (DES) enhance the frequency of intracellular calcium ($[Ca^{2+}]_i$) oscillations in pancreatic β cells within intact islets of Langerhans in wild type mice (Nadal et al., 2000). Further studies performed in βERKO islets by Soriano et al. showed that ERβ is implicated in the rapid regulation of K_{ATP} channel during insulin secretion. E2 blocks K_{ATP} channel in pancreatic β cells from βERKO and this effect is not shown in β cells from wild type and αERKO (Soriano et al., 2009).

In accordance with the study carried out by Le May et al. in αERKO mode (Le May et al., 2006), another study published in Diabetes in 2009 reveals that E2 protects β cells from apoptosis mediated through ERα, GPR30 and also ERβ (Liu et al., 2009).

4.1.1.3 αβERKO

Studies in αERKO and βERKO have shown that estrogens can modulate different or synergistic signaling pathways that have an impact on the physiology and pathophysiology of pancreatic β cells. In order to identify the signaling pathways that regulate each type of receptor, complementary studies have been carried out in the double knock-out mouse model for both receptors: αβERKO.

In 2009 a study led by Mauvais-Jarvis using this model in complementation with the previous ones, revealed that ERβ plays a minor cytoprotective role compared to ERα in the context of islet survival protection. Both ERα and ERβ favor islet survival using different cellular pathways (Liu et al., 2009). βERKO mice presented a medium predisposition to β cell destruction by STZ treatment and αβERKO islets treated with E2 partially prevented STZ-induced β cell destruction, supporting an alternative pathway of E2 in β cell survival. In the absence of both α and β ERs, E2 is able to protect β cells from apoptosis and promote survival, this evidence supports the fact that there must be an alternative pathway of estrogen signaling not dependent on these two classical receptors. In addition, this study also demonstrates that GPERKO mice (see next Section 4.1.1.4) are susceptible to STZ-induced β cell apoptosis, suggesting that the alternative pathway of E2 signaling in survival islet promotion it is mediated by GPER.

4.1.1.4 GPR30KO/GPERKO

GPR30KO mouse model was first developed and used in 2009 by Mårtensson et al. in order to investigate the metabolic role of this estrogen receptor in endocrine pancreas function. This study reveals that GPER is necessary for physiological insulin production and glucose homeostasis in females. Additionally, this research reported defects in skeletal growth and impaired blood pressure related with global GPER deficiency (Martensson et al., 2009). GPER is required for E2 stimulated insulin release in both female and male mice but is essential for normal glucose homeostasis only in females. GPER deletion affects first-phase insulin release. First-phase insulin release is decreased in pancreatic islets from GPR30KO due to, at least in part, a decreased in insulin gene expression and in consequence a decreased in global glucose-stimulated insulin secretion. This effect is reflect in impaired glucose tolerance in vivo in female mice, showing hyperglycemia and glucose intolerance at 6 months of age (Martensson et al., 2009). However, this interesting effect in glucose homeostasis has been not showed in other GPER knock-out models (Liu et al., 2009). There is great controversy regarding the use of this model in regulation of glucose homeostasis because important phenotypic differences have been found depending on the origin. The studies reported in the bibliography are not homogeneous and reach different conclusions (Langer et al., 2010; Ropero et al., 2012).

The mentioned role of GPER in insulin secretion was subsequently investigated by other research groups who observed that GPER activation of the epidermal growth factor receptor and ERK in response to E2 treatment plays a critical role in the secretion of insulin from β cells. Using GPER-selective ligands (G1 and G15) in islets of wild type mice compare with GPERKO islets the authors found that E2 and GPER-selective agonist G-1 mediates calcium mobilization, ERK activation and further modulated by PI3K activity promoting insulin secretion both under low and high glucose stimulation. In addition to this, islet pre-treatment with GPER-selective antagonist G15 inhibited this effect on insulin secretion (Sharma and Prossnitz, 2011). The results of both studies suggest that the activation of downstream signaling pathways by the GPER-selective ligand G-1 could represent a novel therapeutic strategy in the treatment of diabetes.

Regarding the specific regulation of GPER in β cell survival, it has been reported to be involved in the protection from apoptosis and promotion of cell survival alongside ERα and ERβ (Liu et al., 2009).

4.1.2 Tissue-specific knock-out

4.1.2.1 Pancreatic islet-specific ERα knock-out mice (PERαKO)

The most specific model that allows us to evaluate the impact of estrogen signaling on the pancreatic β cell is PERαKO (pancreatic islet-specific ERα knock-out). Several studies carry out in this model reveals that lack of ERα signaling leads directly with regulation of insulin gene transcription, a reduction in lipid synthesis, and apoptosis promotion in pancreatic islets. As previously indicated, E2 is able to up-regulate insulin gene expression and in consequence, insulin content in pancreatic islets. In concordance, E2 insulin regulation is abolished in PERαKO mouse model (Wong et al., 2010). ERα stimulates NeuroD1 binding to the insulin promoter and favors insulin synthesis.

Regulation of lipid metabolism entails islet lipid accumulation and impaired β cell function in response to high fat diet (Tiano et al., 2011). ERα signaling deficiency results in chronically activation of ER stress via suppression of fatty acid synthase expression and activity. The suppression occurs through non-classical pathway and is Stat3-activated dependent. Furthermore, ERα inhibition promotes β cell apoptosis in vivo. ER stress by alloxan exposure promotes β cell destruction and insulin deficient diabetes in PERαKO mouse. Interestingly, ERα mRNA expression was increased by hyperglycemia both in vivo and in vitro, independently of direct insulin action. Thus, induction of ERα expression protects β cells against oxidative injury (Kilic et al., 2014).

Regarding the conclusions in different studies conducted in global αERKO, ERα promotes insulin secretion and pancreatic β cell survival. In addition to this, recent research carried out on a pancreatic β cell tissue specific PERαKO mouse model, concludes that this protection and insulin secretion promotion is due to maintenance of mitochondrial fission-fusion-mitophagy dynamics (Zhou et al., 2018).

ERα protects β cells from apoptosis by direct transcriptional repression of C/EBP homologous protein (*Chop*) and metalloendopeptidase *Oma1* expression regulating mitochondria function, morphology and autophagy and reducing endoplasmic reticulum stress. Both are induce by endoplasmic reticulum stress, in one hand, *Chop* induces apoptosis through mitochondria-dependent pathway and in the other, *Oma1* is an important regulator of inner membrane in the mitochondria. This peptidase controls mitochondrial fusion under stress conditions and is essential for the correct energy balance in cells. Thus, E2 is able to protect β cell from endoplasmic reticulum stress and apoptosis via mitochondrial regulation, this finding is very important in the context of pharmacological approaches against diabetes (Zhou et al., 2018).

In conclusion, this tissue-specific mouse model has allow to deepen in the mechanisms involved in β cell protection and survival in pancreatic β cells previously showed in global αERKO.

4.1.3 Estrogen deficiency, the aromatase knock-out (ArKO) mice

Another interesting study model to take into account as a tool in the research of the impact of sex hormones on the physiology and pathophysiology of pancreatic β cells is the aromatase knock-out mice: ArKO.

ArKO mouse model was generated by targeted disruption of the *cyp19* gene (Fisher et al., 1998) because estrogens production from C19 steroids are catalyzed by aromatase P450 due to *cyp 19* gene expression. ArKO mouse cannot synthesize endogenous estrogens by this aromatase deficiency. This model of complete estrogens ablation due to the inability to convert androgens in estrogens, develops metabolic syndrome, specifically ArKO male mice develops hepatic insulin resistance and glucose intolerance and ArKO female mice becomes glucose intolerance but without insulin resistance (Van Sinderen et al., 2017).

ArKO mice have been used in complementation with αERKO mice to elucidate if female sex steroids protect from β cell damage, as it is suggest by the lower prevalence of diabetes in females. In this way, it has been observed that ArKO mice are vulnerable to β cell apoptosis and prone to insulin-deficient diabetes after exposure to acute oxidative stress with STZ (Le May et al., 2006). E2 post-STZ treatment rescues STZ-induced β cell apoptosis, helps to improve insulin production in residual β cells and prevents diabetes development. These observations, in addition to the observing E2 apoptosis protection in β cells and islets treated with an ERα antagonist and in αERKO islets, reinforce that E2 protects from oxidative injury in pancreatic β cells, at least in part, via ERα in both sexes.

4.2 Knock-in mouse models

As a modification of knock-out technique, knock-in mouse models are also useful tools for the study of proteins of interest and regulation of signaling pathways in tissues and cells. Gene knock-in technology alters the genetic locus of interest via one-for-one DNA sequence information adding information that is not found on select genetic locus.

Using this technique novel ERα knock-in mouse model was generated in order to further investigate estrogens signaling and regulation in physiology and physiopathology. In complementation of PERαKO mouse model, ERα knock-in has been used for the study of E2 regulation of insulin synthesis in vivo (Wong et al., 2010). Using ERα knock-in eliminating estrogen response elements (EREs), it has been described that E2 enhanced insulin synthesis is independent of ERE. Extranuclear ERα activates SRC, to enhanced nuclear location and binding to the insulin promoter of the transcription factor NeuroD1.

Focusing on cell survival, it has been demonstrated thanks to this model that E2 signaling through to ERα protects endocrine islets survival independently of the classical ERE-mediated pathway (Liu et al., 2009).

As we have already seen in previous sections of this chapter, the canonical model for estrogens regulation in physiology is mediated by intracellular hormone-specific ERs. E2 activates ERα, as a dimer an ERE or a non-ERE tethered promoter, to initiate gene transcription. These two regulatory pathways can be divided as direct regulation of gene expression through ERE or indirect regulation of gene expression non-mediated by ERE (non-ERE) (Bjornstrom and Sjoberg, 2005).

In this way, it has been described, using knock-in mouse model with ERE signaling lacking, that, in vivo, the ER cytoprotection role in the islets survival promotion is not mediated by direct ERE binding DNA. E2 favors β cell survival and protect from apoptosis by activation of extranuclear. To investigate whether an ER-ERE or non-ERE signaling mechanism protects β cell survival in vivo, Liu et al. (2009) used an ER knock-in mouse model with a deletion in ER DNA-binding domain.

In conclusion, because estrogens can carry out their physiological and metabolic functions by activating different specific-receptors that will lead to the activation of different signaling pathways, knock-out and knock-in models have been a crucial tool in understanding the impact of E2 on the physiology and pathophysiology of pancreatic β cells. It has also served to identify new therapeutic targets that help control and develop T2DM. Thanks to the studies carried out on these models with specific genetic deletions, we have been able to elucidate the direct effect that the absence of estrogens can have on β cell function as well as the specific pathway by which this effect occurs (Table 3).

Table 3 Summary of the genetic modify mouse models.

Mouse model	Phenotype	B cell regulation
αERKO		Insulin biosynthesis
	Hyperglycaemia	Insulin secretion
	Obesity predisposal	ER stress
	Insulin resistance	Apoptosis
		Differentiation
		Proliferation
βERKO		Insulin secretion
		ER stress
		Apoptosis
GPR30KO	Obesity	Insulin secretion (ERK pathway)
	Insulin resistance	
	Glucose intolerance	Insulin gene expression
	Skeletal growth defects	
	Enhance blood pressure	

Table 3 Summary of the genetic modify mouse models.—cont'd

Mouse model	Phenotype	B cell regulation
PERαKO		Insulin gene expression
		Insulin secretion under high fat diet
		Apoptosis
		Lipid synthesis
		ER stress
ERα knock-in	Estrogen response elements (EREs) elimination	Insulin biosynthesis
		Apoptosis
ArKO	Metabolic syndrome	
	Glucose intolerance (both genders)	
	Hepatic insulin resistance (males)	

5. Menopause, hormone replacement therapy and diabetes

Aging is an important risk factor in the development of insulin resistance and T2DM, both in men and women (Andersson et al., 1997; Zethelius et al., 2004). Furthermore, prevalence of T2DM increases in post-menopausal women. This phenomenon occurs with the lack of estrogens in summation with the subsequence increased in body fat mass that leading to a worsening of insulin resistance. Estrogens and weight loss by exercise has a potential benefits in order to improve glucose homeostasis, reduce insulin resistance and improve β cell function (Andersson et al., 1997; Friday et al., 2001; Gonzalez et al., 2002).

The role of estrogen on endocrine pancreas function takes great relevance in menopausal women. Menopause is a natural physiological event that occurs when the ovaries stop producing hormones, mainly estrogen and progesterone. Normally, this hormonal change appears in women over 50 years old. It is well established that metabolic changes occurring in pre-menopause and menopause are associated with increased risk of cardiovascular disease, osteoporosis, and diabetes (Matthews et al., 1989; Mauvais-Jarvis et al., 2017; Rosen, 2021). Although the association between menopause, insulin resistance and risk of T2DM is apparently clear, the mechanisms by which this increased risk of disease onset at this metabolic stage occurs are controversial, and the direct influence of estrogen on the human β cell is unclear. During menopause circulatory androgen and abdominal fat mass increases which leads with peripheral insulin resistance. This environment leads the β cell to functional failure that evokes in chronic hyperinsulinemia and dyslipidemia. Therefore, menopause risk factor for T2DM development is the sum of impaired insulin secretion and peripheral insulin resistance (Mauvais-Jarvis et al., 2017).

As we have seen in the previous section, many studies in animal models, mainly knock-out and ovariectomized (OVX) rodent models (see Section 5.1), conclude that estrogen deficiency has deleterious effects on insulin-secreting pancreatic β cell function and survival. Extrapolating these conclusions to the human menopause context is not so simple, menopause involves more complex events than the simple lack of estrogen signaling on cells and aging plays an important role of this process. Even so, animal models are the best preclinical approach to be able to elucidate the physiological repercussions that the lack of estrogens has on women after menopause.

HRT is the therapeutic treatment, based in estrogens or/and progesterone, administered in pre-menopausal and menopausal women to counteract the adverse metabolic effects of ovarian non-function that occurs at this stage. HRT had its beginning in the 1960s but it was only in the mid-1990s when it got its higher popularity. But after the publication of the study of Women's Health Initiative (WHI) in 2002 (Rossouw et al., 2002), where it was shown that the HRT use had more harmful than beneficial effects, its use was reduced. The reanalysis of study WHI, which was inadequately designed, evaluated, and reported, along with new studies (reviewed in Cagnacci and Venier, 2019) showed that HRT is highly beneficial when given to symptomatic women within 10 years since the onset of menopause or to symptomatic women that are under 60 years of age. Showing that HRT had a beneficial effect on the cardiovascular system, reducing coronary disease (Cagnacci and Venier, 2019) beside attenuate bone loss (Keck and Taylor, 2018; Newton et al., 1997; Sullivan et al., 1990). In fact, current scientific evidence suggests that the benefits of HRT outweigh the risks in symptomatic menopausal women younger than age 60 or within 10 years of menopause (Santen et al., 2010). The effects of menopause/HRT on the pancreatic β cell function were described in the following section.

5.1 Menopause, hormone replacement therapy and pancreatic β cell function

Focus on the direct effect of lack of estrogens during menopause in humans, the studies are limited and difficult to interpret. Aging is a factor that plays a very important pathophysiological role during this phase, making it difficult to discern the direct impact of lack estrogens on pancreatic β cell function per se.

A study from Walton et al. in 1993, comparing a premenopausal women group with a group of postmenopausal women, in order to elucidate the effect of menopause by discriminating the effect of aging on insulin secretion, concluded that transition from premenopausal to postmenopausal status is associated with reduced pancreatic insulin secretion. This study showed differences in hepatic insulin clearance, which was significantly decreased in postmenopausal women. However, the circulating insulin measured during an IVGTT (intravenous glucose tolerance test) turned out to be the same in the two experimental groups. These results showed that the differences found in insulin clearance should be caused by adaptation of pancreatic β cells function, which would be secreting less insulin into the bloodstream. Decreased insulin clearance in the liver would be a peripheral compensatory mechanism to ensure that the same amount of circulating insulin is maintained despite failure of insulin secretion (Walton et al., 1993).

A good approach for the study of menopause and its pathophysiological consequences is the use of OVX rodent models. The OVX models allows for evaluation of the impact of ovarian hormones independent of age-related changes. OVX mice develop obesity, glucose intolerance and insulin resistance, as a consequence of loss of hormonal signal (Abdel-Hakim et al., 2014; da Silva et al., 2014; Posa et al., 2015; Rogers et al., 2009). It has been reported since 1980 that Langerhans islets from OVX mice showed reduced glucose-stimulated insulin secretion ex vivo, and this defect is possible to restore by E2 administration (El Seifi et al., 1981). Furthermore, ovariectomy has been shown to reduce the adaptive capacity of β cells to respond to insulin resistance, accelerating its failure and the progression of glucose intolerance to T2DM (Vogel et al., 2013). In addition to this, post-ovariectomy treatment with estrogens, have shown improvements in glycemia, glucose tolerance, insulinemia and insulin secretion, independently of body weight and adiposity, suggesting that dysregulated glucose homeostasis is secondary to loss E2 (Santos et al., 2016). Delving into the regulatory mechanism of insulin secretion in the endocrine pancreas it has been observed that first and second phases of insulin secretion are lower in OVX during hyperglycemic clamp, which are improved by estrogen replacement and exercise. The endocrine pancreas of OVX mice shows a marked decrease in β cell mass. The β cell mass of these mice can be restored to physiological values with the administration of estrogen therapy and/or exercise by increased in proliferation and decreased in apoptosis. The increase in cell proliferation is associated with the induction of insulin receptor substrate-2 (IRS2) and PDX-1 via the activation of the cAMP response element binding protein (Choi et al., 2005). Elevated levels of IRS2 can potentiated IGF-I/insulin signaling pathway and increased β cell proliferation with an association in reduced β cell apoptosis.

OVX mouse islets showed reduced protein levels of exocytosis SNAREs complex. These protein complex directly regulate the exocytosis of insulin granules and carry out the vesicle-membrane interaction necessary for the physiological insulin release under high glucose stimulation. E2 administration were able to restore SNAREs complex protein levels to control values, demonstrating the direct involvement of the estrogen signal in insulin secretion (Santos et al., 2016).

In brief, menopause in women increases the risk and prevalence of T2DM. In animal studies, ovariectomy is associated with increased risk of diabetes at least in part, due to decreased insulin secretion in pancreatic β cells. It has been shown that estrogen administration protects against diabetes and increases the insulin response to glucose. Hormone replacement therapy may have benefits in protecting pancreatic β cell function in postmenopausal women, mediated by the protection that estrogens seem to provide against lipoglucotoxic stress environments associated with these physiological stages. Although further studies on the mechanism of action are needed, estrogens may protect pancreatic β function by increasing IRS2 levels and activating the IGF-I/insulin pathway, leading to increased proliferation and survival of β cells as well as a reduction in apoptosis.

6. Summary and concluding remarks

Estrogen plays an important role in the regulation of glucose homeostasis and energy metabolism and it is well established that men and women present differences in the incidence of T2DM. The highest incidence of this endocrine disease in women appears after menopause, associated with an estrogen deficiency. These observations have led multiple research groups to try to elucidate the role of estrogen signaling on pancreatic β cells and its relationship with T2DM onset. In this way, it has been described that pancreatic β cells express ERα, ERβ and GPER receptors through which estrogens can modulate the endocrine function of this cell type. Regarding insulin secretion, the rapid insulinotropic effect of E2 in insulin secretion is mediated by ERβ and GPER, although the implication of ERα has not been formally excluded. On the other hand, insulin biosynthesis appear to be regulated mainly through ERα activation by non-classical pathway with genomic action and GPER is also involved in mentioned regulation, however its mechanism is still unknown. In addition to this, E2 plays an important role in the protection of β cell due to enhanced proliferation, regeneration and survival by these three pathways. In this sense, great efforts have been made to reveal the mechanisms involved in this protection, since they could serve to establish new therapies in the treatment of diabetes. A fundamental tool to understand the metabolic basis of estrogen regulation of insulin-secreting cells have been the generation of genetic modification mouse models. These models have served to delve into the mechanism of cellular action and identify the pathways that are activated in estrogen regulation, this type of study would have been impossible to carry out in humans. Finally, hormone replacement therapy as a therapeutic treatment based in estrogens protects from type 2 diabetes development in post-menopausal stage and could be a strategy in the prevention of this important metabolic disease.

Acknowledgment

B.M. is supported by EFSD/Novo Nordisk.

References

Abdel-Hakim, S.M., Ibrahim, M.Y., Ibrahim, H.M., Ibrahim, M.M., 2014. The effect of ghrelin antagonist (D-Lys3) GHRP-6 on ovariectomy-induced obesity in adult female albino rats. Endocr. Regul. 48, 126–134.

Acconcia, F., Ascenzi, P., Fabozzi, G., Visca, P., Marino, M., 2004. S-palmitoylation modulates human estrogen receptor-alpha functions. Biochem. Biophys. Res. Commun. 316, 878–883.

Adlanmerini, M., Solinhac, R., Abot, A., Fabre, A., Raymond-Letron, I., Guihot, A.L., Boudou, F., Sautier, L., Vessieres, E., Kim, S.H., Liere, P., Fontaine, C., Krust, A., Chambon, P., Katzenellenbogen, J.A., Gourdy, P., Shaul, P.W., Henrion, D., Arnal, J.F., Lenfant, F., 2014. Mutation of the palmitoylation site of estrogen receptor alpha in vivo reveals tissue-specific roles for membrane versus nuclear actions. Proc. Natl. Acad. Sci. U. S. A. 111, E283–E290.

Albanito, L., Madeo, A., Lappano, R., Vivacqua, A., Rago, V., Carpino, A., Oprea, T.I., Prossnitz, E.R., Musti, A.M., Ando, S., Maggiolini, M., 2007. G protein-coupled receptor 30 (GPR30) mediates gene expression changes and growth response to 17 beta-estradiol and selective GPR30 ligand G-1 in ovarian cancer cells. Cancer Res. 67, 1859–1866.

Allard, C., Morford, J.J., Xu, B., Salwen, B., Xu, W., Desmoulins, L., Zsombok, A., Kim, J.K., Levin, E.R., Mauvais-Jarvis, F., 2019. Loss of nuclear and membrane estrogen receptor-alpha differentially impairs insulin secretion and action in male and female mice. Diabetes 68, 490–501.

Alonso-Magdalena, P., Laribi, O., Ropero, A.B., Fuentes, E., Ripoll, C., Soria, B., Nadal, A., 2005. Low doses of bisphenol A and diethylstilbestrol impair Ca^{2+} signals in pancreatic alpha-cells through a nonclassical membrane estrogen receptor within intact islets of Langerhans. Environ. Health Perspect. 113, 969–977.

Alonso-Magdalena, P., Morimoto, S., Ripoll, C., Fuentes, E., Nadal, A., 2006. The estrogenic effect of bisphenol A disrupts pancreatic beta-cell function in vivo and induces insulin resistance. Environ. Health Perspect. 114, 106–112.

Alonso-Magdalena, P., Ropero, A.B., Carrera, M.P., Cederroth, C.R., Baquie, M., Gauthier, B.R., Nef, S., Stefani, E., Nadal, A., 2008. Pancreatic insulin content regulation by the estrogen receptor ER alpha. PLoS One 3, e2069.

Alonso-Magdalena, P., Ropero, A.B., Garcia-Arevalo, M., Soriano, S., Quesada, I., Muhammed, S.J., Salehi, A., Gustafsson, J.A., Nadal, A., 2013. Antidiabetic actions of an estrogen receptor beta selective agonist. Diabetes 62, 2015–2025.

Andersson, B., Mattsson, L.A., Hahn, L., Marin, P., Lapidus, L., Holm, G., Bengtsson, B.A., Bjorntorp, P., 1997. Estrogen replacement therapy decreases hyperandrogenicity and improves glucose homeostasis and plasma lipids in postmenopausal women with noninsulin-dependent diabetes mellitus. J. Clin. Endocrinol. Metab. 82, 638–643.

Aronica, S.M., Kraus, W.L., Katzenellenbogen, B.S., 1994. Estrogen action via the cAMP signaling pathway: stimulation of adenylate cyclase and cAMP-regulated gene transcription. Proc. Natl. Acad. Sci. U. S. A. 91, 8517–8521.

Ashcroft, F.M., Rorsman, P., 1990. ATP-sensitive K+ channels: a link between B-cell metabolism and insulin secretion. Biochem. Soc. Trans. 18, 109–111.

Balhuizen, A., Kumar, R., Amisten, S., Lundquist, I., Salehi, A., 2010. Activation of G protein-coupled receptor 30 modulates hormone secretion and counteracts cytokine-induced apoptosis in pancreatic islets of female mice. Mol. Cell. Endocrinol. 320, 16–24.

Beeson, P.B., 1994. Age and sex associations of 40 autoimmune diseases. Am. J. Med. 96, 457–462.

Berg, J.M., 1989. DNA binding specificity of steroid receptors. Cell 57, 1065–1068.

Berne, R.M., Levy, M.N., Koeppen, B.M., Stanton, B.A., 2009. Physiology. Elsevier Health Sciences.

Bian, C., Bai, B., Gao, Q., Li, S., Zhao, Y., 2019. 17beta-estradiol regulates glucose metabolism and insulin secretion in rat Islet beta cells through GPER and Akt/mTOR/GLUT2 pathway. Front. Endocrinol. (Lausanne) 10, 531.

Bjornstrom, L., Sjoberg, M., 2005. Mechanisms of estrogen receptor signaling: convergence of genomic and nongenomic actions on target genes. Mol. Endocrinol. 19, 833–842.

Bjornstrom, L., Kilic, E., Norman, M., Parker, M.G., Sjoberg, M., 2001. Cross-talk between Stat5b and estrogen receptor-alpha and -beta in mammary epithelial cells. J. Mol. Endocrinol. 27, 93–106.

Blohme, G., Nystrom, L., Arnqvist, H.J., Lithner, F., Littorin, B., Olsson, P.O., Schersten, B., Wibell, L., Ostman, J., 1992. Male predominance of type 1 (insulin-dependent) diabetes mellitus in young adults: results from a 5-year prospective nationwide study of the 15–34-year age group in Sweden. Diabetologia 35, 56–62.

Boland, B.B., Rhodes, C.J., Grimsby, J.S., 2017. The dynamic plasticity of insulin production in beta-cells. Mol. Metab. 6, 958–973.

Bologa, C.G., Revankar, C.M., Young, S.M., Edwards, B.S., Arterburn, J.B., Kiselyov, A.S., Parker, M.A., Tkachenko, S.E., Savchuck, N.P., Sklar, L.A., Oprea, T.I., Prossnitz, E.R., 2006. Virtual and biomolecular screening converge on a selective agonist for GPR30. Nat. Chem. Biol. 2, 207–212.

Bryzgalova, G., Gao, H., Ahren, B., Zierath, J.R., Galuska, D., Steiler, T.L., Dahlman-Wright, K., Nilsson, S., Gustafsson, J.A., Efendic, S., Khan, A., 2006. Evidence that oestrogen receptor-alpha plays an important role in the regulation of glucose homeostasis in mice: insulin sensitivity in the liver. Diabetologia 49, 588–597.
Burks, D.J., Font De Mora, J., Schubert, M., Withers, D.J., Myers, M.G., Towery, H.H., Altamuro, S.L., Flint, C.L., White, M.F., 2000. IRS-2 pathways integrate female reproduction and energy homeostasis. Nature 407, 377–382.
Cagnacci, A., Venier, M., 2019. The controversial history of hormone replacement therapy. Medicina (Kaunas) 55, 602.
Carmeci, C., Thompson, D.A., Ring, H.Z., Francke, U., Weigel, R.J., 1997. Identification of a gene (GPR30) with homology to the G-protein-coupled receptor superfamily associated with estrogen receptor expression in breast cancer. Genomics 45, 607–617.
Carr, M.C., 2003. The emergence of the metabolic syndrome with menopause. J. Clin. Endocrinol. Metab. 88, 2404–2411.
Castoria, G., Migliaccio, A., Bilancio, A., Di Domenico, M., De Falco, A., Lombardi, M., Fiorentino, R., Varricchio, L., Barone, M.V., Auricchio, F., 2001. PI3-kinase in concert with Src promotes the S-phase entry of oestradiol-stimulated MCF-7 cells. EMBO J. 20, 6050–6059.
Castro-Rivera, E., Samudio, I., Safe, S., 2001. Estrogen regulation of cyclin D1 gene expression in ZR-75 breast cancer cells involves multiple enhancer elements. J. Biol. Chem. 276, 30853–30861.
Chambliss, K.L., Yuhanna, I.S., Mineo, C., Liu, P., German, Z., Sherman, T.S., Mendelsohn, M.E., Anderson, R.G., Shaul, P.W., 2000. Estrogen receptor alpha and endothelial nitric oxide synthase are organized into a functional signaling module in caveolae. Circ. Res. 87, E44–E52.
Chen, Z., Yuhanna, I.S., Galcheva-Gargova, Z., Karas, R.H., Mendelsohn, M.E., Shaul, P.W., 1999. Estrogen receptor alpha mediates the nongenomic activation of endothelial nitric oxide synthase by estrogen. J. Clin. Invest. 103, 401–406.
Cheng, S.B., Dong, J., Pang, Y., Larocca, J., Hixon, M., Thomas, P., Filardo, E.J., 2014. Anatomical location and redistribution of G protein-coupled estrogen receptor-1 during the estrus cycle in mouse kidney and specific binding to estrogens but not aldosterone. Mol. Cell. Endocrinol. 382, 950–959.
Choi, S.B., Jang, J.S., Park, S., 2005. Estrogen and exercise may enhance beta-cell function and mass via insulin receptor substrate 2 induction in ovariectomized diabetic rats. Endocrinology 146, 4786–4794.
Chuang, J.C., Cha, J.Y., Garmey, J.C., Mirmira, R.G., Repa, J.J., 2008. Research resource: nuclear hormone receptor expression in the endocrine pancreas. Mol. Endocrinol. 22, 2353–2363.
Claessens, F., Gewirth, D.T., 2004. DNA recognition by nuclear receptors. Essays Biochem. 40, 59–72.
Contreras, J.L., Smyth, C.A., Bilbao, G., Young, C.J., Thompson, J.A., Eckhoff, D.E., 2002. 17beta-estradiol protects isolated human pancreatic islets against proinflammatory cytokine-induced cell death: molecular mechanisms and islet functionality. Transplantation 74, 1252–1259.
Costrini, N.V., Kalkhoff, R.K., 1971. Relative effects of pregnancy, estradiol, and progesterone on plasma insulin and pancreatic islet insulin secretion. J. Clin. Invest. 50, 992–999.
Couse, J.F., Curtis Hewitt, S., Korach, K.S., 2000. Receptor null mice reveal contrasting roles for estrogen receptor alpha and beta in reproductive tissues. J. Steroid Biochem. Mol. Biol. 74, 287–296.
Da Silva, R.P., Zampieri, T.T., Pedroso, J.A., Nagaishi, V.S., Ramos-Lobo, A.M., Furigo, I.C., Camara, N.O., Frazao, R., Donato, J., Jr., 2014. Leptin resistance is not the primary cause of weight gain associated with reduced sex hormone levels in female mice. Endocrinology 155, 4226–4236.

Darimont, B.D., Wagner, R.L., Apriletti, J.W., Stallcup, M.R., Kushner, P.J., Baxter, J.D., Fletterick, R.J., Yamamoto, K.R., 1998. Structure and specificity of nuclear receptor-coactivator interactions. Genes Dev. 12, 3343–3356.

De Francesco, E.M., Pellegrino, M., Santolla, M.F., Lappano, R., Ricchio, E., Abonante, S., Maggiolini, M., 2014. GPER mediates activation of HIF1alpha/VEGF signaling by estrogens. Cancer Res. 74, 4053–4064.

Ding, E.L., Song, Y., Manson, J.E., Hunter, D.J., Lee, C.C., Rifai, N., Buring, J.E., Gaziano, J.M., Liu, S., 2009. Sex hormone-binding globulin and risk of type 2 diabetes in women and men. N. Engl. J. Med. 361, 1152–1163.

Dong, W.H., Chen, J.C., He, Y.L., Xu, J.J., Mei, Y.A., 2013. Resveratrol inhibits K(v)2.2 currents through the estrogen receptor GPR30-mediated PKC pathway. Am. J. Physiol. Cell Physiol. 305, C547–C557.

Driggers, P.H., Segars, J.H., 2002. Estrogen action and cytoplasmic signaling pathways. Part II: the role of growth factors and phosphorylation in estrogen signaling. Trends Endocrinol. Metab. 13, 422–427.

Duan, R., Porter, W., Safe, S., 1998. Estrogen-induced c-fos protooncogene expression in MCF-7 human breast cancer cells: role of estrogen receptor Sp1 complex formation. Endocrinology 139, 1981–1990.

DuPont, S., Krust, A., Gansmuller, A., Dierich, A., Chambon, P., Mark, M., 2000. Effect of single and compound knockouts of estrogen receptors alpha (ERalpha) and beta (ERbeta) on mouse reproductive phenotypes. Development 127, 4277–4291.

Eckhoff, D.E., Smyth, C.A., Eckstein, C., Bilbao, G., Young, C.J., Thompson, J.A., Contreras, J.L., 2003. Suppression of the c-Jun N-terminal kinase pathway by 17beta-estradiol can preserve human islet functional mass from proinflammatory cytokine-induced destruction. Surgery 134, 169–179.

Eckhoff, D.E., Eckstein, C., Smyth, C.A., Vilatoba, M., Bilbao, G., Rahemtulla, F.G., Young, C.J., Anthony Thompson, J., Chaudry, I.H., Contreras, J.L., 2004. Enhanced isolated pancreatic islet recovery and functionality in rats by 17beta-estradiol treatment of brain death donors. Surgery 136, 336–345.

Edwards, D.P., 2005. Regulation of signal transduction pathways by estrogen and progesterone. Annu. Rev. Physiol. 67, 335–376.

Efrat, S., 1991. Sexual dimorphism of pancreatic beta-cell degeneration in transgenic mice expressing an insulin-ras hybrid gene. Endocrinology 128, 897–901.

Efrat, S., Fleischer, N., Hanahan, D., 1990. Diabetes induced in male transgenic mice by expression of human H-ras oncoprotein in pancreatic beta cells. Mol. Cell. Biol. 10, 1779–1783.

Egea, P.F., Klaholz, B.P., Moras, D., 2000. Ligand-protein interactions in nuclear receptors of hormones. FEBS Lett. 476, 62–67.

El Seifi, S., Green, I.C., Perrin, D., 1981. Insulin release and steroid-hormone binding in isolated islets of Langerhans in the rat: effects of ovariectomy. J. Endocrinol. 90, 59–67.

Endoh, H., Sasaki, H., Maruyama, K., Takeyama, K., Waga, I., Shimizu, T., Kato, S., Kawashima, H., 1997. Rapid activation of MAP kinase by estrogen in the bone cell line. Biochem. Biophys. Res. Commun. 235, 99–102.

Engin, F., Nguyen, T., Yermalovich, A., Hotamisligil, G.S., 2014. Aberrant islet unfolded protein response in type 2 diabetes. Sci. Rep. 4, 4054.

Enmark, E., Pelto-Huikko, M., Grandien, K., Lagercrantz, S., Lagercrantz, J., Fried, G., Nordenskjold, M., Gustafsson, J.A., 1997. Human estrogen receptor beta-gene structure, chromosomal localization, and expression pattern. J. Clin. Endocrinol. Metab. 82, 4258–4265.

Faulds, M.H., Pettersson, K., Gustafsson, J.A., Haldosen, L.A., 2001. Cross-talk between ERs and signal transducer and activator of transcription 5 is E2 dependent and involves two functionally separate mechanisms. Mol. Endocrinol. 15, 1929–1940.

Figtree, G.A., McDonald, D., Watkins, H., Channon, K.M., 2003. Truncated estrogen receptor alpha 46-kDa isoform in human endothelial cells: relationship to acute activation of nitric oxide synthase. Circulation 107, 120–126.

Filardo, E.J., 2002. Epidermal growth factor receptor (EGFR) transactivation by estrogen via the G-protein-coupled receptor, GPR30: a novel signaling pathway with potential significance for breast cancer. J. Steroid Biochem. Mol. Biol. 80, 231–238.

Filardo, E.J., Thomas, P., 2005. GPR30: a seven-transmembrane-spanning estrogen receptor that triggers EGF release. Trends Endocrinol. Metab. 16, 362–367.

Filardo, E.J., Quinn, J.A., Bland, K.I., Frackelton, A.R., Jr., 2000. Estrogen-induced activation of Erk-1 and Erk-2 requires the G protein-coupled receptor homolog, GPR30, and occurs via trans-activation of the epidermal growth factor receptor through release of HB-EGF. Mol. Endocrinol. 14, 1649–1660.

Filardo, E.J., Quinn, J.A., Frackelton, A.R., Jr., Bland, K.I., 2002. Estrogen action via the G protein-coupled receptor, GPR30: stimulation of adenylyl cyclase and cAMP-mediated attenuation of the epidermal growth factor receptor-to-MAPK signaling axis. Mol. Endocrinol. 16, 70–84.

Fisher, C.R., Graves, K.H., Parlow, A.F., Simpson, E.R., 1998. Characterization of mice deficient in aromatase (ArKO) because of targeted disruption of the cyp19 gene. Proc. Natl. Acad. Sci. U. S. A. 95, 6965–6970.

Foryst-Ludwig, A., Clemenz, M., Hohmann, S., Hartge, M., Sprang, C., Frost, N., Krikov, M., Bhanot, S., Barros, R., Morani, A., Gustafsson, J.A., Unger, T., Kintscher, U., 2008. Metabolic actions of estrogen receptor beta (ERbeta) are mediated by a negative cross-talk with PPARgamma. PLoS Genet. 4, e1000108.

Friday, K.E., Dong, C., Fontenot, R.U., 2001. Conjugated equine estrogen improves glycemic control and blood lipoproteins in postmenopausal women with type 2 diabetes. J. Clin. Endocrinol. Metab. 86, 48–52.

Gale, E.A., Gillespie, K.M., 2001. Diabetes and gender. Diabetologia 44, 3–15.

Gambineri, A., Pelusi, C., 2019. Sex hormones, obesity and type 2 diabetes: is there a link? Endocr. Connect. 8, R1–R9.

García-Arévalo, M., Lorza-Gil, E., Leite, N., Brunetto, S., Boschero, A.C., Carneiro, E.M., 2019. The 17-beta-estradiol improves insulin sensitivity in a rapid estrogen receptor alpha-dependent manner in an animal model of malnourishment. J. Endocrinol. Metab. 9, 133–146.

Garris, D.R., Garris, B.L., 2005. Estrogenic restoration of functional pancreatic islet cytoarchitecture in diabetes (db/db) mutant C57BL/KsJ mice: relationship to estradiol localization, systemic glycemia, and persistent hyperinsulinemia. Cell Tissue Res. 319, 231–242.

Gaub, M.P., Bellard, M., Scheuer, I., Chambon, P., Sassone-Corsi, P., 1990. Activation of the ovalbumin gene by the estrogen receptor involves the fos-jun complex. Cell 63, 1267–1276.

Geisler, J.G., Zawalich, W., Zawalich, K., Lakey, J.R., Stukenbrok, H., Milici, A.J., Soeller, W.C., 2002. Estrogen can prevent or reverse obesity and diabetes in mice expressing human islet amyloid polypeptide. Diabetes 51, 2158–2169.

Godsland, I.F., 2005. Oestrogens and insulin secretion. Diabetologia 48, 2213–2220.

Gonzalez, C., Alonso, A., Grueso, N.A., Diaz, F., Esteban, M.M., Fernandez, S., Patterson, A.M., 2002. Role of 17beta-estradiol administration on insulin sensitivity in the rat: implications for the insulin receptor. Steroids 67, 993–1005.

Goodman, M.N., Hazelwood, R.L., 1974. Short-term effects of oestradiol benzoate in normal, hypophysectomized and alloxan-diabetic male rats. J. Endocrinol. 62, 439–449.

Gottlicher, M., Heck, S., Herrlich, P., 1998. Transcriptional cross-talk, the second mode of steroid hormone receptor action. J. Mol. Med. (Berl) 76, 480–489.

Green, S., Chambon, P., 1988. Nuclear receptors enhance our understanding of transcription regulation. Trends Genet. 4, 309–314.

Gunsalus, G.L., Musto, N.A., Bardin, W., 1978. Immunoassay of androgen binding protein in blood: a new approach for study of the seminiferous tubule. Science 200, 65–66.

Gurlo, T., Rivera, J.F., Butler, A.E., Cory, M., Hoang, J., Costes, S., Butler, P.C., 2016. CHOP contributes to, but is not the only mediator of, IAPP induced beta-cell apoptosis. Mol. Endocrinol. 30, 446–454.

Gustafsson, J.A., Strom, A., Warner, M., 2019. Update on ERbeta. J. Steroid Biochem. Mol. Biol. 191, 105312.

Halban, P.A., 1991. Structural domains and molecular lifestyles of insulin and its precursors in the pancreatic beta cell. Diabetologia 34, 767–778.

Ham, J., Parker, M.G., 1989. Regulation of gene expression by nuclear hormone receptors. Curr. Opin. Cell Biol. 1, 503–511.

Heine, P.A., Taylor, J.A., Iwamoto, G.A., Lubahn, D.B., Cooke, P.S., 2000. Increased adipose tissue in male and female estrogen receptor-alpha knockout mice. Proc. Natl. Acad. Sci. U. S. A. 97, 12729–12734.

Heldring, N., Pike, A., Andersson, S., Matthews, J., Cheng, G., Hartman, J., Tujague, M., Strom, A., Treuter, E., Warner, M., Gustafsson, J.A., 2007. Estrogen receptors: how do they signal and what are their targets. Physiol. Rev. 87, 905–931.

Hevener, A.L., Clegg, D.J., Mauvais-Jarvis, F., 2015. Impaired estrogen receptor action in the pathogenesis of the metabolic syndrome. Mol. Cell. Endocrinol. 418 (Pt. 3), 306–321.

Hirosawa, M., Minata, M., Harada, K.H., Hitomi, T., Krust, A., Koizumi, A., 2008. Ablation of estrogen receptor alpha (ERalpha) prevents upregulation of POMC by leptin and insulin. Biochem. Biophys. Res. Commun. 371, 320–323.

Huang, X.F., Arvan, P., 1995. Intracellular transport of proinsulin in pancreatic beta-cells. Structural maturation probed by disulfide accessibility. J. Biol. Chem. 270, 20417–20423.

Improta-Brears, T., Whorton, A.R., Codazzi, F., York, J.D., Meyer, T., McDonnell, D.P., 1999. Estrogen-induced activation of mitogen-activated protein kinase requires mobilization of intracellular calcium. Proc. Natl. Acad. Sci. U. S. A. 96, 4686–4691.

Inada, A., Hamamoto, Y., Tsuura, Y., Miyazaki, J., Toyokuni, S., Ihara, Y., Nagai, K., Yamada, Y., Bonner-Weir, S., Seino, Y., 2004. Overexpression of inducible cyclic AMP early repressor inhibits transactivation of genes and cell proliferation in pancreatic beta cells. Mol. Cell. Biol. 24, 2831–2841.

Inada, A., Arai, H., Nagai, K., Miyazaki, J., Yamada, Y., Seino, Y., Fukatsu, A., 2007. Gender difference in ICER Igamma transgenic diabetic mouse. Biosci. Biotechnol. Biochem. 71, 1920–1926.

Inada, A., Inada, O., Fujii, N.L., Fujishima, K., Inai, T., Fujii, H., Sueishi, K., Kurachi, K., 2014. beta-cell induction in vivo in severely diabetic male mice by changing the circulating levels and pattern of the ratios of estradiol to androgens. Endocrinology 155, 3829–3842.

Inoue, S., Hoshino, S., Miyoshi, H., Akishita, M., Hosoi, T., Orimo, H., Ouchi, Y., 1996. Identification of a novel isoform of estrogen receptor, a potential inhibitor of estrogen action, in vascular smooth muscle cells. Biochem. Biophys. Res. Commun. 219, 766–772.

Jakacka, M., Ito, M., Weiss, J., Chien, P.Y., Gehm, B.D., Jameson, J.L., 2001. Estrogen receptor binding to DNA is not required for its activity through the nonclassical AP1 pathway. J. Biol. Chem. 276, 13615–13621.

Jakacka, M., Ito, M., Martinson, F., Ishikawa, T., Lee, E.J., Jameson, J.L., 2002. An estrogen receptor (ER)alpha deoxyribonucleic acid-binding domain knock-in mutation provides evidence for nonclassical ER pathway signaling in vivo. Mol. Endocrinol. 16, 2188–2201.

Janson, J., Soeller, W.C., Roche, P.C., Nelson, R.T., Torchia, A.J., Kreutter, D.K., Butler, P.C., 1996. Spontaneous diabetes mellitus in transgenic mice expressing human islet amyloid polypeptide. Proc. Natl. Acad. Sci. U. S. A. 93, 7283–7288.

Jensen, E.V., Desombre, E.R., 1972. Mechanism of action of the female sex hormones. Annu. Rev. Biochem. 41, 203–230.

Jensen, E.V., Jordan, V.C., 2003. The estrogen receptor: a model for molecular medicine. Clin. Cancer Res. 9, 1980–1989.

Jessop, H.L., Sjoberg, M., Cheng, M.Z., Zaman, G., Wheeler-Jones, C.P., Lanyon, L.E., 2001. Mechanical strain and estrogen activate estrogen receptor alpha in bone cells. J. Bone Miner. Res. 16, 1045–1055.

Joseph, D.R., Hall, S.H., French, F.S., 1987. Rat androgen-binding protein: evidence for identical subunits and amino acid sequence homology with human sex hormone-binding globulin. Proc. Natl. Acad. Sci. U. S. A. 84, 339–343.

Kahn, S.E., Andrikopoulos, S., Verchere, C.B., Wang, F., Hull, R.L., Vidal, J., 2000. Oophorectomy promotes islet amyloid formation in a transgenic mouse model of Type II diabetes. Diabetologia 43, 1309–1312.

Kanaya, A.M., Herrington, D., Vittinghoff, E., Lin, F., Grady, D., Bittner, V., Cauley, J.A., Barrett-Connor, E., Heart & Estrogen/Progestin Replacement Study, 2003. Glycemic effects of postmenopausal hormone therapy: the Heart and Estrogen/progestin Replacement Study. A randomized, double-blind, placebo-controlled trial. Ann. Intern. Med. 138, 1–9.

Kang, L., Zhang, X., Xie, Y., Tu, Y., Wang, D., Liu, Z., Wang, Z.Y., 2010. Involvement of estrogen receptor variant ER-alpha36, not GPR30, in nongenomic estrogen signaling. Mol. Endocrinol. 24, 709–721.

Kato, S., Endoh, H., Masuhiro, Y., Kitamoto, T., Uchiyama, S., Sasaki, H., Masushige, S., Gotoh, Y., Nishida, E., Kawashima, H., Metzger, D., Chambon, P., 1995. Activation of the estrogen receptor through phosphorylation by mitogen-activated protein kinase. Science 270, 1491–1494.

Kawano, K., Hirashima, T., Mori, S., Natori, T., 1994. OLETF (Otsuka Long-Evans Tokushima Fatty) rat: a new NIDDM rat strain. Diabetes Res. Clin. Pract. 24 (Suppl), S317–S320.

Keck, C., Taylor, M., 2018. Emerging research on the implications of hormone replacement therapy on coronary heart disease. Curr. Atheroscler. Rep. 20, 57.

Keselman, A., Heller, N., 2015. Estrogen signaling modulates allergic inflammation and contributes to sex differences in asthma. Front. Immunol. 6, 568.

Kilic, G., Alvarez-Mercado, A.I., Zarrouki, B., Opland, D., Liew, C.W., Alonso, L.C., Myers, M.G., Jr., Jonas, J.C., Poitout, V., Kulkarni, R.N., Mauvais-Jarvis, F., 2014. The islet estrogen receptor-alpha is induced by hyperglycemia and protects against oxidative stress-induced insulin-deficient diabetes. PLoS One 9, e87941.

Kim, J., Mauvais-Jarvis, F., 2016. The combination of conjugated equine estrogens with bazedoxifene prevents streptozotocin-induced diabetes in ovariectomized female mice. Matters.

Kooptiwut, S., Mahawong, P., Hanchang, W., Semprasert, N., Kaewin, S., Limjindaporn, T., Yenchitsomanus, P.T., 2014. Estrogen reduces endoplasmic reticulum stress to protect against glucotoxicity induced-pancreatic beta-cell death. J. Steroid Biochem. Mol. Biol. 139, 25–32.

Korach, K.S., Emmen, J.M., Walker, V.R., Hewitt, S.C., Yates, M., Hall, J.M., Swope, D.L., Harrell, J.C., Couse, J.F., 2003. Update on animal models developed for analyses of estrogen receptor biological activity. J. Steroid Biochem. Mol. Biol. 86, 387–391.

Kousteni, S., Bellido, T., Plotkin, L.I., O'brien, C.A., Bodenner, D.L., Han, L., Han, K., Digregorio, G.B., Katzenellenbogen, J.A., Katzenellenbogen, B.S., Roberson, P.K.,

Weinstein, R.S., Jilka, R.L., Manolagas, S.C., 2001. Nongenotropic, sex-nonspecific signaling through the estrogen or androgen receptors: dissociation from transcriptional activity. Cell 104, 719–730.
Krege, J.H., Hodgin, J.B., Couse, J.F., Enmark, E., Warner, M., Mahler, J.F., Sar, M., Korach, K.S., Gustafsson, J.A., Smithies, O., 1998. Generation and reproductive phenotypes of mice lacking estrogen receptor beta. Proc. Natl. Acad. Sci. U. S. A. 95, 15677–15682.
Kuiper, G.G., Enmark, E., Pelto-Huikko, M., Nilsson, S., Gustafsson, J.A., 1996. Cloning of a novel receptor expressed in rat prostate and ovary. Proc. Natl. Acad. Sci. U. S. A. 93, 5925–5930.
Kumar, R., Balhuizen, A., Amisten, S., Lundquist, I., Salehi, A., 2011. Insulinotropic and antidiabetic effects of 17beta-estradiol and the GPR30 agonist G-1 on human pancreatic islets. Endocrinology 152, 2568–2579.
Kushner, P.J., Agard, D.A., Greene, G.L., Scanlan, T.S., Shiau, A.K., Uht, R.M., Webb, P., 2000. Estrogen receptor pathways to AP-1. J. Steroid Biochem. Mol. Biol. 74, 311–317.
Langer, G., Bader, B., Meoli, L., Isensee, J., Delbeck, M., Noppinger, P.R., Otto, C., 2010. A critical review of fundamental controversies in the field of GPR30 research. Steroids 75, 603–610.
Le May, C., Chu, K., Hu, M., Ortega, C.S., Simpson, E.R., Korach, K.S., Tsai, M.J., Mauvais-Jarvis, F., 2006. Estrogens protect pancreatic beta-cells from apoptosis and prevent insulin-deficient diabetes mellitus in mice. Proc. Natl. Acad. Sci. U. S. A. 103, 9232–9237.
Le, T.N., Nestler, J.E., Strauss, J.F., 3rd, Wickham, E.P., 3rd, 2012. Sex hormone-binding globulin and type 2 diabetes mellitus. Trends Endocrinol. Metab. 23, 32–40.
Lee, Y., Hirose, H., Ohneda, M., Johnson, J.H., Mcgarry, J.D., Unger, R.H., 1994. Beta-cell lipotoxicity in the pathogenesis of non-insulin-dependent diabetes mellitus of obese rats: impairment in adipocyte-beta-cell relationships. Proc. Natl. Acad. Sci. U. S. A. 91, 10878–10882.
Leiter, E.H., Chapman, H.D., Coleman, D.L., 1989. The influence of genetic background on the expression of mutations at the diabetes locus in the mouse. V. Interaction between the db gene and hepatic sex steroid sulfotransferases correlates with gender-dependent susceptibility to hyperglycemia. Endocrinology 124, 912–922.
Levin, E.R., 2009. G protein-coupled receptor 30: estrogen receptor or collaborator? Endocrinology 150, 1563–1565.
Levin, E.R., Hammes, S.R., 2016. Nuclear receptors outside the nucleus: extranuclear signalling by steroid receptors. Nat. Rev. Mol. Cell Biol. 17, 783–797.
Li, C., Briggs, M.R., Ahlborn, T.E., Kraemer, F.B., Liu, J., 2001. Requirement of Sp1 and estrogen receptor alpha interaction in 17beta-estradiol-mediated transcriptional activation of the low density lipoprotein receptor gene expression. Endocrinology 142, 1546–1553.
Li, L., Haynes, M.P., Bender, J.R., 2003. Plasma membrane localization and function of the estrogen receptor alpha variant (ER46) in human endothelial cells. Proc. Natl. Acad. Sci. U. S. A. 100, 4807–4812.
Lindsey, S.H., Liu, L., Chappell, M.C., 2014. Vasodilation by GPER in mesenteric arteries involves both endothelial nitric oxide and smooth muscle cAMP signaling. Steroids 81, 99–102.
Lindstedt, G., Lundberg, P.A., Lapidus, L., Lundgren, H., Bengtsson, C., Bjorntorp, P., 1991. Low sex-hormone-binding globulin concentration as independent risk factor for development of NIDDM. 12-yr follow-up of population study of women in Gothenburg, Sweden. Diabetes 40, 123–128.
Liu, S., Mauvais-Jarvis, F., 2009. Rapid, nongenomic estrogen actions protect pancreatic islet survival. Islets 1, 273–275.

Liu, S., Mauvais-Jarvis, F., 2010. Minireview: estrogenic protection of beta-cell failure in metabolic diseases. Endocrinology 151, 859–864.
Liu, M.M., Albanese, C., Anderson, C.M., Hilty, K., Webb, P., Uht, R.M., Price, R.H., Jr., Pestell, R.G., Kushner, P.J., 2002. Opposing action of estrogen receptors alpha and beta on cyclin D1 gene expression. J. Biol. Chem. 277, 24353–24360.
Liu, S., Le May, C., Wong, W.P., Ward, R.D., Clegg, D.J., Marcelli, M., Korach, K.S., Mauvais-JARVIS, F., 2009. Importance of extranuclear estrogen receptor-alpha and membrane G protein-coupled estrogen receptor in pancreatic islet survival. Diabetes 58, 2292–2302.
Losel, R., Wehling, M., 2003. Nongenomic actions of steroid hormones. Nat. Rev. Mol. Cell Biol. 4, 46–56.
Louet, J.F., Smith, S.B., Gautier, J.F., Molokhia, M., Virally, M.L., Kevorkian, J.P., Guillausseau, P.J., Vexiau, P., Charpentier, G., German, M.S., Vaisse, C., Urbanek, M., Mauvais-Jarvis, F., 2008. Gender and neurogenin3 influence the pathogenesis of ketosis-prone diabetes. Diabetes Obes. Metab. 10, 912–920.
Lubahn, D.B., Moyer, J.S., Golding, T.S., Couse, J.F., Korach, K.S., Smithies, O., 1993. Alteration of reproductive function but not prenatal sexual development after insertional disruption of the mouse estrogen receptor gene. Proc. Natl. Acad. Sci. U. S. A. 90, 11162–11166.
Maclaren, N.K., Neufeld, M., McLaughlin, J.V., Taylor, G., 1980. Androgen sensitization of streptozotocin-induced diabetes in mice. Diabetes 29, 710–716.
Margolis, K.L., Bonds, D.E., Rodabough, R.J., Tinker, L., Phillips, L.S., Allen, C., Bassford, T., Burke, G., Torrens, J., Howard, B.V., Women's Health Initiative Investigators, 2004. Effect of oestrogen plus progestin on the incidence of diabetes in postmenopausal women: results from the Women's Health Initiative Hormone Trial. Diabetologia 47, 1175–1187.
Marino, M., Ascenzi, P., 2006. Steroid hormone rapid signaling: the pivotal role of S-palmitoylation. IUBMB Life 58, 716–719.
Marino, M., Acconcia, F., Bresciani, F., Weisz, A., Trentalance, A., 2002. Distinct nongenomic signal transduction pathways controlled by 17beta-estradiol regulate DNA synthesis and cyclin D(1) gene transcription in HepG2 cells. Mol. Biol. Cell 13, 3720–3729.
Marino, M., Ascenzi, P., Acconcia, F., 2006. S-palmitoylation modulates estrogen receptor alpha localization and functions. Steroids 71, 298–303.
Martel, C., Rheaume, E., Takahashi, M., Trudel, C., Couet, J., Luu-The, V., Simard, J., Labrie, F., 1992. Distribution of 17 beta-hydroxysteroid dehydrogenase gene expression and activity in rat and human tissues. J. Steroid Biochem. Mol. Biol. 41, 597–603.
Martensson, U.E., Salehi, S.A., Windahl, S., Gomez, M.F., Sward, K., Daszkiewicz-Nilsson, J., Wendt, A., Andersson, N., Hellstrand, P., Grande, P.O., Owman, C., Rosen, C.J., Adamo, M.L., Lundquist, I., Rorsman, P., Nilsson, B.O., Ohlsson, C., Olde, B., Leeb-Lundberg, L.M., 2009. Deletion of the G protein-coupled receptor 30 impairs glucose tolerance, reduces bone growth, increases blood pressure, and eliminates estradiol-stimulated insulin release in female mice. Endocrinology 150, 687–698.
Matthews, K.A., Meilahn, E., Kuller, L.H., Kelsey, S.F., Caggiula, A.W., Wing, R.R., 1989. Menopause and risk factors for coronary heart disease. N. Engl. J. Med. 321, 641–646.
Mauvais-Jarvis, F., 2011. Estrogen and androgen receptors: regulators of fuel homeostasis and emerging targets for diabetes and obesity. Trends Endocrinol. Metab. 22, 24–33.
Mauvais-Jarvis, F., 2012. Estrogen sulfotransferase: intracrinology meets metabolic diseases. Diabetes 61, 1353–1354.
Mauvais-Jarvis, F., 2016. Role of sex steroids in beta cell function, growth, and survival. Trends Endocrinol. Metab. 27, 844–855.

Mauvais-Jarvis, F., 2018. Gender differences in glucose homeostasis and diabetes. Physiol. Behav. 187, 20–23.

Mauvais-Jarvis, F., Sobngwi, E., Porcher, R., Riveline, J.P., Kevorkian, J.P., Vaisse, C., Charpentier, G., Guillausseau, P.J., Vexiau, P., Gautier, J.F., 2004. Ketosis-prone type 2 diabetes in patients of sub-Saharan African origin: clinical pathophysiology and natural history of beta-cell dysfunction and insulin resistance. Diabetes 53, 645–653.

Mauvais-Jarvis, F., Clegg, D.J., Hevener, A.L., 2013. The role of estrogens in control of energy balance and glucose homeostasis. Endocr. Rev. 34, 309–338.

Mauvais-Jarvis, F., Manson, J.E., Stevenson, J.C., Fonseca, V.A., 2017. Menopausal hormone therapy and type 2 diabetes prevention: evidence, mechanisms, and clinical implications. Endocr. Rev. 38, 173–188.

McCulloch, L.J., Van De Bunt, M., Braun, M., Frayn, K.N., Clark, A., Gloyn, A.L., 2011. GLUT2 (SLC2A2) is not the principal glucose transporter in human pancreatic beta cells: implications for understanding genetic association signals at this locus. Mol. Genet. Metab. 104, 648–653.

Meitzen, J., Luoma, J.I., Boulware, M.I., Hedges, V.L., Peterson, B.M., Tuomela, K., Britson, K.A., Mermelstein, P.G., 2013. Palmitoylation of estrogen receptors is essential for neuronal membrane signaling. Endocrinology 154, 4293–4304.

Mendoza-Hernandez, G., Lopez-Solache, I., Rendon, J.L., Diaz-Sanchez, V., Diaz-Zagoya, J.C., 1988. 17 beta-hydroxysteroid dehydrogenase activity in canine pancreas. Biochem. Biophys. Res. Commun. 152, 376–382.

Meyer, M.R., Field, A.S., Kanagy, N.L., Barton, M., Prossnitz, E.R., 2012. GPER regulates endothelin-dependent vascular tone and intracellular calcium. Life Sci. 91, 623–627.

Migliaccio, A., Di Domenico, M., Castoria, G., De Falco, A., Bontempo, P., Nola, E., Auricchio, F., 1996. Tyrosine kinase/p21ras/MAP-kinase pathway activation by estradiol-receptor complex in MCF-7 cells. EMBO J. 15, 1292–1300.

Migliaccio, A., Piccolo, D., Castoria, G., Di Domenico, M., Bilancio, A., Lombardi, M., Gong, W., Beato, M., Auricchio, F., 1998. Activation of the Src/p21ras/Erk pathway by progesterone receptor via cross-talk with estrogen receptor. EMBO J. 17, 2008–2018.

Morales, A., Cuellar, A., Ramirez, J., Vilchis, F., Diaz-Sanchez, V., 1999. Synthesis of steroids in pancreas: evidence of cytochrome P-450scc activity. Pancreas 19, 39–44.

Munell, F., Suarez-Quian, C.A., Selva, D.M., Tirado, O.M., Reventos, J., 2002. Androgen-binding protein and reproduction: where do we stand? J. Androl. 23, 598–609.

Nadal, A., Ropero, A.B., Laribi, O., Maillet, M., Fuentes, E., Soria, B., 2000. Nongenomic actions of estrogens and xenoestrogens by binding at a plasma membrane receptor unrelated to estrogen receptor alpha and estrogen receptor beta. Proc. Natl. Acad. Sci. U. S. A. 97, 11603–11608.

Nadal, A., Ropero, A.B., Fuentes, E., Soria, B., Ripoll, C., 2004. Estrogen and xenoestrogen actions on endocrine pancreas: from ion channel modulation to activation of nuclear function. Steroids 69, 531–536.

Nanjappa, M.K., Hess, R.A., Medrano, T.I., Locker, S.H., Levin, E.R., Cooke, P.S., 2016. Membrane-localized estrogen receptor 1 is required for normal male reproductive development and function in mice. Endocrinology 157, 2909–2919.

Newton, K.M., Lacroix, A.Z., McKnight, B., Knopp, R.H., Siscovick, D.S., Heckbert, S.R., Weiss, N.S., 1997. Estrogen replacement therapy and prognosis after first myocardial infarction. Am. J. Epidemiol. 145, 269–277.

O'dowd, B.F., Nguyen, T., Marchese, A., Cheng, R., Lynch, K.R., Heng, H.H., Kolakowski, L.F., Jr., George, S.R., 1998. Discovery of three novel G-protein-coupled receptor genes. Genomics 47, 310–313.

Ogawa, S., Inoue, S., Watanabe, T., Orimo, A., Hosoi, T., Ouchi, Y., Muramatsu, M., 1998. Molecular cloning and characterization of human estrogen receptor betacx: a potential inhibitor of estrogen action in human. Nucleic Acids Res. 26, 3505–3512.

Ogishima, T., Mitani, F., Suematsu, M., 2008. Cytochrome P-450(17alpha) in beta-cells of rat pancreas and its local steroidogenesis. J. Steroid Biochem. Mol. Biol. 111, 80–86.
O'lone, R., Frith, M.C., Karlsson, E.K., Hansen, U., 2004. Genomic targets of nuclear estrogen receptors. Mol. Endocrinol. 18, 1859–1875.
Otto, C., Rohde-Schulz, B., Schwarz, G., Fuchs, I., Klewer, M., Brittain, D., Langer, G., Bader, B., Prelle, K., Nubbemeyer, R., Fritzemeier, K.H., 2008. G protein-coupled receptor 30 localizes to the endoplasmic reticulum and is not activated by estradiol. Endocrinology 149, 4846–4856.
Owman, C., Nilsson, C., Lolait, S.J., 1996. Cloning of cDNA encoding a putative chemoattractant receptor. Genomics 37, 187–194.
Paik, S.G., Michelis, M.A., Kim, Y.T., Shin, S., 1982. Induction of insulin-dependent diabetes by streptozotocin. Inhibition by estrogens and potentiation by androgens. Diabetes 31, 724–729.
Papa, F.R., 2012. Endoplasmic reticulum stress, pancreatic beta-cell degeneration, and diabetes. Cold Spring Harb. Perspect. Med. 2, a007666.
Pappas, T.C., Gametchu, B., Watson, C.S., 1995. Membrane estrogen receptors identified by multiple antibody labeling and impeded-ligand binding. FASEB J. 9, 404–410.
Pawlak, M., Lefebvre, P., Staels, B., 2012. General molecular biology and architecture of nuclear receptors. Curr. Top. Med. Chem. 12, 486–504.
Pedram, A., Razandi, M., Sainson, R.C., Kim, J.K., Hughes, C.C., Levin, E.R., 2007. A conserved mechanism for steroid receptor translocation to the plasma membrane. J. Biol. Chem. 282, 22278–22288.
Pedram, A., Razandi, M., Lewis, M., Hammes, S., Levin, E.R., 2014. Membrane-localized estrogen receptor alpha is required for normal organ development and function. Dev. Cell 29, 482–490.
Petrie, W.K., Dennis, M.K., Hu, C., Dai, D., Arterburn, J.B., Smith, H.O., Hathaway, H.J., Prossnitz, E.R., 2013. G protein-coupled estrogen receptor-selective ligands modulate endometrial tumor growth. Obstet. Gynecol. Int. 2013, 472720.
Porter, W., Saville, B., Hoivik, D., Safe, S., 1997. Functional synergy between the transcription factor Sp1 and the estrogen receptor. Mol. Endocrinol. 11, 1569–1580.
Posa, A., Szabo, R., Kupai, K., Csonka, A., Szalai, Z., Veszelka, M., Torok, S., Daruka, L., Varga, C., 2015. Exercise training and calorie restriction influence the metabolic parameters in ovariectomized female rats. Oxid. Med. Cell. Longev. 2015, 787063.
Prossnitz, E.R., Arterburn, J.B., 2015. International union of basic and clinical pharmacology. XCVII. G protein-coupled estrogen receptor and its pharmacologic modulators. Pharmacol. Rev. 67, 505–540.
Prossnitz, E.R., Arterburn, J.B., Smith, H.O., Oprea, T.I., Sklar, L.A., Hathaway, H.J., 2008. Estrogen signaling through the transmembrane G protein-coupled receptor GPR30. Annu. Rev. Physiol. 70, 165–190.
Puah, J.A., Bailey, C.J., 1985. Insulinotropic effect of ovarian steroid hormones in streptozotocin diabetic female mice. Horm. Metab. Res. 17, 216–218.
Pupo, M., Vivacqua, A., Perrotta, I., Pisano, A., Aquila, S., Abonante, S., Gasperi-Campani, A., Pezzi, V., Maggiolini, M., 2013. The nuclear localization signal is required for nuclear GPER translocation and function in breast Cancer-Associated Fibroblasts (CAFs). Mol. Cell. Endocrinol. 376, 23–32.
Puzianowska-Kuznicka, M., Pawlik-Pachucka, E., Owczarz, M., Budzinska, M., Polosak, J., 2013. Small-molecule hormones: molecular mechanisms of action. Int. J. Endocrinol. 2013, 601246.
Quintero, O.L., Amador-Patarroyo, M.J., Montoya-Ortiz, G., Rojas-Villarraga, A., Anaya, J.M., 2012. Autoimmune disease and gender: plausible mechanisms for the female predominance of autoimmunity. J. Autoimmun. 38, J109–J119.

Razandi, M., Pedram, A., Greene, G.L., Levin, E.R., 1999. Cell membrane and nuclear estrogen receptors (ERs) originate from a single transcript: studies of ERalpha and ERbeta expressed in Chinese hamster ovary cells. Mol. Endocrinol. 13, 307–319.

Razandi, M., Oh, P., Pedram, A., Schnitzer, J., Levin, E.R., 2002. ERs associate with and regulate the production of caveolin: implications for signaling and cellular actions. Mol. Endocrinol. 16, 100–115.

Razandi, M., Pedram, A., Merchenthaler, I., Greene, G.L., Levin, E.R., 2004. Plasma membrane estrogen receptors exist and functions as dimers. Mol. Endocrinol. 18, 2854–2865.

Revankar, C.M., Cimino, D.F., Sklar, L.A., Arterburn, J.B., Prossnitz, E.R., 2005. A transmembrane intracellular estrogen receptor mediates rapid cell signaling. Science 307, 1625–1630.

Riant, E., Waget, A., Cogo, H., Arnal, J.F., Burcelin, R., Gourdy, P., 2009. Estrogens protect against high-fat diet-induced insulin resistance and glucose intolerance in mice. Endocrinology 150, 2109–2117.

Ribas, V., Nguyen, M.T., Henstridge, D.C., Nguyen, A.K., Beaven, S.W., Watt, M.J., Hevener, A.L., 2010. Impaired oxidative metabolism and inflammation are associated with insulin resistance in ERalpha-deficient mice. Am. J. Physiol. Endocrinol. Metab. 298, E304–E319.

Ripoll, C., Ropero, A.B., Alonso-Magdalena, P., Quesada, I., Fuentes, E., Nadal, A., 2008. Rapid regulation of pancreatic alpha- and beta-cell signalling systems by estrogens. Infect. Disord. Drug Targets 8, 61–64.

Rivera, J.F., Gurlo, T., Daval, M., Huang, C.J., Matveyenko, A.V., Butler, P.C., Costes, S., 2011. Human-IAPP disrupts the autophagy/lysosomal pathway in pancreatic beta-cells: protective role of p62-positive cytoplasmic inclusions. Cell Death Differ. 18, 415–426.

Rogers, N.H., Perfield, J.W., 2nd, Strissel, K.J., Obin, M.S., Greenberg, A.S., 2009. Reduced energy expenditure and increased inflammation are early events in the development of ovariectomy-induced obesity. Endocrinology 150, 2161–2168.

Ropero, A.B., Fuentes, E., Rovira, J.M., Ripoll, C., Soria, B., Nadal, A., 1999. Non-genomic actions of 17beta-oestradiol in mouse pancreatic beta-cells are mediated by a cGMP-dependent protein kinase. J. Physiol. 521 (Pt. 2), 397–407.

Ropero, A.B., Soria, B., Nadal, A., 2002. A nonclassical estrogen membrane receptor triggers rapid differential actions in the endocrine pancreas. Mol. Endocrinol. 16, 497–505.

Ropero, A.B., Pang, Y., Alonso-Magdalena, P., Thomas, P., Nadal, A., 2012. Role of ERbeta and GPR30 in the endocrine pancreas: a matter of estrogen dose. Steroids 77, 951–958.

Rosen, C.J., 2021. The epidemiology and pathogenesis of osteoporosis. In: Feingold, K.R., Anawalt, B., Boyce, A., Chrousos, G., de Herder, W.W., Dungan, K., Grossman, A., Hershman, J.M., Hofland, J., Kaltsas, G., Koch, C., Kopp, P., Korbonits, M., McLachlan, R., Morley, J.E., New, M., Purnell, J., Singer, F., Stratakis, C.A., Trence, D.L., Wilson, D.P. (Eds.), Endotext [Internet]. MDText.com, Inc., South Dartmouth, MA. PMID: 25905357.

Rossouw, J.E., Anderson, G.L., Prentice, R.L., Lacroix, A.Z., Kooperberg, C., Stefanick, M.L., Jackson, R.D., Beresford, S.A., Howard, B.V., Johnson, K.C., Kotchen, J.M., Ockene, J., Writing Group for the Women's Health Initiative Investigators, 2002. Risks and benefits of estrogen plus progestin in healthy postmenopausal women: principal results From the Women's Health Initiative randomized controlled trial. JAMA 288, 321–333.

Sabbah, M., Courilleau, D., Mester, J., Redeuilh, G., 1999. Estrogen induction of the cyclin D1 promoter: involvement of a cAMP response-like element. Proc. Natl. Acad. Sci. U. S. A. 96, 11217–11222.

Safe, S., 2001. Transcriptional activation of genes by 17 beta-estradiol through estrogen receptor-Sp1 interactions. Vitam. Horm. 62, 231–252.

Sakurai, N., Miki, Y., Suzuki, T., Watanabe, K., Narita, T., Ando, K., Yung, T.M., Aoki, D., Sasano, H., Handa, H., 2006. Systemic distribution and tissue localizations of human 17beta-hydroxysteroid dehydrogenase type 12. J. Steroid Biochem. Mol. Biol. 99, 174–181.
Santen, R.J., Allred, D.C., Ardoin, S.P., Archer, D.F., Boyd, N., Braunstein, G.D., Burger, H.G., Colditz, G.A., Davis, S.R., Gambacciani, M., Gower, B.A., Henderson, V.W., Jarjour, W.N., Karas, R.H., Kleerekoper, M., Lobo, R.A., Manson, J.E., Marsden, J., Martin, K.A., Martin, L., Pinkerton, J.V., Rubinow, D.R., Teede, H., Thiboutot, D.M., Utian, W.H., Endocrine, S., 2010. Postmenopausal hormone therapy: an Endocrine Society scientific statement. J. Clin. Endocrinol. Metab. 95, s1–s66.
Santos, R.S., Batista, T.M., Camargo, R.L., Morato, P.N., Borck, P.C., Leite, N.C., Kurauti, M.A., Wanschel, A.C., Nadal, A., Clegg, D.J., Carneiro, E.M., 2016. Lacking of estradiol reduces insulin exocytosis from pancreatic beta-cells and increases hepatic insulin degradation. Steroids 114, 16–24.
Sattar, N., 2013. Gender aspects in type 2 diabetes mellitus and cardiometabolic risk. Best Pract. Res. Clin. Endocrinol. Metab. 27, 501–507.
Schiffer, L., Kempegowda, P., Arlt, W., O'reilly, M.W., 2017. Mechanisms in endocrinology: the sexually dimorphic role of androgens in human metabolic disease. Eur. J. Endocrinol. 177, R125–R143.
Schmitt, M., Bausero, P., Simoni, P., Queuche, D., Geoffroy, V., Marschal, C., Kempf, J., Quirin-Stricker, C., 1995. Positive and negative effects of nuclear receptors on transcription activation by AP-1 of the human choline acetyltransferase proximal promoter. J. Neurosci. Res. 40, 152–164.
Sharma, G., Prossnitz, E.R., 2011. Mechanisms of estradiol-induced insulin secretion by the G protein-coupled estrogen receptor GPR30/GPER in pancreatic beta-cells. Endocrinology 152, 3030–3039.
Shi, K., Mizuno, A., Sano, T., Ishida, K., Shima, K., 1994. Sexual difference in the incidence of diabetes mellitus in Otsuka-Long-Evans-Tokushima-Fatty rats: effects of castration and sex hormone replacement on its incidence. Metabolism 43, 1214–1220.
Shimizu, H., Ohtani, K.I., Uehara, Y., Abe, Y., Takahashi, H., Tsuchiya, T., Sato, N., Ibuki, Y., Mori, M., 1998. Orchiectomy and response to testosterone in the development of obesity in young Otsuka-Long-Evans-Tokushima Fatty (OLETF) rats. Int. J. Obes. Relat. Metab. Disord. 22, 318–324.
Simoncini, T., Hafezi-Moghadam, A., Brazil, D.P., Ley, K., Chin, W.W., Liao, J.K., 2000. Interaction of oestrogen receptor with the regulatory subunit of phosphatidylinositol-3-OH kinase. Nature 407, 538–541.
Simpson, E.R., Misso, M., Hewitt, K.N., Hill, R.A., Boon, W.C., Jones, M.E., Kovacic, A., Zhou, J., Clyne, C.D., 2005. Estrogen—the good, the bad, and the unexpected. Endocr. Rev. 26, 322–330.
Smith, H.O., Arias-Pulido, H., Kuo, D.Y., Howard, T., Qualls, C.R., Lee, S.J., Verschraegen, C.F., Hathaway, H.J., Joste, N.E., Prossnitz, E.R., 2009. GPR30 predicts poor survival for ovarian cancer. Gynecol. Oncol. 114, 465–471.
Song, R.X., McPherson, R.A., Adam, L., Bao, Y., Shupnik, M., Kumar, R., Santen, R.J., 2002. Linkage of rapid estrogen action to MAPK activation by ERalpha-Shc association and Shc pathway activation. Mol. Endocrinol. 16, 116–127.
Soriano, S., Ropero, A.B., Alonso-Magdalena, P., Ripoll, C., Quesada, I., Gassner, B., Kuhn, M., Gustafsson, J.A., Nadal, A., 2009. Rapid regulation of K(ATP) channel activity by 17{beta}-estradiol in pancreatic {beta}-cells involves the estrogen receptor {beta} and the atrial natriuretic peptide receptor. Mol. Endocrinol. 23, 1973–1982.
Stein, B., Yang, M.X., 1995. Repression of the interleukin-6 promoter by estrogen receptor is mediated by NF-kappa B and C/EBP beta. Mol. Cell. Biol. 15, 4971–4979.

Steiner, D.F., 1982. The biosynthesis of insulin: genetic, evolutionary, and pathophysiologic aspects. Harvey Lect. 78, 191–228.

Strott, C.A., 1996. Steroid sulfotransferases. Endocr. Rev. 17, 670–697.

Sullivan, J.M., Vander Zwaag, R., Hughes, J.P., Maddock, V., Kroetz, F.W., Ramanathan, K.B., Mirvis, D.M., 1990. Estrogen replacement and coronary artery disease. Effect on survival in postmenopausal women. Arch. Intern. Med. 150, 2557–2562.

Sullivan, P.M., Petrusz, P., Szpirer, C., Joseph, D.R., 1991. Alternative processing of androgen-binding protein RNA transcripts in fetal rat liver. Identification of a transcript formed by trans splicing. J. Biol. Chem. 266, 143–154.

Takada, Y., Kato, C., Kondo, S., Korenaga, R., Ando, J., 1997. Cloning of cDNAs encoding G protein-coupled receptor expressed in human endothelial cells exposed to fluid shear stress. Biochem. Biophys. Res. Commun. 240, 737–741.

Teyssier, C., Belguise, K., Galtier, F., Chalbos, D., 2001. Characterization of the physical interaction between estrogen receptor alpha and JUN proteins. J. Biol. Chem. 276, 36361–36369.

Thomas, P., Pang, Y., Filardo, E.J., Dong, J., 2005. Identity of an estrogen membrane receptor coupled to a G protein in human breast cancer cells. Endocrinology 146, 624–632.

Thomas, P., Alyea, R., Pang, Y., Peyton, C., Dong, J., Berg, A.H., 2010. Conserved estrogen binding and signaling functions of the G protein-coupled estrogen receptor 1 (GPER) in mammals and fish. Steroids 75, 595–602.

Tiano, J.P., Mauvais-Jarvis, F., 2012a. Importance of oestrogen receptors to preserve functional beta-cell mass in diabetes. Nat. Rev. Endocrinol. 8, 342–351.

Tiano, J.P., Mauvais-Jarvis, F., 2012b. Molecular mechanisms of estrogen receptors' suppression of lipogenesis in pancreatic beta-cells. Endocrinology 153, 2997–3005.

Tiano, J.P., Delghingaro-Augusto, V., Le May, C., Liu, S., Kaw, M.K., Khuder, S.S., Latour, M.G., Bhatt, S.A., Korach, K.S., Najjar, S.M., Prentki, M., Mauvais-Jarvis, F., 2011. Estrogen receptor activation reduces lipid synthesis in pancreatic islets and prevents beta cell failure in rodent models of type 2 diabetes. J. Clin. Invest. 121, 3331–3342.

Toyoshima, M., Asakawa, A., Fujimiya, M., Inoue, K., Inoue, S., Kinboshi, M., Koizumi, A., 2007. Dimorphic gene expression patterns of anorexigenic and orexigenic peptides in hypothalamus account male and female hyperphagia in Akita type 1 diabetic mice. Biochem. Biophys. Res. Commun. 352, 703–708.

Tremblay, G.B., Tremblay, A., Copeland, N.G., Gilbert, D.J., Jenkins, N.A., Labrie, F., Giguere, V., 1997. Cloning, chromosomal localization, and functional analysis of the murine estrogen receptor beta. Mol. Endocrinol. 11, 353–365.

Tremblay, A., Tremblay, G.B., Labrie, F., Giguere, V., 1999. Ligand-independent recruitment of SRC-1 to estrogen receptor beta through phosphorylation of activation function AF-1. Mol. Cell 3, 513–519.

Umayahara, Y., Kawamori, R., Watada, H., Imano, E., Iwama, N., Morishima, T., Yamasaki, Y., Kajimoto, Y., Kamada, T., 1994. Estrogen regulation of the insulin-like growth factor I gene transcription involves an AP-1 enhancer. J. Biol. Chem. 269, 16433–16442.

Van De Bunt, M., Gloyn, A.L., 2012. A tale of two glucose transporters: how GLUT2 re-emerged as a contender for glucose transport into the human beta cell. Diabetologia 55, 2312–2315.

Van Sinderen, M., Steinberg, G., Jorgensen, S.B., Honeyman, J., Chow, J.D.Y., Simpson, E.R., Jones, M.E.E., Boon, W.C., 2017. Sexual dimorphism in the glucose homeostasis phenotype of the Aromatase Knockout (ArKO) mice. J. Steroid Biochem. Mol. Biol. 170, 39–48.

Vanacker, J.M., Pettersson, K., Gustafsson, J.A., Laudet, V., 1999. Transcriptional targets shared by estrogen receptor- related receptors (ERRs) and estrogen receptor (ER) alpha, but not by ERbeta. EMBO J. 18, 4270–4279.

Varlamov, O., Bethea, C.L., Roberts, C.T., Jr., 2014. Sex-specific differences in lipid and glucose metabolism. Front. Endocrinol. (Lausanne) 5, 241.

Verchere, C.B., D'alessio, D.A., Palmiter, R.D., Weir, G.C., Bonner-Weir, S., Baskin, D.G., Kahn, S.E., 1996. Islet amyloid formation associated with hyperglycemia in transgenic mice with pancreatic beta cell expression of human islet amyloid polypeptide. Proc. Natl. Acad. Sci. U. S. A. 93, 3492–3496.

Vermeulen, A., Verdonck, L., Kaufman, J.M., 1999. A critical evaluation of simple methods for the estimation of free testosterone in serum. J. Clin. Endocrinol. Metab. 84, 3666–3672.

Vogel, H., Mirhashemi, F., Liehl, B., Taugner, F., Kluth, O., Kluge, R., Joost, H.G., Schurmann, A., 2013. Estrogen deficiency aggravates insulin resistance and induces beta-cell loss and diabetes in female New Zealand obese mice. Horm. Metab. Res. 45, 430–435.

Walton, C., Godsland, I.F., Proudler, A.J., Wynn, V., Stevenson, J.C., 1993. The effects of the menopause on insulin sensitivity, secretion and elimination in non-obese, healthy women. Eur. J. Clin. Invest. 23, 466–473.

Wang, Y., Miksicek, R.J., 1991. Identification of a dominant negative form of the human estrogen receptor. Mol. Endocrinol. 5, 1707–1715.

Wang, J., Takeuchi, T., Tanaka, S., Kubo, S.K., Kayo, T., Lu, D., Takata, K., Koizumi, A., Izumi, T., 1999. A mutation in the insulin 2 gene induces diabetes with severe pancreatic beta-cell dysfunction in the Mody mouse. J. Clin. Invest. 103, 27–37.

Wang, Z., Zhang, X., Shen, P., Loggie, B.W., Chang, Y., Deuel, T.F., 2005. Identification, cloning, and expression of human estrogen receptor-alpha36, a novel variant of human estrogen receptor-alpha66. Biochem. Biophys. Res. Commun. 336, 1023–1027.

Wang, Z., Zhang, X., Shen, P., Loggie, B.W., Chang, Y., Deuel, T.F., 2006. A variant of estrogen receptor-{alpha}, hER-{alpha}36: transduction of estrogen- and antiestrogen-dependent membrane-initiated mitogenic signaling. Proc. Natl. Acad. Sci. U. S. A. 103, 9063–9068.

Watters, J.J., Campbell, J.S., Cunningham, M.J., Krebs, E.G., Dorsa, D.M., 1997. Rapid membrane effects of steroids in neuroblastoma cells: effects of estrogen on mitogen activated protein kinase signalling cascade and c-fos immediate early gene transcription. Endocrinology 138, 4030–4033.

Webb, P., Lopez, G.N., Uht, R.M., Kushner, P.J., 1995. Tamoxifen activation of the estrogen receptor/AP-1 pathway: potential origin for the cell-specific estrogen-like effects of antiestrogens. Mol. Endocrinol. 9, 443–456.

Webb, P., Nguyen, P., Valentine, C., Lopez, G.N., Kwok, G.R., Mcinerney, E., Katzenellenbogen, B.S., Enmark, E., Gustafsson, J.A., Nilsson, S., Kushner, P.J., 1999. The estrogen receptor enhances AP-1 activity by two distinct mechanisms with different requirements for receptor transactivation functions. Mol. Endocrinol. 13, 1672–1685.

Wild, S., Roglic, G., Green, A., Sicree, R., King, H., 2004. Global prevalence of diabetes: estimates for the year 2000 and projections for 2030. Diabetes Care 27, 1047–1053.

Withers, D.J., Gutierrez, J.S., Towery, H., Burks, D.J., Ren, J.M., Previs, S., Zhang, Y., Bernal, D., Pons, S., Shulman, G.I., Bonner-Weir, S., White, M.F., 1998. Disruption of IRS-2 causes type 2 diabetes in mice. Nature 391, 900–904.

Wong, W.P., Tiano, J.P., Liu, S., Hewitt, S.C., Le May, C., Dalle, S., Katzenellenbogen, J.A., Katzenellenbogen, B.S., Korach, K.S., Mauvais-Jarvis, F., 2010. Extranuclear estrogen receptor-alpha stimulates NeuroD1 binding to the insulin promoter and favors insulin synthesis. Proc. Natl. Acad. Sci. U. S. A. 107, 13057–13062.

Wu, R.C., Qin, J., Yi, P., Wong, J., Tsai, S.Y., Tsai, M.J., O'malley, B.W., 2004. Selective phosphorylations of the SRC-3/AIB1 coactivator integrate genomic responses to multiple cellular signaling pathways. Mol. Cell 15, 937–949.

Wyckoff, M.H., Chambliss, K.L., Mineo, C., Yuhanna, I.S., Mendelsohn, M.E., Mumby, S.M., Shaul, P.W., 2001. Plasma membrane estrogen receptors are coupled to endothelial nitric-oxide synthase through Galpha(i). J. Biol. Chem. 276, 27071–27076.

Xu, X., D'hoker, J., Stange, G., Bonne, S., De Leu, N., Xiao, X., Van De Casteele, M., Mellitzer, G., Ling, Z., Pipeleers, D., Bouwens, L., Scharfmann, R., Gradwohl, G., Heimberg, H., 2008. Beta cells can be generated from endogenous progenitors in injured adult mouse pancreas. Cell 132, 197–207.

Xu, B., Allard, C., Alvarez-Mercado, A.I., Fuselier, T., Kim, J.H., Coons, L.A., Hewitt, S.C., Urano, F., Korach, K.S., Levin, E.R., Arvan, P., Floyd, Z.E., Mauvais-Jarvis, F., 2018. Estrogens promote misfolded proinsulin degradation to protect insulin production and delay diabetes. Cell Rep. 24, 181–196.

Yoon, K., Kwack, S.J., Kim, H.S., Lee, B.M., 2014. Estrogenic endocrine-disrupting chemicals: molecular mechanisms of actions on putative human diseases. J. Toxicol. Environ. Health B Crit. Rev. 17, 127–174.

Yoshioka, M., Kayo, T., Ikeda, T., Koizumi, A., 1997. A novel locus, Mody4, distal to D7Mit189 on chromosome 7 determines early-onset NIDDM in nonobese C57BL/6 (Akita) mutant mice. Diabetes 46, 887–894.

Yuchi, Y., Cai, Y., Legein, B., De Groef, S., Leuckx, G., Coppens, V., Van Overmeire, E., Staels, W., De Leu, N., Martens, G., Van Ginderachter, J.A., Heimberg, H., Van De Casteele, M., 2015. Estrogen receptor alpha regulates beta-cell formation during pancreas development and following injury. Diabetes 64, 3218–3228.

Zethelius, B., Hales, C.N., Lithell, H.O., Berne, C., 2004. Insulin resistance, impaired early insulin response, and insulin propeptides as predictors of the development of type 2 diabetes: a population-based, 7-year follow-up study in 70-year-old men. Diabetes Care 27, 1433–1438.

Zhang, Q.X., Borg, A., Fuqua, S.A., 1993. An exon 5 deletion variant of the estrogen receptor frequently coexpressed with wild-type estrogen receptor in human breast cancer. Cancer Res. 53, 5882–5884.

Zhao, C., Dahlman-Wright, K., Gustafsson, J.A., 2010. Estrogen signaling via estrogen receptor {beta}. J. Biol. Chem. 285, 39575–39579.

Zhou, Z., Ribas, V., Rajbhandari, P., Drew, B.G., Moore, T.M., Fluitt, A.H., Reddish, B.R., Whitney, K.A., Georgia, S., Vergnes, L., Reue, K., Liesa, M., Shirihai, O., Van Der Bliek, A.M., Chi, N.W., Mahata, S.K., Tiano, J.P., Hewitt, S.C., Tontonoz, P., Korach, K.S., Mauvais-Jarvis, F., Hevener, A.L., 2018. Estrogen receptor alpha protects pancreatic beta-cells from apoptosis by preserving mitochondrial function and suppressing endoplasmic reticulum stress. J. Biol. Chem. 293, 4735–4751.

Zhu, M., Mizuno, A., Kuwajima, M., Ogino, T., Murakami, T., Noma, Y., Sano, T., Shima, K., 1998. Ovarian hormone-induced beta-cell hypertrophy contributes to the homeostatic control of beta-cell mass in OLETF female rat, a model of Type II diabetes. Diabetologia 41, 799–805.

Zhu, L., Brown, W.C., Cai, Q., Krust, A., Chambon, P., McGuinness, O.P., Stafford, J.M., 2013. Estrogen treatment after ovariectomy protects against fatty liver and may improve pathway-selective insulin resistance. Diabetes 62, 424–434.

CHAPTER THREE

Molecular mechanisms of β-cell dysfunction and death in monogenic forms of diabetes

Laura Sanchez Caballero†, Vyron Gorgogietas†, Maria Nicol Arroyo†, and Mariana Igoillo-Esteve*
ULB Center for Diabetes Research (UCDR), Université Libre de Bruxelles, Brussels, Belgium. http://www.ucdr.be/
*Corresponding author: e-mail address: migoillo@ulb.ac.be

Contents

† Shared first authorship.

International Review of Cell and Molecular Biology, Volume 359
ISSN 1937-6448
https://doi.org/10.1016/bs.ircmb.2021.02.005

Abstract

Monogenetic forms of diabetes represent 1%–5% of all diabetes cases and are caused by mutations in a single gene. These mutations, that affect genes involved in pancreatic β-cell development, function and survival, or insulin regulation, may be dominant or recessive, inherited or de novo. Most patients with monogenic diabetes are very commonly misdiagnosed as having type 1 or type 2 diabetes. The severity of their symptoms depends on the nature of the mutation, the function of the affected gene and, in some cases, the influence of additional genetic or environmental factors that modulate severity and penetrance. In some patients, diabetes is accompanied by other syndromic features such as deafness, blindness, microcephaly, liver and intestinal defects, among others. The age of diabetes onset may also vary from neonatal until early adulthood manifestations. Since the different mutations result in diverse clinical presentations, patients usually need different treatments that range from just diet and exercise, to the requirement of exogenous insulin or other hypoglycemic drugs, e.g., sulfonylureas or glucagon-like peptide 1 analogs to control their glycemia. As a consequence, awareness and correct diagnosis are crucial for the proper management and treatment of monogenic diabetes patients. In this chapter, we describe mutations causing different monogenic forms of diabetes associated with inadequate pancreas development or impaired β-cell function and survival, and discuss the molecular mechanisms involved in β-cell demise.

Abbreviations

AFP	alpha-fetoprotein
ATP	adenosine triphosphate
BLK	B-lymphocyte kinase
bHLH	basic helix-loop-helix
ChIP	chromatin immunoprecipitation
CNS	central nervous system
ER	endoplasmic reticulum
FeS	iron-sulfur cluster
FXN	frataxin
FRDA	Friedreich's ataxia
GCK	glucokinase
GIP	glucose-dependent insulinotropic polypeptide
GLP-1	glucagon-like peptide 1
GLIS3	GLI similar 3
GLUT	glucose transporter
GSIS	glucose-stimulated insulin secretion
HD	homodomain
HNF	hepatocyte nuclear factor
IGF	insulin-like growth factor
INS	insulin
iPSC	inducible pluripotent stem cells
$\mathbf{K_{ATP}}$	adenosine triphosphate sensitive potassium channel
KLF11	Kruppel-like factor 11
KO	knockout
MAMs	mitochondrial associated ER membranes
MODY	maturity-onset diabetes of the young
OP	octapeptide
PAX	paired box
PDX	pancreatic duodenal homeobox
PP	pancreatic polypeptide
PP1	protein-phosphatase-1
PTF1A	pancreas associated transcription factor 1A
RFX6	regulatory factor 6
ROS	reactive oxygen species
SNPs	single nucleotide polymorphisms
SST	somatostatin

SUR1	sulfonylurea receptor 1
T1D	type 1 diabetes
T2D	type 2 diabetes
WRS	Wolcott-Rallison syndrome
WS	Wolfram syndrome
WT	wild type

1. Introduction

Monogenetic forms of diabetes result from mutations in a single gene. They account for 1%–5% of all diabetes cases and may be caused by inherited or the novo, dominant or recessive mutations in key genes related with β-cell development, function and survival or in genes controlling insulin action (McCarthy and Hattersley, 2008). Most monogenic diabetes patients are underdiagnosed or initially misdiagnosed as having type 1 (T1D) or type 2 diabetes (T1D) (Guan et al., 2020). The two main types of monogenic diabetes are neonatal diabetes mellitus and maturity-onset diabetes of the young (MODY). However, there are also other less frequent forms of the disease that are often accompanied by syndromic manifestations (Dipple and McCabe, 2000; Hattersley and Patel, 2017; Kim et al., 2003a). The prevalence of the different monogenic forms of diabetes as well as the severity and variety of their clinical manifestations depend on the nature of the mutation and the identity of the affected gene. Accordingly, some monogenic forms of diabetes are accompanied by extrapancreatic manifestations such as deafness, blindness, microcephaly, liver and intestinal defects and, while in some patients diabetes appears at birth, in others the metabolic alterations manifest in childhood, adolescence or early adulthood (Kanakatti Shankar et al., 2013; Pihoker et al., 2013). Patients suffering from monogenic diabetes also require different treatments. While for some of them diet and exercise are sufficient to keep normoglycemia, others require exogenous insulin or other hypoglycaemic medications, e.g., sulfonylureas or glucagon-like peptide 1 (GLP-1) analogs to achieve proper glycemic control (Rubio-Cabezas et al., 2014b). Thus, awareness and correct diagnosis are crucial for the proper management and treatment of the affected patients.

In this chapter, we will discuss the pathogenic mechanisms of monogenic diabetes caused by autosomal dominant or recessive mutations in essential genes needed for proper β-cell development, function and survival.

In particular, we will provide details on the molecular mechanisms of β-cell failure in different monogenic forms of diabetes associated with impaired function or expression of transcription factors and transcriptional regulators, compromised glucose metabolism and insulin secretion, endoplasmic reticulum stress, mitochondrial dysfunction, and impaired tRNA methylation (Table 1).

2. Monogenic forms of diabetes associated with mutations in transcription factors and transcriptional regulators

Transcription factors and transcriptional regulators play an essential role in pancreas development, β-cell function and survival. During human pancreas development, the differentiation fates and the acquisition of endocrine and exocrine cell identity depend on the controlled expression of unique sets of transcription factors (Conrad et al., 2014; Jennings et al., 2013, 2020; Nair and Hebrok, 2015). Indeed, a large variety of them were shown to be crucial for proper human pancreas development (e.g., SOX9, SOX17, GATA4, PDX1, RFX6, NKX6.1, NKX2.2, HNF1B, NEUROG3, PAX4, PAX6, and NEUROD1) and preserved β-cell function (e.g., PDX1, MAFA, GLISS3, GATA6, PAX6) (Conrad et al., 2014; Jennings et al., 2013, 2020; Nair and Hebrok, 2015). In this section we will discuss monogenic forms of diabetes caused by mutations in transcription factors and transcription regulators controlling β-cell function (Fig. 1), β-cell survival, pancreas development and endocrine differentiation.

2.1 *HNF4A* mutations (MODY1)

2.1.1 Genetic alteration and clinical phenotype

MODY1 is an autosomal dominant monogenic form of diabetes caused by heterozygous inactivating mutations in the *HNF4A* gene coding for the transcription factor hepatocyte nuclear factor 4α (HNF-4α) (Fajans et al., 2001; Yamagata et al., 1996). In some populations, single nucleotide polymorphisms (SNPs) in the P2 promoter of *HNF4A* are associated with increased T2D risk (Love-Gregory et al., 2004; Silander et al., 2004; Weedon et al., 2004). MODY1 accounts for 5–10% of all MODY cases (Firdous et al., 2018; Naylor et al., 2018). MODY1 patients have impaired glucose tolerance due to progressive insulin secretory defects which evolves to overt diabetes with time (Byrne et al., 1995; Herman et al., 1997; Ryffel, 2001). The clinical manifestations usually appear in adolescence or early

Table 1 Monogenic forms of diabetes and their genetic cause classified according to the function or the subcellular localization of the causal protein.

Form of diabetes	Mutated gene	Encoded protein	OMIM	Locus	Inheritance
Monogenic forms of diabetes associated with mutations in transcription factors					
MODY1	*HNF4A (HNF4, NR2A1, TCF14)*	Hepatocyte nuclear factor 4-alpha (HNF-4α)	# 125850, * 600281	20q13.12	AD
MODY3	*HNF1A (TCF1)*	Hepatocyte nuclear factor 1-alpha (HNF-1α)	# 600496, * 142410	12q24.31	AD
MODY4 and NDM	*PDX1*	Pancreas/duodenum homeobox protein 1 (PDX1)	# 606392, * 600733	13q12.2	AD/AR
MODY5	*HNF1B (TCF2)*	Hepatocyte nuclear factor 1-beta (HNF-1β)	# 137920, * 189907	17q12	AD
MODY6	*NEUROD1 (BHLHA3)*	Neurogenic differentiation factor 1 (NEUROD1)	# 606394, * 601724	2q31.3	AD/AR
MODY7	*KLF11 (FKLF, TIEG2)*	Kruppel-like factor 11 (KLF11)	# 610508, * 603301	2p25.1	AD
MODY9	*PAX4*	Paired box protein Pax-4	# 612225, * 167413	7q32.1	AD
NDM	*PTF1A (BHLHA29, PTF1P48)*	Pancreas transcription factor 1 subunit alpha	* 607194	10p12.2	AR
Mitchell-Riley syndrome	*RFX6*	DNA binding protein (RFX6)	* 612659	6q22.1	AR/AD
NDM	*GATA4*	Transcription factor (GATA-4)	* 600576	8p23.1	AD
AOD	*GATA6*	Transcription factor (GATA-6)	* 601656	18q11.2	AD
NDM	*GLIS3*	Zinc finger protein (GLIS3)	* 610192	9p24.2	AR

NDM	*MNX1*	Motor neuron and pancreas homeobox protein 1 (MNX1)	* 142994	7q36.3	AR
NDM, YOD	*PAX6*	Paired box protein 6 (PAX6)	* 607108	11p13	AD/AR
NDM	*NEUROG3, NGN3*	Neurogenin-3 (NGN3)	* 604882	10q22.1	AR
NDM	*NKX2.2*	NK2 Homeobox 2 (NKX2.2)	* 604612	20p11.22	AD
NDM	*CNOT1 (CDC39, KIAA1007, NOT1)*	CCR4-NOT transcription complex subunit 1 (CNOT1)	* 604917	16q21	AD
TND	*ZAC1/PLAGL1*	Pleiomorphic adenoma-like protein 1 (PLAG1) (ZAC1)	* 603044	6q24.2	Imprinting disorder
Monogenic forms of diabetes associated with impaired glucose metabolism and insulin secretion					
MODY2, NDM	*GCK*	Glucokinase (GCK)	# 125851, * 138079	7p13	AD/AR
MODY11, T2D	*BLK*	B-lymphocyte kinase (BLK)	# 613375, * 191305	8p23.1	AD
MODY12, NDM	*ABCC8 (HRINS, SUR, SUR1)*	ATP-binding cassette subfamily C member 8	* 600509	11p15.1	AD/AR/de novo
MODY13	*KCNJ11*	ATP-sensitive inward-rectifier potassium channel 11	# 616329, * 600937	11p15.1	AD/de novo
YOD	*PCSK1*	Proprotein convertase 1 (PC1, PC1/3)	* 162150	5q15	AR

Continued

Table 1 Monogenic forms of diabetes and their genetic cause classified according to the function or the subcellular localization of the causal protein.—cont'd

Form of diabetes	Mutated gene	Encoded protein	OMIM	Locus	Inheritance
Fanconi Bickel syndrome	*SLC2A2*	Glucose transporter 2 (GLUT2)	* 138160	3q26.2	AR
MODY14	*APPL1*	Adaptor protein, phosphotyrosine interacting with PH domain and leucine zipper 1 (APPL1)	* 604299	3p14.3	AD
Monogenic forms of diabetes associated with endoplasmic reticulum stress					
NDM	*EIF2S3*	Eukaryotic translation initiation factor 2 subunit 3	* 300161	Xp22.11	X-linked
Wolcott-Rallison syndrome (NDM)	*EIF2AK3*	Eukaryotic translation initiation factor 2-alpha kinase 3	* 604032	2p11.2	AR
Wolcott-Rallison Like syndrome (NDM)	*IER3IP1*	Immediate early response 3 interacting protein 1	* 609382	18q21.1	AR
YOD	*PPP1R15B*	Protein phosphatase 1 regulatory subunit 15B	* 613257	1q32.1	AR
Wolfram syndrome 1	*WFS1*	Wolframin	# 222300, * 606201	4p16.1	AR/AD/de novo
Wolfram syndrome 2	*CISD2 (CDGSH2, ERIS, ZCD2)*	CDGSH iron-sulfur domain-containing protein 2	# 604928, * 611507	4q24	AR
MODY10, NDM	*INS*	Insulin	# 613370, * 176730	11p15.5	AD/AR/de novo
NDM/YOD	*EIF2B1*	Eukaryotic translation initiation factor 2B, subunit alpha (eIF2Bα)	* 606686	12q24.31	De novo

NDM/YOD	*YIPF5, YIP1A*	Yip1 domain family, member 5 (YIPF5)	* 611483	5q31.3	AR
Monogenic forms of diabetes associated with mitochondrial dysfunction					
Friedreich's ataxia	*FXN*	Frataxin	* 606829	9q21.11	AR
Maternally inherited diabetes-deafness syndrome (MIDD)	*MTTL1, MTTE*	Mitochondrially encoded tRNA leucine 1, mitochondrially encoded tRNA glutamic acid	* 590050, * 590025	chrMT:3229-3307	MT
Monogenic forms of diabetes associated with impaired tRNA methylation					
TRM10A diabetes (YOD)	*TRMT10A*	tRNA methyltransferase 10 homolog A	* 616013	4q23	AR

This table reports the genes associated with each monogenic diabetes form discussed in this chapter. The affected gene, name of the protein encoded by the gene, the authoritative compendium of the genes (OMIM), the gene locus and the modus of inheritance are described. The genes are categorized based on molecular pathways or the subcellular organelle in which they participate.

Abbreviations: AD, autosomal dominant; AOD, adult onset diabetes mellitus; AR, autosomal recessive; MIID, maternally inherited diabetes; MODY, maturity-onset diabetes of the young; MT, mitochondrial; NDM, neonatal diabetes mellitus, PNMD, permanent neonatal diabetes mellitus; TNDM, transient neonatal diabetes mellitus, T2D, type 2 diabetes; YOD, young onset diabetes mellitus. OMIM entry numbers preceded by (*) are genes, the ones preceded by (#) its a descriptive entry, in these cases, a phenotype.

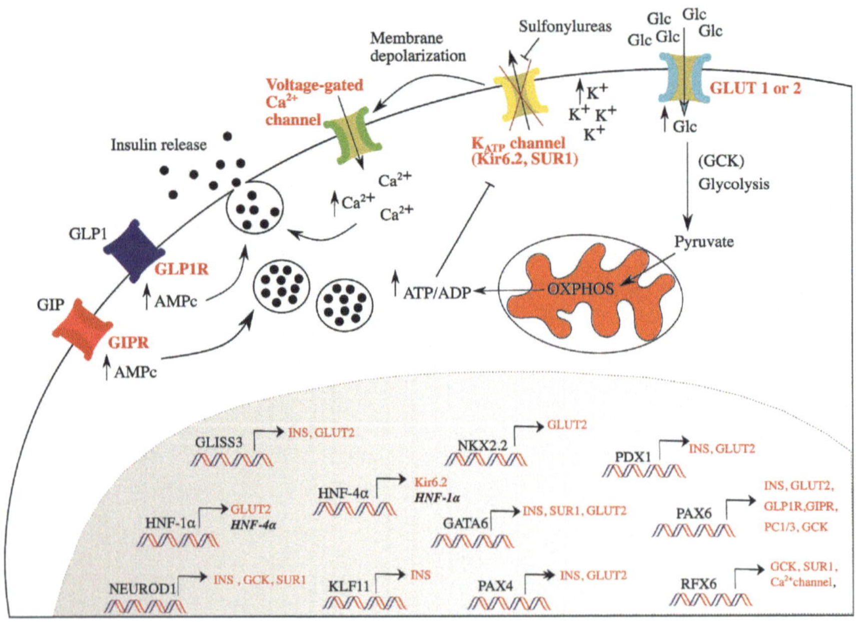

Fig. 1 Molecular mechanism of glucose-induced insulin secretion in pancreatic β-cells. In the nucleus we show transcription factors controlling the expression of genes related with the insulin secretion machinery and whose mutations are associated with monogenic forms of diabetes. AMPc, cyclic AMP; GCK, glucokinase; Glc, glucose; GIP, glucose-dependent insulinotropic polypeptide; GIPR, GIP receptor; GLP1, glucagon-like peptide 1; GLP1R, GLP1 receptor, GLUT, glucose transporter ; INS, insulin; OXPHOS, oxidative phosphorylation.

adulthood (before 25 years) (Yamagata et al., 1996). Most patients respond to sulfonylureas, which is the treatment of choice, but others require a combined therapy of sulfonylureas plus insulin (Byrne et al., 1995). GLP-1 analogs recently showed to be effective in normalizing glycemia in MODY1 patients with lower hypoglycaemic risk (Ostoft et al., 2014). In some patients, *HNF4A* mutations may lead to increased birth weight, macrosomia and hyperinsulinemic hypoglycemia at birth that resolves spontaneously and is followed by diabetes later in life. The exact cause of this phenomenon is currently unknown (Pearson et al., 2007).

2.1.2 Molecular phenotype

HNF-4α is a transcription factor belonging to the nuclear receptor superfamily that is expressed in liver, kidney, intestine and pancreatic islets (Yamagata, 2003). Similar to other transcription factors, it has several functional domains that include a N-terminal transactivation domain, a DNA binding domain, and a complex C-terminal region that forms a ligand-binding domain, a dimerization interface and an additional transactivation domain (Miura et al., 2006). The expression of *HNF4A* is controlled by two promoters, P1 (proximal) and P2 (distal) (Thomas et al., 2001). More than 12 different HNF-4α isoforms exist due to alternative splicing and the use of these two alterative promoters. The different HNF-4α isoforms are tissue specific and are differentially expressed during embryonic development. They are required for the correct development and function of different organs including the pancreas (Huang et al., 2009; Jiang et al., 2003; Ng et al., 2019). The HNF-4α isoforms 7–9 (derived from the P2 promoter) are expressed in pancreatic β-cells (Thomas et al., 2001). In vitro studies performed in the rat β-cell line INS-1 have shown that hnf-4α regulates the expression of β-cell genes involved in glucose metabolism (e.g., insulin (*Ins*), glucose transporter-2 (*Glut2*), L-pyruvate kinase, and aldolase B between others) (Wang et al., 2000). In contrast to humans, heterozygous *Hnf4a* $^{(+/-)}$ mice have normal glucose tolerance (Stoffel and Duncan, 1997) while *Hnf4a* knockout (KO) mice have a defective gastrulation and die at E9, proving the importance of this transcription factor for proper development (Chen et al., 1994). To examine the role of Hnf-4α in pancreatic β-cells, two β-cell-specific *Hnf4a* KO mouse models were generated (Gupta et al., 2005; Miura et al., 2006). Despite the fact that the same RIP-Cre transgene was used to create both models, important differences in their phenotypes were observed, which could potentially be explained by mouse strain differences (Miura et al., 2006). While the mice generated by Miura and colleagues showed impaired glucose tolerance and defective insulin secretion respect to wild-type (WT) mice (Miura et al., 2006), the mouse line generated by Gupta showed a significant decrease in fasting and fed glycemia together with increased plasma insulin levels (Gupta et al., 2005). The β-cell specific *Hnf4a* KO mice generated by Gupta showed significantly reduced Kir6.2 expression (Gupta et al., 2005). However, no changes in the expression of this gene were observed in the mice generated by Miura et al. (2006). The variable phenotype of the *Hnf4a*-deficient mice suggests that rodent models are not ideal to recapitulate human MODY1.

In humans, it has been shown that *HNF4A* haploinsufficiency leads to β-cell dysfunction accompanied by reduced expression of genes involved in early pancreatic development (Gupta et al., 2005; Ng et al., 2019; Rhee et al., 2003). A recent work using inducible pluripotent stem cells (iPSC)-derived hepatocytes and pancreatic β-like cells from members of a MODY1 family with and without a heterozygous *HNF4A* mutation, showed that *HNF4A* haploinsufficiency affects foregut development and negatively influences hepatic and pancreatic β-cell fate (Ng et al., 2019). RNA sequencing of MODY1 hepatopancreatic progenitors showed reduced levels of a subset of pancreas-related genes such as *PDX1*, the *FOXA* gene family, *GATA4*, *RFX6*, *HNF1B*, *KRT19*, and *SOX17*. During β-cell differentiation the MODY1 β-like cells showed reduced levels of both HNF-4α and HNF-1α, as well as reduced expression of PDX1 and other β-cell progenitor markers (Ng et al., 2019).

As mentioned before, HNF-4α controls *HNF-1α* expression in many cell types, including embryonic endoderm, pancreatic β-cells and liver (Gragnoli et al., 1997; Wang et al., 2000). However, it was shown that in differentiated β-cells there is a regulatory loop between HNF-4α and HNF-1α that suggests an interdependency between these two transcription factors that could explain the clinical similarities of MODY1 and MODY3 patients, the latter caused by *HNF1A* mutations (see below) (Ferrer, 2002; Ng et al., 2019).

2.2 *HNF1A* mutations (MODY3)

2.2.1 Genetic alteration and clinical phenotype

MODY3 is the most common form of MODY which accounts for 30–65% of all MODY cases (Firdous et al., 2018; Naylor et al., 2018). Indeed, the proportion of MODY3 patients respect to other MODY individuals varies from country to country (Firdous et al., 2018; Valkovicova et al., 2019). MODY3 is caused by heterozygous germline loss-of-function mutations in the *HNF1A* gene coding for the transcription factor HNF1-α (Bach et al., 1990; Yamagata et al., 1996). More than 1200 different pathogenic and non-pathogenic *HNF1A* variants have been identified (Lek et al., 2016). While some of them are associated with MODY3, e.g., the P291fsinsC mutation in exon 4 that is the most common (Colclough et al., 2013; Ellard and Colclough, 2006), others confer susceptibility for developing T2D, or are benign and neutral (Althari et al., 2020). MODY3 has high penetrance and is a very variable disease, even in siblings presenting severe insulin secretion defects. In some individuals, diabetes appears after an initial phase of transient neonatal hyperinsulinemia and hypoglycemia (Dusatkova et al., 2011; Stanescu et al., 2012). The type and position of the *HNF1A* mutation modulates the age of diabetes onset, partially explaining the variability of the clinical manifestations (Bellanne-Chantelot et al., 2008). It has been proposed that mutations in the different *HNF1A* isoforms (see below) influence the age of disease onset (Valkovicova et al., 2019). Accordingly, the symptoms may appear within a range of 6–25 years old, with some asymptomatic individuals. MODY3 patients are non-obese, show progressive hyperglycemia with impaired β-cell function as the primary cause of diabetes, and have risk of micro- and macrovascular complications alike to T1D and T2D patients (Colclough et al., 2013; Lehto et al., 1997; Steele et al., 2010). Similar to MODY1 patients, MODY3 individuals have initial good response to sulfonylureas and may require insulin supplementation with time (Urakami, 2019).

It has been shown that a subtype of hepatocellular adenoma (HCA) is caused by biallelic somatic *HNF1A* mutations, and some cases of familial liver adenomatosis have been associated with MODY 3 (Fu et al., 2020; Haring et al., 2019). In these patients, one *HNF1A* mutation was germline, whereas the other was somatic and occurred only in tumor cells (Bacq et al., 2003; Fu et al., 2020; Haring et al., 2019; Reznik et al., 2004). The association between HCA and MODY3 is currently recognized and a systematic screening of HCA in MODY 3 patients through non-invasive liver imaging has been advised (Fu et al., 2020; Haring et al., 2019; Nakamura et al., 2012).

2.2.2 Molecular phenotype

HNF-1α is a transcription factor that, together with HNF-1β, belongs to the HNF1 homeobox family. Both proteins have an N-terminal dimerization domain (residues 1–32) through which they form homo- or heterodimers between them; a bipartite DNA binding domain that interacts with the consensus sequence 5′-GTTAATNATTAAC-3′; and a C-terminal transactivation domain (De Simone et al., 1991; Lau et al., 2018). HNF-1α plays a crucial role regulating β-cell proliferation and function (Yamagata, 2014). In MODY3 patients the DNA binding domain of HNF-1α has the highest mutation frequency (Ellard and Colclough, 2006). Three HNF-1α transcriptional isoforms have been identified (isoform A, B and C) which are generated from the same promoter by alternative splicing (Harries et al., 2006). HNF-1α is expressed in the pancreas but also in liver, stomach, small intestine and kidneys. The expression of the different isoforms is tissue specific. Regarding the pancreas, the isoform A is mainly expressed in the fetal tissue, while the isoforms B and C are present in mature pancreas (Harries et al., 2006; Valkovicova et al., 2019). As mentioned before, there is an interdependency between HNF-1α and HNF-4α in pancreatic β-cells where HNF-1α appears as an upstream regulator of HNF-4α in differentiated (mature) β-cells (Odom et al., 2004). HNF-1α plays an important role in embryonic development since it controls cell lineage differentiation (Lussier et al., 2010), but it is also an important transcription factor in adult tissues (Harries et al., 2006). In human pancreatic islets, it controls the expression of at least 106 target genes (Odom et al., 2004).

In contrast to the human situation, *Hnf1a* heterozygous KO mice are normal, while the homozygous KO animals have reduced β-cell mass and abnormal glucose-stimulated insulin secretion together with low levels of insulin and insulin-like growth factor-1 (Igf1) (Dukes et al., 1998; Lee et al., 1998; Pontoglio et al., 1998; Servitja et al., 2009). The reduced glucose-induced insulin secretion observed in the KO animals is the consequence of impaired ATP production. Indeed, electrophysiological analysis of *Hnf1a* KO mouse β-cells, together with fluorescence and luminescence assays for ATP measurement, showed that the ability of glucose to induce the closure of the K_{ATP} channel is significantly reduced in these cells. This is the consequence of impaired-glucose-induced ATP generation resulting from reduced cytoplasmic NADH levels or impaired NADH transfer into the mitochondria (Dukes et al., 1998). In line with that, it was shown that in INS-1 cells, *Hnf1a* dominant-negative mutations impair both, glucose and amino acid-stimulated insulin secretion, in part, as the result of mitochondrial dysfunction (Pongratz et al., 2009). Oligonucleotide array of *Hnf1a* KO mouse islets showed that this transcription factor controls the expression of a large variety of genes. Indeed, *Hnf1a*-deficient islets displayed an important reduction in genes involved in glucose metabolism, such as *Glut2* (Ban et al., 2002), *Foxa3*, *Slc2a2* and *Pklr2* (Boj et al., 2001; Parrizas et al., 2001; Shih et al., 2001), as well as in electron transport and oxidative phosphorylation (Servitja et al., 2009). Genes related with amino acid transport and metabolism, lipid metabolism, endoplasmic retic ulum and Golgi function, among others, were also modified. For many of the genes that are severely downregulated in *Hnf1a* KO islets, it was experimentally demonstrated that they play an important role in controlling β-cell function. Based on that, β-cell demise in MODY3 is probably the result of a large genetic program failure rather than alterations in one isolated biological pathway (Servitja et al., 2009).

Recently, histological analysis of post-mortem pancreas from one individual (originally diagnosed as having T1D) but bearing a disease-causing *HNF1A*$^{(WT/T260M)}$ variant showed normal islet β- and δ-cell mass but slightly increased α-cell number (Haliyur et al., 2019). Dynamic insulin secretion assays in isolated islets from the same donor showed high basal insulin release, impaired glucose-stimulated insulin secretion (GSIS) and, different from *Hnf1a* KO mice (Pontoglio et al., 1998), impaired response to KCl-mediated membrane depolarization (Haliyur et al., 2019). The response of *HNF1A*$^{(WT/T260M)}$ β-cells to high glucose coupled with the phosphodiesterase inhibitor isobutyl methylxanthine (IBMX) was normal. Interestingly, the α-cell-mediated glucagon secretion in response to low glucose (1.7 mM) and epinephrine (1 μM) was also reduced suggesting that *HNF1A*$^{(WT/T260M)}$ affects both β- and α-cells (Haliyur et al., 2019). Donor and control pancreas showed comparable HNF-1α protein expression that was detected in the nucleus of endocrine and exocrine cells. Nevertheless, the DNA binding capacity of HNF-1α^{T260M} protein (assessed by electrophoretic mobility shift assay, EMSA) was importantly reduced respect to HNF-1α^{WT} (Haliyur et al., 2019). Through RNA-sequencing and transcriptional profiling the authors showed that purified *HNF1A*$^{(WT/T260M)}$ β-cells had preserved expression of INS and transcription factors associated with β-cell identity, e.g., PDX1, NKX2.2, and NKX6.1. Nevertheless, the expression of transcription factors related with β-cell function (i.e., MAFA, SIX3, and RFX6) was reduced (Haliyur et al., 2019). This analysis also revealed decreased levels of known (e.g., HNF-4α, PKM, OGDH, G6PC2; Servitja et al., 2009) and new HNF-1α targets such as the K_{ATP} channel subunits SUR1, Kir6.2, and Kir6.1 whose decrease expression probably contributes to the loss of GSIS observed in *HNF1A*$^{(WT/T260M)}$ purified islets (Haliyur et al., 2019). Similar to the *hnf1a* KO mice data (Dukes et al., 1998), pathway analysis of human *HNF1A*$^{(WT/T260M)}$ β-cells unveiled changes in glucose metabolism and ATP production. Approximately 50% of the genes altered by the *HNF1A*$^{(WT/T260M)}$ variant in pancreatic β-cells were also changed in α-cells suggesting that HNF-1α is an important transcription factor for both β- and α-cells. This study, even if based in only one patient, provides insights in potential new therapeutic opportunities for MODY 3 patients. Indeed, it was previously shown that these patients have an increased sensitivity to sulfonylureas which may lead to hypoglycaemic events (Pearson et al., 2000). The work of Haliyur et al. (2019) suggests that this may be the result of impaired glucagon secretion. The preserved insulin secretion of *HNF1A*$^{(WT/T260M)}$ β-cells when cAMP-dependent pathways are stimulated, suggest that GLP-1 agonists may have therapeutic advantages over sulfonylureas in MODY 3 patients lowering the risk of hypoglycemia (Haliyur et al., 2019).

2.3 *PDX1* mutations (MODY4 and neonatal diabetes mellitus)

2.3.1 Genetic alteration and clinical phenotype

Heterozygous missense or frameshifting mutations in Pancreatic Duodenal Homeobox 1 (*PDX1*) gene cause MODY4 (Clocquet et al., 2000; Ellard et al., 2008; Fajans et al., 2010), while loss-of-function homozygous *PDX1* mutations cause pancreatic agenesis and neonatal diabetes mellitus (Kulkarni et al., 2017; Stoffers et al., 1997b; Wright et al., 1993). It was also shown that hypomorphic biallelic mutations in the same gene lead to permanent neonatal diabetes mellitus without pancreatic agenesis (De Franco et al., 2013a; Nicolino et al., 2010). MODY4 is a rare form of MODY that accounts for less than 1% of all MODY cases (Firdous et al., 2018; Naylor et al., 2018). MODY4 patients have glucose intolerance, decreased insulin secretion and β-cell loss (Ellard et al., 2008). Overweight and obesity have also been reported in some MODY4 individuals (Fajans et al., 2010). The disease severity is variable and correlates with the nature and functional consequences of the mutation. The clinical manifestations may appear later than in other types of MODY (Fajans et al., 2001). The treatment for MODY4 patients is based on oral hypoglycaemic agents and insulin (Fajans et al., 2001).

2.3.2 Molecular phenotype

PDX1, also called insulin promoter factor 1 (IPF1) is a key transcription factor that plays an essential role in pancreas development and also controls the function and survival of adult β-cells. It has been shown that Pdx1 is initially expressed in the mouse developing pancreas but it is after restricted to maturing β-cells (Ohlsson et al., 1993). Mice lacking *Pdx1* fail to form a pancreas, which prevents studying the role Pdx1 in mature β-cells (Ahlgren et al., 1996; Jonsson et al., 1994; Offield et al., 1996). To address the role of Pdx1 in adult differentiated β-cells, Ahlgren and colleagues generated β-cell-specific *Pdx1* mutant mice using the CRE–Lox system (Ahlgren et al., 1998). They showed that, while the β-cell specific $pdx1^{-/-}$ mice are overt diabetic, have 40% reduction in insulin-positive cells and 90% reduction in insulin levels, the $pdx1^{+/-}$ mice are normoglycemic but have glucose intolerance (Ahlgren et al., 1998). This and other studies showed that Pdx1 regulates the expression of *Ins*, somatostatin (*Sst*), glucokinase (Gck) and *Glut2* (Ahlgren et al., 1998; Leonard et al., 1993; Ohlsson et al., 1993), suggesting that the loss of *GLUT2* and INS expression may be the cause of hyperglycemia and diabetes in patients with *PDX1* haploinsufficiency (Ahlgren et al., 1998). Recent studies performed with induced pluripotent stem cells (iPSCs) from patients carrying *PDX1* missense mutations or *PDX1* haploinsufficiency, showed that both heterozygous mutations impair the development of the pancreatic lineage and hinder β-cell glucose-responses via deregulation of several PDX1-target genes involved in β-cell development, maturation, and function (Wang et al., 2019). Several in vitro and in vivo studies of partial *PDX1* deficiency in pancreatic β-cells showed that reduced PDX1 expression leads to: (1) increased β-cell apoptosis (although little is known about how PDX1 regulates β-cell survival), (2) enhanced autophagy, and (3) decreased insulin biosynthesis (Fujimoto and Polonsky, 2009).

2.4 *HNF1B* mutations (MODY5 and neonatal diabetes)

2.4.1 Genetic alteration and clinical phenotype

Autosomal dominant loss-of-function mutations in the *HNF1B* gene coding for the transcription factor hepatocyte nuclear factor-1-beta (*HNF-1β*) cause MODY5, also called renal cysts and diabetes syndrome, and accounts for about 5% of all MODY cases (Firdous et al., 2018; Naylor et al., 2018). Indeed, besides having diabetes, MODY 5 patients often present with liver and renal abnormalities and genital tract malformations (Alvelos et al., 2015; Edghill et al., 2006a; Kettunen et al., 2017; Kotalova et al., 2015). Half of the patients also show hypomagnesemia and hypokalemia (Adalat et al., 2009; Faguer et al., 2011), and some of them also present with exocrine pancreas dysfunction (Bellanne-Chantelot et al., 2004). When both manifestations are present, diabetes usually manifests after the renal disease. The age-range for the appearance of the metabolic alterations goes from 10 until to 31 years old (Bingham and Hattersley, 2004). In certain cases, heterozygous *HNF1B* mutations have been associated with neonatal diabetes (Edghill et al., 2006b; Haldorsen et al., 2008; Yorifuji et al., 2004). More than 106 different *HNF1B* mutations have been reported (Alvelos et al., 2015). The onset of diabetes and other associated symptoms varies widely, even among individuals with the same inherited mutation. No clear genotype-phenotype correlation has been demonstrated so far but it was suggested that missense and frameshift mutations may be associated with a higher diabetes and renal disease penetrance, respectively (Chen et al., 2010). MODY5 patients have reduced insulin secretion probably as a result of pancreatic atrophy and the indicated treatment to achieve glycemic control is insulin (Bingham and Hattersley, 2004; Fajans et al., 2001).

2.4.2 Molecular phenotype

As mentioned before, HNF-1β is a transcription factor from the HNF1 homeobox family that has high homology with the HNF-1α, and forms homodimers or heterodimers with HNF-1α. HNF-1β may act as transcriptional activator or repressor and plays an important role in the development and differentiation of several tissues such as liver, kidney, the genitourinary tract and pancreas (Ferre and Igarashi, 2019; Poll et al., 2006). Different from HNF-1α that is expressed in the developing and mature pancreas, HNF-1β is lowly expressed in differentiated β-cells but it is enriched in the pancreatic endoderm. During adulthood, HNF-1β is predominantly expressed in the kidneys (Maestro et al., 2003). Using mouse models of *Hnf1b* deficiency it has been demonstrated that this transcription factor is a marker of early and late pancreatic precursor cells, and an essential regulator of transcriptional networks controlling the pancreas specification, morphogenesis, and differentiation (Maestro et al., 2003). Indeed, it was shown that while $Hnf1b^{+/-}$ mice have no phenotype, $Hnf1b^{-/-}$ embryos had no ventral and dorsal pancreas with no expression of Pdx1, Ptf1, Hlxb9 and Ngn3 (Haumaitre et al., 2005). Selective deletion of *Hnf1b* in mouse pancreatic β-cells leads to glucose intolerance and reduced glucose-induced insulin secretion (Wang et al., 2004b). It has been shown that Hnf-1β activates the expression of Onecut transcription factor 1 (Oct1), also called Hnf-6, in pancreatic precursor cells. This initiates pancreatic cell fate specification by inducing the expression of Pdx1 (Poll et al., 2006). Later during pancreas development Oct1 regulates pancreatic endocrine differentiation by controlling Ngn3 expression (Lau et al., 2018). All together these results suggest that Hnf1β is involved in the regulation of key β-cell transcription factors and that it is necessary for glucose sensing or glycolytic signaling.

2.5 *NEUROD1* mutations (MODY6 and neonatal diabetes mellitus)

2.5.1 Genetic alteration and clinical phenotype

MODY6 is caused by a wide variety of heterozygous mutations (missense, small insertions or deletions) in the *NEUROD1* gene coding for the transcription factor neuronal differentiation 1 (NEUROD1), also known as β-cell *E*-box transactivator 2 (BETA2). MODY 6 is a low penetrant and very rare MODY which accounts for less than 1% of all MODY cases (Firdous et al., 2018; Naylor et al., 2018). To date seven families have been reported since the discovery of the disease (Horikawa et al., 2018). Besides MODY6, heterozygous mutations in *NEUROD1* may cause T2D, while homozygous loss-of-function *NEUROD1* mutations lead to permanent neonatal diabetes mellitus, that is associated with severe neurological abnormalities, e.g., severe cerebellar hypoplasia, learning difficulties and sensorineural deafness together with visual impairment (Gonsorcikova et al., 2008; Malecki et al., 1999; Rubio-Cabezas et al., 2010). MODY6 patients have slightly preserved pancreatic β-cell function (Chapla et al., 2015; Gonsorcikova et al., 2008; Horikawa et al., 2018; Kristinsson et al., 2001; Liu et al., 2007; Malecki et al., 1999; Szopa et al., 2016). The disease severity varies considerably among patients, even in individuals bearing the same mutation, suggesting that other genetic modifying factors, environmental factors or both, influence the clinical manifestations of MODY6 (Horikawa et al., 2018).

2.5.2 Molecular phenotype

NEUROD1 is a basic helix-loop-helix (bHLH) transcriptional activator expressed in pancreatic and neuronal cells, essential for enteroendocrine progenitor differentiation and maturation. The bHLH domain is responsible for DNA binding so mutations in this region disrupt the NEUROD1 recognition of its target genes. NEUROD1 also contains a transactivation domain, which interacts with p300/CBP (CREB binding protein), a cellular coactivator that controls the expression of β-cell specific genes. NEUROD1 is expressed in the developing pancreas and in mature β-cells. By forming heterodimers with the bHLH transcription factor E47, which is ubiquitously expressed and binds to specific *E*-box motifs in the promoter of specific target genes, NEUROD1 triggers the transcription of *INS* (Kristinsson et al., 2001; Naya et al., 1995), the sulfonylurea receptor 1 (*SUR1*) (Kim et al., 2002), glucokinase (*GCK*) (Moates et al., 2003), the islet-specific glucose-6-phosphatase catalytic subunit-related protein (IGRP) (Moates et al., 2003) and *PAX6*, all of them important for glucose homeostasis. Mouse models of *Neurod1* deficiency do not survive more than a week due to acute diabetes mellitus. They present high ketone levels in the urine as a consequence of severe insulinopenia, and also show important neurological disorders characterized by ataxia, a complete loss of the dentate gyrus of the hippocampus and reduced cerebellum's size, suggesting that *Neurod1* plays a crucial role in pancreas and brain development (Miyata et al., 1999; Naya et al., 1997; Schwab et al., 2000). It was also shown that the loss of *Neurod1* in mice causes not only β-cell demise, but also α-cells loss of due to exacerbated apoptosis (Naya et al., 1997). Pancreatic islets from mice lacking *Neurod1* in the β-cells had poor response to glucose, and their metabolic profile resembled the one of immature β-cells with enhanced basal insulin secretion and increased expression of glycolytic genes (Gu et al., 2010). These results indicate NEUROD1 is needed for β-cell maturation and the maintenance of fully functional glucose responsiveness.

2.6 *KLF11* mutations (MODY7)

2.6.1 Genetic alteration and clinical phenotype

Heterozygous mutations in the *KLF11* gene coding for the transcription factor Krüppel-like factor 11 (KLF11) cause MODY7, and mutations in its DNA binding site within the insulin promoter cause neonatal diabetes mellitus (Bonnefond et al., 2011; Lomberk et al., 2013; Neve et al., 2005). Some nonsynonymous *KLF11* variants have also been associated with early onset T2D (Neve et al., 2005). MODY7 is a very rare form of MODY that accounts for less than 1% of all MODY cases (Firdous et al., 2018; Naylor et al., 2018). The clinical features of MODY7 have not been fully established, but it is known that *KLF11* mutations may lead to a wide phenotypic spectrum of diabetes mellitus, similar to the ones associated with *HNF1B*, *INS* and *KCNJ11* mutations (D'Amato et al., 2008; Edghill et al., 2008; Liang et al., 2020). Recently, a candidate gene- or exome-targeted capture sequencing performed in 76 probands of Chinese origin from unrelated families fulfilling clinical diagnostic criteria for MODY showed that a high proportion of the MODY7 patients in this population are overweight or obese (Liang et al., 2020).

2.6.2 Molecular phenotype

KLF11 is a zinc finger transcription factor member of the Sp1/KLF family of transcription factors that is composed of 17 different proteins ubiquitously expressed in human tissues (Lomberk and Urrutia, 2005). KLF11 may act as a transcriptional repressor or activator and has been shown to suppress exocrine pancreas cell growth by binding to GC-rich sequences in the promoter region of its target genes. KLF11contains three zinc finger domains that allow the interaction with its target DNA, and three transcriptional repressor domains through which it interacts with specific cofactors (Cook et al., 1999a, b; Zhang et al., 2001). The KLF11 transcriptional repressor activity is mediated by its association with the Sin3a-histone deacetylase complex, while its transcriptional activity is facilitated by its association with p300 (Fernandez-Zapico et al., 2009). In pancreatic β-cells KLF11 expression is upregulated by high glucose and TGF-β. Chromatin immunoprecipitation experiments (ChIP) experiments showed that this transcription factor is able to bind and activate the insulin promoter in β-cells under high glucose conditions (Bonnefond et al., 2011; Neve et al., 2005), and it was suggested that some *KLF11* mutations may lead to diabetes by affecting insulin production (Neve et al., 2005). In line with that, *Klf11*$^{-/-}$ mice have reduced islet mass and decreased insulin expression (Bonnefond et al., 2011). This mouse model, however, does not completely recapitulate the human phenotype since, despite their altered insulin homeostasis, they do not develop diabetes and show improved glucose tolerance respect to WT animals due to enhanced peripheral insulin sensitivity (Mathison et al., 2015). Fernandez-Zapico and co-workers showed that Klf11 also regulates the expression of *Pdx1* in pancreatic β-cells by interacting with GC-rich elements present in Area II of the *Pdx1* promoter (Fernandez-Zapico et al., 2009). They have also shown that certain *KLF11* mutations alter PDX1 expression. Since proper PDX1 expression is crucial for adequate pancreatic exocrine and endocrine cell development and maintenance of adult β-cell function, preserved KLF11 activity seems to be essential for both, β-cell function and development (Bonnefond et al., 2011; Fernandez-Zapico et al., 2009).

2.7 *PAX4* mutations (MODY9)

2.7.1 Genetic alteration and clinical phenotype

MODY9 is caused by autosomal dominant mutations in paired box4 (*PAX4*) gene, a transcription factor mainly expressed in the endocrine pancreas and required for β-cell differentiation. MODY9 is rare and accounts for less than 1% of all MODY cases (Firdous et al., 2018; Naylor et al., 2018). In MODY9 patients, diabetes appears in childhood or early adulthood (usually before 25 years of age) (Brun and Gauthier, 2008; Miyata et al., 1999; Plengvidhya et al., 2007) as a consequence of impaired insulin secretion. A link between polymorphisms on *PAX4* and a higher susceptibility to develop T2D has also been established (Kanatsuka et al., 2002; Mauvais-Jarvis et al., 2004; Shimajiri et al., 2001; Tokuyama et al., 2006).

2.7.2 Molecular phenotype

PAX4 belongs to a family of evolutionary conserved sequence-specific transcription factors involved in the regulation of β-cell plasticity in mature islets and in embryonic organogenesis (Blake et al., 2008; Blake and Ziman, 2014; Lang et al., 2007; Robson et al., 2006; Wang et al., 2008). The PAX family members are classified in different groups according to the organization of three conserved structural motifs: the DNA binding domain, a complete or truncated DNA interacting homodomain, and the octapeptide motif. PAX4 as well as PAX6 are in the group IV of the PAX family and have very similar binding sites. Indeed, PAX4 has also the capacity to bind PAX6 sites, although with less affinity, allowing competition for the same targets (Fujitani et al., 1999; Ritz-Laser et al., 2002; Smith et al., 1999). A coordinated regulation of gene expression is needed for normal pancreatic endocrine development. Accordingly, PAX4 expression is controlled by NGN3, HNF-1α, HNF-4α, PDX1 and NEUROD1, all of them key transcription factors controlling endocrine lineage differentiation, and whose deficiency has been associated with different MODY forms (Al-Shammari et al., 2011; Napolitano et al., 2015). The correlation between *PAX4* mutations and diabetes is clear, especially for mutations affecting its DNA binding domains, as the vast majority of them cause diabetes (Martin-Montalvo et al., 2017; Sujjitjoon et al., 2016). Indeed, one of the best described human mutations in *PAX4* is the R121W, that alters the DNA binding site leading to β-cell apoptosis (Brun et al., 2004; Hu He et al., 2011; Shimajiri et al., 2001). Pax4 can function as a transcriptional repressor and it is highly expressed in endocrine progenitors, but lowly expressed in adult human and rodent islets (Brun et al., 2004, 2008). In progenitor cells, Pax4 suppresses α-cell differentiation, by repressing the α-cell transcription factor Arx (Collombat et al., 2003), and promotes the differentiation of insulin producing β-cells and somatostatin-producing δ-cells (Greenwood et al., 2007; Lorenzo et al., 2015; Smith et al., 1999; Sosa-Pineda et al., 1997; Wang et al., 2004a). In fact, a co-regulatory loop between Pax4 and Arx has been described (Napolitano et al., 2015). Pax4 is one of the essential players in the production of insulin during embryonic development and also in adults, as it plays a crucial role in pancreatic plasticity (Brun and Gauthier, 2008; Martin-Montalvo et al., 2017; Napolitano et al., 2015; Sosa-Pineda, 2004). *Pax4* deletion in mice causes the complete or near complete absence of β- and δ-cells, together with an increase of α-cells, leading to hyperglycemia that is closely followed by new-born death (Collombat et al., 2005; Sosa-Pineda et al., 1997; Wang et al., 2004a). In the mirror experiment, enhanced expression of *Pax4* in early pancreatic epithelium or in endocrine precursors triggers β-cell differentiation (Collombat et al., 2009). Moreover, ectopic *PAX4* expression in vitro facilitates the generation of β-like cells and glucagon gene repression in α-cells (Petersen et al., 2000; Plengvidhya et al., 2007; Ritz-Laser et al., 2002) an effect that results from the competition between PAX4 and PAX6 for their common targets. In rodents, *Pax4* overexpression was shown to protect β-cells from apoptosis induced by streptozotocin through increased levels of antiapoptotic factors of the intrinsic apoptotic pathway (Brun et al., 2004; Hu He et al., 2011; Lu et al., 2007; Mellado-Gil et al., 2016; Rezende et al., 2007). In line with that, enhanced *Pax4* expression in the mouse insulinoma cell line MIN6, showed protection against apoptosis associated with increased BCL-XL expression (Lu et al., 2007), while *Pax4* silencing in INS-1E cells was associated with decreased BCL-XL levels, increased basal apoptosis and higher sensitivity to cytokine-induced cell death (Brun et al., 2007).

PAX4 also plays a role in proper endoplasmic reticulum (ER) function since it controls the expression of Calreticulin, a calcium binding protein crucial for ER homeostasis preservation. Moreover, it was shown that Pax4 overexpression protects isolated islets from ER stress and apoptosis induced by the synthetic ER stressor thapsigargin (Mellado-Gil et al., 2016).

2.8 *PTF1A* mutations (permanent neonatal diabetes and pancreatic agenesis)

2.8.1 Genetic alteration and clinical phenotype

Homozygous truncating mutations in the coding region of Pancreas Associated Transcription Factor 1A (*PTF1A*) gene cause permanent neonatal diabetes, pancreas agenesis, exocrine pancreas dysfunction and complete cerebellar agenesis (Al-Shammari et al., 2011; Sellick et al., 2004; Tutak et al., 2009). Only four cases with such mutations have been identified up to date. Homozygous missense (hypomorphic) mutations in the same gene cause neonatal diabetes mellitus, associated with pancreatic aplasia/hypoplasia but without neurological manifestations (Houghton et al., 2016). In addition, biallelic mutations in *PTF1A* enhancer regions have been associated with pancreas agenesis without extrapancreatic or neurological manifestations (Demirbilek et al., 2020; Weedon et al., 2014). Indeed, an initial study showed that recessive mutations in a 400 bp distal *PTF1A* enhancer located 25 kb downstream *PTF1A* were causal of isolated pancreas agenesis in 10 independent families (Weedon et al., 2014). A recent genetic and phenotypic follow-up study performed in a very large cohort of patients with *PTF1A* enhancer mutations (30 cases) identified one additional recessive mutation associated with isolated pancreatic agenesis, and provided a very detailed analysis of the clinical characteristics of the patients identifying recurrent extrapancreatic features that accompany the pancreatic manifestations, e.g., hyperferritinemia, anemia, transient mild to severe elevation in liver trans-aminases among others (Demirbilek et al., 2020).

2.8.2 Molecular phenotype

PTF1A is a helix–loop–helix (HLH) class II transcription factor and a component of the trimeric pancreas transcription factor-1 (PTF1) (Krapp et al., 1996). It has been demonstrated that the identified enhancer region located 25 kb downstream the PTF1A gene directly interacts with the *PTF1A* promoter in human pancreatic progenitor cells and plays a critical role in *PFT1A* transcription during pancreatic development (Weedon et al., 2014). In addition, another enhancer region located at 10.8 kb downstream the *Ptf1a* coding sequence, is required for tissue-specific expression of *Ptf1a* in the dorsal neural tube (Jin and Xiang, 2019; Mona et al., 2016).

PTF1A plays an essential role in controlling the development and function of different organs including pancreas, brain, spinal cord and retina, among others (Jin and Xiang, 2019). Indeed, PTF1A is expressed at early stages of the development in pancreatic ducts, exocrine and endocrine progenitors (Adell et al., 2000; Rose et al., 2001). It is especially important in the acquisition of pancreatic fate during organogenesis (Kawaguchi et al., 2002). PTF1A expression during the development is progressively circumscribed from the pancreatic buds (where it is broadly expressed) to the exocrine lineage, and finally to acinar cells (Kawaguchi et al., 2002; Krapp et al., 1996, 1998; Masui et al., 2008). Studies performed in $Ptf1a^{CreER/floxed}$ compound heterozygous mice have shown that this transcription factor is required to maintain acinar cell identity in adults, and that there is a correlation between *Ptf1a* dosage and pancreatic phenotype. Indeed, it was shown that reduced *Ptf1a* mRNA expression below a certain threshold results in reduced pancreatic bud size, impaired specification of the pancreatic precursors, altered exocrine pancreas formation and reduced β-cell to non-β-cell proportion (Fukuda et al., 2008; Houghton et al., 2016). Interestingly, it was demonstrated that partial loss of *Ptf1a* in mouse adult acinar cells leads to acinar cell proliferation and acinar-to-ductal metaplasia (Rodolosse et al., 2004). In addition, *Ptf1a* dosage experiments from the homozygous-to-null status done using conditional *Ptf1a* KO mice, showed pancreatic hypoplasia and glucose intolerance in a dosage-dependent manner (Fukuda et al., 2008). In addition, it was shown that *Ptf1a* null acinar cells have enhanced ER stress and accelerated apoptosis evidenced by increased PERK, eIF2α, ATF4 and ATF6 pathways, suppression of the IRE1 pathways and increased CHOP levels, leading to reduced pancreatic size (Sakikubo et al., 2018). The authors proposed that the shift in cellular identity (acinar to ductal-like) may be a compensatory response to promote cell survival. The mechanism involved, however, remains to be elucidated (Sakikubo et al., 2018). Interestingly, it was shown that transient *Hnf1a* ectopic expression in mouse embryonic stem cells launches pancreatic programs and induces pdx1 expression, thus potentiating endocrine differentiation (Nair et al., 2014; Jin and Xiang, 2019). Surprisingly, *PTF1A* overexpression in human or mouse fibroblasts induces the direct reprogramming of these somatic cells into neurospheres containing neural stem cells able to differentiate into neurons, astrocytes and oligodendrocytes (Jin and Xiang, 2019; Xiao et al., 2018).

2.9 *RFX6* mutations (neonatal diabetes mellitus and MODY)

2.9.1 Genetic alteration and clinical phenotype

Biallelic autosomal recessive truncating or missense mutations in the *RFX6* gene coding for the regulatory factor X-box binding 6 (RFX6) cause a syndromic form of neonatal diabetes mellitus previously described as Mitchell-Riley syndrome (Mitchell et al., 2004; Smith et al., 2010). These patients present with neonatal diabetes mellitus, intestinal atresia, with or without intestinal malrotation, hypoplastic or annular pancreas, and gallbladder hypoplasia or agenesis. The prognosis is poor and most of the patients die within the first months of life. Up to now, eight patients from seven independent families with this syndrome have been described (Chappell et al., 2008; Concepcion et al., 2014; Martinovici et al., 2010; Mitchell et al., 2004; Smith et al., 2010; Spiegel et al., 2011). In 2015, two additional individuals from a same family with compound heterozygous nonsense *RFX6* mutations were described. Different to the previously reported cases, these patients develop diabetes only during childhood but presented the previously reported intestinal phenotype. This differential phenotype was attributed to the expression of a partially truncated version of the wild-type RFX6 protein from one of the mutated alleles. This truncated protein was predicted to have some function given the full preservation of its DNA binding domain (Sansbury et al., 2015). In addition, heterozygous protein truncating *RFX6* variants have been associated with a new form of MODY with reduced penetrance. In these patients, diabetes was not accompanied by any of the intestinal manifestations present in patients with homozygous *RFX6* mutations and the mean age of diagnosis was 32 years (Patel et al., 2017). Interestingly, *RFX6*-MODY patients showed a significant reduction in glucose-dependent insulinotropic polypeptide (GIP) secretion compared to age-matched controls, while the GLP-1 levels were unchanged (Patel et al., 2017). GIP is a gut hormone is secreted by enteroendocrine K cells in response to food intake that enhances insulin biosynthesis and secretion in response to glucose, and exert antiapoptotic effects in pancreatic β-cells (Drucker, 2007). *RFX6*-diabetes is the first human form of diabetes associated with decreased GIP secretion (Patel et al., 2017). The *RFX6* heterozygous truncating mutations were initially identified in 2 unrelated patients and thereafter confirmed in 10 additional unrelated MODY patients in whom the common causes of MODY were excluded. Interestingly, this study unveiled a higher frequency of *RFX6* truncating mutations in Finish respect to non-Finish populations (Patel et al., 2017).

2.9.2 Molecular phenotype

RFX6 belongs to the regulatory factor X-box binding (RFX) family (Aftab et al., 2008), a DNA binding protein family that associates with the X-box DNA motif (Aftab et al., 2008). RFX6 is mainly expressed in pancreas, small intestine and colon, with lower expression in liver (Smith et al., 2010). Studies in mice have demonstrated that Rfx6 is needed for islet differentiation during embryonic pancreas development (Smith et al., 2010; Soyer et al., 2010), and that Rfx6 expression is maintained in adult β-cells. Rfx6 acts downstream of NGN3 and regulates the expression of transcription factors related with β-cell maturation and function. *Rfx6* null mice lack endocrine cells (including β-cells) with the exception of pancreatic polypeptide (PP) cells, responsible of pancreatic polypeptide production (Smith et al., 2010). In humans, RFX6 has been reported to be critical for islet and β-cell formation (Chappell et al., 2008; Concepcion et al., 2014; Martinovici et al., 2010; Mitchell et al., 2004; Smith et al., 2010; Spiegel et al., 2011). The role of *Rfx6* in adult β-cells has been studied via the use of conditional KO mice. These animals showed glucose intolerance and impaired insulin secretion respect to controls. The molecular analysis of *Rfx6*-KO islets showed reduced expression of keys components of the insulin secretion machinery namely, GCK, the K_{ATP} channel SUR1, and the voltage-gated Ca^{2+} channels. In addition, this analysis showed that *Rfx6* inactivation in adult β-cells leads to the re-expression of "disallowed" genes that are normally expressed in most mammalian tissues, but selectively and highly repressed in mouse mature β-cells, suggesting that Rfx6 is needed for β-cell identity preservation (Piccand et al., 2014). Interestingly, in good correlation with the data on RFX6-MODY patients, experiments performed with mouse intestinal cells showed that Rfx6 regulates GIP expression (Suzuki et al., 2013).

2.10 *GATA4* and *GATA6* mutations (neonatal diabetes mellitus and adult onset diabetes)

2.10.1 Genetic alteration and clinical phenotype

Heterozygous mutations in the *GATA4* gene, coding for the transcription factor GATA binding factor 4 (GATA 4) are a rare cause of neonatal diabetes mellitus and pancreatic agenesis. More than 100 different *GATA4* mutations with incomplete penetrance have been described. While most of them cause GATA4 haploinsufficiency leading to dysmorphic features, congenital heart malformations, and mental retardation, only six patients bearing these mutations developed early infancy diabetes with different degrees of exocrine insufficiency (D'Amato et al., 2010; Garg et al., 2003; Nemer et al., 2006; Shaw-Smith et al., 2014; Wat et al., 2009). The factors behind the variability in the phenotypic presentation associated with *GATA4* mutations are not completely understood.

On the other hand, heterozygous mutations in the *GATA6* gene, coding for the transcription GATA binding factor 6 (GATA 6) cause congenital heart malformations (Du et al., 2019; Kodo et al., 2009; Raghuram et al., 2020; Sanchez-Lechuga et al., 2020; Sun et al., 2019b; Yorifuji et al., 2012) and a wide spectrum of diabetic manifestations ranging from pancreatic agenesis to adult onset diabetes with absence of exocrine insufficiency (Allen et al., 2011; De Franco et al., 2013b; Du et al., 2019; Gaisl et al., 2019; Raghuram et al., 2020; Sanchez-Lechuga et al., 2020; Sun et al., 2019b; Yorifuji et al., 2012). Despite the fact that inactivating mutations in GATA6 are the most common cause pancreatic agenesis and hypoplasia (Chao et al., 2015; De Franco et al., 2013b; Lemelman et al., 2018; Sanchez-Lechuga et al., 2020; Yau et al., 2017), until now there is no clear genotype-phenotype correlation allowing to conciliate the broad spectrum of GATA6 mutations with the rainbow of clinical manifestations associated with them (Skoric-Milosavljevic et al., 2019).

2.10.2 Molecular phenotype

GATA4 and GATA6 are closely related members of the GATA transcription factor family that plays a crucial role on cellular differentiation and in the development of endodermal- and mesodermal-derived organs such as pancreas and heart. They contain 1 or 2 zinc finger motifs (Evans et al., 1988) and recognize a DNA consensus sequence known as GATA motif (WGATAR (W = A or T and R = A or G)) (Merika and Orkin, 1993), that is an important cis-element localized in the promoter region of multiple genes (Arceci et al., 1993).

In humans, GATA4 is expressed together with PDX1 in the developing pancreas suggesting that it may be important for pancreas development (Jennings et al., 2013). While *Gata4* heterozygous KO mice are normal, *Gata4* homozygous KO animals present defective extraembryonic tissue formation and premature death. In this last model, when the embryonic lethality was rescued, the mice developed pancreatic agenesis (Watt et al., 2007).

Gata6 deficiency in mouse adult β-cells leads to reduced insulin transcription and insulin secretion, increased number of immature insulin granules, and decreased Snap-25, t-Snare, Sur1 and Glut2 protein expression (Villamayor et al., 2018), together with ultrastructural mitochondrial and ER alterations (Malhotra and Kaufman, 2011). The inactivation of *Gata6* also leads to a diminished expression of transcription factors that are crucial for proper insulin secretion, such as *Pdx1*, *Nkx6.1* and *Mafa* (Iype et al., 2005; Kataoka et al., 2002; Matsuoka et al., 2004).

2.11 *GLIS3* mutations (neonatal diabetes mellitus)

2.11.1 Genetic alteration and clinical phenotype

Homozygous inactivating mutations in the GLI similar 3 (*GLIS3*) gene, coding for the transcription factor GLIS3, cause permanent neonatal diabetes mellitus, that may be accompanied by exocrine pancreas dysfunction and a large variety of non-pancreatic manifestations such as intrauterine growth retardation, congenital hypothyroidism, congenital glaucoma, polycystic kidneys, liver disease, skeletal deformity and bilateral sensorineural deafness (Alghamdi et al., 2017; Dimitri et al., 2011, 2015; Kim et al., 2003b; Senee et al., 2006; Taha et al., 2003). The presence and nature of these extra-pancreatic manifestations are dependent on the type and localization of the *GLIS3* mutation (Dimitri et al., 2011, 2015). In addition, certain variants in *GLIS3* were reported to be a risk factor for the development of both T1D (Barrett et al., 2009) and T2D (Cho et al., 2011; Liu et al., 2011; Rees et al., 2011). Interestingly in 2019, Sun and colleagues reported one rare *GLIS3* variant (p.I28V) found in healthy individuals and associated with reduced plasma glucose levels, suggesting that this variant may have a protective effect against the development of T2D (Sun et al., 2019a).

2.11.2 Molecular phenotype

GLIS3 is an 83.8 kDa transcription factor of a subfamily of Krüppel-like zinc finger proteins that acts both as a transcriptional activator and repressor in different cell types (Kim et al., 2003b). It is expressed in the developing and adult pancreas being particularly enriched in β-cells. GLIS3 is mostly localized in the nucleus in a punctuate pattern, thanks to a nuclear localization signal (NLS) present between Arg489 and Lys505 (Kim et al., 2003b). Both, the C-terminal and N-terminal region are required for its transcriptional activity. Deletions in the C-terminal domain lead to a partial or complete loss of GLIS3 functionality, while deletions in its N-terminal region enhance its transcriptional activity (Kim et al., 2003b). The *GLIS3* gene can be transcribed into two main transcripts: a longer one of 7.5 kb and a smaller of 0.8–2 kb which are differentially expressed in human tissues. The smaller is enriched in liver, kidney, heart and skeletal muscle, while the longer one is mostly located in pancreas, thyroid and kidney (Senee et al., 2006). It was reported that the promoter region of the human *INS* gene and the mouse insulin 2 gene (*Ins2*) have two GLIS-binding sites, and that GLIS3 is able to bind to both of them regulating *INS* and *Ins2* transcription (Kang et al., 2009). Individual mutations in either site partially reduce insulin transcription, the same as deletions in the C-terminal domain of GLIS3. In contrast, deletions in the N-terminal of this transcription factor enhances *Ins2* expression, pointing to the presence of a repressing domain in this area (Kang et al., 2009). Apart from its direct participation in insulin regulation, GLIS3 generates a scaffold that stabilizes the binding of PDX1, NEUROD1, and MAFA to their respective binding sites within the insulin promoter by interacting with p300/CBP (Zeruth et al., 2013). Indeed, it was demonstrated that in the absence of GLIS3 the insulin transcription by these transcription factors is importantly reduced (Zeruth et al., 2013). Yang and colleagues showed that GLIS3 also enhances MAFA expression and downregulates NKX6.1 and PAX6, all direct regulators of insulin expression (Mirmira et al., 2000; Sander et al., 1997; Zhao et al., 2005).

What is more, it was shown that mice lacking *Glis3* had an important decrease in the transcription factors Pdx1, Nkx6.1 and Ngn3, crucial for pancreatic endocrine cell differentiation and β-cell specification during the embryonic development (Nelson et al., 2007; Yang et al., 2011). The regulation of Ngn3 by Glis3 may be the consequence of direct binding to its promoter region (Yang et al., 2011) or by regulating the expression of Hnf6 and Foxa2 (two upstream regulators of Ngn3) (Jacquemin et al., 2000; Lee et al., 2001). Mice expressing an inactive form of Glis3 have hyperglycemia and hypoinsulinemia and die within days after birth (Kang et al., 2009). Despite the fact that the size of the pancreas was similar between WT and mutant mice, the latter showed disorganized islets of reduced size with diminished β-cell and δ-cells number (Kang et al., 2009). In addition, it was shown that *Glis3* deletion in adult mice causes diabetes some weeks after that the KO is induced. The islets of these KO mice were small and disorganized with a complete absence of β-cells and increased levels of apoptotic markers, indicating that *Glis3* is also necessary for β-cell survival (Yang et al., 2013). Based on that, it was proposed that GLIS3 plays an important role not only in pancreatic development during embryogenesis, but also in adults for the maintenance of the β-cell population and preservation of glucose homeostasis (Yang et al., 2013).

Glis3 silencing in the rat β-cell line INS-1E and in primary rat β-cells led to decreased Mafa, Pdx1, Glut2 and Ins2 expression which affected glucose metabolism and insulin secretion (Nogueira et al., 2013). In addition, *Glis3* deficiency in INS-1E cells increased apoptosis in basal condition and sensitized the cells to cytokine-, double stranded RNA- and free-fatty acid-induced apoptosis, a process mediated by activation of the intrinsic pathway of apoptosis, that was evidenced by increased cleaved caspase 9 and 3 levels and cytochrome *c* release into the cytosol (Nogueira et al., 2013). In human islets, *GLIS3* deficiency reduced insulin secretion by 25% compared to controls and increased basal and cytokines-induced apoptosis. In this context, it was shown that one of the mechanisms by which the lack of GLIS3 induces apoptosis is related with a preferential splicing of the BH3-only protein Bim toward its most pro-apoptotic variant, Bim small (BimS) (Nogueira et al., 2013). This phenomenon was associated with the downregulation of SRp55, a previously described splicing factor involved in Bim splicing (Leu et al., 2012). Amin and colleagues showed that *GLIS3*-deficient mouse and human embryonic stem cells fail to differentiate into β-cells and showed increase levels of apoptosis (Amin et al., 2018). The authors carried out a chemical screening to identify FDA-approved drug candidates that could rescue the increased cell death in GLIS3-deficient cells, and identified Galunisertib (LY2157299), a TGF-βR 1 kinase inhibitor that blocks the phosphorylation of SMAD2/3, as a potential drug candidate to prevent *GLIS3* deficiency-mediated β-cell apoptosis both in vivo and in vitro. These results suggest that that activation of the TGF-β pathway is one of the mechanisms contributing to apoptosis in *GLIS3*-deficient adult β-cells (Amin et al., 2018).

2.12 *MNX1* mutations (neonatal diabetes mellitus)

2.12.1 Genetic alteration and clinical phenotype

Homozygous inactivating mutations in the *MNX1* gene, coding for the transcription factor motor neuron and pancreas homeobox 1 (MNX1) cause permanent neonatal diabetes mellitus. Up to now, two unrelated patients carrying these mutations have been identified (Bonnefond et al., 2013a; Flanagan et al., 2014). While both patients developed permanent neonatal diabetes mellitus, one of them had no developmental pancreatic defects or other clinical features (Bonnefond et al., 2013a), while the other had a wide array of morphological and developmental issues including developmental delay, hypoplastic lungs, poorly developed renal cortex and medulla, reduction of myelin in the central nervous system (CNS) and short stature between others (Flanagan et al., 2014).

Interestingly, it has been shown that *MNX1* haploinsufficiency causes Currarino syndrome, characterized by sacral agenesis and the imperforate anus (Currarino et al., 1981; Ross et al., 1998). Notably, Currarino patients do not present with diabetes, suggesting that one functional copy of *MNX1* is sufficient for typical pancreatic development and peripheral glucose homeostasis (Flanagan et al., 2014).

2.12.2 Molecular phenotype

MNX1 is a transcription factor essential for pancreas and other tissues development (Tanabe et al., 1998). In mice, *Mnx1* is expressed in multiple tissue sites, including motor neurons, the notochord, the dorsal endoderm, and the pre-pancreatic endoderm, and global *Mnx1* inactivation leads to agenesis of the dorsal pancreatic bud, although the ventral bud develops normally (Harrison et al., 1999; Li et al., 1999). To assess the mechanism of neonatal diabetes mellitus associated with homozygous *MNX1* mutations, Pan et al completed a systematic and longitudinal analysis of Mnx1 function during murine pancreatic organogenesis (Pan et al., 2015). In their studies Mnx1 protein was detected in nearly all cells from the ventral and dorsal pancreatic buds at E10.5 while excluded completely from cells of the duodenum (Pan et al., 2015). At E12.5, Mnx1 was still expressed, contradicting previous reports (Harrison et al., 1999; Li et al., 1999) however, this expression was restricted to the pancreatic epithelium tip domains. In this area, Mnx1 was found to co-localize with Ptf1A and Cpa1, two proteins involved in pancreatic organogenesis (Kawaguchi et al., 2002). At E14.5, Mnx1 was only found in insulin-positive β-cells, and showed no co-expression with Ngn3, glucagon, somatostatin, or pancreatic polypeptide, indicating a lack of expression in Ngn3-positive endocrine progenitors or α-, δ-, or pancreatic polypeptide cells. This trend was not only found during embryogenesis but also in adult mice β-cells (Harrison et al., 1999; Li et al., 1999; Pan et al., 2015). To assess whether MNX1 expression in humans follows the same profile, Pan and co-workers examined MNX1 expression in human fetal pancreatic sections and juvenile pancreatic tissue. These experiments showed that MNX1 is expressed as early as gestational week 7, which is equivalent to mouse E10.5 (Pan et al., 2015; Pan and Wright, 2011). At gestational week 18, MNX1 was found in insulin-positive β-cells, supporting a major role of MNX1 during pancreatic organogenesis. Additionally, MNX1 was found in the pancreas of a 5-year-old donor, suggesting that MNX1 expression persists in the post-embryonic pancreas, analogous to the mice findings (Pan et al., 2015). In addition, to assess the role of MNX1 during endocrine differentiation, Pan and colleagues generated a mouse line where *Mnx1* expression was ablated in all endocrine cells at E9.5 (Pan et al., 2015). The *Mnx1* mutant pups developed extreme hyperglycemia by 2 days of age, and all died maximum 10 days after birth. This trend suggested that either the β-cell function or the β-cell population was impacted by the *Mnx1* deficiency. Immunofluorescence of mutant pancreas sections revealed a significant reduction in insulin-positive β-cells with a simultaneous increase in somatostatin-positive δ-cells suggesting that Mnx1 promotes β-cell fate while suppressing δ-cell fate during pancreatic organogenesis (Pan et al., 2015). While MNX1 plays an essential role in β-cell lineage commitment, the expression of the protein following differentiation from endocrine precursors into β-cells suggests a continued role of MNX1 in β-cell differentiation and function. Studies in β-cell specific *Mnx1* KO mice revealed that this transcription factor serves as a lineage allocation factor during pancreatic organogenesis, serving to shift Pax6-positive endocrine precursor cells to a β-cell fate as opposed to an α- or δ-cell fate (Pan et al., 2015).

2.13 *PAX6* mutations (young onset diabetes and neonatal diabetes mellitus)

2.13.1 Genetic alteration and clinical phenotype

Heterozygous inactivating mutations in the *PAX6* gene, coding for the transcription factor Paired box protein 6 (PAX6) cause glucose intolerance (in most cases), or young onset diabetes (identified in two unrelated patients (Nishi et al., 2005; Yasuda et al., 2002)). The metabolic alterations are usually accompanied by a large variety of eye malformations, such as Peters anomaly, anophthalmia (the absence of one or the two eyes), microphthalmia (eyes abnormally small), aniridia (absence of the iris), congenital cataracts and retinal colobomas (holes in the retina) (MacDonald et al., 2019; Yasuda et al., 2002). In addition, a 484–630 kb deletion distal to PAX6 demonstrated dominant co-segregation with diabetes and aniridia (MacDonald et al., 2019), and recessive *PAX6* mutations were shown to cause neonatal diabetes mellitus in one individual (Rubio-Cabezas and Ellard, 2013; Solomon et al., 2009). *PAX6* diabetes is characterized by progressive hyperglycemia as a consequence of abnormal proinsulin processing, impaired glucose-stimulated insulin secretion, and reduced β-cell mass (Boese et al., 2020; Gosmain et al., 2012a,b; MacDonald et al., 2019; Motoda et al., 2019; Wen et al., 2009; Yasuda et al., 2002).

2.13.2 Molecular phenotype

PAX6, also known as aniridia type II (AN2) protein, is a transcription factor essential for the normal development and maturation of the central nervous system (eyes and brain) and pancreas (St-Onge et al., 1997; Turque et al., 1994). It promotes neurogenesis during early embryo development, and plays an important role in the development, maturation and function of β-cells and α-cells (St-Onge et al., 1997; Walther and Gruss, 1991). PAX6 regulates the expression of several genes that directly or indirectly participate in insulin biosynthesis, insulin secretion and β-cell metabolism. Accordingly, it has been shown that in rodents, Pax6 directly regulates the expression of *Insulin 1* and *2*, prohormone convertase 1/3 (*Pc1/3*), *Pc2*, *Pdx1*, *cMaf*, *Mafa*, *Glut2*, *Nkx6.1*, *Glp1R*, *Gck* and *Gip-R* in β-cells by binding to specific regions in their promoters (Gosmain et al., 2012b). *Pax6*-silencing in primary rat β-cells results in decreased expression of the aforementioned genes, together with reduced insulin levels, reduced glucose-induced insulin secretion and impaired β-cell response to GLP-1 (Gosmain et al., 2012b). In addition, it was shown that *Pax6*-mutant mice are glucose intolerant, have reduced insulin secretion and show elevated proinsulin/insulin levels as a consequence of reduced expression of Pc1/3 (responsible of proinsulin processing) (Wen et al., 2009). Mitchell and co-workers studied the importance of Pax6 in adult β-cells (Mitchell et al., 2017). For that, they generated a transgenic mouse where *Pax6* expression was selectively ablated in pancreatic β-cells by crossing $\text{Pax6}^{\text{flox/flox}}$ mice with $\text{Pdx1CreERT}^{+/-}$ mice. These experiments demonstrated that, after tamoxifen-mediated Cre induction, the *Pax6*-depleted mice show severe hyperglycemia, that is accompanied by reduced insulin expression and secretion (Mitchell et al., 2017). In addition, an important decrease in genes related with β-cell identity such as *Ins2*, *Slc30a8*, *Mafa*, *Slc2a2*, *Nkx6.1*, *Pdx1* and *Glp1-R*, and increased expression of the disallowed gene *Slc16a1*, encoding monocarboxylate transporter 1 (Mct-1) was observed. Moreover, experiments performed with intact islets of β-cell specific *Pax6* KO mice and controls, showed reduced glucose-induced

ATP/ADP ratio in the mutant cells. This was accompanied by attenuated increases in cytosolic $Ca2^+$ levels, which explained the impaired insulin secretion (Mitchell et al., 2017). Collectively, the results derived from these studies indicate that Pax6 is necessary for proper β-cell development, preserved β-cell identity and function, and proper incretin response.

2.14 *NEUROG3* mutations (neonatal diabetes mellitus, and insulin-dependent diabetes)

2.14.1 Genetic alteration and clinical phenotype

Biallelic hypomorphic and inactivating mutations in the *NEUROG3* gene, also called *ATOH5* or *NGN3*, that codes for the transcription factor Neurogenin 3 (NGN3), cause a rare form of permanent neonatal diabetes mellitus that was identified in two unrelated patients (Pinney et al., 2011; Rubio-Cabezas et al., 2011). In addition, functionally null homozygous *NGN3* mutations, identified in two patients from two independent families of Mexican-American origin, were shown to cause insulin-dependent diabetes that develops in middle childhood (Solorzano-Vargas et al., 2020). Besides the endocrine manifestations the four patients showed also congenital malabsorptive diarrhea as a result of the complete absence of enteroendocrine cells (Pinney et al., 2011; Rubio-Cabezas et al., 2011; Solorzano-Vargas et al., 2020).

2.14.2 Molecular phenotype

NGN3 is a helix-loop-helix transcription factor key for both the development of CNS, and pancreas in humans. Regarding the latter, it participates in the specification of a common pancreatic precursor that gives rise to the four endocrine cell lineages (McGrath et al., 2015). NGN3 directly regulates the expression of a variety of genes important for endocrine cell development such as *RFX6, PAX4, PAX6, MYT* and *NEUROD1* (McGrath et al., 2015). Despite *Ngn3*-deficient mice and pigs show a complete absence of endocrine hormones at birth (Gradwohl et al., 2000; Sheets et al., 2018), and CRISPR/Cas9-mediated complete *NGN3* ablation in human stem cells (hES) results in the absence of pancreatic endocrine cell specification when they are differentiated into pancreatic precursors, individuals bearing homozygous or biallelic *NGN3* mutations are all born with a functional pancreas and a certain degree of endocrine cells, evidenced by detectable circulating C-peptide levels (Jensen et al., 2007; Ohsie et al., 2009; Pinney et al., 2011; Rubio-Cabezas et al., 2011, 2014a; Sayar et al., 2013). This suggests that, since the majority of the reported human mutations are hypomorphic rather than null, residual *NGN3* levels are sufficient for endocrine cell formation (McGrath et al., 2015; Solorzano-Vargas et al., 2020). Accordingly, experiments done in *NGN3*-silenced hESs showed that as little as 11% of normal *NGN3* levels are sufficient for a relatively normal formation of insulin-expressing cells in vitro (McGrath et al., 2015). The recent report of Solorzano-Vargas and colleagues describing two patients with functionally null mutations in *NGN3* that develop diabetes in middle childhood instead of neonatally, imply that, unlike in mice, pancreatic endocrine cell generation in humans may not be entirely dependent on *NGN3* expression (Solorzano-Vargas et al., 2020).

2.15 *NKX2.2* mutations (neonatal diabetes mellitus)

2.15.1 Genetic alteration and clinical phenotype

Homozygous nonsense or frameshift mutations in the *NKX2.2* gene coding for the transcription factor NKX2.2 cause permanent neonatal diabetes mellitus (Flanagan et al., 2014). Up to date, three patients from two independent consanguineous families have been reported. The three patients had very important insulin secretion defects but no alterations in the exocrine pancreas function (Flanagan et al., 2014), and two of them also had several extrapancreatic manifestations including moderate to severe developmental delay affecting motor and intellectual function, hypotonia, bilateral hearing impairment, cortical blindness and short stature among others (Flanagan et al., 2014).

2.15.2 Molecular phenotype

NKX2.2 is a transcription factor member of the family of mammalian NK2 homeobox transcription factors. It is expressed in the CNS and the pancreas and it plays an important role in endocrine pancreas specification (Anderson et al., 2009a,b; Papizan et al., 2011). It has been proposed that during this last process it acts predominantly as a transcriptional repressor through its interaction with the co-repressor protein Groucho-3 (GRg3) (Doyle et al., 2007). Indeed, it was shown that the interaction between Nkx2.2 and GRg3 is not needed for the early commitment of β- versus α-cells, but it is necessary for the adequate balance between β- and ε-cells. This interaction is also required for the maintenance of the β-cell identity. The lack of Nkx2.2 abolishes this interaction negatively affecting β-cell specification (Papizan et al., 2011). Experiments performed in *Nkx2.2* KO mice showed the relevance of this transcription factor for final β-, α- and PP cell differentiation (Papizan et al., 2011). Indeed, it was shown that the *Nkx2.2*

KO mice had a complete absence of pancreatic β-cells, a significant reduction in α-cell number, and an elevated proportion of endocrine cells having some β-cell markers such as Pdx1 and islet amyloid polypeptide, but lacking mature β-cell markers such as Nkx6.1 and Glut2. These mice showed overt hyperglycemia and died very soon after birth (Papizan et al., 2011; Prado et al., 2004). Postnatal depletion of *Nkx2.2* in mouse pancreatic β-cells caused rapid onset diabetes associated with reduced insulin levels, down-regulation of β-cell functional genes, and transdifferentiation of pancreatic β-cells into bihormonal cells which resulted in the loss of β-cell identity (Gutierrez et al., 2017). Likewise, loss of *Nkx2.2* in human islets causes dysregulation of key genes for β-cell development and glucose homeostasis (Churchill et al., 2017).

2.16 *CNOT1* mutations (neonatal diabetes mellitus)

2.16.1 Genetic alteration and clinical phenotype

It was recently shown that a spontaneous heterozygous c.1603C > T; p.Arg535Cys missense mutation in the *CNOT1* gene, coding for the CNOT1 protein is associated with pancreatic agenesis and holoprosencephaly, a disorder in which the forebrain of the embryo fails to develop into two hemispheres (De Franco et al., 2019; Hilbrands et al., 2017; Kruszka et al., 2019). Four unrelated patients bearing the identified mutation and presenting these clinicals manifestations have been identified (De Franco et al., 2019; Hilbrands et al., 2017; Kruszka et al., 2019). The four probands developed diabetes very early requiring insulin and pancreatic enzymes replacement therapy within the first 6 weeks of life, and presented also with holoprosencephaly and other extrapancreatic manifestations (De Franco et al., 2019; Hilbrands et al., 2017; Kruszka et al., 2019). In three of them, very low birth weight was reported consistent with in utero insulin deficiency in the last trimester of pregnancy (De Franco et al., 2019).

It is important to highlight that not all the heterozygous *CNOT1* mutations cause pancreatic agenesis and brain structural malformations, suggesting that these phenotypes are mutation-specific (Firth et al., 2011). Indeed, a very recent genetic study performed on 37 independent families identified a large number of the novo *CNOT1* mutations (Vissers et al., 2020). All these patients showed a broad phenotypic spectrum of neurological manifestations but no diabetes. While some individuals showed important intellectual disability, development delay, motor and speech delay and seizures; others showed borderline normal IQ and no other major clinical manifestations (Vissers et al., 2020). All types of genetic variations were found in these patients ranging from missense to nonsense mutations, splice variants, and partial gene deletions. No particular genotype-phenotype correlation based on the type of variant (truncating versus missense) or the location of the missense variants could be established. The essentially preserved pancreatic phenotype of the patients carrying the c.1603C > T; p.Arg535Cys suggested that mutations affecting the same protein domain could have similar outcomes. This, however, was not confirmed since individuals with mutations in the same domain did not develop holoprosencephaly or diabetes (Vissers et al., 2020) suggesting that the clinical phenotype of the patients could be associated directly with the mutation itself.

2.16.2 Molecular phenotype

CNOT1 encodes for CNOT1 protein that is the central scaffolding subunit of the CCR4-NOT complex, that regulates gene expression and mRNA stability and turnover by controlling RNA deadenylation and protein ubiquitination (Elmen et al., 2020). Indeed, the CCR4-NOT complex is one of the main cellular mRNA deadenylases that shorten the polyA tail in mRNA leading to mRNA degradation (Bartlam and Yamamoto, 2010). It has also been proposed that CNOT1 acts as an independent factor that mediates transcriptional repression and is expressed very early during pancreas development (Winkler et al., 2006). In vitro studies have shown that CNOT1 plays an essential role in maintaining pluripotency in human and mice stem cells by inhibiting primitive endoderm factors such as GATA 4 and 6 (Zheng et al., 2012). In human iPSCs the highest CNOT1 expression is detected in undifferentiated cells and it is reduced during the differentiation of iPSC toward the pancreatic endocrine lineage (Van De Bunt et al., 2016). *Cnot1*-deficient mouse embryos showed increased Shh expression which inhibits endocrine cell differentiation by reducing the expression of Pdx1, Ins, Hnf1β, and Ptf1α. Indeed, it has been shown that for a successful differentiation toward dorsal pancreas, Shh expression needs to be repressed in the dorsal foregut endoderm (Apelqvist et al., 1997; Jennings et al., 2013). During normal pancreas development, pro-differentiating factors repress CNOT1 activity allowing endocrine cell differentiation. It has been proposed that the presence of the p.Arg535Cys mutation, results in a CNOT1 version with preserved inhibitory activity on the primitive endoderm factors resulting in impaired pancreas development (De Franco et al., 2019). Studies done in mice have shown that homozygous *Cnot1* mutations are lethal and that embryos show severe morphological abnormalities such as exencephaly, defects on their eyes, edema and a severe reduction on pancreas size, highlighting the role of CNOT1 in pancreatic development (De Franco et al., 2019).

2.17 *ZAC1* overexpression (transient neonatal diabetes)

2.17.1 Genetic alteration and clinical phenotype

About 2/3 of all cases of transient neonatal diabetes mellitus (TNDM) are the consequence of overexpression of the *ZAC1* (also known PLAGL1) and *HYMA* genes located in the 6q24 locus (Gardner et al., 2000; Hoffmann and Spengler, 2015; MacKay and Temple, 2010; Temple et al., 2000). The expression of these two genes is normally restricted to the paternal allele as a result of maternal DNA methylation of the promoter region, indicating that this form of TNDM is a rare imprinting disorder. Enhanced *ZAC1* and *HYMA* expression may be caused by, e.g., paternal uniparental disomy of chromosome 6, de novo or *inherited* paternal duplication of 6q24, or loss of maternal DNA methylation at the ZAC1 locus (MacKay et al., 2006; MacKay and Temple, 2010). ZAC1-mediated TNDM patients present with uncontrollable hyperglycemia in the first 6 month of life, growth retardation, undetectable or very low insulin levels, and severe dehydration. In addition, it was shown that about 50% of the patients that have low methylation at the 6q24 locus may also display hypomethylation at additional imprinted loci which causes multilocus imprinting disturbances (MLID). In patients having TNDM-MLID diabetes is accompanied by macroglossia and other extrapancreatic symptoms of varying penetrance, e.g., cardiac anomalies, umbilical hernia, facial dysmorphisms, and developmental delay (MacKay and Temple, 2010; Touati et al., 2019). Interestingly, the majority of patients with TNDM-MLID have inactivating mutations in the *ZPF57* gene present in the 6p22.3 locus and coding for the transcription factor of the same name that is involved in the correct maintenance of methylation of imprinted loci. Around 14 different *ZPF57* inactivating mutations have been described so far (Touati et al., 2019).

2.17.2 Molecular phenotype

The exact biological role of ZAC1 is not well understood. It has been classified as a zinc finger containing transcription factor with tumor suppressor activity due to its capacity to regulate cell cycle and apoptosis. *ZAC1* is expressed in diverse types of embryonic and adult human tissues (Abdollahi, 2007). Experiments in mice showed that Zac1 controls the expression of target genes during embryonic development and postpartum contributing to the regulation of developmental pathways. In addition, it was shown that it induces cell cycle arrest and apoptosis through the regulation of the p53 signaling (Huang and Stallcup, 2000; Huang et al., 2001). In good correlation with the patient phenotype, mouse models of *Zac1* overexpression develop neonate hyperglycemia with juvenile remission and adult glucose intolerance. Moreover, Zac1 overexpression in rodent cell lines reduces GSIS (Du et al., 2012; Ma et al., 2004). Histological analysis of Zac1-overexpressing mouse pancreas revealed a reduced number of all endocrine cells including β-cells (Hoffmann and Spengler, 2015). Indeed, it was shown that *Zac1* overexpression leads to reduced β-cell mass as a consequence of cell cycle arrest or apoptosis, mainly in neonatal stages, with the subsequent reduction in insulin production and secretion. In the mirror situation, reduced *ZAC1* expression has been associated with the incidence and progression of several human tumors, including cervical cancer, breast cancer, ovarian cancer, pituitary tumors, and basal cell carcinoma (Abdollahi, 2007; Su et al., 2020).

3. Monogenic forms of diabetes associated with impaired glucose metabolism and insulin secretion

Preserved glucose sensing, transport and metabolism, together with a functional insulin secretion machinery are crucial to maintain normoglycemia (Fig. 1). Indeed, insulin secretion depends on proper insulin sensing and metabolism and on the preserved activity of the ATP-sensitive potassium (K_{ATP}) channel and the voltage-gated Ca^{2+} channels involved in insulin secretion. The K_{ATP} channel is a hetero-octameric channel located in the plasma membrane of the pancreatic β-cells. It is constituted by four inward-rectifier potassium ion subunits (Kir6.2) forming the pore, and four outer SUR1 subunits (McTaggart et al., 2010). Closure of the K_{ATP} channel is required for insulin secretion, while its opening, inhibits this process. At low (sub-stimulatory) glucose concentrations, the K_{ATP} channels stay open allowing the K^+ efflux from the cell. This leads to the hyperpolarization of the β-cell plasma membrane to a potential of around −70 mV which keeps voltage-gated Ca^{2+} channels closed. When the glycemia increases, the glucose uptake and metabolism enhance the intracellular ATP/ADP ratio which induces the closure of the K_{ATP} channel. This results in the depolarization of the plasma membrane (to around −30 mV) which opens the voltage-gated Ca^{2+} channels inducing calcium entry and the concomitant insulin release (Fig. 1) (Beltrand et al., 2020; Demirbilek and Hussain, 2017). In this section we will discuss on monogenic forms of diabetes associated with impaired glucose sensing, metabolism and transport, reduced insulin secretion and proinsulin processing.

3.1 *GCK* mutations (MODY2 and neonatal diabetes mellitus)

3.1.1 Genetic alteration and clinical phenotype

Heterozygous inactivating mutations in the *GCK* gene coding for the enzyme glucokinase (GCK) cause MODY2, while homozygous loss-of-function mutations on the same gene cause permanent neonatal diabetes mellitus (Bennett et al., 2011; Froguel et al., 1992; Hattersley et al., 1992, 1993; Njolstad et al., 2003; Osbak et al., 2009; Rubio-Cabezas et al., 2008; Vionnet et al., 1992). MODY2 is one of the most common reported forms of MODY which accounts for 15–50% of all MODY cases (Firdous et al., 2018; Naylor et al., 2018). More than 600 *GCK* mutations have been described in more than 1400 families (Osbak et al., 2009).

MODY2 patients have mild fasting hyperglycemia that, different from other MODYs, may appear at birth but many patients are generally asymptomatic and hyperglycemia is often discovered in routine medical checks. Indeed, MODY2 patients have relatively preserved insulin secretion and moderate glucose excursions after meals; their β-cell function only slightly deteriorates with increasing age. Diabetes-related complications are very uncommon (Naylor et al., 2018; Steele et al., 2014). Due to the mild phenotypic presentation, MODY2 patients are often misdiagnosed with early T1D (if they are children) or T2D when they are adults, and sometimes are mistreated with insulin. In general, MODY2 patients do not require a particular treatment besides some exercise and diet. However, women with MODY2 often develop diabetes during pregnancy and in this case, exogenous insulin administration is required to avoid the deleterious effect of the mother's hyperglycemia on the fetus (Hattersley et al., 1992, 1993).

It was also shown that heterozygous activating *GCK* mutations may cause a mild form of familial hyperinsulinemic hypoglycemia or persistent hyperinsulinemic hypoglycemia. Up to date, 20 cases of GCK heterozygous activating mutations have been reported (Beer et al., 2011; Christesen et al., 2008; Cuesta-Munoz et al., 2004; Glaser et al., 1998; Gloyn et al., 2003; Jannin et al., 2018; Martinez et al., 2017; Sayed et al., 2009; Wabitsch et al., 2007; Yamagata et al., 1996). 19 of the 20 mutations identified were missense, and the 1 remaining was an insertion mutation. One third of the mutations identified were the novo, while the rest were autosomal dominant. The age of onset varied between birth until 44 years, but most patients developed hypoglycemia before 10 years old (Ping et al., 2019). The severity of symptoms associated with the hypoglycemia ranged from asymptomatic to epilepsy, coma and death (Cuesta-Munoz et al., 2004).

3.1.2 Molecular phenotype

GCK catalyzes the first reaction of the glycolytic pathway and acts as the primary glucose sensor in the body. By using different mouse models of *Gck* deficiency or overexpression in pancreatic β-cells and liver, it was demonstrated that *Gck* expression in both β-cells and hepatocytes independently contributes to the maintenance of blood glucose homeostasis (Desai et al., 2001; Ferre et al., 1996; Grupe et al., 1995; Hariharan et al., 1997; Matschinsky, 1990; Pino et al., 2007; Terauchi et al., 1995). The regulation of GCK expression in β-cells and liver is different (Matschinsky and Wilson, 2019). The aminoacid changes derived from the inactivating or activating mutations identified affect either the function, the glucose affinity or the folding of the enzyme (Capuano et al., 2012; Negahdar et al., 2012). As mentioned above, in pancreatic β-cells GCK works as a glucose sensor by controlling the rate of entry of glucose into the glycolytic pathway and its catabolism (Matschinsky and Wilson, 2019). Indeed, GCK phosphorylates glucose to glucose-6 phosphate, which is then further metabolized with the concomitant ATP production. Since ATP is required for insulin secretion, preserved GCK activity is crucial for maintaining the normal insulin secretory capacity of pancreatic β-cells. The glucose affinity of GCK is the main determinant of the glucose threshold for GSIS. This threshold is tightly maintained in human β-cells to close to 5 mmoL/L (Matschinsky, 2002). In patients with heterozygous inactivating mutations in GCK (MODY2) the glucose threshold for GSIS may be only moderately increased (to around 7 mmoL/L) while in the case homozygous inactivating mutations in the gene, causing permanent neonatal diabetes mellitus, this threshold is infinitely increased preventing insulin secretion (Matschinsky, 2002; Matschinsky and Wilson, 2019). *GCK* mutations leading hyperinsulinism increase the maximal GCK activity or its glucose affinity. The severity of the clinical manifestations associated with these mutations are directly correlated with the increase in these two parameters. Different from mouse models in which homozygous activating *Gck* mutations are tolerated (Pino et al., 2007), in humans, all cases described up to now of *GCK*-associated hyperinsulinism involve heterozygous mutations suggesting that activation of both alleles is lethal for the human embryos (Matschinsky and Wilson, 2019).

3.2 BLK mutations (MODY 11)

3.2.1 Genetic alteration and clinical phenotype

It has been suggested that heterozygous loss-of-function mutations in the *BLK* gene (located in 8p23) and coding for B-lymphocyte kinase (BLK) cause MODY11. MODY11 is a very rare form of MODY which accounts for less than 1% of all MODY cases (Firdous et al., 2018). Kim et al. were the first to identify a new MODY locus on 8p23 that was related with diabetes in a good proportion of MODY cases unlinked to known MODY genes (Kim et al., 2004). Later on, Borowiec and colleagues identified five different *BLK* mutations (one coding (p.Ala71Thr, c.211G > A) and four non-coding) in 21 diabetic individuals from three unrelated families (Borowiec et al., 2009). The phenotype of the affected subjects was similar to the one of MODY6 patients characterized by overweight and a relative, rather than an absolute, impaired insulin secretion (Borowiec et al., 2009). Despite this initial association, an additional study aiming to reassess the potential role of BLK-pA71T loss-of-function mutation in MODY11 and T2D failed to find such association. Indeed, in this study the mutation of interest was found in 52 normoglycemic individuals but was absent in the 64 unelucidated MODY samples analyzed. Only a nominal association between this variant and increased T2D risk was found with an enrichment of the mutation in obese patients (Bonnefond et al., 2013b).

Moreover, the genome databases show that the frequency of BLK p. Ala71Thr (c.211G > A) mutation is very low suggesting that this *BLK* variant may act as benign or a pathological factor depending on if it is combined with other diabetic risk factors (Du et al., 2020).

Recently, the *BLK* Ala71Thr mutation was detected in a diabetic patient together with a hemizygous 1.58 Mb deletion in Chr17q12. The affected individual had insulinopenic diabetes and bilateral renal cists without additional severe phenotypes. The contribution of the BLK mutation to diabetes development in this individual could not be assessed (Du et al., 2020). Altogether, these data indicate that the contribution of *BKL* variants to the development of MODY still requires validation.

3.2.2 Molecular phenotype

BLK is a non-receptor tyrosine kinase of the src family of proto-oncogenes. Besides being expressed in B-lymphocytes (Dymecki et al., 1990), it has been demonstrated that it is expressed in human islets were it exclusively co-localizes with insulin, suggesting that it is only expressed in pancreatic β-cells (Borowiec et al., 2009). Whole body *Blk* KO mice have been generated and studied in the context of B-lymphocyte physiology (23). During the time-frame of the study (8 weeks) no disturbances in glucose homeostasis were reported in these animals. Nevertheless, their metabolic response after exposure to a diabetogenic environment, e.g., a high-fat diet, has not been studied (Texido et al., 2000). Conversely, experiments of *Blk* silencing or overexpression in the mouse β-cell line MIN6 cells showed that it is an essential regulator of β-cell function, acting as an inducer of insulin synthesis and secretion after glucose stimulation (Borowiec et al., 2009). The actions described previously appeared to be mediated by an upregulation of Pdx1 and Nkx6.1, both key regulators of β-cell function and MODY genes (Dutta et al., 1998; Edlund, 2001; Pedersen et al., 2005; Schisler et al., 2008; Stoffers et al., 1997a).

3.3 ABCC8 mutations (MODY12 and neonatal diabetes mellitus)

3.3.1 Genetic alteration and clinical phenotype

Dominant missense gain-of-function mutations in the *ABCC8* gene, coding for the sulfonylurea receptor 1 (SUR1) may cause MODY12 (a rare form of MODY) that accounts for less than 1% of all MODY cases (Firdous et al., 2018) and transient or permanent neonatal diabetes mellitus (Babenko et al., 2006; Cattoni et al., 2019; Fantes et al., 1995; Gloyn et al., 2006; Henwood et al., 2005; Johnson et al., 2018; Laukkanen et al., 2004; Meirhaeghe et al., 2001; Mohan et al., 2018; Proks et al., 2006; Reis et al., 2000; Shima et al., 2018; Thornton et al., 2003; Verkarre et al., 1998; Voevoda et al., 2016). Recently, recessive *ABCC8* gain-of-function mutations were also reported as causal of neonatal diabetes in two unrelated patients expanding the mode of inheritance of this pathology (Flanagan et al., 2017). The metabolic phenotype associated with the *ABCC8* mutations is largely variable ranging from glucose intolerance to overt diabetes that may manifest from birth until adulthood. This may be explained by the location and type of the *ABCC8* mutation or by the influence of other genetic or environmental factors (Reilly et al., 2020). Interestingly, it has been demonstrated that the mutated K_{ATP} channels remain sensitive to sulfonylureas in 90% of cases (Beltrand et al., 2020). This indicates that sulfonylureas can be used to control glycemia in patients with MODY 12 or neonatal diabetes caused by both ABCC8 and KCNJ11 mutations (see below). Indeed, it was shown that, similar to MODY1 and MODY3 patients, MODY12 individuals or patients with transient neonatal diabetes associated with *ABCC8* mutations may be treated with low doses of sulfonylureas rather than insulin (Reilly et al., 2020). However, patients with permanent neonatal diabetes caused by *ABCC8* mutations require high doses of sulfonylureas, insulin or a combination of both (Misra et al., 2018). In these individuals the sulfonylureas therapy is safe and showed to be successful in most patients, inducing insulin secretion in response to a meal. Garcin and co-workers showed that the success of the treatment depends on the pharmacogenetics, pharmacokinetics, treatment compliance, maximal dose used, and patient's age (Beltrand et al., 2020; Garcin et al., 2020).

In addition to diabetes associated *ABCC8* mutations, it has been shown that recessive and autosomal dominant *inactivating* mutations in the same gene are causal of medically responsive and unresponsive congenital hyperinsulinism (Arya et al., 2014; De Lonlay et al., 1997; Giurgea et al., 2006; Glaser et al., 1994, 1999; Goksel et al., 1998; Gutgold et al., 2017; Lord et al., 2013; MacMullen et al., 2011; Nestorowicz et al., 1996; Otonkoski et al., 1999; Pinney et al., 2008; Ryan et al., 1998; Thomas et al., 1995; Thornton et al., 1998; Tornovsky et al., 2004).

3.3.2 Molecular phenotype

As mentioned above, SUR1 is a subunit of the K_{ATP} channel involved in insulin secretion. The closure of the K_{ATP} channel is required for insulin release while its opening inhibits this process.

In line with that, the diabetes-causing gain-of-function mutations in *ABCC8* prevent the closure of the K_{ATP} channel which causes cell hyperpolarization (even under high glucose conditions). This keeps the voltage-gated Ca^{2+} channels closed preventing Ca^{2+} entry into the β-cell and insulin secretion (Aittoniemi et al., 2009). The SUR1 subunits are the binding site of sulfonylureas (that induce insulin secretion by having an inhibitory effect on the K_{ATP} channel) and diazoxide (that abolishes insulin release by opening the K_{ATP} channel) (Ashcroft, 2005; Ashcroft and Gribble, 2000; Gribble and Reimann, 2003). As previously discussed, despite the presence of the activating SUR1 mutations, the mutated K_{ATP} channel can still be closed in response to sulfonylureas, indicating that most patients with *ABCC8* mutations may be treated with this oral antidiabetic medication rather than insulin (Beltrand et al., 2020).

More than 100 mutations associated with *ABCC8*-mediated hyperinsulinism have been identified which can be divided in two classes. Class I mutations that cause impaired SUR1 trafficking to the plasma membrane, and Class II mutations that lead to a dysfunctional channel (insensitive to

MgADP) that remains permanently closed, even at low glucose concentrations. This results in permanent membrane depolarization and uncontrolled insulin release (Ashcroft, 2005). The majority of the Class II mutations are located in the NBD of SUR1 (Gloyn et al., 2006). Interestingly, it has been shown than some patients with hyperinsulinism caused by certain *ABCC8* mutations may progress to T2D later in life (Abdulhadi-Atwan et al., 2008; Ashcroft, 2005; Isik et al., 2019). A quick diagnosis and immediate management of hyperinsulinemic hypoglycemia is essential to avoid hypoglycaemic brain injury and long-term neurological complications in children (Demirbilek and Hussain, 2017). Patients with mutations causing mild congenital hyperinsulinism may be treated by diet or diazoxide. However, patients with more severe forms of the disease associated with Class I mutations (that prevent the channel from reaching the plasma membrane), fail to respond to diazoxide therapy and are usually treated by sub-total pancreatectomy (Dekel et al., 2002) .

3.4 KCNJ11 mutations (MODY13 and neonatal diabetes mellitus)

3.4.1 Genetic alteration and clinical phenotype

Heterozygous activating mutations in the *KCNJ11* gene encoding Kir6.2 are the most common cause of permanent neonatal diabetes mellitus in a large variety of ethnic groups and populations accounting for 1/3 to 1/2 of all neonatal diabetes cases (Gloyn et al., 2004, 2006). *KCNJ11* mutations were also shown to cause a new syndrome of neonatal diabetes, developmental delay, muscle weakness and epilepsy crisis. Additionally, *KCNJ11* mutations have been associated with transient neonatal diabetes mellitus and MODY 13 (Bonnefond et al., 2012; Flanagan et al., 2009; Gloyn et al., 2004; Hattersley and Ashcroft, 2005; Liu et al., 2013; Massa et al., 2005; Sagen et al., 2004; Vaxillaire et al., 2004; Yorifuji et al., 2005). The latter is a very rare form of MODY which accounts for less than 1% of all MODY cases (Firdous et al., 2018). Similar to ABCC8 mutations, the metabolic phenotype associated *KCNJ11* mutations is very variable and may not be totally penetrant. These differences may be explained by the mutation itself and/or by the presence of other genetic or epigenetic changes that may influence the age of onset and severity of the clinical manifestations (Bonnefond et al., 2012). In line with that, it has been shown that there is a marked genotype-phenotype relationship between particular activating *KCNJ11* mutations and the development of either transient neonatal diabetes, permanent neonatal diabetes, or syndromic neonatal diabetes (Hattersley and Ashcroft, 2005).

Patients with neonatal diabetes mellitus often receive insulin replacement therapy. Nevertheless, as mentioned above, patients with K_{ATP} channel mutations respond to sulfonylureas treatment. Related with *KCNJ11* mutations, it has been demonstrated that high doses of sulfonylurea administered from diagnosis is the best therapeutic approach to maintain normoglycemia in patients with permanent neonatal diabetes caused by mutations in this gene (Bowman et al., 2018; Garcin et al., 2020; Pearson et al., 2006). A recent report has shown that this therapy is safe and highly effective, allowing to maintain an excellent glycemic control for at least 10 years (Bowman et al., 2018; Garcin et al., 2020).

Similar to *ABCC8*, loss-of-function mutations in *KCNJ11* have been associated with congenital hyperinsulinism and hypoglycemia. Less than 25 different *KCNJ11* inactivating mutations have been identified (Flanagan et al., 2009; Gloyn et al., 2006).

3.4.2 Molecular phenotype

As mentioned above, Kir6.2 is the pore-forming subunit of the β-cell K_{ATP} channel. Similar to activating *ABCC8* mutations, gain of function mutations in *KCNJ11* importantly reduce GSIS. Studies done in mice have shown that *Kcnj11* activating mutations reduce the ATP sensitivity of the channel resulting in its defective closure in response to high glucose and impaired insulin release. Indeed, it was shown that transgenic mice expressing a β-cell specific mutated Kir6.2 protein (with reduced ATP sensitivity) develop hypoinsulinemia, severe hyperglycemia and ketoacidosis shortly after birth (Hattersley and Ashcroft, 2005; Koster et al., 2000). The severity of the clinical manifestations of patients with *KCNJ11* activating mutations correlates with the ATP sensitivity of the channels that is higher in transient neonatal diabetes than in permanent neonatal diabetes. In syndromic neonatal diabetes caused by *KCNJ11* mutations, the ATP sensitivity of the channel is further reduced (Hattersley and Ashcroft, 2005).

On the other hand, the inactivating *KCNJ11* mutations associated with congenital hyperinsulinism either reduce or completely abolish the K_{ATP} channel activity, or impair its traffic to the plasma membrane, thus leading to permanent membrane polarization and insulin release (Flanagan et al., 2009).

3.5 *PCSK1* mutations (young onset diabetes)

3.5.1 Genetic alteration and clinical phenotype

Biallelic inactivating mutations in the *PCSK1*gene, coding for proprotein convertase (PC1), also known as neuroendocrine convertase 1 or prohormone convertase 1/3 (PC1/3), lead to a large variety of symptoms.

These may include severe malabsorptive diarrhea that generally manifests before the age of 1 month, obesity; and several endocrine alterations, such as, young onset diabetes insipidus (characterized by hyperproinsulinemia) that generally manifest before the age of 3 years, hypogonadism, central hypothyroidism and mild central hypocortisolism. The clinical manifestations associated with PC1/3 deficiency may occur simultaneously or sequentially. PC1/3 deficiency is a rare disorder, as only 26 cases of PC1/3-deficiency, caused by 24 different missense, nonsense, splice-modifying or frameshifting *PCSK1* mutations have been reported up to date (Frank et al., 2013; Harter et al., 2016; Jackson et al., 1997; Martin et al., 2013; Muhsin et al., 2020; O'Rahilly et al., 1995; Pepin et al., 2019; Yourshaw et al., 2013).

3.5.2 Molecular phenotype

PC1/3 is expressed in a variety of tissues such as the endocrine cells in the gut, β-cells of the pancreas (where it plays a crucial role in proinsulin processing), monocytes, and the arcuate and paraventricular nuclei of the hypothalamus (Harter et al., 2016). PC1 is a 753 amino acids calcium- and pH-dependent endoprotease located in secretory granules that cleaves substrates immediately after a pair of dibasic residues such as RR or KR. PC1/3 is responsible of cleavage-mediated protein maturation and, as mentioned above, in pancreatic β-cells it is responsible of the conversion of proinsulin into insulin (Chen et al., 2018; Martin et al., 2013). PC1/3 expression is positively regulated by the transcription factors PAX6 and MAFA and negatively regulated by glycated hemoglobin c levels (HbA1c) (Ahlqvist et al., 2012; Wang et al., 2007; Wen et al., 2009). Patients with decreased PC1/3 protein levels have an increased ratio of proinsulin/insulin, as a result of impaired proinsulin processing (Heni et al., 2010; Teitelman, 2019). In addition, the failure of enteroendocrine cells to produce functional hormones results in the generalized malabsorption seen in these patients, suggesting that PC1/3 is also needed for the processing of several enteric hormones (Martin et al., 2013). High proinsulin levels accompanied by less mature secretory granules were also seen in β-cells from *Psk1*-KO mice (Zhu et al., 2002).

3.6 *SLC2A2* mutations (Fanconi Bickel Syndrome)

3.6.1 Genetic alteration and clinical phenotype

Fanconi Bickel Syndrome is a rare clinical entity, that follows an autosomal recessive mode of inheritance. It is caused by missense, nonsense, in-frame indels, frameshift, splice site, and compound heterozygous variants in the

SLC2A2 gene coding for the glucose transporter GLUT2. Up to now around 70 different *SLC2A2* mutations present in a 88 families have been identified (Abbasi et al., 2015; Al-Haggar et al., 2012; Kehar et al., 2014; Matsuura et al., 2002; Mohandas Nair et al., 2012; Peduto et al., 2004; Riva et al., 2004; Sakamoto et al., 2000; Sansbury et al., 2012; Sharari et al., 2020). This syndrome is characterized by proximal renal tubular dysfunction, hepatorenal glycogen accumulation, and impaired utilization of glucose and galactose (Manz et al., 1987). The clinical manifestations appear during the first months of life. Patients present with short stature, hepatomegaly, glucosuria, polyuria, failure to thrive, and dysglycemia that can be very variable ranging from fasting hypoglycemia, postprandial hyperglycemia, glucose intolerance, transient neonatal diabetes or gestational diabetes to overt diabetes mellitus. No association between the type of mutation and the clinical manifestations were found (Sharari et al., 2020). The clinical management of these patients is challenging and there are currently no specific drugs to treat this condition.

3.6.2 Molecular phenotype

GLUT2 is a member of the facilitative glucose transporter family expressed in hepatocytes, pancreatic β-cells, and the basolateral membranes of intestinal, renal tubular epithelial cells, some neurons and astrocytes (Kellett and Brot-Laroche, 2005; Leturque et al., 2005; McCulloch et al., 2011; Mueckler et al., 1994; Sharari et al., 2020). It is a low affinity and high Km facilitative glucose transporter that plays an important role in regulating glucose homeostasis. It controls glucose uptake and release in the liver in response to the fasting or feeding state, and glucose uptake in pancreatic β-cells proportionally to the circulating blood glucose levels (Thorens, 2015). Its role as a β-cell glucose transporter is clear in rodents in which Glut2 is highly expressed. However, its function in human β-cells is still controversial since adult human β-cells mainly express GLUT1 and 3 instead of GLUT2. Low GLUT2 expression levels have, however, been detected in fetal islets during human pancreatic development (De Vos et al., 1995; McCulloch et al., 2011; Richardson et al., 2007). Only few cases of patients with Fanconi Bickel syndrome presenting impaired GSIS or overt diabetes have been described (Chesney et al., 1980; Garty et al., 1974; Manz et al., 1987) and, as mentioned before, some of them have transient neonatal diabetes mellitus (Bahillo-Curieses et al., 2018; Khandelwal et al., 2018; Riva et al., 2004). This suggests that GLUT2 may play an important role in controlling β-cell physiology and insulin secretion during development and in the new-born period (Khandelwal et al., 2018; Sansbury et al., 2012). This is further reinforced by the fact that nearly all Fanconi Bickel Syndrome have low weight at birth pointing to a potential defect in insulin production during the three last months of gestation. Diabetes remission in Fanconi Bickel Syndrome after birth may be the consequence of a compensatory mechanism relying on, for example, GLUT1 or GLUT3 expression that take over the GLUT2 function (Sharari et al., 2020). Further studies are needed in human pancreatic β-cells, liver and kidney to understand the molecular mechanisms underlying dysglycemia caused by *SLC2A2* mutations.

3.7 *APPL1* mutations (MODY14)

3.7.1 Genetic alteration and clinical phenotype

Dominant mutations in the *APPL1* gene, coding for Adaptor Protein, Phosphotyrosine Interacting With PH Domain And Leucine Zipper 1 (APPL1) were identified as causal of MODY14 in two large families with high diabetes prevalence (Prudente et al., 2015). Ten members of the first family, having diabetes or pre-diabetes, showed a heterozygous nonsense mutation (c.1655T>A, p.L552X) in the *APPL1* gene. Several unaffected carriers of the mutation were also identified but they were all below 38 years old suggesting that the metabolic manifestations could still appear with age. In the second family, a missense mutation (c.280G>A; p.D94N) in *APPL1* was identified in five of the seven diabetic members. Both mutations were predicted to be loss-of-function (Prudente et al., 2015). A recent genetic analysis of a large cohort of Russian patients having MODY-type diabetes of unidentified origin failed to identify novel or the previously reported *APPL1* mutations. This study, however, unveiled one single nucleotide variant (rs11544593) in the *APPL1* gene as potentially associated with earlier onset of carbohydrate metabolism disorders. This association still needs to be replicated in a larger study population (Ivanoshchuk et al., 2020) .

3.7.2 Molecular phenotype

APPL1 is an Akt/protein kinase B-binding protein involved in signal transduction and membrane trafficking pathways. In humans APPL1 is highly expressed in different tissues such as pancreas, liver, adipose tissue, brain and muscles (Ivanoshchuk et al., 2020). APPL1 is a 710 aminoacid protein constituted of 5 different functional domains: one NH2-terminal Bin/Amphiphysin/Rvs domain (BAR), a central pleckstrin homology domain (PH), a BPP domain, a COOH-terminal phosphotyrosine-binding domain (PTB) and a coiled-coil (CC domain) (Cheng et al., 2012) through which it interacts with a large number of proteins including critical components of the insulin-signaling pathway. Accordingly, it has been demonstrated that APPL1 is an AKT2 interactor and a key regulator of insulin signaling that plays an important role in the regulation of insulin secretion (Cheng et al., 2012). It was also shown that APPL1 interacts with different signaling receptors, proteins and lipids mediating the crosstalk between adiponectin and insulin signaling (Habermann, 2004; Ivanoshchuk et al., 2020; Li et al., 2007; Takei et al., 1999). The role of Appl1 in insulin secretion has been supported by different mouse models of *Appl1* deficiency and overexpression. In fact, it has been shown that *Appl1* KO mice develop glucose intolerance and impaired GSIS, while mice overexpressing *Appl1* are protected from high-fat diet-induced glucose intolerance due to elevated GSIS (Cheng et al., 2012). *Appl1* deficiency in rat β-cells and pancreatic islets led to reduced expression of SNARE proteins from the secretory machinery, which was accompanied by reduced exocytosis and insulin-stimulated Akt activation (Cheng et al., 2012). In the liver, APPL1 potentiates the inhibitory effect of insulin on hepatic gluconeogenesis through activation of AKT2. Indeed, it was shown that if *Appl1* is overexpressed in the liver of diabetic mice it normalizes hyperglycemia (Cheng et al., 2009). In skeletal muscle and adipocytes, APPL1 regulates insulin-stimulated glucose uptake by controlling the translocation of the glucose transporter 4 (GLUT4) from the cytosol to the plasma membrane (Cheng et al., 2009).

4. Monogenic forms of diabetes associated with endoplasmic reticulum stress

The ER is an essential organelle for secretory cells such as pancreatic β-cells and neurons. The majority of the secretory proteins or proteins residing within the ER are translated on ER membrane-associated ribosomes and transported into the ER lumen, which highlights the importance of this organelle for proper secretory function (Cnop et al., 2017). Pancreatic β-cells rely on the ER for proper insulin processing, folding, and release (Liu et al., 2018). It has been extensively shown that impaired ER function induces ER stress which leads to β-cell demise and diabetes in humans (Cnop et al., 2017; Lin and Popko, 2009; Marciniak and Ron, 2006). ER stress is the result of an imbalance between the protein load into the ER and the folding capacity of the organelle which leads to the accumulation of unfolded or misfolded proteins within ER lumen. Under this condition, and to try to restore ER homeostasis, the cells trigger what is called the unfolded protein response (UPR) or ER stress response which is an adaptive mechanism aiming to (1) increase the folding capacity of the ER (by enhancing the ER chaperone expression and increasing the size of the organelle), (2) decrease the amount of newly synthesized proteins arriving into ER by reducing protein translation, and (3) enhancing the degradation of misfolded ER proteins. The ER stress response is initiated mainly via the activation of three ER stress transducers (PERK, ATF6 and IRE1) located in the ER membrane (Fig. 2). PERK activation leads to phosphorylation of the eukaryotic translation initiation factor 2 alpha (P-eIF2α) which reduces general translation initiation while favoring the translation of selected transcripts such as ATF4, a transcription factor controlling the expression of genes related with aminoacid transport and metabolism and antioxidant responses. ATF4 also regulates the expression of ATF3 and CHOP, two pro-apoptotic transcription factors controlling the expression of UPR-related genes (Baird and Wek, 2012; Harding et al., 2003; Ron and Walter, 2007) (Fig. 2). ATF6 activation results in enhanced ER chaperone levels, increased expression of genes involved in lipid biosynthesis, and ER-associated degradation (ERAD), a mechanism consisting in the retro-translocation, polyubiquitination and proteasomal degradation of misfolded proteins from the secretory pathway (Meusser et al., 2005) (Fig. 2). IRE1 activation results in mRNA splicing of X-box binding protein 1 (XBP1) producing spliced XBP1 (sXBP1) a transcription factor that, the same as ATF6, upregulates ER chaperones and ERAD genes, but also leads to the IRE1-dependent decay of ER-associated mRNAs, contributing to reduce the ER protein load. IRE1 activation also induces JNK expression which contributes to β-cell demise (Hollien and Weissman, 2006; Pirot et al., 2007; Ron and Walter, 2007) (Fig. 2).

The UPR response is expected to be beneficial. However, when prolonged in time, it leads to β-cell dysfunction and death (Eizirik and Cnop, 2010). Here below we present different monogenic forms of diabetes caused by mutations in ER stress-associated proteins.

4.1 *EIF2S3* mutations (MEHMO syndrome)

4.1.1 Genetic alteration and clinical phenotype

Mutations in the *EIF2S3* gene, located within Xp21.1-p22.13 and coding for eukaryotic translation initiation factor 2 subunit 3 (eIF2S3 or eIF2γ) have been associated with the MEHMO syndrome characterized by Mental retardation, Epileptic seizures, Hypogenitalism, Microcephaly and Obesity that is often accompanied by non-autoimmune neonatal or

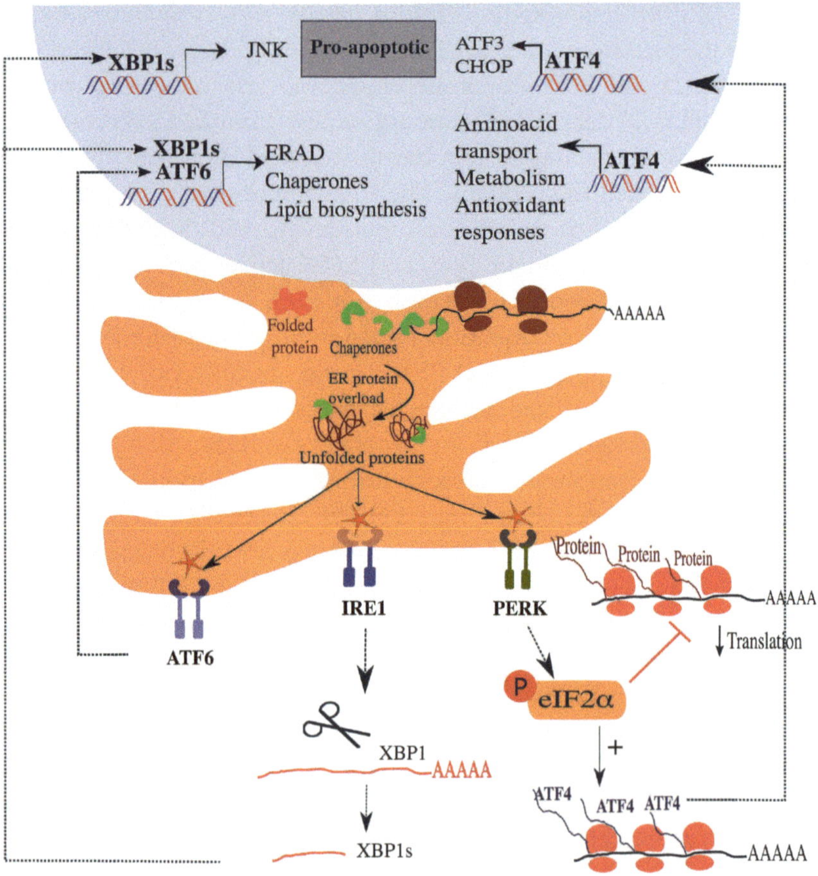

Fig. 2 Endoplasmic reticulum stress signaling. An imbalance between the protein load into the ER and the folding capacity of the organelle leads to the accumulation of unfolded or misfolded proteins within ER lumen resulting in the activation of the ER stress transducers PERK, IRE1 and ATF6. Activated PERK phosphorylates eIF2α leading to a general repression in translation. However, eIF2α phosphorylation also enhances ATF4 translation that induces the transcription of proteins related with aminoacid transport, metabolism and antioxidant responses, together with enhanced expression of the pro-apoptotic proteins ATF3 and CHOP. IRE1 activation causes the splicing of XBP1 mRNA. Spliced XBP1 (XBP1s) is a transcription factor that enhances the expression of chaperones, components of the ERAD machinery, and lipid biosynthesis as well as the pro-apoptotic protein JNK. Activated ATF6 translocates to the nucleus where it enhances chaperone expression.

infant-onset diabetes (Barbetti and D'Annunzio, 2018; Skopkova et al., 2017; Steinmuller et al., 1998). Mutations in *EIF2S3* have also been correlated with several additional endocrine manifestations such as growth hormone deficiency and hypoglycemia in addition to MEHMO's phenotype (Moortgat et al., 2016). Up to now, 9 independent families with different missense and frameshift *EIF2S3* mutations have been reported (Borck et al., 2012; Gregory et al., 2019; Moortgat et al., 2016; Skopkova et al., 2017; Steinmuller et al., 1998). The location and nature of the mutations have been associated with the clinical severity and variability of the symptoms (Borck et al., 2012; Gregory et al., 2019; Maghnie and Barbetti, 2019; Skopkova et al., 2017).

4.1.2 Molecular phenotype

eIF2γ is one of the three subunits that form the eukaryotic translation initiation factor 2 (eIF2 protein) essential for protein synthesis. Indeed, eIF2 is a heterotrimeric factor composed by the eIF2α, eIF2β and eIF2γ subunits. eIF2 forms a ternary complex with GTP and the initiator methionine tRNA ($tRNA_i^{Met}$) that binds to the small 40S ribosomal subunit along with other factors to form the 43S preinitiation complex, which binds to the mRNA, searches for a start codon and initiates translation (Hinnebusch, 1985, 2005). eIF2 is regulated by guanine nucleotide exchange and phosphorylation of eIF2α subunit. In situ hybridization studies revealed that *EIF2S3* mRNA is strongly expressed in the fetal pancreas suggesting a role of this protein in endocrine pancreas development (Maghnie and Barbetti, 2019). It has been proposed that the *EIF2S3* mutations may alter eIF2γ expression or stability or affect its binding to eIF2α and eIF2β, resulting in altered eIF2 binding to GPT and/or RNA_i^{Met} which compromises protein synthesis (Skopkova et al., 2017; Young-Baird et al., 2019). Initial experiments in yeast and human cells showed that the phenotype of eIF2γ -deficient cells is similar to the one caused by enhanced eIF2α phosphorylation resulting in impaired eIF2 function, reduced protein expression and translation fidelity and increased ATF4 and CHOP translation (Skopkova et al., 2017). In addition, *EIF2S3* silencing in a hybrid human β-cell line led to enhanced basal and cytokine-induced caspase activity and apoptosis (Gregory et al., 2019). Recently, a newly identified Pro432Ser mutation in *EIF2S3* has been associated with a novel phenotype of mild learning difficulties, hypopituitarism and early hypoglycemia followed by later hyperglycemia (Gregory et al., 2019). Mutagenesis studies performed in a humanized yeast eIF2γ showed that the Pro432Ser mutation results in either a defect in the ternary complex formation or altered delivery of $tRNA_i^{Met}$ to the ribosome, thus confirming previous findings that suggested that the *EIF2S3* mutations alter the fidelity of the translation start site selection (Gregory et al., 2019; Skopkova et al., 2017).

4.2 *EIF2AK3* mutations (Wolcott-Rallison syndrome)

4.2.1 Genetic alteration and clinical phenotype

Mutations in the eukaryotic translation initiation factor 2-alpha kinase 3 (*EIF2AK3)* gene, encoding PERK, cause Wolcott-Rallison syndrome (WRS) which is a rare syndromic form of diabetes with a prevalence of 1/1.000.000 (Delepine et al., 2000) but recognized as the most common cause of permanent neonatal diabetes in consanguineous families (Rubio-Cabezas et al., 2009). While frameshift or nonsense *EIF2AK3* mutations are present in the majority of the patients (65%), missense mutations appear in less than 30% of the cases. All these mutations are located in the kinase domains of PERK (Julier and Nicolino, 2010; Senee et al., 2004). WRS is characterized by non-autoimmune permanent neonatal or early infancy diabetes mellitus and exocrine pancreas atrophy, hepatic dysfunction (which includes recurrent hepatitis), skeletal abnormalities (epiphyseal, pelvis and spine dysplasia), osteopenia, microcephaly and intellectual disability (Iida et al., 2007; Julier and Nicolino, 2010; Li et al., 2003; Zhang et al., 2002, 2006). Growth retardation and developmental delays are very common in WRS patients and several cases of WRS are accompanied by hypothyroidism, simplified gyral patterns, epilepsy, and dental discoloration (Iida et al., 2007; Li et al., 2003; Zhang et al., 2002, 2006).

4.2.2 Molecular phenotype

PERK is a protein kinase located in the ER membrane which, as mentioned above, acts as an ER stress sensor and is one of the three master regulators that control the UPR. It has been demonstrated that the onset of neonatal diabetes mellitus in WRS is caused by insulin insufficiency which is associated with reduced β-cell mass. Indeed, post-mortem pancreas analysis of two WRS patients showed a selective loss of pancreatic β-cells (Collardeau-Frachon et al., 2015). Studies in *Eif2ak3 KO* mice, that mirror the WRS complex phenotype, showed that proper Perk function is required for normal β-cell differentiation and proliferation during the fetal period which is essential to develop a normal β-cell mass and normal insulin synthesis and secretion (Harding et al., 2001; Zhang et al., 2002, 2006). Interestingly, it was shown that Perk-deficient rodent β-cells do not have an uncontrolled protein synthesis (due to the lack of eIF2α phosphorylation) but an accumulation of proinsulin within the ER. This is due to a defect in ER to Golgi anterograde trafficking, a reduced retrotranslocation of misfolded proteins from the ER into de cytosol and decreased proteasomal activity (Gupta et al., 2010). It was shown that PERK-deficient β-cells are more sensitive to ER stress-induced apoptosis since their capacity to phosphorylate eIF2α and repress protein synthesis is importantly diminished. Indeed, deficient and excessive eIF2α phosphorylation (see below) is poorly tolerated by pancreatic β-cells pointing to the need of finely-tuned P-eIF2α levels for β-cell preservation (Cnop et al., 2017).

4.3 *EIF2B1* mutations (neonatal/early onset diabetes and transient liver dysfunction)

4.3.1 Genetic alteration and clinical phenotype

Recently, five heterozygous de novo missense mutations in the *EIF2B1* gene (encoding the α subunit of the eIF2B complex) haven been associated with permanent neonatal/early onset neonatal diabetes and transient liver dysfunction in five patients of different origins (De Franco et al., 2020a). All patients developed permanent diabetes mellitus between the age of 4 and 56 weeks and required insulin replacement therapy to manage diabetes. Low birth weight, consistent with reduced in utero insulin secretion, was reported in four of them while no data was available for the remaining one. During childhood four out of five patients suffered from episodes of hepatitis-like liver dysfunction of variable severity that was fatal in one case. The three remaining patients recovered from liver disease. No severe neurological defects were detected in any of the patients (De Franco et al., 2020a).

4.3.2 Molecular phenotype

As mentioned above, *EIF2B1* encodes for the α subunit of the eIF2B complex which regulates the activity of eIF2 during translation initiation. As previously mentioned protein translation is regulated by the eIF2-GTP-Met-tRNAi ternary complex that, together with eIF5, eIF2B and the 40S subunit of the ribosome bind and scan the mRNA until detection of the translation start site (Hinnebusch, 1985, 2005). At this moment, the GTP bound to eIF2γ is hydrolyzed by eIF5 resulting in the release of eIF5 and eIF2 from the 40S subunit of the ribosome. This allows the 60S subunit of the ribosome to bind and participate in protein elongation (Hinnebusch, 1985, 2005). To be engaged in new rounds of translation initiation, the GDP attached to the released eIF2γ needs to be exchanged for GTP, a reaction that is catalyzed by eIF2B (Wortham and Proud, 2015). Indeed, it was recently shown that when eIF2α is non-phosphorylated, it makes contact with the β-γ and δ subunits of eIF2B, but it does not interact with eIF2Bα. On the contrary, when eIF2α is phosphorylated at Ser 51 (as occurs during ER stress) it binds tightly to eIF2Bα resulting in eIF2B inhibition which prevents GTP reloading in the ternary complex with the concomitant blockage in protein synthesis (Kashiwagi et al., 2019).

An in silico protein analysis, performed to assess the impact of the identified *IEF2B1* de novo mutations on protein function, revealed that the mutated residues in eIF2Bα were highly conserved across species and were localized at the same protein surface that is normally occupied by phosphorylated eIF2α (De Franco et al., 2020a). This suggested that the *IEF2B1* mutations may render the eIF2B complex insensitive to eIF2α phosphorylation under stress conditions (De Franco et al., 2020a). As discussed above, *EIF2AK3* mutations, causing Wolcott-Rallison syndrome, result in reduced eIF2α phosphorylation leading to exacerbated UPR under stress conditions which results in β-cell death. Since *EIF2AK3* and e *IEF2B1* act in different steps of the same pathway it is likely that the failure of eIF2Bα to sense eIF2α phosphorylation will have the same deleterious outcome in pancreatic β-cells than *EIF2AK3* mutations. This, however, still needs to be experimentally demonstrated.

4.4 *IER3IP1*-mutations (neonatal diabetes mellitus)

4.4.1 Genetic alteration and clinical phenotype

Loss-of-function mutations in the *IER3IP1* gene, coding immediate early response 3 interacting protein 1 (IER3IP1) cause a Wolcott-Rallison-Like syndrome characterized by permanent neonatal diabetes mellitus, microcephaly, severe epileptic encephalopathy and simplified gyral patterns. Some patients also developed cortical blindness, developmental delay and numerous dysmorphic features (Poulton et al., 2011; Shalev et al., 2014). Until now, three different *IER3IP1* mutations have been reported that were present in six unrelated families: two missense mutations (Val21Gly and Leu78Pro) and a frameshift mutation Phe27fsSer2 all predicted to promote IER3IP1 loss-of-function (Abdel-Salam et al., 2012; Poulton et al., 2011; Rubio-Cabezas et al., 2009; Shalev et al., 2014; Valenzuela et al., 2017).

4.4.2 Molecular phenotype

IER3IP1 is an 82 amino acids long protein localized in the ER through its C-terminal transmembrane domain. It is highly conserved among species and it is ubiquitously expressed, even though its highest expression levels were seen in the developing brain and pancreas (Poulton et al., 2011). Post-mortem analysis of these tissues from IER3IP1-deficient patients unveiled increased neuronal apoptosis and reduced β-cell mass (Poulton et al., 2011). The current knowledge of IER3IP1 function is relatively limited but it has been suggested that it is involved in the protein transport between the ER and the Golgi apparatus and in the regulation of apoptosis in cortical neurons and pancreatic β-cells (Sun and Ren, 2017). Indeed, it was shown that IER3IP1 deficiency leads to a dysregulation in ER stress response mechanisms and β-cell apoptosis. More specifically, Sun and Ren (2017) showed that *Ier3ip1*-silencing in the mouse β-cell line MIN6 leads to increased caspase 3 cleavage, accompanied by upregulation of the pro-apoptotic proteins Bim and Bax and decreased levels of the anti-apoptotic protein Bcl-xl (Sun and Ren, 2017). They also confirmed that *Ier3ip1*deficiency impairs the Ire1α/-sXbp-1, and Perk-eif2α/Atf4 arms of the UPR, suggesting that the lack of *Ier3ip1* induces β-cell apoptosis due to a failure in the UPR regulation (Sun and Ren, 2017).

4.5 *PPP1R15B* mutations (young onset diabetes and microcephaly)

4.5.1 Genetic alteration and clinical phenotype

In 2015 Abdulkarim et al. reported a homozygous missense mutation (c.1972C>T; p.R658C) in Protein Phosphatase 1 Regulatory subunit 15B (*PPP1R15B)* gene as causal of a new syndrome of juvenile onset diabetes, microcephaly and short stature in two patients from a consanguineous family (Abdulkarim et al., 2015). On the same year Kernohan et al. reported on two additional patients from an independent family bearing the same mutation and having microcephaly, short stature and intellectual disability but not diabetes (Kernohan et al., 2015). In addition, compound heterozygous inactivating mutations in *PPP1R15B*, predicted to result in a very short and inactive form of CReP, were detected in two additional patients from a non-consanguineous family (Mohammad et al., 2016). These patients had microcephaly, short stature, intellectual disability, hypothyroidism and severe liver disease (not present in the other two families) but no diabetes (Mohammad et al., 2016). This indicates that while microcephaly, short stature and neurological alterations are general features of *PPP1R15B* mutations, the presence of additional non-neurological manifestations is more variable and may depend on the presence of still unidentified additional genetic or environmental factors.

What is more, mutations in *PPP1R15A*, have been linked with obesity, fatty liver, liver cirrhosis and pancreatic β-cell dysfunction leading to T2D (Patel et al., 2019).

4.5.2 Molecular phenotype

PPP1R15B encodes the constitutive repressor of eIF2α phosphorylation (CReP), while *PPP1R15A* codes for GADD34. CReP and GADD34 are two regulatory subunits of protein-phosphatase-1 (PP1), a holophosphatase involved in eIF2α dephosphorylation. While CreP is constitutively expressed and represses eIF2α phosphorylation in basal condition (Jousse et al., 2003), GADD34 regulates eIF2α phosphorylation under stress. Indeed, GADD34 is activated by eIF2α phosphorylation and contributes to recover protein synthesis upon stress activation (Novoa et al., 2001).

Abdulkarim et al., showed that a R658C mutation in *PPP1R15B* leads to destabilization of the CreP-PP1 complex, with a consequent decrease in eIF2α dephosphorylation (Abdulkarim et al., 2015). Accordingly, siRNA-mediated *Ppp1r15b* silencing in INS-1E cells enhanced basal eIF2α phosphorylation that was accompanied by reduced insulin content, and altered normal β-cell function; this was evidenced by enhanced basal insulin secretion and GSIS, pointing to an important role of *Ppp1r15b* in exocytosis (Abdulkarim et al., 2015). Indeed, the involvement of this protein in exocytosis and membrane traffic had been previously reported in other cell types (Kloft et al., 2012). Different from other tissues, in which *PPP1R15B* expression is constitutive and it is not modulated by stress, in pancreatic β-cells *PPP1R15B* expression is enhanced by ER stress in a PERK-dependent manner (Abdulkarim et al., 2015). *Ppp1r15b* deficiency in clonal and primary rat β-cells, besides affecting cell function, induced apoptosis in control condition and sensitized β-cells to free-fatty acid and ER stress-mediated cell death. The *Ppp1r15b*-deficient β-cells showed enhanced cytochrome *c* release and cleaved caspase 9, together with induction of the pro-apoptotic Bh3-only proteins Dp5, Puma, and Bim indicating that the lack of this protein activates the intrinsic pathway of apoptosis (Abdulkarim et al., 2015). In good correlation with that, it was previously shown that excessive eIF2α phosphorylation is deleterious for pancreatic β-cells and leads to the activation of the intrinsic apoptosis pathway (Cunha et al., 2012). Indeed, it was demonstrated that besides *PPP1R15B* deficiency, Salubrinal, a chemical inhibitor of eIF2a dephosphorylation that prevents GADD34/PP1 and CReP/PPI complexes formation (Boyce et al., 2005), and Guanabenz, an α 2-adrenergic receptor agonist that binds to GADD34 inhibiting eIF2a dephosphorylation (Tsaytler et al., 2011), exacerbate ER stress-induced β-cell apoptosis and β-cell dysfunction in vitro and in vivo (Abdulkarim et al., 2017; Cnop et al., 2012; Ladriere et al., 2010).

4.6 *WFS1* and *CISD2* mutations (Wolfram syndrome 1 and 2)

4.6.1 Genetic alteration and clinical phenotype

Wolfram syndrome (WS) is a rare autosomal life-threatening syndrome with a frequency of 1/160.000 to 1/770.000 individuals (in the USA and UK, respectively) (Barrett et al., 1995; Fraser and Gunn, 1977). Two classes of WS exist: WS-1, caused by mutations in the *WFS1* gene, and WS-2, caused by mutations in the *CISD2* gene. The majority of the WS patients present biallelic homozygous or compound heterozygous mutations in the *WFS1* gene (Inoue et al., 1998). Nevertheless autosomal dominant inherited or *the novo* mutations in the same gene have also been reported (Bespalova et al., 2001; Bonnycastle et al., 2013; De Franco et al., 2017; Eiberg et al., 2006; Hogewind et al., 2010). More than 100 mutations have been identified that are essentially located within the exon 8 (Cryns et al., 2003; Gomez-Zaera et al., 2001; Hansen et al., 2005; Hardy et al., 1999; Khanim et al., 2001). Most patients with autosomal recessive WFS1 mutations present with young onset diabetes mellitus (that apereas in the first decade of life, usually before the age of 6 years), optic atrophy, diabetes insipidus, hearing loss and neurodegeneration that manifests principally as cerebellar ataxia, gait abnormalities, memory loss, speech difficulties, dysphagia, nystagmus, anxiety and depression (Barrett et al., 1995; Kinsley et al., 1995; Marshall et al., 2013; Rando et al., 1992). Accordingly, post-mortem neuropathological and neuroimaging studies have shown cerebellar, hypothalamus, hippocampus, pons and brain stem degeneration in WS patients respect to controls (Barrett and Bundey, 1997; D'Annunzio et al., 2008; Hardy et al., 1999; Hershey et al., 2012; Hilson et al., 2009; Jackson et al., 1994; Lugar et al., 2016, 2019; Scolding et al., 1996). The main clinical manifestations of WS-1 (diabetes mellitus, optic atrophy, hearing loss and neuropathy) are also present in WS-2 (El-Shanti et al., 2000; Rondinelli et al., 2015) that is also characterized by defective platelet aggregation that results in peptic ulcer bleeding (Mozzillo et al., 2014). Most autosomal dominant *WFS1* mutations have been associated with milder clinical manifestations that appear in an isolated manner such as adult onset diabetes, optic atrophy, cataracts and hearing impairment (Bespalova et al., 2001; Bonnycastle et al., 2013; Eiberg et al., 2006; Hogewind et al., 2010). However, De Franco et al. have recently identified three dominant de novo *WFS1* mutations as causal of a severe congenital syndrome of neonatal/infancy-onset diabetes, congenital cataracts, and congenital sensorineural deafness (De Franco et al., 2017). This highlights the large variability in the clinical presentations associated with WFS1 mutations. While it is known that Wolfram syndrome patients carrying recessive and missense variants tend to have milder manifestations, the exact genotype-phenotype correlation of the different recessive and dominant WFS1 mutations identified is not completely understood (De Heredia et al., 2013; Stone et al., 2021).

4.6.2 Molecular phenotype

WFS1 encodes wolframin, a ubiquitously transmembrane protein located in the ER. The pathogenesis of WS-1 has been investigated using in vitro and in vivo models for the disease. These studies showed that *WFS1* deficiency results in ER stress leading to neurodegeneration and pancreatic β-cell dysfunction and death (Fonseca et al., 2005, 2011, 2010; Ishihara et al., 2004; Riggs et al., 2005). Indeed, it was demonstrated that WFS1 plays an important role in modulating the UPR via its interaction with ATF6 (Fonseca et al., 2010), and was also shown to regulates cellular and ER Ca^{2+} levels (Lu et al., 2014). The Ca^{2+} concentration in the cytoplasm is primarily regulated by the Ca^{2+} transport in and out of the ER. It has been shown that wolframin-deficient β-cells and neurons have increased cytosolic Ca^{2+}, probably due to altered ER Ca^{2+} homeostasis leading to calpain activation and cell death (Lu et al., 2014). Besides its association with ER stress, WFS1 deficiency has recently been linked to impaired granule acidification in β-cells and neurons (Gharanei et al., 2013; Hatanaka et al., 2011; Zatyka et al., 2008). In β-cells, acidification of secretory granules is required for proper insulin processing and release (Hatanaka et al., 2011; Sudhof and Rizo, 2011). Experiments performed in wolframin-deficient rodent neurons showed that the lack of this protein leads to alterations in mitochondrial dynamics, inhibition in mitochondrial trafficking and increased mitophagy (Cagalinec et al., 2016). Recent studies by Angebault and colleagues in WS-1 patient fibroblasts (Angebault et al., 2018) suggest that wolframin deficiency may cause mitochondrial dysfunction through reduced expression of the Ca^{2+} sensor NCS1 (Angebault et al., 2018). The authors showed that the ER-mitochondrial interactions through mitochondrial associated ER membranes (MAMs) are reduced in *WFS1*-deficient fibroblasts leading to reduced $Ca2^{+}$ transfer from the ER to the mitochondria and mitochondrial failure (Angebault et al., 2018). Since granule exocytosis is an energy-dependent process, mitochondrial dysfunction will directly affect the functionality of pancreatic β-cells. In line with that, a recent work by Maxwell and colleagues using iPSC-derived β-cells from one Wolfram syndrome patient and one CRISPR/Cas-9 mediated isogenic control demonstrated that, besides being dysfunctional and showing signs of ER stress, the patient-derived β-cells have smaller and fragmented mitochondria with signs of fission, suggesting that mitochondrial failure is also a characteristic of wolframin-deficient β-cells (Maxwell et al., 2020).

CISD2 encodes a high-conserved zinc finger Fe-S cluster small protein called CISD2, Miner1, NAF-1, NOXp70 or Eris. This protein is localized in the ER and in MAMs (Amr et al., 2007). The specific function of CISD2 is yet unknown, but it has been proposed that it plays an important role in iron donation to the mitochondria, reactive oxygen species (ROS) regulation, and preservation of mitochondrial and ER Ca^{2+} homeostasis (Wang et al., 2014b; Wiley et al., 2013). Indeed, it was demonstrated that *CISD2* mutations alter the Ca^{2+} flux between the ER and the mitochondria (Wang et al., 2014a). In humans, *CISD2* associates BCL-2, BCL-XL and the IP3 receptor. The interaction of CISD2 with BCL-2 in the ER is necessary to inhibit autophagy through Beclin1 (Chang et al., 2010, 2012; Chen et al., 2009a; Pattingre et al., 2005). Mice models of *Cids2* deficiency reproduce several of the clinical manifestations of WS-2 (Chen et al., 2009b). This model contributed to demonstrate that *Cisd2* downregulation leads to decreased mitochondrial function and integrity, together with an upregulation of autophagy and apoptotic factors (Chen et al., 2009b; Danielpur et al., 2016; Holt et al., 2016; Sohn et al., 2013). Although the precise mechanism is yet to be known, experiments performed with induced pluripotent stem cells from WS-2 patients and *Cisd2* silencing in the mouse neuronal NSC34 cells and in the rat and mouse β-cell lines INS1 832/13 and MIN6 cells, showed that *Cisd2* deficiency leads to neuronal and β-cell death a process probably mediated by calpain activation (Lu et al., 2014).

4.7 *INS* mutations (MODY10 and permanent neonatal diabetes mellitus)

4.7.1 Genetic alteration and clinical phenotype

Autosomal dominant inherited or *the novo* missense mutations in the insulin gene *INS* are the second most common cause of permanent neonatal diabetes mellitus and also a rare form of MODY-type diabetes (MODY10) (Colombo et al., 2008; Edghill et al., 2008; Stoy et al., 2007, 2010). At least 23 different dominant missense *INS* mutations have been identified as causal of permanent neonatal diabetes (Colombo et al., 2008; Edghill et al., 2008; Stoy et al., 2007). The majority of the mutations were located within the A or B chains of insulin (see below), and were predicted to interfere with disulfide bond formation affecting proinsulin folding (Edghill et al., 2008; Stoy et al., 2007).

As mentioned above, dominant *INS* missense mutations have also been associated with MODY10. Indeed, the Arg6Cys, Arg46Gln, Leu30Met, Arg55Cys and Arg6His *INS* mutations have been linked with MODY diabetes in five independent families (Edghill et al., 2008; Meur et al., 2010; Molven et al., 2008). In addition, the Leu68Met and Arg55Cys *INS* mutations were associated with T2D and non-autoimmune T1D respectively, in two independent families (Edghill et al., 2008; Molven et al., 2008). Garin and co-workers identified 10 different recessive *INS* mutations as causal of neonatal diabetes in 15 probands from consanguineous families (Garin et al., 2010). Patients with dominant *INS* mutations have severe hyperglycemia at diagnosis and marked insulin deficiency that needs to be treated with insulin replacement therapy. At the beginning, they may have detectable or high C-peptide levels, suggestive of residual β-cell mass and function, but this is rapidly lost probably due to increased β-cell death (Colombo et al., 2008; Edghill et al., 2008; Stoy et al., 2007, 2010). Patients with recessive *INS* mutations have reduced birth weight and lower age of diabetes onset than the ones with dominant *INS* mutations (Garin et al., 2010). Moreover, some patients with recessive *INS* mutations had transient neonatal diabetes mellitus, a feature not reported for dominant mutations (Garin et al., 2010). To add to the complexity of the variable phenotype associated with *INS* mutations, some of them were reported to cause hyperproinsulinemia, hyperinsulinemia, and mild diabetes (Haneda et al., 1984; Tager et al., 1979; Yano et al., 1992). The broad spectrum of clinical phenotypes and age of diabetes onset associated with *INS* mutations seem to be related with the type of mutation on itself (see below). In addition, still unidentified environmental and/or additional genetic factors may explain the phenotypic variabilities within and between families bearing the same INS mutation (Edghill et al., 2008; Meur et al., 2010).

4.7.2 Molecular phenotype

The insulin molecule is composed by two different polypeptide chains, peptide A and B, which are bound by two disulfide bonds and are generated after proinsulin proteolytic cleavage (Steiner and Oyer, 1967). The first evidence of diabetes associated with mutations in the insulin gene arise from the Akita mice. These animals have a heterozygous missense mutation (Cys96Tyr) in the Ins2 gene which disrupts one the interchain disulfide bonds in the insulin molecule. This leads to proinsulin misfolding within the ER causing hyperglycemia, impaired β-cell function and β-cell death within the first 4 weeks of age (Wang et al., 1999). The Akita mice are considered the prototype of ER stress related diabetes since it was shown that the accumulation of misfolded proinsulin induces ER stress and activates downstream apoptotic signals that lead to β-cell dysfunction and death (Wang et al., 1999).

As mentioned before, in humans *INS* mutations may cause variable phenotypes ranging from mild diabetes with hyperproinsulinemia or hyperinsulinemia, to MODY10 or neonatal diabetes. The severity of the clinical manifestations seems to be, at least in part, associated with the nature of the *INS* mutation. Accordingly, the development of mild diabetes with hyperproinsulinemia or hyperinsulinemia seem to be linked with mutations that result in impaired proinsulin cleavage or the production of a structurally defective insulin (Haneda et al., 1984; Yano et al., 1992). On the other hand, mutations causing MODY10 are predicted to reduce the insulin stability or induce a partial misfolding of the protein which causes its retention within the ER causing mild ER stress (Edghill et al., 2008; Meur et al., 2010; Molven et al., 2008). Interestingly, while recessive *INS* mutations were shown to cause neonatal diabetes as a result of reduced insulin biosynthesis through distinct mechanisms, such as altered mRNA stability due to the lack of polyadenylation signal, lack of the translation initiation signal, gene deletion, or abnormal *INS* transcription (Garin et al., 2010), most dominant *INS* mutations causing permanent neonatal diabetes are predicted to induce proinsulin misfolding due to impaired disulfide bond formation, proinsulin ER retention and ER stress (Colombo et al., 2008; Liu et al., 2015; Wang et al., 1999). Indeed, 60% of the *INS* mutations identified either lead to a cysteine substitution, e.g., Cys96Tyr, or the creation of an additional cysteine, e.g., Arg89Cys, that alike Akita insulin, result in impaired or aberrant disulfide bond formation and proinsulin misfolding (Liu et al., 2015). The rest of the dominant *INS* mutations identified generate either a premature stop codon or alter the secondary structure of the proinsulin causing its misfolding (Colombo et al., 2008). All these alterations result in ER stress that triggers β-cell apoptosis at least in part via CHOP induction highlighting the deleterious impact of prolonged ER stress in pancreatic β-cells (Oyadomari et al., 2002). A recent study by Balboa et al. (2018) using iPSC-derived β-like cells from patients bearing *INS* mutations, showed proinsulin accumulation within the ER, ER stress, impaired cell proliferation, mitochondrial dysfunction but no apoptosis in the INS mutant β-cells. These results suggest that *INS* mutations associated with neonatal diabetes are already pathogenic during pancreas development due to failure of neonatal β-cell expansion (Balboa et al., 2018).

4.8 *YIPF5* mutations (neonatal/early onset diabetes, microcephaly and epilepsy)

4.8.1 Genetic alteration and clinical phenotype

Homozygous mutations in the *YIPF5* gene were recently identified as causal of a new syndrome of neonatal/early onset diabetes, severe microcephaly, and epilepsy in six patients from five consanguineous families (De Franco et al., 2020b). Five of the six patients carried homozygous missense mutations in *YIPF5* while the remaining one had an *in-frame* deletion in the same gene. All patients had severe microcephaly, epilepsy (that was diagnosed between 1 and 7 months of age) and neonatal/early onset diabetes (diagnosed between 4 weeks to 20 months). The six showed reduced birth weight, consistent with reduced insulin secretion in utero, and all required full insulin replacement to manage diabetes (De Franco et al., 2020b).

4.8.2 Molecular phenotype

YIPF5 is a transmembrane protein that cycles between the ER and the Golgi. It is localized in the exit site of the ER, in the ER-Golgi intermediate compartment, and in the cis-Golgi (Kranjc et al., 2017; Yoshida et al., 2008). Studies in HeLa cells suggested that it may play a role in ER to Golgi trafficking (Dykstra et al., 2010). The mutations identified in the six patients reported by De Franco et al. were located in different regions of the YIPF5 protein and, even if they were predicted to be pathogenic it is plausible that YIPF5 function was not completely loss. In vitro analysis of *YIPF5* expression in human tissues showed that it is ubiquitously expressed, but it is enriched in pancreatic tissue, islets, β-cells, and brain (De Franco et al., 2020b). In good correlation with a potential role of *YIPF5* in the development of the cerebral cortex, in situ hybridization studies showed that *YIPF5* is broadly expressed in the developing human cortex at all the embryonic stages analyzed. Studies of *YIPF5* silencing in the human β-cell line EndoC-βH1 and human islets showed that partial *YIPF5* deficiency does not affect β-cell function, proliferation or survival in basal conditions, but it sensitizes β-cells to ER stress-induced apoptosis (De Franco et al., 2020b). Moreover, it was demonstrated that *YIPF5* deficiency induces ER stress and enhances the expression of the pro-apoptotic BH3 only proteins Puma and DP5 (De Franco et al., 2020b). Interestingly, the complete absence of *YIPF5* in stem cell-derived islet cells, resulted in proinsulin retention within the ER (consistent with proinsulin not being transported into the Golgi), pronounced ER stress, and β-cell failure, while *YIPF5*-KO α-cells did not show signs of ER stress or impaired survival.

A differentiation of iPSCs from two patients having a same YIPF5 mutations (Ile98Ser) into pancreatic islet cells showed that the phenotype observed in the differentiated cells was milder than the one observed in YIPF5 KO cells, suggesting that the YIPF5 protein harboring the Ile98Ser mutation still had some residual activity (De Franco et al., 2020b). All together, these findings highlight the importance of *YIPF5* for preserved β-cell function and survival and identified impaired ER-Golgi trafficking as a novel cause of monogenic diabetes.

5. Monogenic forms of diabetes associated with mitochondrial dysfunction

All secretory cells including pancreatic β-cells strongly depend on functional mitochondria for proper function and survival. Indeed, mitochondria not only provide the necessary ATP and antioxidants/metabolites that the β-cells need for their normal function but also regulate cell apoptosis controlling β-cell mass (Cnop et al., 2012). Therefore, mitochondrial failure will not only cause β-cell dysfunction but it may also induce β-cell death. In this section we discuss monogenic forms of diabetes caused by mutations in one mitochondrial protein and two mitochondrial transfer RNAs (mt-tRNAs) leading to impaired mitochondrial function.

5.1 *FXN* mutations (Friedreich's ataxia)

5.1.1 Genetic alteration and clinical phenotype

Friedreich's ataxia (FRDA) is a rare neurodegenerative disorder caused by reduced expression of the mitochondrial protein frataxin. Most FRDA patients have biallelic GAA trinucleotide-repeat-expansions (70–1700) in the first intron of the frataxin (*FXN*) gene, while a small proportion of them have an expanded GAA repeat allele and a *FXN* loss-of-function mutation in the other (Campuzano et al., 1996). The prevalence of FRDA is estimated at 1/20.000–1/50.000 (Vankan, 2013). The syndrome is characterized by severe neurological problems: clumsiness, gait ataxia, upper-limp ataxia instability coupled with dysphagia, hearing difficulties and visual loss (Marmolino, 2011; Selvadurai et al., 2018). As FRDA progresses, the patients suffer from spasticity and hypertrophic cardiopathy (Parkinson et al., 2013; Payne and Wagner, 2012; Raman et al., 2011), and 30% of them suffer from diabetes mellitus as a consequence of β-cell dysfunction and death that is accompanied by insulin resistance (Cnop et al., 2012). Accordingly, post-mortem pancreas analysis of FRDA patients showed an important reduction in β-cell mass (Cnop et al., 2012). In several cases ketoacidosis has been reported indicating absolute insulin deficiency as a result of β-cell loss (Bird et al., 1978).

5.1.2 Molecular phenotype

Frataxin is a 210 amino acid long mitochondrial protein that has been associated with cell iron metabolism. The exact role of frataxin still needs to be elucidated but it has been proposed that it plays an important role in iron homeostasis regulation, iron storage, iron-sulfur cluster (FeS) and heme biosynthesis and iron carriage (Adamec et al., 2000; Adinolfi et al., 2002, 2009; Becker et al., 2002; Condo et al., 2010; Emond et al., 2000; Gerber et al., 2003; Paupe et al., 2009; Schulz et al., 2000; Tsai and Barondeau, 2010; Yoon and Cowan, 2003, 2004). Accordingly, it was shown that frataxin deficiency leads to reduced biosynthesis and activity of heme and several FeS-containing enzymes such as proteins of the respiratory complex I, II, III and catalase resulting in decreased mitochondrial functionality, reduced ATP production and oxidative stress (Lin et al., 2017; Lu and Cortopassi, 2007; Marmolino et al., 2010; Poburski et al., 2016; Rotig et al., 1997; Salehi et al., 2014). The increased GAA repeat expansion within the first intron of the *FXN* gene reduces *FXN* transcription by heterochromatin formation resulting in 65%–95% reduction on normal frataxin protein levels which is central in the disease pathogenesis (Herman et al., 2006; Saveliev et al., 2003; Silva et al., 2015). In pancreatic β-cells frataxin deficiency leads to reduced ATP production, impaired glucose-induced insulin secretion and enhanced basal and stress-induced apoptosis (Cnop et al., 2012). It was shown that frataxin deficiency causes mitochondrial dysfunction as a consequence of mitochondrial iron overload. Indeed, FRDA patient cells and mouse models of partial frataxin deficiency showed activation of the iron regulatory proteins Irp1 and Irp2 which results in the upregulation of the transferrin receptor 1 (Tfr1) and the mitochondrial iron importer Mfrn2 together with a downregulation of the iron exporter-ferroportin and iron storage proteins (ferritins and transferrins) (Huang et al., 2011; Koeppen et al., 2009; Martelli and Puccio, 2014; Richardson and Ponka, 1997; Whitnall et al., 2008). In several cell models of frataxin deficiency, including pancreatic β-cells, it was demonstrated that the mitochondrial iron accumulation causes oxidative stress through the Fenton reaction leading to DNA damage, protein and lipid oxidation (Bradshaw et al., 2016; Eaton and Qian, 2002) and apoptosis (Igoillo-Esteve et al., 2015; Park et al., 2002). Related to the later, iPSC-derived neurons from

FRDA patients and frataxin-deficient rodent and human β-cells showed that apoptosis in FRDA is the consequence of mitochondrial-failure-mediated enhanced oxidative stress that leads to activation of the intrinsic pathway of apoptosis (Igoillo-Esteve et al., 2015, 2020). The pro-apoptotic BCL-2 family members BAD, BIM and DP5 were identified as key mediators of this process (Igoillo-Esteve et al., 2015). It has also been proposed that the increased ROS levels seen in frataxin-deficient cells is the consequence of reduced antioxidant capacity that was evidenced by impaired NRF2 signaling (Paupe et al., 2009; Shan et al., 2013), and increased GSSG/GSH ratio (Piermarini et al., 2016). Due to their capacity to counteract heterochromatin formation and enhance frataxin expression it has been proposed that histone deacetylase (HDAC) inhibitors may have therapeutic potential for FRDA (Herman et al., 2006; Soragni et al., 2014; Soragni and Gottesfeld, 2016). GLP-1 analogs were also pointed as potential novel therapeutics for FRDA since, besides enhancing frataxin protein levels in vitro and in vivo (Igoillo-Esteve et al., 2020), they showed to be effective in reducing oxidative stress, improving mitochondrial function, and cell viability in neuronal and β-cell models of frataxin deficiency (Cnop et al., 2012; Igoillo-Esteve et al., 2015, 2020).

5.2 *MTTL1* and *MTTE* mutations (maternally inherited diabetes and deafness, MIDD)

5.2.1 Genetic alteration and clinical phenotype

Maternally inherited diabetes and deafness (MIDD) is caused by various mitochondrial DNA mutations. The most frequent one is the 3243A > G in mitochondrial *MTTL1* gene coding for mitochondrial transfer RNA Leucine (mt-tRNA$^{Leu(UUR)}$) (Finsterer, 2007; Maassen et al., 2005; Murphy et al., 2008; Tsukuda et al., 1997). Mutations in the mitochondrial *MTTE* gene that encodes mt-tRNA^{Glu} are also cause of MIDD but are much less frequent (Hao et al., 1995; Shoffner et al., 1990). As the name implies, MIDD is predominantly inherited from the mother due to the maternal-inheritance pattern of mitochondrial DNA (Giles et al., 1980). The genetic presentation of each patient is highly variable and depend on the type and predominance (heteroplasmy) of the mtDNA mutation that varies among patients and within tissues (Koga et al., 2000; Murphy et al., 2008). The first clinical manifestations may appear at any age, but the disease is generally diagnosed in young adults with deafness appearing before diabetes. Diabetes is caused by an age dependent deterioration of β-cell function, with or without insulin resistance (Maassen et al., 2004).

Interestingly, the mutation is also the cause of MELAS, Mitochondrial encephalopathy, lactic acidosis, and stroke-like episodes syndrome, a much more severe phenotype of a neuromuscular disease with abnormalities in mitochondrial structure in muscles, that is usually accompanied by a proportion of mutated mitochondrial genome of 56–95% in muscle and lower levels in blood (Ciafaloni et al., 1992). Besides diabetes and sensorineural deafness MIDD patients present with a wide array of symptoms such as abnormalities in brain scans, psychiatric disorders, macular retinal dystrophy and pigmentary lesions, cardiomyopathy, kidney disease (Guillausseau et al., 2001), short stature, malabsorption and cynical vomiting syndrome and muscular abnormalities, cramps or weakness (Murphy et al., 2008).

5.2.2 Molecular phenotype

Cytosolic and mt-tRNAs are essential molecules for protein translation. In physiological conditions tRNAs undergo several post-transcriptional modifications that regulate their folding, stability and function (Phizicky and Hopper, 2010). At a molecular level, the *MTTL1* mutation affects the stability of the $tRNA_{(UUR)}{}^{Leu}$ and as a consequence, the steady state amounts of this tRNA are decreased (Yasukawa et al., 2000). Additionally, the mutant $tRNA_{(UUR)}{}^{Leu}$ lack a taurine modification in the uridine present in the anticodon wobble position (τm5U) (Yasukawa et al., 2000). The absence of this modification prevents the tRNA from properly binding the rRNA at the A-site to decode the UUG codon resulting in defective mitochondrial protein translation and mitochondrial failure (Kirino et al., 2004). At a cellular level, the 3243A > G *MTTL1* mutation results in dysregulation of the mitochondrial bioenergetics, with reduced activity of most mitochondrial enzymes such as cytochrome *c* oxidase, NADHQ1 oxidoreductase, and NADH O2 oxidoreductase (Van Den Ouweland et al., 1992) and a general decrease in mitochondrial respiratory chain activity (King et al., 1992). Due to the decrease in mitochondrial function, the mutant cells only depend on their glycolytic metabolism for ATP production (De Andrade et al., 2006). Accordingly, cybrid mutant cells (containing only mutant mitochondria) showed high ROS levels and an increase in lactate production compared to wild-type cells where pyruvate gets oxidized (Wang et al., 2018). The lack of ATP production after the addition of pyruvate to the medium in mutant cells compared to control, proves that the formers rely completely on cytosolic ATP production. Last but not least, the mutant cells presented altered calcium homeostasis (De Andrade et al., 2006). In pancreatic β-cells, which are a highly metabolic, the mitochondrial dysregulation caused by *MTTL1* mutations leads to β-cell dysfunction, decreased insulin production and loss of β-cell mass, and (Murphy et al., 2008). As mentioned before, *MTTE* mutations also cause MIDD but with less frequency. These mutations impair mt-tRNA^{Glu} pseudouridination at position 55 resulting in mt-tRNA^{Glu} degradation. This negatively impacts the translation of glutamate-rich proteins within the mitochondria, e.g., some proteins of the respiratory complexes I and IV, resulting in impaired mitochondrial function, reduced ATP production, and oxidative stress (Wang et al., 2016).

6. Monogenic forms of diabetes associated with impaired tRNA methylation

Transfer RNAs (tRNA) work as adaptor molecules between the mRNA and the aminoacids during protein synthesis (Chan and Lowe, 2016). Besides this canonical function tRNAs also play an important role in the regulation of cell homeostasis (Phizicky and Hopper, 2010). As mentioned above, tRNAs are heavily post-transcriptionally modified. The presence or absence of each of these modifications affects the 3D structure, stability and function of the tRNAs (Pan, 2018). Genetic and environmentally-mediated perturbations in tRNA metabolism have been associated with various human diseases such as neurodegenerative diseases (Blanco et al., 2014; Delaunay et al., 2016; Huang et al., 2018; Karlsborn et al., 2014; Martinez et al., 2012; Simpson et al., 2009) mitochondrial-linked disorders (Abbott et al., 2014; Asano et al., 2018; Latkany et al., 1999; Shoffner et al., 1990; Torres et al., 2014) and more recently also to the development of diabetes mellitus (Cosentino et al., 2018; Igoillo-Esteve et al., 2013; Steinthorsdottir et al., 2007; Zhou et al., 2019). tRNAs molecules may undergo fragmentation. This is a controlled process modulated by stress conditions, environmental factors, or changes in tRNA modifications (Chen et al., 2016; Fu et al., 2009; Ivanov et al., 2011). tRNA fragmentation gives origin to small tRNA-derived fragments (tRFs) and tRNA-derived halves that act as small non-coding RNAs that modulate several important cell processes such as apoptosis, protein synthesis and cell proliferation among others (Chen et al., 2016; Fu et al., 2009; Ivanov et al., 2011). The generation of tRNA fragments has been associated with human diseases such as cancer, viral infections and monogenic diabetes (see below) (Cosentino et al., 2018, 2019; Igoillo-Esteve et al., 2013; Nientiedt et al., 2016; Pereira et al., 2018; Ruggero et al., 2014;

Soares and Santos, 2017). Here below we will discuss a monogenic form of diabetes associated with impaired tRNA methylation and enhanced tRNA fragmentation.

6.1 *TRMT10A* mutations (young onset diabetes)

6.1.1 Genetic alteration and clinical phenotype

Inactivating autosomal recessive mutations in the *TRMT10A* gene, coding for the tRNA methyltransferase TRMT10A, have been associated with young onset diabetes, microcephaly and intellectual disability (Igoillo-Esteve et al., 2013). The first reported cases were three siblings born from consanguineous parents of Moroccan origin (Igoillo-Esteve et al., 2013). The patients presented with insulin-dependent diabetes, that was diagnosed in adolescence, together with short stature, microcephaly and intellectual disability. They were negative for anti-insulin, anti-GAD65, anti-IA2 antibodies and islet cell autoantibodies (Igoillo-Esteve et al., 2013). A homozygous nonsense c.379G>A; p.Arg127Stop mutation in *TRMT10A* was identified as causal of this new syndrome (Igoillo-Esteve et al., 2013). After this first report, eight additional patients from four independent families all having different nonsense, missense or deletion mutations in *TRMT10A* and similar clinical manifestations were described (Gillis et al., 2014; Narayanan et al., 2015; Yew et al., 2016; Zung et al., 2015).

6.1.2 Molecular phenotype

TRMT10A is a tRNA modifying enzyme that methylates cytosolic tRNAs at their guanosine in position 9 (m^1G_9) (Cosentino et al., 2018; Howell et al., 2019; Vilardo et al., 2020). TRMT10A protein is mostly located in the nucleus, a subcellular localization that correlates with its function since cytosolic tRNAs can be modified in the nucleus or in the cytoplasm (Igoillo-Esteve et al., 2013). TRMT10A is ubiquitously expressed in adult tissues but it is enriched in pancreatic islets and brain, the two main tissues affected by TRMT10A deficiency (Igoillo-Esteve et al., 2013). TRMT10A is also expressed in human fetal brain suggesting that it may play a role in brain development human (Igoillo-Esteve et al., 2013). All the pathogenic TRMT10A mutations identified are inactivating and lead either to the complete absence of TRMT10A or to the formation of truncated (inactive) form of the protein (Igoillo-Esteve et al., 2013; Narayanan et al., 2015; Yew et al., 2016; Zung et al., 2015). The missense mutation described by Gillis et al. (2014) results in the generation of a full length non-functional protein that retains its ability to bind tRNAs but lacks the methyl transferase activity suggesting that the main 3D structure of TRMT10A is conserved but the binding site to the donor methyl group is affected (Gillis et al., 2014). TRMT10A deficiency leads to the absence of m^1G_9 modification in certain cytosolic tRNAs: e.g., $tRNA^{iMet}$ and in $tRNA^{Gln}$ (Cosentino et al., 2018), as well as $tRNA^{Gly}$, $tRNA^{Trp}$, $tRNA^{Asn}$ and $tRNA^{Arg}$ (Howell et al., 2019). Human and rodent models of β-cell TRMT10A-deficiency showed that the absence of TRMT10A does not impair total protein synthesis or insulin secretion in control condition (Igoillo-Esteve et al., 2013). However, the lack of TRMT10A sensitizes rat β-cells and human islets, to free-fatty acid-, endoplasmic reticulum stress-, and high glucose-induced apoptosis (Igoillo-Esteve et al., 2013). Additionally, in both human and rat β-cells, TRMT10A deficiency leads to increased ROS that is accompanied by activation of the intrinsic pathway of apoptosis (Cosentino et al., 2018). Interestingly, Cosentino and co-workers showed that the absence of m^1G_9 tRNA modification leads to the $tRNA^{Gln}$ fragmentation, both in lymphoblasts and TRMT10A-deficient iPSC-derived β-cells from TRMT10A patients (Cosentino et al., 2018) and identified the 5'$tRNA^{Gln}$ fragments as key mediators apoptosis in TRMT10A-deficient human β-cells (Cosentino et al., 2018). In the haploid human cells HAP1, TRMT10A deficiency did not lead to $tRNA^{Gln}$ fragmentation (Vilardo et al., 2020), and the absence of m^1G_9 modification in $tRNA^{iMet}$ was accompanied by a decrease in the abundance of this tRNA, an effect not seen in TRMT10A-deficient β-cells (Cosentino et al., 2018; Vilardo et al., 2020). Altogether, this suggests that the outcome of impaired tRNA methylation is probably tissue and cell specific.

A more recent study done in HEK293T cells showed that TRMT10A plays also a role in the regulation of mRNA methylation via binding and enhancing the activity of the Fat Mass and obesity-associated protein (FTO) (Ontiveros et al., 2020). FTO is a demethylase that removes m^6A modification from certain mRNAs (Jia et al., 2011). The binding of TRMT10A and FTO is facilitated by the presence of FTO mRNA targets. The m^6A modification is an important regulator of mRNA stability (Wang et al., 2014c). Indeed, the presence of m^6A in the mRNA leads to its degradation by YTDHF (Wang et al., 2014c). It was shown that the absence of TRMT10A leads to a consequential decrease in the functionality of FTO, resulting in enhanced m^6A modification, and concomitant degradation of certain mRNAs. Interestingly, the tRNA methyltransferase activity of TRMT10A is not fundamental to the binding of FTO and its function in mRNA modifications regulation (Ontiveros et al., 2020). Whether TRMT10A affects FTO function in pancreatic β-cells and its implications in gene expression regulation remains to be elucidated.

7. Concluding remarks

In this chapter we have provided a wide overview on the genetic cause, clinical features, treatment and molecular phenotype of different monogenic forms of diabetes caused by mutations leading to impaired pancreas development and insulin secretion, mitochondrial dysfunction, ER stress and tRNA hypomethylation, all associated with pancreatic β-cell demise. Despite the advances in genetic testing, monogenic diabetes patients are still underdiagnosed or misdiagnosed as having T1D or T2D and are often wrongly treated with insulin. Therefore, awareness about the genetic, clinical and molecular characteristics of these rare forms of diabetes is crucial for a proper diagnosis and treatment. A careful study of the clinical phenotype and family history of each patient, together with genetic testing, is essential to identify and understand the disease, predict its evolution and provide tailored treatments. The current advances in next generation sequencing, allowing to test multiple genes at the same time, makes genetic testing more accessible than before in many countries but not yet worldwide. Despite the very large number of genes associated with monogenic forms of diabetes that have already been identified, there is still a proportion of patients suspected to have monogenic diabetes for which the causal mutation is still unknown. Additional efforts are then needed to identify the mutated genes, assess the pathogenicity of the identified variants, and the pathogenic mechanisms associated with them. Indeed, determining the pathogenicity of the variants and their genotype-phenotype association at clinical and molecular level is a very big challenge, but has great importance from a clinical perspective, since it will directly influence the patient diagnosis and management. As mentioned before, some monogenic forms of diabetes have very variable clinical presentations and penetrance, and some common genetic variants in genes linked with monogenic diabetes have been associated with increased T2D or T1D risk. This large variability in the clinical manifestations associated with mutations in a same gene have often been attributed to the type of mutation or to the influence of still unidentified extra genetic or environmental factors. Identifying such factors is then an additional unmet need that has important clinical implications. Studying the molecular mechanisms underlying monogenic forms of diabetes, by using in vitro and in vivo disease models, is of outmost importance; not only to understand the pathogenesis of these rare forms of diabetes and identify potential therapeutic targets, but also to understand the pathophysiology of more common forms of the disease such as T1D and T2D. The recent development of iPSC-derived β-cell models offers the unevaluable possibility to study mutation effects directly in patient cells. Current protocols for iPSC generation from peripheral blood mononuclear cells (PBMCs) have facilitated the development of such models. Thus, clinician awareness on the importance of getting informed-consent blood samples from monogenic diabetes patient for iPSC generation is essential to foster mechanistic studies in these rare forms of diabetes.

Author contributions

L.S.C., V.G., M.N.A., and M.I.-E. performed the bibliographic research and wrote the manuscript. M.I.-E. is the guarantor of this work. All authors revised the manuscript and approved the final version.

Conflicts of interest

All the authors declare to have no conflict of interest.

Funds

This work was supported by the Brussels Region Innoviris project DiaType, the EFSD/Novo Nordisk Program for Diabetes Research in Europe, the actions de Recherche Concertées de la Communauté Française (ARC). M.N.A. is a FRIA-FNRS fellow.

References

Abbasi, F., Azizi, F., Javaheri, M., Mosallanejad, A., Ebrahim-Habibi, A., Ghafouri-Fard, S., 2015. Segregation of a novel homozygous 6 nucleotide deletion in GLUT2 gene in a Fanconi-Bickel syndrome family. Gene 557, 103–105.

Abbott, J.A., Francklyn, C.S., Robey-Bond, S.M., 2014. Transfer RNA and human disease. Front. Genet. 5, 158.

Abdel-Salam, G.M., Schaffer, A.E., Zaki, M.S., Dixon-Salazar, T., Mostafa, I.S., Afifi, H.H., Gleeson, J.G., 2012. A homozygous IER3IP1 mutation causes microcephaly with simplified gyral pattern, epilepsy, and permanent neonatal diabetes syndrome (MEDS). Am. J. Med. Genet. A 158A, 2788–2796.

Abdollahi, A., 2007. LOT1 (ZAC1/PLAGL1) and its family members: mechanisms and functions. J. Cell. Physiol. 210, 16–25.

Abdulhadi-Atwan, M., Bushman, J., Tornovsky-Babaey, S., Perry, A., Abu-Libdeh, A., Glaser, B., Shyng, S.L., Zangen, D.H., 2008. Novel de novo mutation in sulfonylurea receptor 1 presenting as hyperinsulinism in infancy followed by overt diabetes in early adolescence. Diabetes 57, 1935–1940.

Abdulkarim, B., Nicolino, M., Igoillo-Esteve, M., Daures, M., Romero, S., Philippi, A., Senee, V., Lopes, M., Cunha, D.A., Harding, H.P., Derbois, C., Bendelac, N., Hattersley, A.T., Eizirik, D.L., Ron, D., Cnop, M., Julier, C., 2015. A missense mutation in PPP1R15B causes a syndrome including diabetes, short stature, and microcephaly. Diabetes 64, 3951–3962.

Abdulkarim, B., Hernangomez, M., Igoillo-Esteve, M., Cunha, D.A., Marselli, L., Marchetti, P., Ladriere, L., Cnop, M., 2017. Guanabenz sensitizes pancreatic beta cells to lipotoxic endoplasmic reticulum stress and apoptosis. Endocrinology 158, 1659–1670.

Adalat, S., Woolf, A.S., Johnstone, K.A., Wirsing, A., Harries, L.W., Long, D.A., Hennekam, R.C., Ledermann, S.E., Rees, L., Van't Hoff, W., Marks, S.D., Trompeter, R.S., Tullus, K., Winyard, P.J., Cansick, J., Mushtaq, I., Dhillon, H.K., Bingham, C., Edghill, E.L., Shroff, R., Stanescu, H., Ryffel, G.U., Ellard, S., Bockenhauer, D., 2009. HNF1B mutations associate with hypomagnesemia and renal magnesium wasting. J. Am. Soc. Nephrol. 20, 1123–1131.

Adamec, J., Rusnak, F., Owen, W.G., Naylor, S., Benson, L.M., Gacy, A.M., Isaya, G., 2000. Iron-dependent self-assembly of recombinant yeast frataxin: implications for Friedreich ataxia. Am. J. Hum. Genet. 67, 549–562.

Adell, T., Gomez-Cuadrado, A., Skoudy, A., Pettengill, O.S., Longnecker, D.S., Real, F.X., 2000. Role of the basic helix-loop-helix transcription factor p48 in the differentiation phenotype of exocrine pancreas cancer cells. Cell Growth Differ. 11, 137–147.

Adinolfi, S., Trifuoggi, M., Politou, A.S., Martin, S., Pastore, A., 2002. A structural approach to understanding the iron-binding properties of phylogenetically different frataxins. Hum. Mol. Genet. 11, 1865–1877.

Adinolfi, S., Iannuzzi, C., Prischi, F., Pastore, C., Iametti, S., Martin, S.R., Bonomi, F., Pastore, A., 2009. Bacterial frataxin CyaY is the gatekeeper of iron-sulfur cluster formation catalyzed by IscS. Nat. Struct. Mol. Biol. 16, 390–396.

Aftab, S., Semenec, L., Chu, J.S., Chen, N., 2008. Identification and characterization of novel human tissue-specific RFX transcription factors. BMC Evol. Biol. 8, 226.

Ahlgren, U., Jonsson, J., Edlund, H., 1996. The morphogenesis of the pancreatic mesenchyme is uncoupled from that of the pancreatic epithelium in IPF1/PDX1-deficient mice. Development 122, 1409–1416.

Ahlgren, U., Jonsson, J., Jonsson, L., Simu, K., Edlund, H., 1998. Beta-cell-specific inactivation of the mouse Ipf1/Pdx1 gene results in loss of the beta-cell phenotype and maturity onset diabetes. Genes Dev. 12, 1763–1768.

Ahlqvist, E., Turrini, F., Lang, S.T., Taneera, J., Zhou, Y., Almgren, P., Hansson, O., Isomaa, B., Tuomi, T., Eriksson, K., Eriksson, J.G., Lyssenko, V., Groop, L., 2012. A common variant upstream of the PAX6 gene influences islet function in man. Diabetologia 55, 94–104.

Aittoniemi, J., Fotinou, C., Craig, T.J., De Wet, H., Proks, P., Ashcroft, F.M., 2009. Review. SUR1: a unique ATP-binding cassette protein that functions as an ion channel regulator. Philos. Trans. R. Soc. Lond. B Biol. Sci. 364, 257–267.

Alghamdi, K.A., Alsaedi, A.B., Aljasser, A., Altawil, A., Kamal, N.M., 2017. Extended clinical features associated with novel Glis3 mutation: a case report. BMC Endocr. Disord. 17, 14.

Al-Haggar, M., Sakamoto, O., Shaltout, A., Al-Hawari, A., Wahba, Y., Abdel-Hadi, D., 2012. Mutation analysis of the GLUT2 gene in three unrelated Egyptian families with Fanconi-Bickel syndrome: revisited gene atlas for renumbering. Clin. Exp. Nephrol. 16, 604–610.

Allen, H.L., Flanagan, S.E., Shaw-Smith, C., De Franco, E., Akerman, I., Caswell, R., International Pancreatic Agenesis Consortium, Ferrer, J., Hattersley, A.T., Ellard, S., 2011. GATA6 haploinsufficiency causes pancreatic agenesis in humans. Nat. Genet. 44, 20–22.

Al-Shammari, M., Al-Husain, M., Al-Kharfy, T., Alkuraya, F.S., 2011. A novel PTF1A mutation in a patient with severe pancreatic and cerebellar involvement. Clin. Genet. 80, 196–198.
Althari, S., Najmi, L.A., Bennett, A.J., Aukrust, I., Rundle, J.K., Colclough, K., Molnes, J., Kaci, A., Nawaz, S., Van Der Lugt, T., Hassanali, N., Mahajan, A., Molven, A., Ellard, S., Mccarthy, M.I., Bjorkhaug, L., Njolstad, P.R., Gloyn, A.L., 2020. Unsupervised clustering of missense variants in HNF1A using multidimensional functional data aids clinical interpretation. Am. J. Hum. Genet. 107, 670–682.
Alvelos, M.I., Rodrigues, M., Lobo, L., Medeira, A., Sousa, A.B., Simao, C., Lemos, M.C., 2015. A novel mutation of the HNF1B gene associated with hypoplastic glomerulocystic kidney disease and neonatal renal failure: a case report and mutation update. Medicine (Baltimore) 94, e469.
Amin, S., Cook, B., Zhou, T., Ghazizadeh, Z., Lis, R., Zhang, T., Khalaj, M., Crespo, M., Perera, M., Xiang, J.Z., Zhu, Z., Tomishima, M., Liu, C., Naji, A., Evans, T., Huangfu, D., Chen, S., 2018. Discovery of a drug candidate for GLIS3-associated diabetes. Nat. Commun. 9, 2681.
Amr, S., Heisey, C., Zhang, M., Xia, X.J., Shows, K.H., Ajlouni, K., Pandya, A., Satin, L.S., El-Shanti, H., Shiang, R., 2007. A homozygous mutation in a novel zinc-finger protein, ERIS, is responsible for Wolfram syndrome 2. Am. J. Hum. Genet. 81, 673–683.
Anderson, K.R., Torres, C.A., Solomon, K., Becker, T.C., Newgard, C.B., Wright, C.V., Hagman, J., Sussel, L., 2009a. Cooperative transcriptional regulation of the essential pancreatic islet gene NeuroD1 (beta2) by Nkx2.2 and neurogenin 3. J. Biol. Chem. 284, 31236–31248.
Anderson, K.R., White, P., Kaestner, K.H., Sussel, L., 2009b. Identification of known and novel pancreas genes expressed downstream of Nkx2.2 during development. BMC Dev. Biol. 9, 65.
Angebault, C., Fauconnier, J., Patergnani, S., Rieusset, J., Danese, A., Affortit, C.A., Jagodzinska, J., Megy, C., Quiles, M., Cazevieille, C., Korchagina, J., Bonnet-Wersinger, D., Milea, D., Hamel, C., Pinton, P., Thiry, M., Lacampagne, A., Delprat, B., Delettre, C., 2018. ER-mitochondria cross-talk is regulated by the Ca(2+) sensor NCS1 and is impaired in Wolfram syndrome. Sci. Signal. 11. eaaq1380.
Apelqvist, A., Ahlgren, U., Edlund, H., 1997. Sonic hedgehog directs specialised mesoderm differentiation in the intestine and pancreas. Curr. Biol. 7, 801–804.
Arceci, R.J., King, A.A., Simon, M.C., Orkin, S.H., Wilson, D.B., 1993. Mouse GATA-4: a retinoic acid-inducible GATA-binding transcription factor expressed in endodermally derived tissues and heart. Mol. Cell. Biol. 13, 2235–2246.
Arya, V.B., Guemes, M., Nessa, A., Alam, S., Shah, P., Gilbert, C., Senniappan, S., Flanagan, S.E., Ellard, S., Hussain, K., 2014. Clinical and histological heterogeneity of congenital hyperinsulinism due to paternally inherited heterozygous ABCC8/KCNJ11 mutations. Eur. J. Endocrinol. 171, 685–695.
Asano, K., Suzuki, T., Saito, A., Wei, F.Y., Ikeuchi, Y., Numata, T., Tanaka, R., Yamane, Y., Yamamoto, T., Goto, T., Kishita, Y., Murayama, K., Ohtake, A., Okazaki, Y., Tomizawa, K., Sakaguchi, Y., Suzuki, T., 2018. Metabolic and chemical regulation of tRNA modification associated with taurine deficiency and human disease. Nucleic Acids Res. 46, 1565–1583.
Ashcroft, F.M., 2005. ATP-sensitive potassium channelopathies: focus on insulin secretion. J. Clin. Invest. 115, 2047–2058.
Ashcroft, F.M., Gribble, F.M., 2000. New windows on the mechanism of action of K(ATP) channel openers. Trends Pharmacol. Sci. 21, 439–445.
Babenko, A.P., Polak, M., Cave, H., Busiah, K., Czernichow, P., Scharfmann, R., Bryan, J., Aguilar-Bryan, L., Vaxillaire, M., Froguel, P., 2006. Activating mutations in the ABCC8 gene in neonatal diabetes mellitus. N. Engl. J. Med. 355, 456–466.

Bach, I., Galcheva-Gargova, Z., Mattei, M.G., Simon-Chazottes, D., Guenet, J.L., Cereghini, S., Yaniv, M., 1990. Cloning of human hepatic nuclear factor 1 (HNF1) and chromosomal localization of its gene in man and mouse. Genomics 8, 155–164.

Bacq, Y., Jacquemin, E., Balabaud, C., Jeannot, E., Scotto, B., Branchereau, S., Laurent, C., Bourlier, P., Pariente, D., De Muret, A., Fabre, M., Bioulac-Sage, P., Zucman-Rossi, J., 2003. Familial liver adenomatosis associated with hepatocyte nuclear factor 1alpha inactivation. Gastroenterology 125, 1470–1475.

Bahillo-Curieses, M.P., Garrote-Molpeceres, R., Minambres-Rodriguez, M., Del Real-Llorente, M.R., Tobar-Mideros, C., Rellan-Rodriguez, S., 2018. Glycosuria and hyperglycemia in the neonatal period as the first clinical sign of Fanconi-Bickel syndrome. Pediatr. Diabetes 19, 180–183.

Baird, T.D., Wek, R.C., 2012. Eukaryotic initiation factor 2 phosphorylation and translational control in metabolism. Adv. Nutr. 3, 307–321.

Balboa, D., Saarimäki-Vire, J., Borshagovski, D., Survila, M., Lindholm, P., Galli, E., Eurola, S., Ustinov, J., Grym, H., Huopio, H., Partanen, J., Wartiovaara, K., Otonkoski, T., 2018. Insulin mutations impair beta-cell development in a patient-derived iPSC model of neonatal diabetes. eLife 7, e38519.

Ban, N., Yamada, Y., Someya, Y., Miyawaki, K., Ihara, Y., Hosokawa, M., Toyokuni, S., Tsuda, K., Seino, Y., 2002. Hepatocyte nuclear factor-1alpha recruits the transcriptional co-activator p300 on the GLUT2 gene promoter. Diabetes 51, 1409–1418.

Barbetti, F., D'Annunzio, G., 2018. Genetic causes and treatment of neonatal diabetes and early childhood diabetes. Best Pract. Res. Clin. Endocrinol. Metab. 32, 575–591.

Barrett, T.G., Bundey, S.E., 1997. Wolfram (DIDMOAD) syndrome. J. Med. Genet. 34, 838–841.

Barrett, T.G., Bundey, S.E., Macleod, A.F., 1995. Neurodegeneration and diabetes: UK nationwide study of Wolfram (DIDMOAD) syndrome. Lancet 346, 1458–1463.

Barrett, J.C., Clayton, D.G., Concannon, P., Akolkar, B., Cooper, J.D., Erlich, H.A., Julier, C., Morahan, G., Nerup, J., Nierras, C., Plagnol, V., Pociot, F., Schuilenburg, H., Smyth, D.J., Stevens, H., Todd, J.A., Walker, N.M., Rich, S.S., Type 1 Diabetes Genetics Consortium, 2009. Genome-wide association study and meta-analysis find that over 40 loci affect risk of type 1 diabetes. Nat. Genet. 41, 703–707.

Bartlam, M., Yamamoto, T., 2010. The structural basis for deadenylation by the CCR4-NOT complex. Protein Cell 1, 443–452.

Becker, E.M., Greer, J.M., Ponka, P., Richardson, D.R., 2002. Erythroid differentiation and protoporphyrin IX down-regulate frataxin expression in friend cells: characterization of frataxin expression compared to molecules involved in iron metabolism and hemoglobinization. Blood 99, 3813–3822.

Beer, N.L., Van De Bunt, M., Colclough, K., Lukacs, C., Arundel, P., Chik, C.L., Grimsby, J., Ellard, S., Gloyn, A.L., 2011. Discovery of a novel site regulating glucokinase activity following characterization of a new mutation causing hyperinsulinemic hypoglycemia in humans. J. Biol. Chem. 286, 19118–19126.

Bellanne-Chantelot, C., Chauveau, D., Gautier, J.F., Dubois-Laforgue, D., Clauin, S., Beaufils, S., Wilhelm, J.M., Boitard, C., Noel, L.H., Velho, G., Timsit, J., 2004. Clinical spectrum associated with hepatocyte nuclear factor-1beta mutations. Ann. Intern. Med. 140, 510–517.

Bellanne-Chantelot, C., Carette, C., Riveline, J.P., Valero, R., Gautier, J.F., Larger, E., Reznik, Y., Ducluzeau, P.H., Sola, A., Hartemann-Heurtier, A., Lecomte, P., Chaillous, L., Laloi-Michelin, M., Wilhem, J.M., Cuny, P., Duron, F., Guerci, B., Jeandidier, N., Mosnier-Pudar, H., Assayag, M., Dubois-Laforgue, D., Velho, G., Timsit, J., 2008. The type and the position of HNF1A mutation modulate age at diagnosis of diabetes in patients with maturity-onset diabetes of the young (MODY)-3. Diabetes 57, 503–508.

Beltrand, J., Busiah, K., Vaivre-Douret, L., Fauret, A.L., Berdugo, M., Cave, H., Polak, M., 2020. Neonatal diabetes mellitus. Front. Pediatr. 8, 540718.
Bennett, K., James, C., Mutair, A., Al-Shaikh, H., Sinani, A., Hussain, K., 2011. Four novel cases of permanent neonatal diabetes mellitus caused by homozygous mutations in the glucokinase gene. Pediatr. Diabetes 12, 192–196.
Bespalova, I.N., Van Camp, G., Bom, S.J., Brown, D.J., Cryns, K., Dewan, A.T., Erson, A.E., Flothmann, K., Kunst, H.P., Kurnool, P., Sivakumaran, T.A., Cremers, C.W., Leal, S.M., Burmeister, M., Lesperance, M.M., 2001. Mutations in the Wolfram syndrome 1 gene (WFS1) are a common cause of low frequency sensorineural hearing loss. Hum. Mol. Genet. 10, 2501–2508.
Bingham, C., Hattersley, A.T., 2004. Renal cysts and diabetes syndrome resulting from mutations in hepatocyte nuclear factor-1beta. Nephrol. Dial. Transplant. 19, 2703–2708.
Bird, T.D., Turner, J.L., Sumi, S.M., Bierman, E.L., 1978. Abnormal function of endocrine pancreas and anterior pituitary in Friedreich's ataxia. Studies in a family. Ann. Intern. Med. 88, 478–481.
Blake, J.A., Ziman, M.R., 2014. Pax genes: regulators of lineage specification and progenitor cell maintenance. Development 141, 737–751.
Blake, J.A., Thomas, M., Thompson, J.A., White, R., Ziman, M., 2008. Perplexing pax: from puzzle to paradigm. Dev. Dyn. 237, 2791–2803.
Blanco, S., Dietmann, S., Flores, J.V., Hussain, S., Kutter, C., Humphreys, P., Lukk, M., Lombard, P., Treps, L., Popis, M., Kellner, S., Holter, S.M., Garrett, L., Wurst, W., Becker, L., Klopstock, T., Fuchs, H., Gailus-Durner, V., Hrabe De Angelis, M., Karadottir, R.T., Helm, M., Ule, J., Gleeson, J.G., Odom, D.T., Frye, M., 2014. Aberrant methylation of tRNAs links cellular stress to neuro-developmental disorders. EMBO J. 33, 2020–2039.
Boese, E.A., Tollefson, M.R., Schnieders, M.J., Darbro, B.W., Alward, W.L.M., Fingert, J.H., 2020. Novel intragenic PAX6 deletion in a pedigree with aniridia, morbid obesity, and diabetes. Curr. Eye Res. 45, 91–96.
Boj, S.F., Parrizas, M., Maestro, M.A., Ferrer, J., 2001. A transcription factor regulatory circuit in differentiated pancreatic cells. Proc. Natl. Acad. Sci. U. S. A. 98, 14481–14486.
Bonnefond, A., Lomberk, G., Buttar, N., Busiah, K., Vaillant, E., Lobbens, S., Yengo, L., Dechaume, A., Mignot, B., Simon, A., Scharfmann, R., Neve, B., Tanyolac, S., Hodoglugil, U., Pattou, F., Cave, H., Iovanna, J., Stein, R., Polak, M., Vaxillaire, M., Froguel, P., Urrutia, R., 2011. Disruption of a novel Kruppel-like transcription factor p300-regulated pathway for insulin biosynthesis revealed by studies of the c.-331 INS mutation found in neonatal diabetes mellitus. J. Biol. Chem. 286, 28414–28424.
Bonnefond, A., Philippe, J., Durand, E., Dechaume, A., Huyvaert, M., Montagne, L., Marre, M., Balkau, B., Fajardy, I., Vambergue, A., Vatin, V., Delplanque, J., Le Guilcher, D., De Graeve, F., Lecoeur, C., Sand, O., Vaxillaire, M., Froguel, P., 2012. Whole-exome sequencing and high throughput genotyping identified KCNJ11 as the thirteenth MODY gene. PLoS One 7, e37423.
Bonnefond, A., Vaillant, E., Philippe, J., Skrobek, B., Lobbens, S., Yengo, L., Huyvaert, M., Cave, H., Busiah, K., Scharfmann, R., Polak, M., Abdul-Rasoul, M., Froguel, P., Vaxillaire, M., 2013a. Transcription factor gene MNX1 is a novel cause of permanent neonatal diabetes in a consanguineous family. Diabetes Metab. 39, 276–280.
Bonnefond, A., Yengo, L., Philippe, J., Dechaume, A., Ezzidi, I., Vaillant, E., Gjesing, A.P., Andersson, E.A., Czernichow, S., Hercberg, S., Hadjadj, S., Charpentier, G., Lantieri, O., Balkau, B., Marre, M., Pedersen, O., Hansen, T., Froguel, P., Vaxillaire, M., 2013b. Reassessment of the putative role of BLK-p.A71T loss-of-function mutation in MODY and type 2 diabetes. Diabetologia 56, 492–496.
Bonnycastle, L.L., Chines, P.S., Hara, T., Huyghe, J.R., Swift, A.J., Heikinheimo, P., Mahadevan, J., Peltonen, S., Huopio, H., Nuutila, P., Narisu, N., Goldfeder, R.L.,

Stitzel, M.L., Lu, S., Boehnke, M., Urano, F., Collins, F.S., Laakso, M., 2013. Autosomal dominant diabetes arising from a Wolfram syndrome 1 mutation. Diabetes 62, 3943–3950.

Borck, G., Shin, B.S., Stiller, B., Mimouni-Bloch, A., Thiele, H., Kim, J.R., Thakur, M., Skinner, C., Aschenbach, L., Smirin-Yosef, P., Har-Zahav, A., Nurnberg, G., Altmuller, J., Frommolt, P., Hofmann, K., Konen, O., Nurnberg, P., Munnich, A., Schwartz, C.E., Gothelf, D., Colleaux, L., Dever, T.E., Kubisch, C., Basel-Vanagaite, L., 2012. eIF2gamma mutation that disrupts eIF2 complex integrity links intellectual disability to impaired translation initiation. Mol. Cell 48, 641–646.

Borowiec, M., Liew, C.W., Thompson, R., Boonyasrisawat, W., Hu, J., Mlynarski, W.M., El Khattabi, I., Kim, S.H., Marselli, L., Rich, S.S., Krolewski, A.S., Bonner-Weir, S., Sharma, A., Sale, M., Mychaleckyj, J.C., Kulkarni, R.N., Doria, A., 2009. Mutations at the BLK locus linked to maturity onset diabetes of the young and beta-cell dysfunction. Proc. Natl. Acad. Sci. U. S. A. 106, 14460–14465.

Bowman, P., Sulen, A., Barbetti, F., Beltrand, J., Svalastoga, P., Codner, E., Tessmann, E.H., Juliusson, P.B., Skrivarhaug, T., Pearson, E.R., Flanagan, S.E., Babiker, T., Thomas, N.J., Shepherd, M.H., Ellard, S., Klimes, I., Szopa, M., Polak, M., Iafusco, D., Hattersley, A.T., Njolstad, P.R., Neonatal Diabetes International Collaborative Group, 2018. Effectiveness and safety of long-term treatment with sulfonylureas in patients with neonatal diabetes due to KCNJ11 mutations: an international cohort study. Lancet Diabetes Endocrinol. 6, 637–646.

Boyce, M., Bryant, K.F., Jousse, C., Long, K., Harding, H.P., Scheuner, D., Kaufman, R.J., Ma, D., Coen, D.M., Ron, D., Yuan, J., 2005. A selective inhibitor of eIF2alpha dephosphorylation protects cells from ER stress. Science 307, 935–939.

Bradshaw, T.Y., Romano, L.E., Duncan, E.J., Nethisinghe, S., Abeti, R., Michael, G.J., Giunti, P., Vermeer, S., Chapple, J.P., 2016. A reduction in Drp1-mediated fission compromises mitochondrial health in autosomal recessive spastic ataxia of Charlevoix Saguenay. Hum. Mol. Genet. 25, 3232–3244.

Brun, T., Gauthier, B.R., 2008. A focus on the role of Pax4 in mature pancreatic islet beta-cell expansion and survival in health and disease. J. Mol. Endocrinol. 40, 37–45.

Brun, T., Franklin, I., St-Onge, L., Biason-Lauber, A., Schoenle, E.J., Wollheim, C.B., Gauthier, B.R., 2004. The diabetes-linked transcription factor PAX4 promotes {beta}-cell proliferation and survival in rat and human islets. J. Cell Biol. 167, 1123–1135.

Brun, T., Duhamel, D.L., Hu He, K.H., Wollheim, C.B., Gauthier, B.R., 2007. The transcription factor PAX4 acts as a survival gene in INS-1E insulinoma cells. Oncogene 26, 4261–4271.

Brun, T., Hu He, K.H., Lupi, R., Boehm, B., Wojtusciszyn, A., Sauter, N., Donath, M., Marchetti, P., Maedler, K., Gauthier, B.R., 2008. The diabetes-linked transcription factor Pax4 is expressed in human pancreatic islets and is activated by mitogens and GLP-1. Hum. Mol. Genet. 17, 478–489.

Byrne, M.M., Sturis, J., Fajans, S.S., Ortiz, F.J., Stoltz, A., Stoffel, M., Smith, M.J., Bell, G.I., Halter, J.B., Polonsky, K.S., 1995. Altered insulin secretory responses to glucose in subjects with a mutation in the MODY1 gene on chromosome 20. Diabetes 44, 699–704.

Cagalinec, M., Liiv, M., Hodurova, Z., Hickey, M.A., Vaarmann, A., Mandel, M., Zeb, A., Choubey, V., Kuum, M., Safiulina, D., Vasar, E., Veksler, V., Kaasik, A., 2016. Role of mitochondrial dynamics in neuronal development: mechanism for Wolfram syndrome. PLoS Biol. 14, e1002511.

Campuzano, V., Montermini, L., Molto, M.D., Pianese, L., Cossee, M., Cavalcanti, F., Monros, E., Rodius, F., Duclos, F., Monticelli, A., Zara, F., Canizares, J., Koutnikova, H., Bidichandani, S.I., Gellera, C., Brice, A., Trouillas, P., De Michele, G., Filla, A., De Frutos, R., Palau, F., Patel, P.I., Di Donato, S., Mandel, J.L., Cocozza, S., Koenig, M., Pandolfo, M., 1996. Friedreich's ataxia: autosomal recessive disease caused by an intronic GAA triplet repeat expansion. Science 271, 1423–1427.

Capuano, M., Garcia-Herrero, C.M., Tinto, N., Carluccio, C., Capobianco, V., Coto, I., Cola, A., Iafusco, D., Franzese, A., Zagari, A., Navas, M.A., Sacchetti, L., 2012. Glucokinase (GCK) mutations and their characterization in MODY2 children of southern Italy. PLoS One 7, e38906.

Cattoni, A., Jackson, C., Bain, M., Houghton, J., Wei, C., 2019. Phenotypic variability in two siblings with monogenic diabetes due to the same ABCC8 gene mutation. Pediatr. Diabetes 20, 482–485.

Chan, P.P., Lowe, T.M., 2016. GtRNAdb 2.0: an expanded database of transfer RNA genes identified in complete and draft genomes. Nucleic Acids Res. 44, D184–D189.

Chang, N.C., Nguyen, M., Germain, M., Shore, G.C., 2010. Antagonism of beclin 1-dependent autophagy by BCL-2 at the endoplasmic reticulum requires NAF-1. EMBO J. 29, 606–618.

Chang, N.C., Nguyen, M., Bourdon, J., Risse, P.A., Martin, J., Danialou, G., Rizzuto, R., Petrof, B.J., Shore, G.C., 2012. Bcl-2-associated autophagy regulator Naf-1 required for maintenance of skeletal muscle. Hum. Mol. Genet. 21, 2277–2287.

Chao, C.S., Mcknight, K.D., Cox, K.L., Chang, A.L., Kim, S.K., Feldman, B.J., 2015. Novel GATA6 mutations in patients with pancreatic agenesis and congenital heart malformations. PLoS One 10, e0118449.

Chapla, A., Mruthyunjaya, M.D., Asha, H.S., Varghese, D., Varshney, M., Vasan, S.K., Venkatesan, P., Nair, V., Mathai, S., Paul, T.V., Thomas, N., 2015. Maturity onset diabetes of the young in India—a distinctive mutation pattern identified through targeted next-generation sequencing. Clin. Endocrinol. (Oxf) 82, 533–542.

Chappell, L., Gorman, S., Campbell, F., Ellard, S., Rice, G., Dobbie, A., Crow, Y., 2008. A further example of a distinctive autosomal recessive syndrome comprising neonatal diabetes mellitus, intestinal atresias and gall bladder agenesis. Am. J. Med. Genet. A 146A, 1713–1717.

Chen, W.S., Manova, K., Weinstein, D.C., Duncan, S.A., Plump, A.S., Prezioso, V.R., Bachvarova, R.F., Darnell Jr., J.E., 1994. Disruption of the HNF-4 gene, expressed in visceral endoderm, leads to cell death in embryonic ectoderm and impaired gastrulation of mouse embryos. Genes Dev. 8, 2466–2477.

Chen, Y.F., Kao, C.H., Chen, Y.T., Wang, C.H., Wu, C.Y., Tsai, C.Y., Liu, F.C., Yang, C.W., Wei, Y.H., Hsu, M.T., Tsai, S.F., Tsai, T.F., 2009a. Cisd2 deficiency drives premature aging and causes mitochondria-mediated defects in mice. Genes Dev. 23, 1183–1194.

Chen, Y.F., Kao, C.H., Kirby, R., Tsai, T.F., 2009b. Cisd2 mediates mitochondrial integrity and life span in mammals. Autophagy 5, 1043–1045.

Chen, Y.Z., Gao, Q., Zhao, X.Z., Chen, Y.Z., Bennett, C.L., Xiong, X.S., Mei, C.L., Shi, Y.Q., Chen, X.M., 2010. Systematic review of TCF2 anomalies in renal cysts and diabetes syndrome/maturity onset diabetes of the young type 5. Chin Med J (Engl) 123, 3326–3333.

Chen, Q., Yan, M., Cao, Z., Li, X., Zhang, Y., Shi, J., Feng, G.H., Peng, H., Zhang, X., Zhang, Y., Qian, J., Duan, E., Zhai, Q., Zhou, Q., 2016. Sperm tsRNAs contribute to intergenerational inheritance of an acquired metabolic disorder. Science 351, 397–400.

Chen, Y.C., Taylor, A.J., Verchere, C.B., 2018. Islet prohormone processing in health and disease. Diabetes Obes. Metab. 20 (Suppl. 2), 64–76.

Cheng, K.K., Iglesias, M.A., Lam, K.S., Wang, Y., Sweeney, G., Zhu, W., Vanhoutte, P.M., Kraegen, E.W., Xu, A., 2009. APPL1 potentiates insulin-mediated inhibition of hepatic glucose production and alleviates diabetes via Akt activation in mice. Cell Metab. 9, 417–427.

Cheng, K.K., Lam, K.S., Wu, D., Wang, Y., Sweeney, G., Hoo, R.L., Zhang, J., Xu, A., 2012. APPL1 potentiates insulin secretion in pancreatic beta cells by enhancing protein kinase Akt-dependent expression of SNARE proteins in mice. Proc. Natl. Acad. Sci. U. S. A. 109, 8919–8924.

Chesney, R.W., Kaplan, B.S., Colle, E., Scriver, C.R., Mcinnes, R.R., Dupont, C.H., Drummond, K.N., 1980. Abnormalities of carbohydrate metabolism in idiopathic Fanconi syndrome. Pediatr. Res. 14, 209–215.

Cho, Y.S., Chen, C.H., Hu, C., Long, J., Ong, R.T., Sim, X., Takeuchi, F., Wu, Y., Go, M.J., Yamauchi, T., Chang, Y.C., Kwak, S.H., Ma, R.C., Yamamoto, K., Adair, L.S., Aung, T., Cai, Q., Chang, L.C., Chen, Y.T., Gao, Y., Hu, F.B., Kim, H.L., Kim, S., Kim, Y.J., Lee, J.J., Lee, N.R., Li, Y., Liu, J.J., Lu, W., Nakamura, J., Nakashima, E., Ng, D.P., Tay, W.T., Tsai, F.J., Wong, T.Y., Yokota, M., Zheng, W., Zhang, R., Wang, C., So, W.Y., Ohnaka, K., Ikegami, H., Hara, K., Cho, Y.M., Cho, N.H., Chang, T.J., Bao, Y., Hedman, A.K., Morris, A.P., Mccarthy, M.I., Consortium, D., Mu, T.C., Takayanagi, R., Park, K.S., Jia, W., Chuang, L.M., Chan, J.C., Maeda, S., Kadowaki, T., Lee, J.Y., Wu, J.Y., Teo, Y.Y., Tai, E.S., Shu, X.O., Mohlke, K.L., Kato, N., Han, B.G., Seielstad, M., 2011. Meta-analysis of genome-wide association studies identifies eight new loci for type 2 diabetes in east Asians. Nat. Genet. 44, 67–72.

Christesen, H.B., Tribble, N.D., Molven, A., Siddiqui, J., Sandal, T., Brusgaard, K., Ellard, S., Njolstad, P.R., Alm, J., Brock Jacobsen, B., Hussain, K., Gloyn, A.L., 2008. Activating glucokinase (GCK) mutations as a cause of medically responsive congenital hyperinsulinism: prevalence in children and characterisation of a novel GCK mutation. Eur. J. Endocrinol. 159, 27–34.

Churchill, A.J., Gutierrez, G.D., Singer, R.A., Lorberbaum, D.S., Fischer, K.A., Sussel, L., 2017. Genetic evidence that Nkx2.2 acts primarily downstream of Neurog3 in pancreatic endocrine lineage development. Elife 6, e20010.

Ciafaloni, E., Ricci, E., Shanske, S., Moraes, C.T., Silvestri, G., Hirano, M., Simonetti, S., Angelini, C., Donati, M.A., Garcia, C., Al, E., 1992. MELAS: clinical features, biochemistry, and molecular genetics. Ann. Neurol. 31, 391–398.

Clocquet, A.R., Egan, J.M., Stoffers, D.A., Muller, D.C., Wideman, L., Chin, G.A., Clarke, W.L., Hanks, J.B., Habener, J.F., Elahi, D., 2000. Impaired insulin secretion and increased insulin sensitivity in familial maturity-onset diabetes of the young 4 (insulin promoter factor 1 gene). Diabetes 49, 1856–1864.

Cnop, M., Igoillo-Esteve, M., Rai, M., Begu, A., Serroukh, Y., Depondt, C., Musuaya, A.E., Marhfour, I., Ladriere, L., Moles Lopez, X., Lefkaditis, D., Moore, F., Brion, J.P., Cooper, J.M., Schapira, A.H., Clark, A., Koeppen, A.H., Marchetti, P., Pandolfo, M., Eizirik, D.L., Fery, F., 2012. Central role and mechanisms of beta-cell dysfunction and death in friedreich ataxia-associated diabetes. Ann. Neurol. 72, 971–982.

Cnop, M., Toivonen, S., Igoillo-Esteve, M., Salpea, P., 2017. Endoplasmic reticulum stress and eIF2alpha phosphorylation: the Achilles heel of pancreatic beta cells. Mol. Metab. 6, 1024–1039.

Colclough, K., Bellanne-Chantelot, C., Saint-Martin, C., Flanagan, S.E., Ellard, S., 2013. Mutations in the genes encoding the transcription factors hepatocyte nuclear factor 1 alpha and 4 alpha in maturity-onset diabetes of the young and hyperinsulinemic hypoglycemia. Hum. Mutat. 34, 669–685.

Collardeau-Frachon, S., Vasiljevic, A., Jouvet, A., Bouvier, R., Senee, V., Nicolino, M., 2015. Microscopic and ultrastructural features in Wolcott-Rallison syndrome, a permanent neonatal diabetes mellitus: about two autopsy cases. Pediatr. Diabetes 16, 510–520.

Collombat, P., Mansouri, A., Hecksher-Sorensen, J., Serup, P., Krull, J., Gradwohl, G., Gruss, P., 2003. Opposing actions of Arx and Pax4 in endocrine pancreas development. Genes Dev. 17, 2591–2603.

Collombat, P., Hecksher-Sorensen, J., Broccoli, V., Krull, J., Ponte, I., Mundiger, T., Smith, J., Gruss, P., Serup, P., Mansouri, A., 2005. The simultaneous loss of Arx and Pax4 genes promotes a somatostatin-producing cell fate specification at the expense of the alpha- and beta-cell lineages in the mouse endocrine pancreas. Development 132, 2969–2980.

Collombat, P., Xu, X., Ravassard, P., Sosa-Pineda, B., Dussaud, S., Billestrup, N., Madsen, O.D., Serup, P., Heimberg, H., Mansouri, A., 2009. The ectopic expression of Pax4 in the mouse pancreas converts progenitor cells into alpha and subsequently beta cells. Cell 138, 449–462.

Colombo, C., Porzio, O., Liu, M., Massa, O., Vasta, M., Salardi, S., Beccaria, L., Monciotti, C., Toni, S., Pedersen, O., Hansen, T., Federici, L., Pesavento, R., Cadario, F., Federici, G., Ghirri, P., Arvan, P., Iafusco, D., Barbetti, F., Early Onset Diabetes Study Group of the Italian Society of Pediatric Endocrinology & Diabetes, 2008. Seven mutations in the human insulin gene linked to permanent neonatal/infancy-onset diabetes mellitus. J. Clin. Invest. 118, 2148–2156.

Concepcion, J.P., Reh, C.S., Daniels, M., Liu, X., Paz, V.P., Ye, H., Highland, H.M., Hanis, C.L., Greeley, S.A., 2014. Neonatal diabetes, gallbladder agenesis, duodenal atresia, and intestinal malrotation caused by a novel homozygous mutation in RFX6. Pediatr. Diabetes 15, 67–72.

Condo, I., Malisan, F., Guccini, I., Serio, D., Rufini, A., Testi, R., 2010. Molecular control of the cytosolic aconitase/IRP1 switch by extramitochondrial frataxin. Hum. Mol. Genet. 19, 1221–1229.

Conrad, E., Stein, R., Hunter, C.S., 2014. Revealing transcription factors during human pancreatic beta cell development. Trends Endocrinol. Metab. 25, 407–414.

Cook, T., Gebelein, B., Belal, M., Mesa, K., Urrutia, R., 1999a. Three conserved transcriptional repressor domains are a defining feature of the TIEG subfamily of Sp1-like zinc finger proteins. J. Biol. Chem. 274, 29500–29504.

Cook, T., Gebelein, B., Urrutia, R., 1999b. Sp1 and its likes: biochemical and functional predictions for a growing family of zinc finger transcription factors. Ann. N. Y. Acad. Sci. 880, 94–102.

Cosentino, C., Toivonen, S., Diaz Villamil, E., Atta, M., Ravanat, J.L., Demine, S., Schiavo, A.A., Pachera, N., Deglasse, J.P., Jonas, J.C., Balboa, D., Otonkoski, T., Pearson, E.R., Marchetti, P., Eizirik, D.L., Cnop, M., Igoillo-Esteve, M., 2018. Pancreatic beta-cell tRNA hypomethylation and fragmentation link TRMT10A deficiency with diabetes. Nucleic Acids Res. 46, 10302–10318.

Cosentino, C., Cnop, M., Igoillo-Esteve, M., 2019. The tRNA epitranscriptome and diabetes: emergence of trna hypomodifications as a cause of pancreatic beta-cell failure. Endocrinology 160, 1262–1274.

Cryns, K., Sivakumaran, T.A., Van Den Ouweland, J.M., Pennings, R.J., Cremers, C.W., Flothmann, K., Young, T.L., Smith, R.J., Lesperance, M.M., Van Camp, G., 2003. Mutational spectrum of the WFS1 gene in Wolfram syndrome, nonsyndromic hearing impairment, diabetes mellitus, and psychiatric disease. Hum. Mutat. 22, 275–287.

Cuesta-Munoz, A.L., Huopio, H., Otonkoski, T., Gomez-Zumaquero, J.M., Nanto-Salonen, K., Rahier, J., Lopez-Enriquez, S., García-Gimeno, M.A., Sanz, P., Soriguer, F.C., Laakso, M., 2004. Severe persistent hyperinsulinemic hypoglycemia due to a de novo glucokinase mutation. Diabetes 53, 2164–2168.

Cunha, D.A., Igoillo-Esteve, M., Gurzov, E.N., Germano, C.M., Naamane, N., Marhfour, I., Fukaya, M., Vanderwinden, J.M., Gysemans, C., Mathieu, C., Marselli, L., Marchetti, P., Harding, H.P., Ron, D., Eizirik, D.L., Cnop, M., 2012. Death protein 5 and p53-upregulated modulator of apoptosis mediate the endoplasmic reticulum stress-mitochondrial dialog triggering lipotoxic rodent and human beta-cell apoptosis. Diabetes 61, 2763–2775.

Currarino, G., Coln, D., Votteler, T., 1981. Triad of anorectal, sacral, and presacral anomalies. Am. J. Roentgenol. 137, 395–398.

D'Amato, E., Tammaro, P., Craig, T.J., Tosi, A., Giorgetti, R., Lorini, R., Ashcroft, F.M., 2008. Variable phenotypic spectrum of diabetes mellitus in a family carrying a novel KCNJ11 gene mutation. Diabet. Med. 25, 651–656.

D'Amato, E., Giacopelli, F., Giannattasio, A., D'annunzio, G., Bocciardi, R., Musso, M., Lorini, R., Ravazzolo, R., 2010. Genetic investigation in an Italian child with an unusual association of atrial septal defect, attributable to a new familial GATA4 gene mutation, and neonatal diabetes due to pancreatic agenesis. Diabet. Med. 27, 1195–1200.

Danielpur, L., Sohn, Y.S., Karmi, O., Fogel, C., Zinger, A., Abu-Libdeh, A., Israeli, T., Riahi, Y., Pappo, O., Birk, R., Zangen, D.H., Mittler, R., Cabantchik, Z.I., Cerasi, E., Nechushtai, R., Leibowitz, G., 2016. GLP-1-RA corrects mitochondrial labile iron accumulation and improves beta-cell function in type 2 Wolfram syndrome. J. Clin. Endocrinol. Metab. 101, 3592–3599.

D'Annunzio, G., Minuto, N., D'amato, E., De Toni, T., Lombardo, F., Pasquali, L., Lorini, R., 2008. Wolfram syndrome (diabetes insipidus, diabetes, optic atrophy, and deafness): clinical and genetic study. Diabetes Care 31, 1743–1745.

De Andrade, P.B., Rubi, B., Frigerio, F., Van Den Ouweland, J.M., Maassen, J.A., Maechler, P., 2006. Diabetes-associated mitochondrial DNA mutation A3243G impairs cellular metabolic pathways necessary for beta cell function. Diabetologia 49, 1816–1826.

De Franco, E., Shaw-Smith, C., Flanagan, S.E., Edghill, E.L., Wolf, J., Otte, V., Ebinger, F., Varthakavi, P., Vasanthi, T., Edvardsson, S., Hattersley, A.T., Ellard, S., 2013a. Biallelic PDX1 (insulin promoter factor 1) mutations causing neonatal diabetes without exocrine pancreatic insufficiency. Diabet. Med. 30, e197–e200.

De Franco, E., Shaw-Smith, C., Flanagan, S.E., Shepherd, M.H., International NDM Consortium, Hattersley, A.T., Ellard, S., 2013b. GATA6 mutations cause a broad phenotypic spectrum of diabetes from pancreatic agenesis to adult-onset diabetes without exocrine insufficiency. Diabetes 62, 993–997.

De Franco, E., Flanagan, S.E., Yagi, T., Abreu, D., Mahadevan, J., Johnson, M.B., Jones, G., Acosta, F., Mulaudzi, M., Lek, N., Oh, V., Petz, O., Caswell, R., Ellard, S., Urano, F., Hattersley, A.T., 2017. Dominant ER stress-inducing WFS1 mutations underlie a genetic syndrome of neonatal/infancy-onset diabetes, congenital sensorineural deafness, and congenital cataracts. Diabetes 66, 2044–2053.

De Franco, E., Watson, R.A., Weninger, W.J., Wong, C.C., Flanagan, S.E., Caswell, R., Green, A., Tudor, C., Lelliott, C.J., Geyer, S.H., Maurer-Gesek, B., Reissig, L.F., Lango Allen, H., Caliebe, A., Siebert, R., Holterhus, P.M., Deeb, A., Prin, F., Hilbrands, R., Heimberg, H., Ellard, S., Hattersley, A.T., Barroso, I., 2019. A specific CNOT1 mutation results in a Novel syndrome of pancreatic agenesis and holoprosencephaly through impaired pancreatic and neurological development. Am. J. Hum. Genet. 104, 985–989.

De Franco, E., Caswell, R., Johnson, M.B., Wakeling, M.N., Zung, A., Dung, V.C., Bich Ngoc, C.T., Goonetilleke, R., Vivanco Jury, M., El-Khateeb, M., Ellard, S., Flanagan, S.E., Ron, D., Hattersley, A.T., 2020a. De novo mutations in EIF2B1 affecting eIF2 signaling cause neonatal/early-onset diabetes and transient hepatic dysfunction. Diabetes 69, 477–483.

De Franco, E., Lytrivi, M., Ibrahim, H., Montaser, H., Wakeling, M.N., Fantuzzi, F., Patel, K., Demarez, C., Cai, Y., Igoillo-Esteve, M., Cosentino, C., Lithovius, V., Vihinen, H., Jokitalo, E., Laver, T.W., Johnson, M.B., Sawatani, T., Shakeri, H., Pachera, N., Haliloglu, B., Ozbek, M.N., Unal, E., Yildirim, R., Godbole, T., Yildiz, M., Aydin, B., Bilheu, A., Suzuki, I., Flanagan, S.E., Vanderhaeghen, P., Senee, V., Julier, C., Marchetti, P., Eizirik, D.L., Ellard, S., Saarimaki-Vire, J., Otonkoski, T., Cnop, M., Hattersley, A.T., 2020b. YIPF5 mutations cause neonatal diabetes and microcephaly through endoplasmic reticulum stress. J. Clin. Invest. 130, 6338–6353.

De Heredia, M.L., Cleries, R., Nunes, V., 2013. Genotypic classification of patients with Wolfram syndrome: insights into the natural history of the disease and correlation with phenotype. Genet. Med. 15, 497–506.

De Lonlay, P., Fournet, J.C., Rahier, J., Gross-Morand, M.S., Poggi-Travert, F., Foussier, V., Bonnefont, J.P., Brusset, M.C., Brunelle, F., Robert, J.J., Nihoul-Fekete, C., Saudubray, J.M., Junien, C., 1997. Somatic deletion of the imprinted 11p15 region in sporadic persistent hyperinsulinemic hypoglycemia of infancy is specific of focal adenomatous hyperplasia and endorses partial pancreatectomy. J. Clin. Invest. 100, 802–807.

De Simone, V., De Magistris, L., Lazzaro, D., Gerstner, J., Monaci, P., Nicosia, A., Cortese, R., 1991. LFB3, a heterodimer-forming homeoprotein of the LFB1 family, is expressed in specialized epithelia. EMBO J. 10, 1435–1443.

De Vos, A., Heimberg, H., Quartier, E., Huypens, P., Bouwens, L., Pipeleers, D., Schuit, F., 1995. Human and rat beta cells differ in glucose transporter but not in glucokinase gene expression. J. Clin. Invest. 96, 2489–2495.

Dekel, B., Lubin, D., Modan-Moses, D., Quint, J., Glaser, B., Meyerovitch, J., 2002. Compound heterozygosity for the common sulfonylurea receptor mutations can cause mild diazoxide-sensitive hyperinsulinism. Clin. Pediatr. (Phila) 41, 183–186.

Delaunay, S., Rapino, F., Tharun, L., Zhou, Z., Heukamp, L., Termathe, M., Shostak, K., Klevernic, I., Florin, A., Desmecht, H., Desmet, C.J., Nguyen, L., Leidel, S.A., Willis, A.E., Buttner, R., Chariot, A., Close, P., 2016. Elp3 links tRNA modification to IRES-dependent translation of LEF1 to sustain metastasis in breast cancer. J. Exp. Med. 213, 2503–2523.

Delepine, M., Nicolino, M., Barrett, T., Golamaully, M., Lathrop, G.M., Julier, C., 2000. EIF2AK3, encoding translation initiation factor 2-alpha kinase 3, is mutated in patients with Wolcott-Rallison syndrome. Nat. Genet. 25, 406–409.

Demirbilek, H., Hussain, K., 2017. Congenital hyperinsulinism: diagnosis and treatment update. J. Clin. Res. Pediatr. Endocrinol. 9, 69–87.

Demirbilek, H., Cayir, A., Flanagan, S.E., Yildirim, R., Kor, Y., Gurbuz, F., Haliloglu, B., Yildiz, M., Baran, R.T., Akbas, E.D., Demiral, M., Unal, E., Arslan, G., Vuralli, D., Buyukyilmaz, G., Al-Khawaga, S., Saeed, A., Al Maadheed, M., Khalifa, A., Onal, H., Yuksel, B., Ozbek, M.N., Bereket, A., Hattersley, A.T., Hussain, K., De Franco, E., 2020. Clinical characteristics and long-term follow-up of patients with diabetes due to PTF1A enhancer mutations. J. Clin. Endocrinol. Metab. 105. e4351-e4359.

Desai, U.J., Slosberg, E.D., Boettcher, B.R., Caplan, S.L., Fanelli, B., Stephan, Z., Gunther, V.J., Kaleko, M., Connelly, S., 2001. Phenotypic correction of diabetic mice by adenovirus-mediated glucokinase expression. Diabetes 50, 2287–2295.

Dimitri, P., Warner, J.T., Minton, J.A., Patch, A.M., Ellard, S., Hattersley, A.T., Barr, S., Hawkes, D., Wales, J.K., Gregory, J.W., 2011. Novel GLIS3 mutations demonstrate an extended multisystem phenotype. Eur. J. Endocrinol. 164, 437–443.

Dimitri, P., Habeb, A.M., Gurbuz, F., Millward, A., Wallis, S., Moussa, K., Akcay, T., Taha, D., Hogue, J., Slavotinek, A., Wales, J.K., Shetty, A., Hawkes, D., Hattersley, A.T., Ellard, S., De Franco, E., 2015. Expanding the clinical spectrum associated with GLIS3 mutations. J. Clin. Endocrinol. Metab. 100, E1362–E1369.

Dipple, K.M., McCabe, E.R., 2000. Phenotypes of patients with "simple" Mendelian disorders are complex traits: thresholds, modifiers, and systems dynamics. Am. J. Hum. Genet. 66, 1729–1735.

Doyle, M.J., Loomis, Z.L., Sussel, L., 2007. Nkx2.2-repressor activity is sufficient to specify alpha-cells and a small number of beta-cells in the pancreatic islet. Development 134, 515–523.

Drucker, D.J., 2007. The role of gut hormones in glucose homeostasis. J. Clin. Invest. 117, 24–32.

Du, X., Ounissi-Benkalha, H., Loder, M.K., Rutter, G.A., Polychronakos, C., 2012. Overexpression of ZAC impairs glucose-stimulated insulin translation and secretion in clonal pancreatic beta-cells. Diabetes Metab. Res. Rev. 28, 645–653.

Du, Y.T., Moore, L., Poplawski, N.K., De Sousa, S.M.C., 2019. Familial GATA6 mutation causing variably expressed diabetes mellitus and cardiac and renal abnormalities. Endocrinol. Diabetes Metab. Case Rep. 2019. 19-0022.
Du, T., Zeng, N., Wen, X., Zhu, P., Li, W., 2020. A rare combination of MODY5 and duodenal atresia in a patient: a case report. BMC Med. Genet. 21, 24.
Dukes, I.D., Sreenan, S., Roe, M.W., Levisetti, M., Zhou, Y.P., Ostrega, D., Bell, G.I., Pontoglio, M., Yaniv, M., Philipson, L., Polonsky, K.S., 1998. Defective pancreatic beta-cell glycolytic signaling in hepatocyte nuclear factor-1alpha-deficient mice. J. Biol. Chem. 273, 24457–24464.
Dusatkova, P., Pruhova, S., Sumnik, Z., Kolouskova, S., Obermannova, B., Cinek, O., Lebl, J., 2011. HNF1A mutation presenting with fetal macrosomia and hypoglycemia in childhood prior to onset of overt diabetes. J. Pediatr. Endocrinol. Metab. 24, 187–189.
Dutta, S., Bonner-Weir, S., Montminy, M., Wright, C., 1998. Regulatory factor linked to late-onset diabetes? Nature 392, 560.
Dykstra, K.M., Pokusa, J.E., Suhan, J., Lee, T.H., 2010. Yip1A structures the mammalian endoplasmic reticulum. Mol. Biol. Cell 21, 1556–1568.
Dymecki, S.M., Niederhuber, J.E., Desiderio, S.V., 1990. Specific expression of a tyrosine kinase gene, blk, in B lymphoid cells. Science 247, 332–336.
Eaton, J.W., Qian, M., 2002. Molecular bases of cellular iron toxicity. Free Radic. Biol. Med. 32, 833–840.
Edghill, E.L., Bingham, C., Ellard, S., Hattersley, A.T., 2006a. Mutations in hepatocyte nuclear factor-1beta and their related phenotypes. J. Med. Genet. 43, 84–90.
Edghill, E.L., Bingham, C., Slingerland, A.S., Minton, J.A., Noordam, C., Ellard, S., Hattersley, A.T., 2006b. Hepatocyte nuclear factor-1 beta mutations cause neonatal diabetes and intrauterine growth retardation: support for a critical role of HNF-1beta in human pancreatic development. Diabet. Med. 23, 1301–1306.
Edghill, E.L., Flanagan, S.E., Patch, A.M., Boustred, C., Parrish, A., Shields, B., Shepherd, M.H., Hussain, K., Kapoor, R.R., Malecki, M., Macdonald, M.J., Stoy, J., Steiner, D.F., Philipson, L.H., Bell, G.I., Neonatal Diabetes International Collaborative Group, Hattersley, A.T., Ellard, S., 2008. Insulin mutation screening in 1,044 patients with diabetes: mutations in the INS gene are a common cause of neonatal diabetes but a rare cause of diabetes diagnosed in childhood or adulthood. Diabetes 57, 1034–1042.
Edlund, H., 2001. Factors controlling pancreatic cell differentiation and function. Diabetologia 44, 1071–1079.
Eiberg, H., Hansen, L., Kjer, B., Hansen, T., Pedersen, O., Bille, M., Rosenberg, T., Tranebjaerg, L., 2006. Autosomal dominant optic atrophy associated with hearing impairment and impaired glucose regulation caused by a missense mutation in the WFS1 gene. J. Med. Genet. 43, 435–440.
Eizirik, D.L., Cnop, M., 2010. ER stress in pancreatic beta cells: the thin red line between adaptation and failure. Sci. Signal. 3, pe7.
Ellard, S., Colclough, K., 2006. Mutations in the genes encoding the transcription factors hepatocyte nuclear factor 1 alpha (HNF1A) and 4 alpha (HNF4A) in maturity-onset diabetes of the young. Hum. Mutat. 27, 854–869.
Ellard, S., Bellanne-Chantelot, C., Hattersley, A.T., European Molecular Genetics Quality Network MODY Group, 2008. Best practice guidelines for the molecular genetic diagnosis of maturity-onset diabetes of the young. Diabetologia 51, 546–553.
Elmen, L., Volpato, C.B., Kervadec, A., Pineda, S., Kalvakuri, S., Alayari, N.N., Foco, L., Pramstaller, P.P., Ocorr, K., Rossini, A., Cammarato, A., Colas, A.R., Hicks, A.A., Bodmer, R., 2020. Silencing of CCR4-NOT complex subunits affects heart structure and function. Dis. Model. Mech. 13. dmm044727.

El-Shanti, H., Lidral, A.C., Jarrah, N., Druhan, L., Ajlouni, K., 2000. Homozygosity mapping identifies an additional locus for Wolfram syndrome on chromosome 4q. Am. J. Hum. Genet. 66, 1229–1236.
Emond, M., Lepage, G., Vanasse, M., Pandolfo, M., 2000. Increased levels of plasma malondialdehyde in Friedreich ataxia. Neurology 55, 1752–1753.
Evans, T., Reitman, M., Felsenfeld, G., 1988. An erythrocyte-specific DNA-binding factor recognizes a regulatory sequence common to all chicken globin genes. Proc. Natl. Acad. Sci. U. S. A. 85, 5976–5980.
Faguer, S., Decramer, S., Chassaing, N., Bellanne-Chantelot, C., Calvas, P., Beaufils, S., Bessenay, L., Lengele, J.P., Dahan, K., Ronco, P., Devuyst, O., Chauveau, D., 2011. Diagnosis, management, and prognosis of HNF1B nephropathy in adulthood. Kidney Int. 80, 768–776.
Fajans, S.S., Bell, G.I., Polonsky, K.S., 2001. Molecular mechanisms and clinical pathophysiology of maturity-onset diabetes of the young. N. Engl. J. Med. 345, 971–980.
Fajans, S.S., Bell, G.I., Paz, V.P., Below, J.E., Cox, N.J., Martin, C., Thomas, I.H., Chen, M., 2010. Obesity and hyperinsulinemia in a family with pancreatic agenesis and MODY caused by the IPF1 mutation Pro63fsX60. Transl. Res. 156, 7–14.
Fantes, J.A., Oghene, K., Boyle, S., Danes, S., Fletcher, J.M., Bruford, E.A., Williamson, K., Seawright, A., Schedl, A., Hanson, I., Al, E., 1995. A high-resolution integrated physical, cytogenetic, and genetic map of human chromosome 11: distal p13 to proximal p15.1. Genomics 25, 447–461.
Fernandez-Zapico, M.E., Van Velkinburgh, J.C., Gutierrez-Aguilar, R., Neve, B., Froguel, P., Urrutia, R., Stein, R., 2009. MODY7 gene, KLF11, is a novel p300-dependent regulator of Pdx-1 (MODY4) transcription in pancreatic islet beta cells. J. Biol. Chem. 284, 36482–36490.
Ferre, S., Igarashi, P., 2019. New insights into the role of HNF-1beta in kidney (patho)physiology. Pediatr. Nephrol. 34, 1325–1335.
Ferre, T., Riu, E., Bosch, F., Valera, A., 1996. Evidence from transgenic mice that glucokinase is rate limiting for glucose utilization in the liver. FASEB J. 10, 1213–1218.
Ferrer, J., 2002. A genetic switch in pancreatic beta-cells: implications for differentiation and haploinsufficiency. Diabetes 51, 2355–2362.
Finsterer, J., 2007. Genetic, pathogenetic, and phenotypic implications of the mitochondrial A3243G tRNALeu(UUR) mutation. Acta Neurol. Scand. 116, 1–14.
Firdous, P., Nissar, K., Ali, S., Ganai, B.A., Shabir, U., Hassan, T., Masoodi, S.R., 2018. Genetic testing of maturity-onset diabetes of the young current status and future perspectives. Front. Endocrinol. (Lausanne) 9, 253.
Firth, H.V., Wright, C.F., Study, D.D.D., 2011. The deciphering developmental disorders (DDD) study. Dev. Med. Child Neurol. 53, 702–703.
Flanagan, S.E., Clauin, S., Bellanne-Chantelot, C., De Lonlay, P., Harries, L.W., Gloyn, A.L., Ellard, S., 2009. Update of mutations in the genes encoding the pancreatic beta-cell K(ATP) channel subunits Kir6.2 (KCNJ11) and sulfonylurea receptor 1 (ABCC8) in diabetes mellitus and hyperinsulinism. Hum. Mutat. 30, 170–180.
Flanagan, S.E., De Franco, E., Lango Allen, H., Zerah, M., Abdul-Rasoul, M.M., Edge, J.A., Stewart, H., Alamiri, E., Hussain, K., Wallis, S., De Vries, L., Rubio-Cabezas, O., Houghton, J.A., Edghill, E.L., Patch, A.M., Ellard, S., Hattersley, A.T., 2014. Analysis of transcription factors key for mouse pancreatic development establishes NKX2-2 and MNX1 mutations as causes of neonatal diabetes in man. Cell Metab. 19, 146–154.
Flanagan, S.E., Dung, V.C., Houghton, J.a.L., De Franco, E., Ngoc, C.T.B., Damhuis, A., Ashcroft, F.M., Harries, L.W., Ellard, S., 2017. An ABCC8 nonsense mutation causing neonatal diabetes through altered transcript expression. J. Clin. Res. Pediatr. Endocrinol. 9, 260–264.

Fonseca, S.G., Fukuma, M., Lipson, K.L., Nguyen, L.X., Allen, J.R., Oka, Y., Urano, F., 2005. WFS1 is a novel component of the unfolded protein response and maintains homeostasis of the endoplasmic reticulum in pancreatic beta-cells. J. Biol. Chem. 280, 39609–39615.

Fonseca, S.G., Ishigaki, S., Oslowski, C.M., Lu, S., Lipson, K.L., Ghosh, R., Hayashi, E., Ishihara, H., Oka, Y., Permutt, M.A., Urano, F., 2010. Wolfram syndrome 1 gene negatively regulates ER stress signaling in rodent and human cells. J. Clin. Invest. 120, 744–755.

Fonseca, S.G., Gromada, J., Urano, F., 2011. Endoplasmic reticulum stress and pancreatic beta-cell death. Trends Endocrinol. Metab. 22, 266–274.

Frank, G.R., Fox, J., Candela, N., Jovanovic, Z., Bochukova, E., Levine, J., Papenhausen, P.R., O'rahilly, S., Farooqi, I.S., 2013. Severe obesity and diabetes insipidus in a patient with PCSK1 deficiency. Mol. Genet. Metab. 110, 191–194.

Fraser, F.C., Gunn, T., 1977. Diabetes mellitus, diabetes insipidus, and optic atrophy. an autosomal recessive syndrome? J. Med. Genet. 14, 190–193.

Froguel, P., Vaxillaire, M., Sun, F., Velho, G., Zouali, H., Butel, M.O., Lesage, S., Vionnet, N., Clement, K., Fougerousse, F., Al, E., 1992. Close linkage of glucokinase locus on chromosome 7p to early-onset non-insulin-dependent diabetes mellitus. Nature 356, 162–164.

Fu, H., Feng, J., Liu, Q., Sun, F., Tie, Y., Zhu, J., Xing, R., Sun, Z., Zheng, X., 2009. Stress induces tRNA cleavage by angiogenin in mammalian cells. FEBS Lett. 583, 437–442.

Fu, J., Wang, T., Zhai, X., Xiao, X., 2020. Primary hepatocellular adenoma due to biallelic HNF1A mutations and its co-occurrence with MODY 3: case-report and review of the literature. Endocrine 67, 544–551.

Fujimoto, K., Polonsky, K.S., 2009. Pdx1 and other factors that regulate pancreatic beta-cell survival. Diabetes Obes. Metab. 11 (Suppl. 4), 30–37.

Fujitani, Y., Kajimoto, Y., Yasuda, T., Matsuoka, T.A., Kaneto, H., Umayahara, Y., Fujita, N., Watada, H., Miyazaki, J.I., Yamasaki, Y., Hori, M., 1999. Identification of a portable repression domain and an E1A-responsive activation domain in Pax4: a possible role of Pax4 as a transcriptional repressor in the pancreas. Mol. Cell. Biol. 19, 8281–8291.

Fukuda, A., Kawaguchi, Y., Furuyama, K., Kodama, S., Horiguchi, M., Kuhara, T., Kawaguchi, M., Terao, M., Doi, R., Wright, C.V., Hoshino, M., Chiba, T., Uemoto, S., 2008. Reduction of Ptf1a gene dosage causes pancreatic hypoplasia and diabetes in mice. Diabetes 57, 2421–2431.

Gaisl, O., Konrad, D., Joset, P., Lang-Muritano, M., 2019. A novel GATA6 variant in a boy with neonatal diabetes and diaphragmatic hernia: a familial case with a review of the literature. J. Pediatr. Endocrinol. Metab. 32, 1027–1030.

Garcin, L., Mericq, V., Fauret-Amsellem, A.L., Cave, H., Polak, M., Beltrand, J., 2020. Neonatal diabetes due to potassium channel mutation: response to sulfonylurea according to the genotype. Pediatr. Diabetes 21, 932–941.

Gardner, R.J., Mackay, D.J., Mungall, A.J., Polychronakos, C., Siebert, R., Shield, J.P., Temple, I.K., Robinson, D.O., 2000. An imprinted locus associated with transient neonatal diabetes mellitus. Hum. Mol. Genet. 9, 589–596.

Garg, V., Kathiriya, I.S., Barnes, R., Schluterman, M.K., King, I.N., Butler, C.A., Rothrock, C.R., Eapen, R.S., Hirayama-Yamada, K., Joo, K., Matsuoka, R., Cohen, J.C., Srivastava, D., 2003. GATA4 mutations cause human congenital heart defects and reveal an interaction with TBX5. Nature 424, 443–447.

Garin, I., Edghill, E.L., Akerman, I., Rubio-Cabezas, O., Rica, I., Locke, J.M., Maestro, M.A., Alshaikh, A., Bundak, R., Del Castillo, G., Deeb, A., Deiss, D., Fernandez, J.M., Godbole, K., Hussain, K., O'connell, M., Klupa, T., Kolouskova, S., Mohsin, F., Perlman, K., Sumnik, Z., Rial, J.M., Ugarte, E., Vasanthi, T., Neonatal Diabetes

International Group, Johnstone, K., Flanagan, S.E., Martinez, R., Castano, C., Patch, A.M., Fernandez-Rebollo, E., Raile, K., Morgan, N., Harries, L.W., Castano, L., Ellard, S., Ferrer, J., Perez De Nanclares, G., Hattersley, A.T., 2010. Recessive mutations in the INS gene result in neonatal diabetes through reduced insulin biosynthesis. Proc. Natl. Acad. Sci. U. S. A. 107, 3105–3110.

Garty, R., Cooper, M., Tabachnik, E., 1974. The Fanconi syndrome associated with hepatic glycogenosis and abnormal metabolism of galactose. J. Pediatr. 85, 821–823.

Gerber, J., Muhlenhoff, U., Lill, R., 2003. An interaction between frataxin and Isu1/Nfs1 that is crucial for Fe/S cluster synthesis on Isu1. EMBO Rep. 4, 906–911.

Gharanei, S., Zatyka, M., Astuti, D., Fenton, J., Sik, A., Nagy, Z., Barrett, T.G., 2013. Vacuolar-type H+-ATPase V1A subunit is a molecular partner of Wolfram syndrome 1 (WFS1) protein, which regulates its expression and stability. Hum. Mol. Genet. 22, 203–217.

Giles, R.E., Blanc, H., Cann, H.M., Wallace, D.C., 1980. Maternal inheritance of human mitochondrial DNA. Proc. Natl. Acad. Sci. U. S. A. 77, 6715–6719.

Gillis, D., Krishnamohan, A., Yaacov, B., Shaag, A., Jackman, J.E., Elpeleg, O., 2014. TRMT10A dysfunction is associated with abnormalities in glucose homeostasis, short stature and microcephaly. J. Med. Genet. 51, 581–586.

Giurgea, I., Sempoux, C., Bellanne-Chantelot, C., Ribeiro, M., Hubert, L., Boddaert, N., Saudubray, J.M., Robert, J.J., Brunelle, F., Rahier, J., Jaubert, F., Nihoul-Fekete, C., De Lonlay, P., 2006. The Knudson's two-hit model and timing of somatic mutation may account for the phenotypic diversity of focal congenital hyperinsulinism. J. Clin. Endocrinol. Metab. 91, 4118–4123.

Glaser, B., Chiu, K.C., Anker, R., Nestorowicz, A., Landau, H., Ben-Bassat, H., Shlomai, Z., Kaiser, N., Thornton, P.S., Stanley, C.A., Al, E., 1994. Familial hyperinsulinism maps to chromosome 11p14-15.1, 30 cM centromeric to the insulin gene. Nat. Genet. 7, 185–188.

Glaser, B., Kesavan, P., Heyman, M., Davis, E., Cuesta, A., Buchs, A., Stanley, C.A., Thornton, P.S., Permutt, M.A., Matschinsky, F.M., Herold, K.C., 1998. Familial hyperinsulinism caused by an activating glucokinase mutation. N. Engl. J. Med. 338, 226–230.

Glaser, B., Furth, J., Stanley, C.A., Baker, L., Thornton, P.S., Landau, H., Permutt, M.A., 1999. Intragenic single nucleotide polymorphism haplotype analysis of SUR1 mutations in familial hyperinsulinism. Hum. Mutat. 14, 23–29.

Gloyn, A.L., Noordam, K., Willemsen, M.A., Ellard, S., Lam, W.W., Campbell, I.W., Midgley, P., Shiota, C., Buettger, C., Magnuson, M.A., Matschinsky, F.M., Hattersley, A.T., 2003. Insights into the biochemical and genetic basis of glucokinase activation from naturally occurring hypoglycemia mutations. Diabetes 52, 2433–2440.

Gloyn, A.L., Pearson, E.R., Antcliff, J.F., Proks, P., Bruining, G.J., Slingerland, A.S., Howard, N., Srinivasan, S., Silva, J.M., Molnes, J., Edghill, E.L., Frayling, T.M., Temple, I.K., Mackay, D., Shield, J.P., Sumnik, Z., Van Rhijn, A., Wales, J.K., Clark, P., Gorman, S., Aisenberg, J., Ellard, S., Njolstad, P.R., Ashcroft, F.M., Hattersley, A.T., 2004. Activating mutations in the gene encoding the ATP-sensitive potassium-channel subunit Kir6.2 and permanent neonatal diabetes. N. Engl. J. Med. 350, 1838–1849.

Gloyn, A.L., Siddiqui, J., Ellard, S., 2006. Mutations in the genes encoding the pancreatic beta-cell KATP channel subunits Kir6.2 (KCNJ11) and SUR1 (ABCC8) in diabetes mellitus and hyperinsulinism. Hum. Mutat. 27, 220–231.

Goksel, D.L., Fischbach, K., Duggirala, R., Mitchell, B.D., Aguilar-Bryan, L., Blangero, J., Stern, M.P., O'connell, P., 1998. Variant in sulfonylurea receptor-1 gene is associated with high insulin concentrations in non-diabetic Mexican Americans: SUR-1 gene variant and hyperinsulinemia. Hum. Genet. 103, 280–285.

Gomez-Zaera, M., Strom, T.M., Rodriguez, B., Estivill, X., Meitinger, T., Nunes, V., 2001. Presence of a major WFS1 mutation in Spanish Wolfram syndrome pedigrees. Mol. Genet. Metab. 72, 72–81.

Gonsorcikova, L., Pruhova, S., Cinek, O., Ek, J., Pelikanova, T., Jorgensen, T., Eiberg, H., Pedersen, O., Hansen, T., Lebl, J., 2008. Autosomal inheritance of diabetes in two families characterized by obesity and a novel H241Q mutation in NEUROD1. Pediatr. Diabetes 9, 367–372.

Gosmain, Y., Cheyssac, C., Masson, M.H., Guerardel, A., Poisson, C., Philippe, J., 2012a. Pax6 is a key component of regulated glucagon secretion. Endocrinology 153, 4204–4215.

Gosmain, Y., Katz, L.S., Masson, M.H., Cheyssac, C., Poisson, C., Philippe, J., 2012b. Pax6 is crucial for beta-cell function, insulin biosynthesis, and glucose-induced insulin secretion. Mol. Endocrinol. 26, 696–709.

Gradwohl, G., Dierich, A., Lemeur, M., Guillemot, F., 2000. neurogenin3 is required for the development of the four endocrine cell lineages of the pancreas. Proc. Natl. Acad. Sci. U. S. A. 97, 1607–1611.

Gragnoli, C., Lindner, T., Cockburn, B.N., Kaisaki, P.J., Gragnoli, F., Marozzi, G., Bell, G.I., 1997. Maturity-onset diabetes of the young due to a mutation in the hepatocyte nuclear factor-4 alpha binding site in the promoter of the hepatocyte nuclear factor-1 alpha gene. Diabetes 46, 1648–1651.

Greenwood, A.L., Li, S., Jones, K., Melton, D.A., 2007. Notch signaling reveals developmental plasticity of Pax4(+) pancreatic endocrine progenitors and shunts them to a duct fate. Mech. Dev. 124, 97–107.

Gregory, L.C., Ferreira, C.B., Young-Baird, S.K., Williams, H.J., Harakalova, M., Van Haaften, G., Rahman, S.A., Gaston-Massuet, C., Kelberman, D., Gosgene, Qasim, W., Camper, S.A., Dever, T.E., Shah, P., Robinson, I., Dattani, M.T., 2019. Impaired EIF2S3 function associated with a novel phenotype of X-linked hypopituitarism with glucose dysregulation. EBioMedicine 42, 470–480.

Gribble, F.M., Reimann, F., 2003. Sulphonylurea action revisited: the post-cloning era. Diabetologia 46, 875–891.

Grupe, A., Hultgren, B., Ryan, A., Ma, Y.H., Bauer, M., Stewart, T.A., 1995. Transgenic knockouts reveal a critical requirement for pancreatic beta cell glucokinase in maintaining glucose homeostasis. Cell 83, 69–78.

Gu, C., Stein, G.H., Pan, N., Goebbels, S., Hornberg, H., Nave, K.A., Herrera, P., White, P., Kaestner, K.H., Sussel, L., Lee, J.E., 2010. Pancreatic beta cells require NeuroD to achieve and maintain functional maturity. Cell Metab. 11, 298–310.

Guan, Y., Maloney, K.A., Pollin, T.I., 2020. Patient perspectives on the diagnostic journey to a monogenic diabetes diagnosis: barriers and facilitators. J. Genet. Couns. 29, 1106–1113.

Guillausseau, P.J., Massin, P., Dubois-Laforgue, D., Timsit, J., Virally, M., Gin, H., Bertin, E., Blickle, J.F., Bouhanick, B., Cahen, J., Caillat-Zucman, S., Charpentier, G., Chedin, P., Derrien, C., Ducluzeau, P.H., Grimaldi, A., Guerci, B., Kaloustian, E., Murat, A., Olivier, F., Paques, M., Paquis-Flucklinger, V., Porokhov, B., Samuel-Lajeunesse, J., Vialettes, B., 2001. Maternally inherited diabetes and deafness: a multicenter study. Ann. Intern. Med. 134, 721–728.

Gupta, R.K., Vatamaniuk, M.Z., Lee, C.S., Flaschen, R.C., Fulmer, J.T., Matschinsky, F.M., Duncan, S.A., Kaestner, K.H., 2005. The MODY1 gene HNF-4alpha regulates selected genes involved in insulin secretion. J. Clin. Invest. 115, 1006–1015.

Gupta, S., Mcgrath, B., Cavener, D.R., 2010. PERK (EIF2AK3) regulates proinsulin trafficking and quality control in the secretory pathway. Diabetes 59, 1937–1947.

Gutgold, A., Gross, D.J., Glaser, B., Szalat, A., 2017. Diagnosis of ABCC8 congenital hyperinsulinism of infancy in a 20-year-old man evaluated for factitious hypoglycemia. J. Clin. Endocrinol. Metab. 102, 345–349.

Gutierrez, G.D., Bender, A.S., Cirulli, V., Mastracci, T.L., Kelly, S.M., Tsirigos, A., Kaestner, K.H., Sussel, L., 2017. Pancreatic beta cell identity requires continual repression of non-beta cell programs. J. Clin. Invest. 127, 244–259.

Habermann, B., 2004. The BAR-domain family of proteins: a case of bending and binding? EMBO Rep. 5, 250–255.

Haldorsen, I.S., Vesterhus, M., Raeder, H., Jensen, D.K., Sovik, O., Molven, A., Njolstad, P.R., 2008. Lack of pancreatic body and tail in HNF1B mutation carriers. Diabet. Med. 25, 782–787.

Haliyur, R., Tong, X., Sanyoura, M., Shrestha, S., Lindner, J., Saunders, D.C., Aramandla, R., Poffenberger, G., Redick, S.D., Bottino, R., Prasad, N., Levy, S.E., Blind, R.D., Harlan, D.M., Philipson, L.H., Stein, R.W., Brissova, M., Powers, A.C., 2019. Human islets expressing HNF1A variant have defective beta cell transcriptional regulatory networks. J. Clin. Invest. 129, 246–251.

Haneda, M., Polonsky, K.S., Bergenstal, R.M., Jaspan, J.B., Shoelson, S.E., Blix, P.M., Chan, S.J., Kwok, S.C., Wishner, W.B., Zeidler, A., Al, E., 1984. Familial hyperinsulinemia due to a structurally abnormal insulin. Definition of an emerging new clinical syndrome. N. Engl. J. Med. 310, 1288–1294.

Hansen, L., Eiberg, H., Barrett, T., Bek, T., Kjaersgaard, P., Tranebjaerg, L., Rosenberg, T., 2005. Mutation analysis of the WFS1 gene in seven Danish Wolfram syndrome families; four new mutations identified. Eur. J. Hum. Genet. 13, 1275–1284.

Hao, H., Bonilla, E., Manfredi, G., Dimauro, S., Moraes, C.T., 1995. Segregation patterns of a novel mutation in the mitochondrial tRNA glutamic acid gene associated with myopathy and diabetes mellitus. Am. J. Hum. Genet. 56, 1017–1025.

Harding, H.P., Zeng, H., Zhang, Y., Jungries, R., Chung, P., Plesken, H., Sabatini, D.D., Ron, D., 2001. Diabetes mellitus and exocrine pancreatic dysfunction in perk-/- mice reveals a role for translational control in secretory cell survival. Mol. Cell 7, 1153–1163.

Harding, H.P., Zhang, Y., Zeng, H., Novoa, I., Lu, P.D., Calfon, M., Sadri, N., Yun, C., Popko, B., Paules, R., Stojdl, D.F., Bell, J.C., Hettmann, T., Leiden, J.M., Ron, D., 2003. An integrated stress response regulates amino acid metabolism and resistance to oxidative stress. Mol. Cell 11, 619–633.

Hardy, C., Khanim, F., Torres, R., Scott-Brown, M., Seller, A., Poulton, J., Collier, D., Kirk, J., Polymeropoulos, M., Latif, F., Barrett, T., 1999. Clinical and molecular genetic analysis of 19 Wolfram syndrome kindreds demonstrating a wide spectrum of mutations in WFS1. Am. J. Hum. Genet. 65, 1279–1290.

Hariharan, N., Farrelly, D., Hagan, D., Hillyer, D., Arbeeny, C., Sabrah, T., Treloar, A., Brown, K., Kalinowski, S., Mookhtiar, K., 1997. Expression of human hepatic glucokinase in transgenic mice liver results in decreased glucose levels and reduced body weight. Diabetes 46, 11–16.

Haring, M.P.D., Vriesendorp, T.M., Klein Wassink-Ruiter, J.S., De Haas, R.J., Gouw, A.S.H., De Meijer, V.E., 2019. Diagnosis of hepatocellular adenoma in men before onset of diabetes in HNF1A-MODY: watch out for winkers. Liver Int. 39, 2042–2045.

Harries, L.W., Ellard, S., Stride, A., Morgan, N.G., Hattersley, A.T., 2006. Isomers of the TCF1 gene encoding hepatocyte nuclear factor-1 alpha show differential expression in the pancreas and define the relationship between mutation position and clinical phenotype in monogenic diabetes. Hum. Mol. Genet. 15, 2216–2224.

Harrison, K.A., Thaler, J., Pfaff, S.L., Gu, H., Kehrl, J.H., 1999. Pancreas dorsal lobe agenesis and abnormal islets of Langerhans in Hlxb9-deficient mice. Nat. Genet. 23, 71–75.

Harter, B., Fuchs, I., Muller, T., Akbulut, U.E., Cakir, M., Janecke, A.R., 2016. Early clinical diagnosis of PC1/3 deficiency in a patient with a novel homozygous PCSK1 splice-site mutation. J. Pediatr. Gastroenterol. Nutr. 62, 577–580.
Hatanaka, M., Tanabe, K., Yanai, A., Ohta, Y., Kondo, M., Akiyama, M., Shinoda, K., Oka, Y., Tanizawa, Y., 2011. Wolfram syndrome 1 gene (WFS1) product localizes to secretory granules and determines granule acidification in pancreatic beta-cells. Hum. Mol. Genet. 20, 1274–1284.
Hattersley, A.T., Ashcroft, F.M., 2005. Activating mutations in Kir6.2 and neonatal diabetes: new clinical syndromes, new scientific insights, and new therapy. Diabetes 54, 2503–2513.
Hattersley, A.T., Patel, K.A., 2017. Precision diabetes: learning from monogenic diabetes. Diabetologia 60, 769–777.
Hattersley, A.T., Turner, R.C., Permutt, M.A., Patel, P., Tanizawa, Y., Chiu, K.C., O'rahilly, S., Watkins, P.J., Wainscoat, J.S., 1992. Linkage of type 2 diabetes to the glucokinase gene. Lancet 339, 1307–1310.
Hattersley, A.T., Saker, P.J., Patel, P., Lo, Y.M., Page, R., O'rahilly, S., Turner, R.C., Permutt, M.A., Wainscoat, J.S., 1993. Linkage of maturity-onset diabetes of the young to the glucokinase gene—evidence of genetic heterogeneity. Biochem. Soc. Trans. 21, 24S.
Haumaitre, C., Barbacci, E., Jenny, M., Ott, M.O., Gradwohl, G., Cereghini, S., 2005. Lack of TCF2/vHNF1 in mice leads to pancreas agenesis. Proc. Natl. Acad. Sci. U. S. A. 102, 1490–1495.
Heni, M., Haupt, A., Schafer, S.A., Ketterer, C., Thamer, C., Machicao, F., Stefan, N., Staiger, H., Haring, H.U., Fritsche, A., 2010. Association of obesity risk SNPs in PCSK1 with insulin sensitivity and proinsulin conversion. BMC Med. Genet. 11, 86.
Henwood, M.J., Kelly, A., Macmullen, C., Bhatia, P., Ganguly, A., Thornton, P.S., Stanley, C.A., 2005. Genotype-phenotype correlations in children with congenital hyperinsulinism due to recessive mutations of the adenosine triphosphate-sensitive potassium channel genes. J. Clin. Endocrinol. Metab. 90, 789–794.
Herman, W.H., Fajans, S.S., Smith, M.J., Polonsky, K.S., Bell, G.I., Halter, J.B., 1997. Diminished insulin and glucagon secretory responses to arginine in nondiabetic subjects with a mutation in the hepatocyte nuclear factor-4alpha/MODY1 gene. Diabetes 46, 1749–1754.
Herman, D., Jenssen, K., Burnett, R., Soragni, E., Perlman, S.L., Gottesfeld, J.M., 2006. Histone deacetylase inhibitors reverse gene silencing in Friedreich's ataxia. Nat. Chem. Biol. 2, 551–558.
Hershey, T., Lugar, H.M., Shimony, J.S., Rutlin, J., Koller, J.M., Perantie, D.C., Paciorkowski, A.R., Eisenstein, S.A., Permutt, M.A., Washington University Wolfram Study Group, 2012. Early brain vulnerability in Wolfram syndrome. PLoS One 7, e40604.
Hilbrands, R., Keymolen, K., Michotte, A., Marichal, M., Cools, F., Goossens, A., Veld, P.I., De Schepper, J., Hattersley, A., Heimberg, H., 2017. Pancreas and gallbladder agenesis in a newborn with semilobar holoprosencephaly, a case repor. BMC Med. Genet. 18, 57.
Hilson, J.B., Merchant, S.N., Adams, J.C., Joseph, J.T., 2009. Wolfram syndrome: a clinicopathologic correlation. Acta Neuropathol. 118, 415–428.
Hinnebusch, A.G., 1985. A hierarchy of trans-acting factors modulates translation of an activator of amino acid biosynthetic genes in Saccharomyces cerevisiae. Mol. Cell. Biol. 5, 2349–2360.
Hinnebusch, A.G., 2005. Translational regulation of GCN4 and the general amino acid control of yeast. Annu. Rev. Microbiol. 59, 407–450.
Hoffmann, A., Spengler, D., 2015. Role of ZAC1 in transient neonatal diabetes mellitus and glucose metabolism. World J. Biol. Chem. 6, 95–109.

Hogewind, B.F., Pennings, R.J., Hol, F.A., Kunst, H.P., Hoefsloot, E.H., Cruysberg, J.R., Cremers, C.W., 2010. Autosomal dominant optic neuropathy and sensorineual hearing loss associated with a novel mutation of WFS1. Mol. Vis. 16, 26–35.
Hollien, J., Weissman, J.S., 2006. Decay of endoplasmic reticulum-localized mRNAs during the unfolded protein response. Science 313, 104–107.
Holt, S.H., Darash-Yahana, M., Sohn, Y.S., Song, L., Karmi, O., Tamir, S., Michaeli, D., Luo, Y., Paddock, M.L., Jennings, P.A., Onuchic, J.N., Azad, R.K., Pikarsky, E., Cabantchik, I.Z., Nechushtai, R., Mittler, R., 2016. Activation of apoptosis in NAF-1-deficient human epithelial breast cancer cells. J. Cell Sci. 129, 155–165.
Horikawa, Y., Enya, M., Mabe, H., Fukushima, K., Takubo, N., Ohashi, M., Ikeda, F., Hashimoto, K.I., Watada, H., Takeda, J., 2018. NEUROD1-deficient diabetes (MODY6): Identification of the first cases in Japanese and the clinical features. Pediatr. Diabetes 19, 236–242.
Houghton, J.A., Swift, G.H., Shaw-Smith, C., Flanagan, S.E., De Franco, E., Caswell, R., Hussain, K., Mohamed, S., Abdulrasoul, M., Hattersley, A.T., Macdonald, R.J., Ellard, S., 2016. Isolated pancreatic aplasia due to a hypomorphic PTF1A mutation. Diabetes 65, 2810–2815.
Howell, N.W., Jora, M., Jepson, B.F., Limbach, P.A., Jackman, J.E., 2019. Distinct substrate specificities of the human tRNA methyltransferases TRMT10A and TRMT10B. RNA 25, 1366–1376.
Hu He, K.H., Lorenzo, P.I., Brun, T., Jimenez Moreno, C.M., Aeberhard, D., Vallejo Ortega, J., Cornu, M., Thorel, F., Gjinovci, A., Thorens, B., Herrera, P.L., Meda, P., Wollheim, C.B., Gauthier, B.R., 2011. In vivo conditional Pax4 overexpression in mature islet beta-cells prevents stress-induced hyperglycemia in mice. Diabetes 60, 1705–1715.
Huang, S.M., Stallcup, M.R., 2000. Mouse Zac1, a transcriptional coactivator and repressor for nuclear receptors. Mol. Cell. Biol. 20, 1855–1867.
Huang, S.M., Schonthal, A.H., Stallcup, M.R., 2001. Enhancement of p53-dependent gene activation by the transcriptional coactivator Zac1. Oncogene 20, 2134–2143.
Huang, J., Levitsky, L.L., Rhoads, D.B., 2009. Novel P2 promoter-derived HNF4alpha isoforms with different N-terminus generated by alternate exon insertion. Exp. Cell Res. 315, 1200–1211.
Huang, M.L., Lane, D.J., Richardson, D.R., 2011. Mitochondrial mayhem: the mitochondrion as a modulator of iron metabolism and its role in disease. Antioxid. Redox Signal. 15, 3003–3019.
Huang, S.Q., Sun, B., Xiong, Z.P., Shu, Y., Zhou, H.H., Zhang, W., Xiong, J., Li, Q., 2018. The dysregulation of tRNAs and tRNA derivatives in cancer. J. Exp. Clin. Cancer Res. 37, 101.
Igoillo-Esteve, M., Genin, A., Lambert, N., Desir, J., Pirson, I., Abdulkarim, B., Simonis, N., Drielsma, A., Marselli, L., Marchetti, P., Vanderhaeghen, P., Eizirik, D.L., Wuyts, W., Julier, C., Chakera, A.J., Ellard, S., Hattersley, A.T., Abramowicz, M., Cnop, M., 2013. tRNA methyltransferase homolog gene TRMT10A mutation in young onset diabetes and primary microcephaly in humans. PLoS Genet. 9, e1003888.
Igoillo-Esteve, M., Gurgul-Convey, E., Hu, A., Romagueira Bichara Dos Santos, L., Abdulkarim, B., Chintawar, S., Marselli, L., Marchetti, P., Jonas, J.C., Eizirik, D.L., Pandolfo, M., Cnop, M., 2015. Unveiling a common mechanism of apoptosis in beta-cells and neurons in Friedreich's ataxia. Hum. Mol. Genet. 24, 2274–2286.
Igoillo-Esteve, M., Oliveira, A.F., Cosentino, C., Fantuzzi, F., Demarez, C., Toivonen, S., Hu, A., Chintawar, S., Lopes, M., Pachera, N., Cai, Y., Abdulkarim, B., Rai, M., Marselli, L., Marchetti, P., Tariq, M., Jonas, J.C., Boscolo, M., Pandolfo, M., Eizirik, D.L., Cnop, M., 2020. Exenatide induces frataxin expression and improves mitochondrial function in Friedreich ataxia. JCI Insight 5, e134221.

Iida, K., Li, Y., Mcgrath, B.C., Frank, A., Cavener, D.R., 2007. PERK eIF2 alpha kinase is required to regulate the viability of the exocrine pancreas in mice. BMC Cell Biol. 8, 38.

Inoue, H., Tanizawa, Y., Wasson, J., Behn, P., Kalidas, K., Bernal-Mizrachi, E., Mueckler, M., Marshall, H., Donis-Keller, H., Crock, P., Rogers, D., Mikuni, M., Kumashiro, H., Higashi, K., Sobue, G., Oka, Y., Permutt, M.A., 1998. A gene encoding a transmembrane protein is mutated in patients with diabetes mellitus and optic atrophy (Wolfram syndrome). Nat. Genet. 20, 143–148.

Ishihara, H., Takeda, S., Tamura, A., Takahashi, R., Yamaguchi, S., Takei, D., Yamada, T., Inoue, H., Soga, H., Katagiri, H., Tanizawa, Y., Oka, Y., 2004. Disruption of the WFS1 gene in mice causes progressive beta-cell loss and impaired stimulus-secretion coupling in insulin secretion. Hum. Mol. Genet. 13, 1159–1170.

Isik, E., Demirbilek, H., Houghton, J.A., Ellard, S., Flanagan, S.E., Hussain, K., 2019. Congenital hyperinsulinism and evolution to sulfonylurearesponsive diabetes later in life due to a novel homozygous p.L171F ABCC8 mutation. J. Clin. Res. Pediatr. Endocrinol. 11, 82–87.

Ivanoshchuk, D.E., Shakhtshneider, E.V., Rymar, O.D., Ovsyannikova, A.K., Mikhailova, S.V., Orlov, P.S., Ragino, Y.I., Voevoda, M.I., 2020. Analysis of APPL1 gene polymorphisms in patients with a phenotype of maturity onset diabetes of the young. J. Pers. Med. 10, 100.

Ivanov, P., Emara, M.M., Villen, J., Gygi, S.P., Anderson, P., 2011. Angiogenin-induced tRNA fragments inhibit translation initiation. Mol. Cell 43, 613–623.

Iype, T., Francis, J., Garmey, J.C., Schisler, J.C., Nesher, R., Weir, G.C., Becker, T.C., Newgard, C.B., Griffen, S.C., Mirmira, R.G., 2005. Mechanism of insulin gene regulation by the pancreatic transcription factor Pdx-1: application of pre-mRNA analysis and chromatin immunoprecipitation to assess formation of functional transcriptional complexes. J. Biol. Chem. 280, 16798–16807.

Jackson, M.J., Bindoff, L.A., Weber, K., Wilson, J.N., Ince, P., Alberti, K.G., Turnbull, D.M., 1994. Biochemical and molecular studies of mitochondrial function in diabetes insipidus, diabetes mellitus, optic atrophy, and deafness. Diabetes Care 17, 728–733.

Jackson, R.S., Creemers, J.W., Ohagi, S., Raffin-Sanson, M.L., Sanders, L., Montague, C.T., Hutton, J.C., O'rahilly, S., 1997. Obesity and impaired prohormone processing associated with mutations in the human prohormone convertase 1 gene. Nat. Genet. 16, 303–306.

Jacquemin, P., Durviaux, S.M., Jensen, J., Godfraind, C., Gradwohl, G., Guillemot, F., Madsen, O.D., Carmeliet, P., Dewerchin, M., Collen, D., Rousseau, G.G., Lemaigre, F.P., 2000. Transcription factor hepatocyte nuclear factor 6 regulates pancreatic endocrine cell differentiation and controls expression of the proendocrine gene ngn3. Mol. Cell. Biol. 20, 4445–4454.

Jannin, A., Espiard, S., Douillard, C., Pasquier, F., Bellanne-Chantelot, C., Vantyghem, M.C., 2018. Hyperinsulinemic hypoglycemia without insulinoma: think of activating glucokinase mutation. Presse Med. 47, 595–597.

Jennings, R.E., Berry, A.A., Kirkwood-Wilson, R., Roberts, N.A., Hearn, T., Salisbury, R.J., Blaylock, J., Piper Hanley, K., Hanley, N.A., 2013. Development of the human pancreas from foregut to endocrine commitment. Diabetes 62, 3514–3522.

Jennings, R.E., Scharfmann, R., Staels, W., 2020. Transcription factors that shape the mammalian pancreas. Diabetologia 63, 1974–1980.

Jensen, J.N., Rosenberg, L.C., Hecksher-Sorensen, J., Serup, P., 2007. Mutant neurogenin-3 in congenital malabsorptive diarrhea. N. Engl. J. Med. 356, 1781–1782. author reply 1782.

Jia, G., Fu, Y., Zhao, X., Dai, Q., Zheng, G., Yang, Y., Yi, C., Lindahl, T., Pan, T., Yang, Y.G., He, C., 2011. N6-methyladenosine in nuclear RNA is a major substrate of the obesity-associated FTO. Nat. Chem. Biol. 7, 885–887.

Jiang, S., Tanaka, T., Iwanari, H., Hotta, H., Yamashita, H., Kumakura, J., Watanabe, Y., Uchiyama, Y., Aburatani, H., Hamakubo, T., Kodama, T., Naito, M., 2003. Expression and localization of P1 promoter-driven hepatocyte nuclear factor-4alpha (HNF4alpha) isoforms in human and rats. Nucl. Recept. 1, 5.

Jin, K., Xiang, M., 2019. Transcription factor Ptf1a in development, diseases and reprogramming. Cell. Mol. Life Sci. 76, 921–940.

Johnson, S.R., Leo, P., Conwell, L.S., Harris, M., Brown, M.A., Duncan, E.L., 2018. Clinical usefulness of comprehensive genetic screening in maturity onset diabetes of the young (MODY): a novel ABCC8 mutation in a previously screened family. J. Diabetes 10, 764–767.

Jonsson, J., Carlsson, L., Edlund, T., Edlund, H., 1994. Insulin-promoter-factor 1 is required for pancreas development in mice. Nature 371, 606–609.

Jousse, C., Oyadomari, S., Novoa, I., Lu, P., Zhang, Y., Harding, H.P., Ron, D., 2003. Inhibition of a constitutive translation initiation factor 2alpha phosphatase, CReP, promotes survival of stressed cells. J. Cell Biol. 163, 767–775.

Julier, C., Nicolino, M., 2010. Wolcott-Rallison syndrome. Orphanet J. Rare Dis. 5, 29.

Kanakatti Shankar, R., Pihoker, C., Dolan, L.M., Standiford, D., Badaru, A., Dabelea, D., Rodriguez, B., Black, M.H., Imperatore, G., Hattersley, A., Ellard, S., Gilliam, L.K., SEARCH for Diabetes in Youth Study Group, 2013. Permanent neonatal diabetes mellitus: prevalence and genetic diagnosis in the SEARCH for Diabetes in Youth Study. Pediatr. Diabetes 14, 174–180.

Kanatsuka, A., Tokuyama, Y., Nozaki, O., Matsui, K., Egashira, T., 2002. Beta-cell dysfunction in late-onset diabetic subjects carrying homozygous mutation in transcription factors NeuroD1 and Pax4. Metabolism 51, 1161–1165.

Kang, H.S., Kim, Y.S., Zeruth, G., Beak, J.Y., Gerrish, K., Kilic, G., Sosa-Pineda, B., Jensen, J., Pierreux, C.E., Lemaigre, F.P., Foley, J., Jetten, A.M., 2009. Transcription factor Glis3, a novel critical player in the regulation of pancreatic beta-cell development and insulin gene expression. Mol. Cell. Biol. 29, 6366–6379.

Karlsborn, T., Tukenmez, H., Chen, C., Bystrom, A.S., 2014. Familial dysautonomia (FD) patients have reduced levels of the modified wobble nucleoside mcm(5)s(2)U in tRNA. Biochem. Biophys. Res. Commun. 454, 441–445.

Kashiwagi, K., Yokoyama, T., Nishimoto, M., Takahashi, M., Sakamoto, A., Yonemochi, M., Shirouzu, M., Ito, T., 2019. Structural basis for eIF2B inhibition in integrated stress response. Science 364, 495–499.

Kataoka, K., Han, S.I., Shioda, S., Hirai, M., Nishizawa, M., Handa, H., 2002. MafA is a glucose-regulated and pancreatic beta-cell-specific transcriptional activator for the insulin gene. J. Biol. Chem. 277, 49903–49910.

Kawaguchi, Y., Cooper, B., Gannon, M., Ray, M., Macdonald, R.J., Wright, C., 2002. The role of the transcriptional regulator Ptf1a in converting intestinal to pancreatic progenitors. Nat. Genet. 32, 128–134.

Kehar, M., Bijarnia, S., Ellard, S., Houghton, J., Saxena, R., Verma, I.C., Wadhwa, N., 2014. Fanconi-Bickel syndrome—mutation in SLC2A2 gene. Indian J. Pediatr. 81, 1237–1239.

Kellett, G.L., Brot-Laroche, E., 2005. Apical GLUT2: a major pathway of intestinal sugar absorption. Diabetes 54, 3056–3062.

Kernohan, K.D., Tetreault, M., Liwak-Muir, U., Geraghty, M.T., Qin, W., Venkateswaran, S., Davila, J., Care4rare Canada Consortium, Holcik, M., Majewski, J., Richer, J., Boycott, K.M., 2015. Homozygous mutation in the eukaryotic translation initiation factor 2alpha phosphatase gene, PPP1R15B, is associated with severe microcephaly, short stature and intellectual disability. Hum. Mol. Genet. 24, 6293–6300.

Kettunen, J.L.T., Parviainen, H., Miettinen, P.J., Farkkila, M., Tamminen, M., Salonen, P., Lantto, E., Tuomi, T., 2017. Biliary anomalies in patients with HNF1B diabetes. J. Clin. Endocrinol. Metab. 102, 2075–2082.

Khandelwal, P., Sinha, A., Jain, V., Houghton, J., Hari, P., Bagga, A., 2018. Fanconi syndrome and neonatal diabetes: phenotypic heterogeneity in patients with GLUT2 defects. CEN Case Rep. 7, 1–4.

Khanim, F., Kirk, J., Latif, F., Barrett, T.G., 2001. WFS1/wolframin mutations, Wolfram syndrome, and associated diseases. Hum. Mutat. 17, 357–367.

Kim, J.W., Seghers, V., Cho, J.H., Kang, Y., Kim, S., Ryu, Y., Baek, K., Aguilar-Bryan, L., Lee, Y.D., Bryan, J., Suh-Kim, H., 2002. Transactivation of the mouse sulfonylurea receptor I gene by BETA2/NeuroD. Mol. Endocrinol. 16, 1097–1107.

Kim, S.H., Ma, X., Klupa, T., Powers, C., Pezzolesi, M., Warram, J.H., Rich, S.S., Krolewski, A.S., Doria, A., 2003a. Genetic modifiers of the age at diagnosis of diabetes (MODY3) in carriers of hepatocyte nuclear factor-1alpha mutations map to chromosomes 5p15, 9q22, and 14q24. Diabetes 52, 2182–2186.

Kim, Y.S., Nakanishi, G., Lewandoski, M., Jetten, A.M., 2003b. GLIS3, a novel member of the GLIS subfamily of Kruppel-like zinc finger proteins with repressor and activation functions. Nucleic Acids Res. 31, 5513–5525.

Kim, S.H., Ma, X., Weremowicz, S., Ercolino, T., Powers, C., Mlynarski, W., Bashan, K.A., Warram, J.H., Mychaleckyj, J., Rich, S.S., Krolewski, A.S., Doria, A., 2004. Identification of a locus for maturity-onset diabetes of the young on chromosome 8p23. Diabetes 53, 1375–1384.

King, M.P., Koga, Y., Davidson, M., Schon, E.A., 1992. Defects in mitochondrial protein synthesis and respiratory chain activity segregate with the tRNA(Leu(UUR)) mutation associated with mitochondrial myopathy, encephalopathy, lactic acidosis, and strokelike episodes. Mol. Cell. Biol. 12, 480–490.

Kinsley, B.T., Swift, M., Dumont, R.H., Swift, R.G., 1995. Morbidity and mortality in the Wolfram syndrome. Diabetes Care 18, 1566–1570.

Kirino, Y., Yasukawa, T., Ohta, S., Akira, S., Ishihara, K., Watanabe, K., Suzuki, T., 2004. Codon-specific translational defect caused by a wobble modification deficiency in mutant tRNA from a human mitochondrial disease. Proc. Natl. Acad. Sci. U. S. A. 101, 15070–15075.

Kloft, N., Neukirch, C., Von Hoven, G., Bobkiewicz, W., Weis, S., Boller, K., Husmann, M., 2012. A subunit of eukaryotic translation initiation factor 2alpha-phosphatase (CreP/PPP1R15B) regulates membrane traffic. J. Biol. Chem. 287, 35299–35317.

Kodo, K., Nishizawa, T., Furutani, M., Arai, S., Yamamura, E., Joo, K., Takahashi, T., Matsuoka, R., Yamagishi, H., 2009. GATA6 mutations cause human cardiac outflow tract defects by disrupting semaphorin-plexin signaling. Proc. Natl. Acad. Sci. U. S. A. 106, 13933–13938.

Koeppen, A.H., Morral, J.A., Davis, A.N., Qian, J., Petrocine, S.V., Knutson, M.D., Gibson, W.M., Cusack, M.J., Li, D., 2009. The dorsal root ganglion in Friedreich's ataxia. Acta Neuropathol. 118, 763–776.

Koga, Y., Akita, Y., Takane, N., Sato, Y., Kato, H., 2000. Heterogeneous presentation in A3243G mutation in the mitochondrial tRNA(Leu(UUR)) gene. Arch. Dis. Child. 82, 407–411.

Koster, J.C., Marshall, B.A., Ensor, N., Corbett, J.A., Nichols, C.G., 2000. Targeted overactivity of beta cell K(ATP) channels induces profound neonatal diabetes. Cell 100, 645–654.

Kotalova, R., Dusatkova, P., Cinek, O., Dusatkova, L., Dedic, T., Seeman, T., Lebl, J., Pruhova, S., 2015. Hepatic phenotypes of HNF1B gene mutations: a case of neonatal cholestasis requiring portoenterostomy and literature review. World J. Gastroenterol. 21, 2550–2557.

Kranjc, T., Dempsey, E., Cagney, G., Nakamura, N., Shields, D.C., Simpson, J.C., 2017. Functional characterisation of the YIPF protein family in mammalian cells. Histochem. Cell Biol. 147, 439–451.
Krapp, A., Knofler, M., Frutiger, S., Hughes, G.J., Hagenbuchle, O., Wellauer, P.K., 1996. The p48 DNA-binding subunit of transcription factor PTF1 is a new exocrine pancreas-specific basic helix-loop-helix protein. EMBO J. 15, 4317–4329.
Krapp, A., Knofler, M., Ledermann, B., Burki, K., Berney, C., Zoerkler, N., Hagenbuchle, O., Wellauer, P.K., 1998. The bHLH protein PTF1-p48 is essential for the formation of the exocrine and the correct spatial organization of the endocrine pancreas. Genes Dev. 12, 3752–3763.
Kristinsson, S.Y., Thorolfsdottir, E.T., Talseth, B., Steingrimsson, E., Thorsson, A.V., Helgason, T., Hreidarsson, A.B., Arngrimsson, R., 2001. MODY in Iceland is associated with mutations in HNF-1alpha and a novel mutation in NeuroD1. Diabetologia 44, 2098–2103.
Kruszka, P., Berger, S.I., Weiss, K., Everson, J.L., Martinez, A.F., Hong, S., Anyane-Yeboa, K., Lipinski, R.J., Muenke, M., 2019. A CCR4-NOT transcription complex, subunit 1, CNOT1, variant associated with holoprosencephaly. Am. J. Hum. Genet. 104, 990–993.
Kulkarni, A., Sharma, V.K., Nabi, F., 2017. PDX1 gene mutation with permanent neonatal diabetes mellitus with annular pancreas, duodenal atresia, hypoplastic gall bladder and exocrine pancreatic insufficiency. Indian Pediatr. 54, 1052–1053.
Ladriere, L., Igoillo-Esteve, M., Cunha, D.A., Brion, J.P., Bugliani, M., Marchetti, P., Eizirik, D.L., Cnop, M., 2010. Enhanced signaling downstream of ribonucleic acid-activated protein kinase-like endoplasmic reticulum kinase potentiates lipotoxic endoplasmic reticulum stress in human islets. J. Clin. Endocrinol. Metab. 95, 1442–1449.
Lang, D., Powell, S.K., Plummer, R.S., Young, K.P., Ruggeri, B.A., 2007. PAX genes: roles in development, pathophysiology, and cancer. Biochem. Pharmacol. 73, 1–14.
Latkany, P., Ciulla, T.A., Cacchillo, P.F., Malkoff, M.D., 1999. Mitochondrial maculopathy: geographic atrophy of the macula in the MELAS associated A to G 3243 mitochondrial DNA point mutation. Am. J. Ophthalmol. 128, 112–114.
Lau, H.H., Ng, N.H.J., Loo, L.S.W., Jasmen, J.B., Teo, A.K.K., 2018. The molecular functions of hepatocyte nuclear factors—in and beyond the liver. J. Hepatol. 68, 1033–1048.
Laukkanen, O., Pihlajamaki, J., Lindstrom, J., Eriksson, J., Valle, T.T., Hamalainen, H., Ilanne-Parikka, P., Keinanen-Kiukaanniemi, S., Tuomilehto, J., Uusitupa, M., Laakso, M., Finnish Diabetes Prevention Study Group, 2004. Polymorphisms of the SUR1 (ABCC8) and Kir6.2 (KCNJ11) genes predict the conversion from impaired glucose tolerance to type 2 diabetes. The Finnish Diabetes Prevention Study. J. Clin. Endocrinol. Metab. 89, 6286–6290.
Lee, Y.H., Sauer, B., Gonzalez, F.J., 1998. Laron dwarfism and non-insulin-dependent diabetes mellitus in the Hnf-1alpha knockout mouse. Mol. Cell. Biol. 18, 3059–3068.
Lee, J.C., Smith, S.B., Watada, H., Lin, J., Scheel, D., Wang, J., Mirmira, R.G., German, M.S., 2001. Regulation of the pancreatic pro-endocrine gene neurogenin3. Diabetes 50, 928–936.
Lehto, M., Tuomi, T., Mahtani, M.M., Widen, E., Forsblom, C., Sarelin, L., Gullstrom, M., Isomaa, B., Lehtovirta, M., Hyrkko, A., Kanninen, T., Orho, M., Manley, S., Turner, R.C., Brettin, T., Kirby, A., Thomas, J., Duyk, G., Lander, E., Taskinen, M.R., Groop, L., 1997. Characterization of the MODY3 phenotype. Early-onset diabetes caused by an insulin secretion defect. J. Clin. Invest. 99, 582–591.
Lek, M., Karczewski, K.J., Minikel, E.V., Samocha, K.E., Banks, E., Fennell, T., O'donnell-Luria, A.H., Ware, J.S., Hill, A.J., Cummings, B.B., Tukiainen, T., Birnbaum, D.P., Kosmicki, J.A., Duncan, L.E., Estrada, K., Zhao, F., Zou, J., Pierce-Hoffman, E., Berghout, J., Cooper, D.N., Deflaux, N., Depristo, M., Do, R.,

Flannick, J., Fromer, M., Gauthier, L., Goldstein, J., Gupta, N., Howrigan, D., Kiezun, A., Kurki, M.I., Moonshine, A.L., Natarajan, P., Orozco, L., Peloso, G.M., Poplin, R., Rivas, M.A., Ruano-Rubio, V., Rose, S.A., Ruderfer, D.M., Shakir, K., Stenson, P.D., Stevens, C., Thomas, B.P., Tiao, G., Tusie-Luna, M.T., Weisburd, B., Won, H.H., Yu, D., Altshuler, D.M., Ardissino, D., Boehnke, M., Danesh, J., Donnelly, S., Elosua, R., Florez, J.C., Gabriel, S.B., Getz, G., Glatt, S.J., Hultman, C.M., Kathiresan, S., Laakso, M., Mccarroll, S., Mccarthy, M.I., Mcgovern, D., Mcpherson, R., Neale, B.M., Palotie, A., Purcell, S.M., Saleheen, D., Scharf, J.M., Sklar, P., Sullivan, P.F., Tuomilehto, J., Tsuang, M.T., Watkins, H.C., Wilson, J.G., Daly, M.J., Macarthur, D.G., Exome Aggregation Consortium, 2016. Analysis of protein-coding genetic variation in 60,706 humans. Nature 536, 285–291.

Lemelman, M.B., Letourneau, L., Greeley, S.a.W., 2018. Neonatal diabetes mellitus: an update on diagnosis and management. Clin. Perinatol. 45, 41–59.

Leonard, J., Peers, B., Johnson, T., Ferreri, K., Lee, S., Montminy, M.R., 1993. Characterization of somatostatin transactivating factor-1, a novel homeobox factor that stimulates somatostatin expression in pancreatic islet cells. Mol. Endocrinol. 7, 1275–1283.

Leturque, A., Brot-Laroche, E., Le Gall, M., Stolarczyk, E., Tobin, V., 2005. The role of GLUT2 in dietary sugar handling. J. Physiol. Biochem. 61, 529–537.

Leu, S., Lin, Y.M., Wu, C.H., Ouyang, P., 2012. Loss of Pnn expression results in mouse early embryonic lethality and cellular apoptosis through SRSF1-mediated alternative expression of Bcl-xS and ICAD. J. Cell Sci. 125, 3164–3172.

Li, H., Arber, S., Jessell, T.M., Edlund, H., 1999. Selective agenesis of the dorsal pancreas in mice lacking homeobox gene Hlxb9. Nat. Genet. 23, 67–70.

Li, Y., Iida, K., O'neil, J., Zhang, P., Li, S., Frank, A., Gabai, A., Zambito, F., Liang, S.H., Rosen, C.J., Cavener, D.R., 2003. PERK eIF2alpha kinase regulates neonatal growth by controlling the expression of circulating insulin-like growth factor-I derived from the liver. Endocrinology 144, 3505–3513.

Li, J., Mao, X., Dong, L.Q., Liu, F., Tong, L., 2007. Crystal structures of the BAR-PH and PTB domains of human APPL1. Structure 15, 525–533.

Liang, H., Zhang, Y., Li, M., Yan, J., Yang, D., Luo, S., Zheng, X., Yang, G., Li, Z., Xu, W., Groop, L., Weng, J., 2020. Recognition of maturity-onset diabetes of the young in China. J. Diabetes Investig., 1–9. https://doi.org/10.1111/jdi.13378.

Lin, W., Popko, B., 2009. Endoplasmic reticulum stress in disorders of myelinating cells. Nat. Neurosci. 12, 379–385.

Lin, H., Magrane, J., Rattelle, A., Stepanova, A., Galkin, A., Clark, E.M., Dong, Y.N., Halawani, S.M., Lynch, D.R., 2017. Early cerebellar deficits in mitochondrial biogenesis and respiratory chain complexes in the KIKO mouse model of Friedreich ataxia. Dis. Model. Mech. 10, 1343–1352.

Liu, L., Furuta, H., Minami, A., Zheng, T., Jia, W., Nanjo, K., Xiang, K., 2007. A novel mutation, Ser159Pro in the NeuroD1/BETA2 gene contributes to the development of diabetes in a Chinese potential MODY family. Mol. Cell. Biochem. 303, 115–120.

Liu, C., Li, H., Qi, L., Loos, R.J., Qi, Q., Lu, L., Gan, W., Lin, X., 2011. Variants in GLIS3 and CRY2 are associated with type 2 diabetes and impaired fasting glucose in Chinese Hans. PLoS One 6, e21464.

Liu, L., Nagashima, K., Yasuda, T., Liu, Y., Hu, H.R., He, G., Feng, B., Zhao, M., Zhuang, L., Zheng, T., Friedman, T.C., Xiang, K., 2013. Mutations in KCNJ11 are associated with the development of autosomal dominant, early-onset type 2 diabetes. Diabetologia 56, 2609–2618.

Liu, M., Sun, J., Cui, J., Chen, W., Guo, H., Barbetti, F., Arvan, P., 2015. INS-gene mutations: from genetics and beta cell biology to clinical disease. Mol. Aspects Med. 42, 3–18.

Liu, M., Weiss, M.A., Arunagiri, A., Yong, J., Rege, N., Sun, J., Haataja, L., Kaufman, R.J., Arvan, P., 2018. Biosynthesis, structure, and folding of the insulin precursor protein. Diabetes Obes. Metab. 20 (Suppl. 2), 28–50.

Lomberk, G., Urrutia, R., 2005. The family feud: turning off Sp1 by Sp1-like KLF proteins. Biochem. J. 392, 1–11.

Lomberk, G., Grzenda, A., Mathison, A., Escande, C., Zhang, J.S., Calvo, E., Miller, L.J., Iovanna, J., Chini, E.N., Fernandez-Zapico, M.E., Urrutia, R., 2013. Kruppel-like factor 11 regulates the expression of metabolic genes via an evolutionarily conserved protein interaction domain functionally disrupted in maturity onset diabetes of the young. J. Biol. Chem. 288, 17745–17758.

Lord, K., Dzata, E., Snider, K.E., Gallagher, P.R., De Leon, D.D., 2013. Clinical presentation and management of children with diffuse and focal hyperinsulinism: a review of 223 cases. J. Clin. Endocrinol. Metab. 98, E1786–E1789.

Lorenzo, P.I., Fuente-Martin, E., Brun, T., Cobo-Vuilleumier, N., Jimenez-Moreno, C.M., I, G.H.G., Lopez Noriega, L., Mellado-Gil, J.M., Martin-Montalvo, A., Soria, B., Gauthier, B.R., 2015. PAX4 defines an expandable beta-cell subpopulation in the adult pancreatic islet. Sci. Rep. 5, 15672.

Love-Gregory, L.D., Wasson, J., Ma, J., Jin, C.H., Glaser, B., Suarez, B.K., Permutt, M.A., 2004. A common polymorphism in the upstream promoter region of the hepatocyte nuclear factor-4 alpha gene on chromosome 20q is associated with type 2 diabetes and appears to contribute to the evidence for linkage in an Ashkenazi Jewish population. Diabetes 53, 1134–1140.

Lu, C., Cortopassi, G., 2007. Frataxin knockdown causes loss of cytoplasmic iron-sulfur cluster functions, redox alterations and induction of heme transcripts. Arch. Biochem. Biophys. 457, 111–122.

Lu, J., Li, G., Lan, M.S., Zhang, S., Fan, W., Wang, H., Lu, D., 2007. Pax4 paired domain mediates direct protein transduction into mammalian cells. Endocrinology 148, 5558–5565.

Lu, S., Kanekura, K., Hara, T., Mahadevan, J., Spears, L.D., Oslowski, C.M., Martinez, R., Yamazaki-Inoue, M., Toyoda, M., Neilson, A., Blanner, P., Brown, C.M., Semenkovich, C.F., Marshall, B.A., Hershey, T., Umezawa, A., Greer, P.A., Urano, F., 2014. A calcium-dependent protease as a potential therapeutic target for Wolfram syndrome. Proc. Natl. Acad. Sci. U. S. A. 111, E5292–E5301.

Lugar, H.M., Koller, J.M., Rutlin, J., Marshall, B.A., Kanekura, K., Urano, F., Bischoff, A.N., Shimony, J.S., Hershey, T., Washington University Wolfram Syndrome Research Study Group, 2016. Neuroimaging evidence of deficient axon myelination in Wolfram syndrome. Sci. Rep. 6, 21167.

Lugar, H.M., Koller, J.M., Rutlin, J., Eisenstein, S.A., Neyman, O., Narayanan, A., Chen, L., Shimony, J.S., Hershey, T., 2019. Evidence for altered neurodevelopment and neurodegeneration in Wolfram syndrome using longitudinal morphometry. Sci. Rep. 9, 6010.

Lussier, C.R., Brial, F., Roy, S.A., Langlois, M.J., Verdu, E.F., Rivard, N., Perreault, N., Boudreau, F., 2010. Loss of hepatocyte-nuclear-factor-1alpha impacts on adult mouse intestinal epithelial cell growth and cell lineages differentiation. PLoS One 5, e12378.

Ma, D., Shield, J.P., Dean, W., Leclerc, I., Knauf, C., Burcelin, R.R., Rutter, G.A., Kelsey, G., 2004. Impaired glucose homeostasis in transgenic mice expressing the human transient neonatal diabetes mellitus locus, TNDM. J. Clin. Invest. 114, 339–348.

Maassen, J.A., Lm, T.H., Van Essen, E., Heine, R.J., Nijpels, G., Jahangir Tafrechi, R.S., Raap, A.K., Janssen, G.M., Lemkes, H.H., 2004. Mitochondrial diabetes: molecular mechanisms and clinical presentation. Diabetes 53 (Suppl. 1), S103–S109.

Maassen, J.A., Janssen, G.M., Hart, T., M, L., 2005. Molecular mechanisms of mitochondrial diabetes (MIDD). Ann. Med. 37, 213–221.

MacDonald, G.C., Hesselson, S.E., Chan, J.Y., Jenkins, A.B., Laybutt, D.R., Hesselson, D., Campbell, L., 2019. Deletion distal to the PAX6 coding region reveals a novel basis for familial cosegregation of aniridia and diabetes mellitus. Diabetes Res. Clin. Pract. 148, 64–71.

MacKay, D.J., Temple, I.K., 2010. Transient neonatal diabetes mellitus type 1. Am. J. Med. Genet. C Semin. Med. Genet. 154C, 335–342.

MacKay, D.J., Boonen, S.E., Clayton-Smith, J., Goodship, J., Hahnemann, J.M., Kant, S.G., Njolstad, P.R., Robin, N.H., Robinson, D.O., Siebert, R., Shield, J.P., White, H.E., Temple, I.K., 2006. A maternal hypomethylation syndrome presenting as transient neonatal diabetes mellitus. Hum. Genet. 120, 262–269.

MacMullen, C.M., Zhou, Q., Snider, K.E., Tewson, P.H., Becker, S.A., Aziz, A.R., Ganguly, A., Shyng, S.L., Stanley, C.A., 2011. Diazoxide-unresponsive congenital hyperinsulinism in children with dominant mutations of the beta-cell sulfonylurea receptor SUR1. Diabetes 60, 1797–1804.

Maestro, M.A., Boj, S.F., Luco, R.F., Pierreux, C.E., Cabedo, J., Servitja, J.M., German, M.S., Rousseau, G.G., Lemaigre, F.P., Ferrer, J., 2003. Hnf6 and Tcf2 (MODY5) are linked in a gene network operating in a precursor cell domain of the embryonic pancreas. Hum. Mol. Genet. 12, 3307–3314.

Maghnie, M., Barbetti, F., 2019. MEHMO syndrome and the link between brain, pituitary and pancreas. EBioMedicine 42, 26–27.

Malecki, M.T., Jhala, U.S., Antonellis, A., Fields, L., Doria, A., Orban, T., Saad, M., Warram, J.H., Montminy, M., Krolewski, A.S., 1999. Mutations in NEUROD1 are associated with the development of type 2 diabetes mellitus. Nat. Genet. 23, 323–328.

Malhotra, J.D., Kaufman, R.J., 2011. ER stress and its functional link to mitochondria: role in cell survival and death. Cold Spring Harb. Perspect. Biol. 3, a004424.

Manz, F., Bickel, H., Brodehl, J., Feist, D., Gellissen, K., Gescholl-Bauer, B., Gilli, G., Harms, E., Helwig, H., Nutzenadel, W., Al, E., 1987. Fanconi-Bickel syndrome. Pediatr. Nephrol. 1, 509–518.

Marciniak, S.J., Ron, D., 2006. Endoplasmic reticulum stress signaling in disease. Physiol. Rev. 86, 1133–1149.

Marmolino, D., 2011. Friedreich's ataxia: past, present and future. Brain Res. Rev. 67, 311–330.

Marmolino, D., Manto, M., Acquaviva, F., Vergara, P., Ravella, A., Monticelli, A., Pandolfo, M., 2010. PGC-1alpha down-regulation affects the antioxidant response in Friedreich's ataxia. PLoS One 5, e10025.

Marshall, B.A., Permutt, M.A., Paciorkowski, A.R., Hoekel, J., Karzon, R., Wasson, J., Viehover, A., White, N.H., Shimony, J.S., Manwaring, L., Austin, P., Hullar, T.E., Hershey, T., Washington University Wolfram Study Group, 2013. Phenotypic characteristics of early Wolfram syndrome. Orphanet J. Rare Dis. 8, 64.

Martelli, A., Puccio, H., 2014. Dysregulation of cellular iron metabolism in Friedreich ataxia: from primary iron-sulfur cluster deficit to mitochondrial iron accumulation. Front. Pharmacol. 5, 130.

Martin, M.G., Lindberg, I., Solorzano-Vargas, R.S., Wang, J., Avitzur, Y., Bandsma, R., Sokollik, C., Lawrence, S., Pickett, L.A., Chen, Z., Egritas, O., Dalgic, B., Albornoz, V., De Ridder, L., Hulst, J., Gok, F., Aydogan, A., Al-Hussaini, A., Gok, D.E., Yourshaw, M., Wu, S.V., Cortina, G., Stanford, S., Georgia, S., 2013. Congenital proprotein convertase 1/3 deficiency causes malabsorptive diarrhea and other endocrinopathies in a pediatric cohort. Gastroenterology 145, 138–148.

Martinez, F.J., Lee, J.H., Lee, J.E., Blanco, S., Nickerson, E., Gabriel, S., Frye, M., Al-Gazali, L., Gleeson, J.G., 2012. Whole exome sequencing identifies a splicing mutation in NSUN2 as a cause of a Dubowitz-like syndrome. J. Med. Genet. 49, 380–385.

Martinez, R., Gutierrez-Nogues, A., Fernandez-Ramos, C., Velayos, T., Vela, A., Spanish Congenital Hyperinsulinism Group, Navas, M.A., Castano, L., 2017. Heterogeneity in phenotype of hyperinsulinism caused by activating glucokinase mutations: a novel mutation and its functional characterization. Clin. Endocrinol. (Oxf) 86, 778–783.

Martin-Montalvo, A., Lorenzo, P.I., Lopez-Noriega, L., Gauthier, B.R., 2017. Targeting pancreatic expressed PAX genes for the treatment of diabetes mellitus and pancreatic neuroendocrine tumors. Expert Opin. Ther. Targets 21, 77–89.

Martinovici, D., Ransy, V., Vanden Eijnden, S., Ridremont, C., Pardou, A., Cassart, M., Avni, F., Donner, C., Lingier, P., Mathieu, A., Gulbis, B., De Brouckere, V., Cnop, M., Abramowicz, M., Desir, J., 2010. Neonatal hemochromatosis and Martinez-Frias syndrome of intestinal atresia and diabetes mellitus in a consanguineous newborn. Eur. J. Med. Genet. 53, 25–28.

Massa, O., Iafusco, D., D'amato, E., Gloyn, A.L., Hattersley, A.T., Pasquino, B., Tonini, G., Dammacco, F., Zanette, G., Meschi, F., Porzio, O., Bottazzo, G., Crino, A., Lorini, R., Cerutti, F., Vanelli, M., Barbetti, F., Early Onset Diabetes Study Group of the Italian Society of Pediatric Endocrinology & Diabetology, 2005. KCNJ11 activating mutations in Italian patients with permanent neonatal diabetes. Hum. Mutat. 25, 22–27.

Masui, T., Swift, G.H., Hale, M.A., Meredith, D.M., Johnson, J.E., Macdonald, R.J., 2008. Transcriptional autoregulation controls pancreatic Ptf1a expression during development and adulthood. Mol. Cell. Biol. 28, 5458–5468.

Mathison, A., Escande, C., Calvo, E., Seo, S., White, T., Salmonson, A., Faubion Jr., W.A., Buttar, N., Iovanna, J., Lomberk, G., Chini, E.N., Urrutia, R., 2015. Phenotypic characterization of mice carrying homozygous deletion of KLF11, a gene in which mutations cause human neonatal and MODY VII diabetes. Endocrinology 156, 3581–3595.

Matschinsky, F.M., 1990. Glucokinase as glucose sensor and metabolic signal generator in pancreatic beta-cells and hepatocytes. Diabetes 39, 647–652.

Matschinsky, F.M., 2002. Regulation of pancreatic beta-cell glucokinase: from basics to therapeutics. Diabetes 51 (Suppl. 3), S394–S404.

Matschinsky, F.M., Wilson, D.F., 2019. The central role of glucokinase in glucose homeostasis: a perspective 50 years after demonstrating the presence of the enzyme in islets of langerhans. Front. Physiol. 10, 148.

Matsuoka, T.A., Artner, I., Henderson, E., Means, A., Sander, M., Stein, R., 2004. The MafA transcription factor appears to be responsible for tissue-specific expression of insulin. Proc. Natl. Acad. Sci. U. S. A. 101, 2930–2933.

Matsuura, T., Tamura, T., Chinen, Y., Ohta, T., 2002. A novel mutation (N32K) of GLUT2 gene in a Japanese patient with Fanconi-Bickel syndrome. Clin. Genet. 62, 255–256.

Mauvais-Jarvis, F., Smith, S.B., Le May, C., Leal, S.M., Gautier, J.F., Molokhia, M., Riveline, J.P., Rajan, A.S., Kevorkian, J.P., Zhang, S., Vexiau, P., German, M.S., Vaisse, C., 2004. PAX4 gene variations predispose to ketosis-prone diabetes. Hum. Mol. Genet. 13, 3151–3159.

Maxwell, K.G., Augsornworawat, P., Velazco-Cruz, L., Kim, M.H., Asada, R., Hogrebe, N.J., Morikawa, S., Urano, F., Millman, J.R., 2020. Gene-edited human stem cell-derived beta cells from a patient with monogenic diabetes reverse preexisting diabetes in mice. Sci. Transl. Med. 12. eaax9106.

McCarthy, M.I., Hattersley, A.T., 2008. Learning from molecular genetics: novel insights arising from the definition of genes for monogenic and type 2 diabetes. Diabetes 57, 2889–2898.

McCulloch, L.J., Van De Bunt, M., Braun, M., Frayn, K.N., Clark, A., Gloyn, A.L., 2011. GLUT2 (SLC2A2) is not the principal glucose transporter in human pancreatic beta cells: implications for understanding genetic association signals at this locus. Mol. Genet. Metab. 104, 648–653.

McGrath, P.S., Watson, C.L., Ingram, C., Helmrath, M.A., Wells, J.M., 2015. The basic helix-loop-helix transcription factor NEUROG3 is required for development of the human endocrine pancreas. Diabetes 64, 2497–2505.

McTaggart, J.S., Clark, R.H., Ashcroft, F.M., 2010. The role of the KATP channel in glucose homeostasis in health and disease: more than meets the islet. J. Physiol. 588, 3201–3209.

Meirhaeghe, A., Helbecque, N., Cottel, D., Arveiler, D., Ruidavets, J.B., Haas, B., Ferrieres, J., Tauber, J.P., Bingham, A., Amouyel, P., 2001. Impact of sulfonylurea receptor 1 genetic variability on non-insulin-dependent diabetes mellitus prevalence and treatment: a population study. Am. J. Med. Genet. 101, 4–8.

Mellado-Gil, J.M., Jimenez-Moreno, C.M., Martin-Montalvo, A., Alvarez-Mercado, A.I., Fuente-Martin, E., Cobo-Vuilleumier, N., Lorenzo, P.I., Bru-Tari, E., Herrera-Gomez Ide, G., Lopez-Noriega, L., Perez-Florido, J., Santoyo-Lopez, J., Spyrantis, A., Meda, P., Boehm, B.O., Quesada, I., Gauthier, B.R., 2016. PAX4 preserves endoplasmic reticulum integrity preventing beta cell degeneration in a mouse model of type 1 diabetes mellitus. Diabetologia 59, 755–765.

Merika, M., Orkin, S.H., 1993. DNA-binding specificity of GATA family transcription factors. Mol. Cell. Biol. 13, 3999–4010.

Meur, G., Simon, A., Harun, N., Virally, M., Dechaume, A., Bonnefond, A., Fetita, S., Tarasov, A.I., Guillausseau, P.J., Boesgaard, T.W., Pedersen, O., Hansen, T., Polak, M., Gautier, J.F., Froguel, P., Rutter, G.A., Vaxillaire, M., 2010. Insulin gene mutations resulting in early-onset diabetes: marked differences in clinical presentation, metabolic status, and pathogenic effect through endoplasmic reticulum retention. Diabetes 59, 653–661.

Meusser, B., Hirsch, C., Jarosch, E., Sommer, T., 2005. ERAD: the long road to destruction. Nat. Cell Biol. 7, 766–772.

Mirmira, R.G., Watada, H., German, M.S., 2000. Beta-cell differentiation factor Nkx6.1 contains distinct DNA binding interference and transcriptional repression domains. J. Biol. Chem. 275, 14743–14751.

Misra, S., Vedovato, N., Cliff, E., De Franco, E., Hattersley, A.T., Ashcroft, F.M., Oliver, N.S., 2018. Permanent neonatal diabetes: combining sulfonylureas with insulin may be an effective treatment. Diabet. Med. 35, 1291–1296.

Mitchell, J., Punthakee, Z., Lo, B., Bernard, C., Chong, K., Newman, C., Cartier, L., Desilets, V., Cutz, E., Hansen, I.L., Riley, P., Polychronakos, C., 2004. Neonatal diabetes, with hypoplastic pancreas, intestinal atresia and gall bladder hypoplasia: search for the aetiology of a new autosomal recessive syndrome. Diabetologia 47, 2160–2167.

Mitchell, R.K., Nguyen-Tu, M.S., Chabosseau, P., Callingham, R.M., Pullen, T.J., Cheung, R., Leclerc, I., Hodson, D.J., Rutter, G.A., 2017. The transcription factor Pax6 is required for pancreatic beta cell identity, glucose-regulated ATP synthesis, and Ca(2+) dynamics in adult mice. J. Biol. Chem. 292, 8892–8906.

Miura, A., Yamagata, K., Kakei, M., Hatakeyama, H., Takahashi, N., Fukui, K., Nammo, T., Yoneda, K., Inoue, Y., Sladek, F.M., Magnuson, M.A., Kasai, H., Miyagawa, J., Gonzalez, F.J., Shimomura, I., 2006. Hepatocyte nuclear factor-4alpha is essential for glucose-stimulated insulin secretion by pancreatic beta-cells. J. Biol. Chem. 281, 5246–5257.

Miyata, T., Maeda, T., Lee, J.E., 1999. NeuroD is required for differentiation of the granule cells in the cerebellum and hippocampus. Genes Dev. 13, 1647–1652.

Moates, J.M., Nanda, S., Cissell, M.A., Tsai, M.J., Stein, R., 2003. BETA2 activates transcription from the upstream glucokinase gene promoter in islet beta-cells and gut endocrine cells. Diabetes 52, 403–408.

Mohammad, S., Wolfe, L.A., Stobe, P., Biskup, S., Wainwright, M.S., Melin-Aldana, H., Malladi, P., Muenke, M., Gahl, W.A., Whitington, P.F., 2016. Infantile cirrhosis,

growth impairment, and neurodevelopmental anomalies associated with deficiency of PPP1R15B. J. Pediatr. 179, 144–149 e2.
Mohan, V., Radha, V., Nguyen, T.T., Stawiski, E.W., Pahuja, K.B., Goldstein, L.D., Tom, J., Anjana, R.M., Kong-Beltran, M., Bhangale, T., Jahnavi, S., Chandni, R., Gayathri, V., George, P., Zhang, N., Murugan, S., Phalke, S., Chaudhuri, S., Gupta, R., Zhang, J., Santhosh, S., Stinson, J., Modrusan, Z., Ramprasad, V.L., Seshagiri, S., Peterson, A.S., 2018. Comprehensive genomic analysis identifies pathogenic variants in maturity-onset diabetes of the young (MODY) patients in South India. BMC Med. Genet. 19, 22.
Mohandas Nair, K., Sakamoto, O., Jagadeesh, S., Nampoothiri, S., 2012. Fanconi-Bickel syndrome. Indian J. Pediatr. 79, 112–114.
Molven, A., Ringdal, M., Nordbo, A.M., Raeder, H., Stoy, J., Lipkind, G.M., Steiner, D.F., Philipson, L.H., Bergmann, I., Aarskog, D., Undlien, D.E., Joner, G., Sovik, O., Norwegian Childhood Diabetes Study Group, Bell, G.I., Njolstad, P.R., 2008. Mutations in the insulin gene can cause MODY and autoantibody-negative type 1 diabetes. Diabetes 57, 1131–1135.
Mona, B., Avila, J.M., Meredith, D.M., Kollipara, R.K., Johnson, J.E., 2016. Regulating the dorsal neural tube expression of Ptf1a through a distal 3' enhancer. Dev. Biol. 418, 216–225.
Moortgat, S., Desir, J., Benoit, V., Boulanger, S., Pendeville, H., Nassogne, M.C., Lederer, D., Maystadt, I., 2016. Two novel EIF2S3 mutations associated with syndromic intellectual disability with severe microcephaly, growth retardation, and epilepsy. Am. J. Med. Genet. A 170, 2927–2933.
Motoda, S., Fujita, S., Kozawa, J., Kimura, T., Fukui, K., Ikuno, Y., Imagawa, A., Iwahashi, H., Shimomura, I., 2019. Case of a novel PAX6 mutation with aniridia and insulin-dependent diabetes mellitus. J. Diabetes Investig. 10, 552–553.
Mozzillo, E., Delvecchio, M., Carella, M., Grandone, E., Palumbo, P., Salina, A., Aloi, C., Buono, P., Izzo, A., D'annunzio, G., Vecchione, G., Orrico, A., Genesio, R., Simonelli, F., Franzese, A., 2014. A novel CISD2 intragenic deletion, optic neuropathy and platelet aggregation defect in Wolfram syndrome type 2. BMC Med. Genet. 15, 88.
Mueckler, M., Kruse, M., Strube, M., Riggs, A.C., Chiu, K.C., Permutt, M.A., 1994. A mutation in the Glut2 glucose transporter gene of a diabetic patient abolishes transport activity. J. Biol. Chem. 269, 17765–17767.
Muhsin, N.I.A., Bentley, L., Bai, Y., Goldsworthy, M., Cox, R.D., 2020. A novel mutation in the mouse Pcsk1 gene showing obesity and diabetes. Mamm. Genome 31, 17–29.
Murphy, R., Turnbull, D.M., Walker, M., Hattersley, A.T., 2008. Clinical features, diagnosis and management of maternally inherited diabetes and deafness (MIDD) associated with the 3243A > G mitochondrial point mutation. Diabet. Med. 25, 383–399.
Nair, G., Hebrok, M., 2015. Islet formation in mice and men: lessons for the generation of functional insulin-producing beta-cells from human pluripotent stem cells. Curr. Opin. Genet. Dev. 32, 171–180.
Nair, G.G., Vincent, R.K., Odorico, J.S., 2014. Ectopic Ptf1a expression in murine ESCs potentiates endocrine differentiation and models pancreas development in vitro. Stem Cells 32, 1195–1207.
Nakamura, A., Ishidu, K., Tajima, T., 2012. Early onset of liver steatosis in a Japanese girl with maturity-onset diabetes of the young type 3 (MODY3). J. Clin. Res. Pediatr. Endocrinol. 4, 104–106.
Napolitano, T., Avolio, F., Courtney, M., Vieira, A., Druelle, N., Ben-Othman, N., Hadzic, B., Navarro, S., Collombat, P., 2015. Pax4 acts as a key player in pancreas development and plasticity. Semin. Cell Dev. Biol. 44, 107–114.
Narayanan, M., Ramsey, K., Grebe, T., Schrauwen, I., Szelinger, S., Huentelman, M., Craig, D., Narayanan, V., C4RCD Research Group, 2015. Case report: compound

heterozygous nonsense mutations in TRMT10A are associated with microcephaly, delayed development, and periventricular white matter hyperintensities. F1000Research 4, 912.

Naya, F.J., Stellrecht, C.M., Tsai, M.J., 1995. Tissue-specific regulation of the insulin gene by a novel basic helix-loop-helix transcription factor. Genes Dev. 9, 1009–1019.

Naya, F.J., Huang, H.P., Qiu, Y., Mutoh, H., Demayo, F.J., Leiter, A.B., Tsai, M.J., 1997. Diabetes, defective pancreatic morphogenesis, and abnormal enteroendocrine differentiation in BETA2/neuroD-deficient mice. Genes Dev. 11, 2323–2334.

Naylor, R., Knight Johnson, A., Del Gaudio, D., 2018. Maturity-onset diabetes of the young overview. In: Adam, M.P., Ardinger, H.H., Pagon, R.A., Wallace, S.E., Bean, L.J.H., Stephens, K., Amemiya, A. (Eds.), GeneReviews((R)). University of Washington, Seattle, WA.

Negahdar, M., Aukrust, I., Johansson, B.B., Molnes, J., Molven, A., Matschinsky, F.M., Sovik, O., Kulkarni, R.N., Flatmark, T., Njolstad, P.R., Bjorkhaug, L., 2012. GCK-MODY diabetes associated with protein misfolding, cellular self-association and degradation. Biochim. Biophys. Acta 1822, 1705–1715.

Nelson, S.B., Schaffer, A.E., Sander, M., 2007. The transcription factors Nkx6.1 and Nkx6.2 possess equivalent activities in promoting beta-cell fate specification in Pdx1 + pancreatic progenitor cells. Development 134, 2491–2500.

Nemer, G., Fadlalah, F., Usta, J., Nemer, M., Dbaibo, G., Obeid, M., Bitar, F., 2006. A novel mutation in the GATA4 gene in patients with tetralogy of fallot. Hum. Mutat. 27, 293–294.

Nestorowicz, A., Wilson, B.A., Schoor, K.P., Inoue, H., Glaser, B., Landau, H., Stanley, C.A., Thornton, P.S., Clement, J.P.T., Bryan, J., Aguilar-Bryan, L., Permutt, M.A., 1996. Mutations in the sulonylurea receptor gene are associated with familial hyperinsulinism in Ashkenazi Jews. Hum. Mol. Genet. 5, 1813–1822.

Neve, B., Fernandez-Zapico, M.E., Ashkenazi-Katalan, V., Dina, C., Hamid, Y.H., Joly, E., Vaillant, E., Benmezroua, Y., Durand, E., Bakaher, N., Delannoy, V., Vaxillaire, M., Cook, T., Dallinga-Thie, G.M., Jansen, H., Charles, M.A., Clement, K., Galan, P., Hercberg, S., Helbecque, N., Charpentier, G., Prentki, M., Hansen, T., Pedersen, O., Urrutia, R., Melloul, D., Froguel, P., 2005. Role of transcription factor KLF11 and its diabetes-associated gene variants in pancreatic beta cell function. Proc. Natl. Acad. Sci. U. S. A. 102, 4807–4812.

Ng, N.H.J., Jasmen, J.B., Lim, C.S., Lau, H.H., Krishnan, V.G., Kadiwala, J., Kulkarni, R.N., Raeder, H., Vallier, L., Hoon, S., Teo, A.K.K., 2019. HNF4A haploinsufficiency in MODY1 abrogates liver and pancreas differentiation from patient-derived induced pluripotent stem cells. iScience 16, 192–205.

Nicolino, M., Claiborn, K.C., Senee, V., Boland, A., Stoffers, D.A., Julier, C., 2010. A novel hypomorphic PDX1 mutation responsible for permanent neonatal diabetes with subclinical exocrine deficiency. Diabetes 59, 733–740.

Nientiedt, M., Deng, M., Schmidt, D., Perner, S., Muller, S.C., Ellinger, J., 2016. Identification of aberrant tRNA-halves expression patterns in clear cell renal cell carcinoma. Sci. Rep. 6, 37158.

Nishi, M., Sasahara, M., Shono, T., Saika, S., Yamamoto, Y., Ohkawa, K., Furuta, H., Nakao, T., Sasaki, H., Nanjo, K., 2005. A case of novel de novo paired box gene 6 (PAX6) mutation with early-onset diabetes mellitus and aniridia. Diabet. Med. 22, 641–644.

Njolstad, P.R., Sagen, J.V., Bjorkhaug, L., Odili, S., Shehadeh, N., Bakry, D., Sarici, S.U., Alpay, F., Molnes, J., Molven, A., Sovik, O., Matschinsky, F.M., 2003. Permanent neonatal diabetes caused by glucokinase deficiency: inborn error of the glucose-insulin signaling pathway. Diabetes 52, 2854–2860.

Nogueira, T.C., Paula, F.M., Villate, O., Colli, M.L., Moura, R.F., Cunha, D.A., Marselli, L., Marchetti, P., Cnop, M., Julier, C., Eizirik, D.L., 2013. GLIS3, a susceptibility gene for type 1 and type 2 diabetes, modulates pancreatic beta cell apoptosis via regulation of a splice variant of the BH3-only protein Bim. PLoS Genet. 9, e1003532.

Novoa, I., Zeng, H., Harding, H.P., Ron, D., 2001. Feedback inhibition of the unfolded protein response by GADD34-mediated dephosphorylation of eIF2alpha. J. Cell Biol. 153, 1011–1022.

Odom, D.T., Zizlsperger, N., Gordon, D.B., Bell, G.W., Rinaldi, N.J., Murray, H.L., Volkert, T.L., Schreiber, J., Rolfe, P.A., Gifford, D.K., Fraenkel, E., Bell, G.I., Young, R.A., 2004. Control of pancreas and liver gene expression by HNF transcription factors. Science 303, 1378–1381.

Offield, M.F., Jetton, T.L., Labosky, P.A., Ray, M., Stein, R.W., Magnuson, M.A., Hogan, B.L., Wright, C.V., 1996. PDX-1 is required for pancreatic outgrowth and differentiation of the rostral duodenum. Development 122, 983–995.

Ohlsson, H., Karlsson, K., Edlund, T., 1993. IPF1, a homeodomain-containing transactivator of the insulin gene. EMBO J. 12, 4251–4259.

Ohsie, S., Gerney, G., Gui, D., Kahana, D., Martin, M.G., Cortina, G., 2009. A paucity of colonic enteroendocrine and/or enterochromaffin cells characterizes a subset of patients with chronic unexplained diarrhea/malabsorption. Hum. Pathol. 40, 1006–1014.

Ontiveros, R.J., Shen, H., Stoute, J., Yanas, A., Cui, Y., Zhang, Y., Liu, K.F., 2020. Coordination of mRNA and tRNA methylations by TRMT10A. Proc. Natl. Acad. Sci. U. S. A. 117, 7782–7791.

O'Rahilly, S., Gray, H., Humphreys, P.J., Krook, A., Polonsky, K.S., White, A., Gibson, S., Taylor, K., Carr, C., 1995. Brief report: impaired processing of prohormones associated with abnormalities of glucose homeostasis and adrenal function. N. Engl. J. Med. 333, 1386–1390.

Osbak, K.K., Colclough, K., Saint-Martin, C., Beer, N.L., Bellanne-Chantelot, C., Ellard, S., Gloyn, A.L., 2009. Update on mutations in glucokinase (GCK), which cause maturity-onset diabetes of the young, permanent neonatal diabetes, and hyperinsulinemic hypoglycemia. Hum. Mutat. 30, 1512–1526.

Ostoft, S.H., Bagger, J.I., Hansen, T., Pedersen, O., Faber, J., Holst, J.J., Knop, F.K., Vilsboll, T., 2014. Glucose-lowering effects and low risk of hypoglycemia in patients with maturity-onset diabetes of the young when treated with a GLP-1 receptor agonist: a double-blind, randomized, crossover trial. Diabetes Care 37, 1797–1805.

Otonkoski, T., Ammala, C., Huopio, H., Cote, G.J., Chapman, J., Cosgrove, K., Ashfield, R., Huang, E., Komulainen, J., Ashcroft, F.M., Dunne, M.J., Kere, J., Thomas, P.M., 1999. A point mutation inactivating the sulfonylurea receptor causes the severe form of persistent hyperinsulinemic hypoglycemia of infancy in Finland. Diabetes 48, 408–415.

Oyadomari, S., Koizumi, A., Takeda, K., Gotoh, T., Akira, S., Araki, E., Mori, M., 2002. Targeted disruption of the Chop gene delays endoplasmic reticulum stress-mediated diabetes. J. Clin. Invest. 109, 525–532.

Pan, T., 2018. Modifications and functional genomics of human transfer RNA. Cell Res. 28, 395–404.

Pan, F.C., Wright, C., 2011. Pancreas organogenesis: from bud to plexus to gland. Dev. Dyn. 240, 530–565.

Pan, F.C., Brissova, M., Powers, A.C., Pfaff, S., Wright, C.V., 2015. Inactivating the permanent neonatal diabetes gene Mnx1 switches insulin-producing beta-cells to a delta-like fate and reveals a facultative proliferative capacity in aged beta-cells. Development 142, 3637–3648.

Papizan, J.B., Singer, R.A., Tschen, S.I., Dhawan, S., Friel, J.M., Hipkens, S.B., Magnuson, M.A., Bhushan, A., Sussel, L., 2011. Nkx2.2 repressor complex regulates islet beta-cell specification and prevents beta-to-alpha-cell reprogramming. Genes Dev. 25, 2291–2305.

Park, S., Gakh, O., Mooney, S.M., Isaya, G., 2002. The ferroxidase activity of yeast frataxin. J. Biol. Chem. 277, 38589–38595.

Parkinson, M.H., Boesch, S., Nachbauer, W., Mariotti, C., Giunti, P., 2013. Clinical features of Friedreich's ataxia: classical and atypical phenotypes. J. Neurochem. 126 (Suppl. 1), 103–117.

Parrizas, M., Maestro, M.A., Boj, S.F., Paniagua, A., Casamitjana, R., Gomis, R., Rivera, F., Ferrer, J., 2001. Hepatic nuclear factor 1-alpha directs nucleosomal hyperacetylation to its tissue-specific transcriptional targets. Mol. Cell. Biol. 21, 3234–3243.

Patel, K.A., Kettunen, J., Laakso, M., Stancakova, A., Laver, T.W., Colclough, K., Johnson, M.B., Abramowicz, M., Groop, L., Miettinen, P.J., Shepherd, M.H., Flanagan, S.E., Ellard, S., Inagaki, N., Hattersley, A.T., Tuomi, T., Cnop, M., Weedon, M.N., 2017. Heterozygous RFX6 protein truncating variants are associated with MODY with reduced penetrance. Nat. Commun. 8, 888.

Patel, V., Bidault, G., Chambers, J.E., Carobbio, S., Everden, A.J.T., Garces, C., Dalton, L.E., Gribble, F.M., Vidal-Puig, A., Marciniak, S.J., 2019. Inactivation of Ppp1r15a minimises weight gain and insulin resistance during caloric excess in female mice. Sci. Rep. 9, 2903.

Pattingre, S., Tassa, A., Qu, X., Garuti, R., Liang, X.H., Mizushima, N., Packer, M., Schneider, M.D., Levine, B., 2005. Bcl-2 antiapoptotic proteins inhibit Beclin 1-dependent autophagy. Cell 122, 927–939.

Paupe, V., Dassa, E.P., Goncalves, S., Auchere, F., Lonn, M., Holmgren, A., Rustin, P., 2009. Impaired nuclear Nrf2 translocation undermines the oxidative stress response in Friedreich ataxia. PLoS One 4, e4253.

Payne, R.M., Wagner, G.R., 2012. Cardiomyopathy in Friedreich ataxia: clinical findings and research. J. Child Neurol. 27, 1179–1186.

Pearson, E.R., Liddell, W.G., Shepherd, M., Corrall, R.J., Hattersley, A.T., 2000. Sensitivity to sulphonylureas in patients with hepatocyte nuclear factor-1alpha gene mutations: evidence for pharmacogenetics in diabetes. Diabet. Med. 17, 543–545.

Pearson, E.R., Flechtner, I., Njolstad, P.R., Malecki, M.T., Flanagan, S.E., Larkin, B., Ashcroft, F.M., Klimes, I., Codner, E., Iotova, V., Slingerland, A.S., Shield, J., Robert, J.J., Holst, J.J., Clark, P.M., Ellard, S., Sovik, O., Polak, M., Hattersley, A.T., Neonatal Diabetes International Collaborative Group, 2006. Switching from insulin to oral sulfonylureas in patients with diabetes due to Kir6.2 mutations. N. Engl. J. Med. 355, 467–477.

Pearson, E.R., Boj, S.F., Steele, A.M., Barrett, T., Stals, K., Shield, J.P., Ellard, S., Ferrer, J., Hattersley, A.T., 2007. Macrosomia and hyperinsulinaemic hypoglycaemia in patients with heterozygous mutations in the HNF4A gene. PLoS Med. 4, e118.

Pedersen, J.K., Nelson, S.B., Jorgensen, M.C., Henseleit, K.D., Fujitani, Y., Wright, C.V., Sander, M., Serup, P., Beta Cell Biology Consortium, 2005. Endodermal expression of Nkx6 genes depends differentially on Pdx1. Dev. Biol. 288, 487–501.

Peduto, A., Spada, M., Alluto, A., La Dolcetta, M., Ponzone, A., Santer, R., 2004. A novel mutation in the GLUT2 gene in a patient with Fanconi-Bickel syndrome detected by neonatal screening for galactosaemia. J. Inherit. Metab. Dis. 27, 279–280.

Pepin, L., Colin, E., Tessarech, M., Rouleau, S., Bouhours-Nouet, N., Bonneau, D., Coutant, R., 2019. A new case of PCSK1 pathogenic variant with congenital proprotein convertase 1/3 deficiency and literature review. J. Clin. Endocrinol. Metab. 104, 985–993.

Pereira, M., Francisco, S., Varanda, A.S., Santos, M., Santos, M.A.S., Soares, A.R., 2018. Impact of tRNA modifications and tRNA-modifying enzymes on proteostasis and human disease. Int. J. Mol. Sci. 19, 3738.

Petersen, H.V., Jorgensen, M.C., Andersen, F.G., Jensen, J., Nielsen, T.F., Jorgensen, R., Madsen, O.D., Serup, P., 2000. Pax4 represses pancreatic glucagon gene expression. Mol. Cell Biol. Res. Commun. 3, 249–254.

Phizicky, E.M., Hopper, A.K., 2010. tRNA biology charges to the front. Genes Dev. 24, 1832–1860.

Piccand, J., Strasser, P., Hodson, D.J., Meunier, A., Ye, T., Keime, C., Birling, M.C., Rutter, G.A., Gradwohl, G., 2014. Rfx6 maintains the functional identity of adult pancreatic beta cells. Cell Rep. 9, 2219–2232.

Piermarini, E., Cartelli, D., Pastore, A., Tozzi, G., Compagnucci, C., Giorda, E., D'amico, J., Petrini, S., Bertini, E., Cappelletti, G., Piemonte, F., 2016. Frataxin silencing alters microtubule stability in motor neurons: implications for Friedreich's ataxia. Hum. Mol. Genet. 25, 4288–4301.

Pihoker, C., Gilliam, L.K., Ellard, S., Dabelea, D., Davis, C., Dolan, L.M., Greenbaum, C.J., Imperatore, G., Lawrence, J.M., Marcovina, S.M., Mayer-Davis, E., Rodriguez, B.L., Steck, A.K., Williams, D.E., Hattersley, A.T., for the SEARCH for Diabetes in Youth Study Group, 2013. Prevalence, characteristics and clinical diagnosis of maturity onset diabetes of the young due to mutations in HNF1A, HNF4A, and glucokinase: results from the SEARCH for Diabetes in Youth. J. Clin. Endocrinol. Metab. 98, 4055–4062.

Ping, F., Wang, Z., Xiao, X., 2019. Clinical and enzymatic phenotypes in congenital hyperinsulinemic hypoglycemia due to glucokinase-activating mutations: a report of two cases and a brief overview of the literature. J. Diabetes Investig. 10, 1454–1462.

Pinney, S.E., Macmullen, C., Becker, S., Lin, Y.W., Hanna, C., Thornton, P., Ganguly, A., Shyng, S.L., Stanley, C.A., 2008. Clinical characteristics and biochemical mechanisms of congenital hyperinsulinism associated with dominant KATP channel mutations. J. Clin. Invest. 118, 2877–2886.

Pinney, S.E., Oliver-Krasinski, J., Ernst, L., Hughes, N., Patel, P., Stoffers, D.A., Russo, P., De Leon, D.D., 2011. Neonatal diabetes and congenital malabsorptive diarrhea attributable to a novel mutation in the human neurogenin-3 gene coding sequence. J. Clin. Endocrinol. Metab. 96, 1960–1965.

Pino, M.F., Kim, K.A., Shelton, K.D., Lindner, J., Odili, S., Li, C., Collins, H.W., Shiota, M., Matschinsky, F.M., Magnuson, M.A., 2007. Glucokinase thermolability and hepatic regulatory protein binding are essential factors for predicting the blood glucose phenotype of missense mutations. J. Biol. Chem. 282, 13906–13916.

Pirot, P., Naamane, N., Libert, F., Magnusson, N.E., Orntoft, T.F., Cardozo, A.K., Eizirik, D.L., 2007. Global profiling of genes modified by endoplasmic reticulum stress in pancreatic beta cells reveals the early degradation of insulin mRNAs. Diabetologia 50, 1006–1014.

Plengvidhya, N., Kooptiwut, S., Songtawee, N., Doi, A., Furuta, H., Nishi, M., Nanjo, K., Tantibhedhyangkul, W., Boonyasrisawat, W., Yenchitsomanus, P.T., Doria, A., Banchuin, N., 2007. PAX4 mutations in Thais with maturity onset diabetes of the young. J. Clin. Endocrinol. Metab. 92, 2821–2826.

Poburski, D., Boerner, J.B., Koenig, M., Ristow, M., Thierbach, R., 2016. Time-resolved functional analysis of acute impairment of frataxin expression in an inducible cell model of Friedreich ataxia. Biol. Open 5, 654–661.

Poll, A.V., Pierreux, C.E., Lokmane, L., Haumaitre, C., Achouri, Y., Jacquemin, P., Rousseau, G.G., Cereghini, S., Lemaigre, F.P., 2006. A vHNF1/TCF2-HNF6 cascade regulates the transcription factor network that controls generation of pancreatic precursor cells. Diabetes 55, 61–69.

Pongratz, R.L., Kibbey, R.G., Kirkpatrick, C.L., Zhao, X., Pontoglio, M., Yaniv, M., Wollheim, C.B., Shulman, G.I., Cline, G.W., 2009. Mitochondrial dysfunction contributes to impaired insulin secretion in INS-1 cells with dominant-negative mutations of HNF-1alpha and in HNF-1alpha-deficient islets. J. Biol. Chem. 284, 16808–16821.

Pontoglio, M., Sreenan, S., Roe, M., Pugh, W., Ostrega, D., Doyen, A., Pick, A.J., Baldwin, A., Velho, G., Froguel, P., Levisetti, M., Bonner-Weir, S., Bell, G.I., Yaniv, M., Polonsky, K.S., 1998. Defective insulin secretion in hepatocyte nuclear factor 1alpha-deficient mice. J. Clin. Invest. 101, 2215–2222.

Poulton, C.J., Schot, R., Kia, S.K., Jones, M., Verheijen, F.W., Venselaar, H., De Wit, M.C., De Graaff, E., Bertoli-Avella, A.M., Mancini, G.M., 2011. Microcephaly with simplified gyration, epilepsy, and infantile diabetes linked to inappropriate apoptosis of neural progenitors. Am. J. Hum. Genet. 89, 265–276.

Prado, C.L., Pugh-Bernard, A.E., Elghazi, L., Sosa-Pineda, B., Sussel, L., 2004. Ghrelin cells replace insulin-producing beta cells in two mouse models of pancreas development. Proc. Natl. Acad. Sci. U. S. A. 101, 2924–2929.

Proks, P., Arnold, A.L., Bruining, J., Girard, C., Flanagan, S.E., Larkin, B., Colclough, K., Hattersley, A.T., Ashcroft, F.M., Ellard, S., 2006. A heterozygous activating mutation in the sulphonylurea receptor SUR1 (ABCC8) causes neonatal diabetes. Hum. Mol. Genet. 15, 1793–1800.

Prudente, S., Jungtrakoon, P., Marucci, A., Ludovico, O., Buranasupkajorn, P., Mazza, T., Hastings, T., Milano, T., Morini, E., Mercuri, L., Bailetti, D., Mendonca, C., Alberico, F., Basile, G., Romani, M., Miccinilli, E., Pizzuti, A., Carella, M., Barbetti, F., Pascarella, S., Marchetti, P., Trischitta, V., Di Paola, R., Doria, A., 2015. Loss-of-function mutations in APPL1 in familial diabetes mellitus. Am. J. Hum. Genet. 97, 177–185.

Raghuram, N., Marwaha, A., Greer, M.C., Gauda, E., Chitayat, D., 2020. Congenital hypothyroidism, cardiac defects, and pancreatic agenesis in an infant with GATA6 mutation. Am. J. Med. Genet. A 182, 1496–1499.

Raman, R., Vaitheeswaran, K., Vinita, K., Sharma, T., 2011. Is prevalence of retinopathy related to the age of onset of diabetes? Sankara Nethralaya Diabetic Retinopathy Epidemiology and Molecular Genetic Report No. 5. Ophthalmic Res. 45, 36–41.

Rando, T.A., Horton, J.C., Layzer, R.B., 1992. Wolfram syndrome: evidence of a diffuse neurodegenerative disease by magnetic resonance imaging. Neurology 42, 1220–1224.

Rees, S.D., Hydrie, M.Z., O'hare, J.P., Kumar, S., Shera, A.S., Basit, A., Barnett, A.H., Kelly, M.A., 2011. Effects of 16 genetic variants on fasting glucose and type 2 diabetes in South Asians: ADCY5 and GLIS3 variants may predispose to type 2 diabetes. PLoS One 6, e24710.

Reilly, F., Sanchez-Lechuga, B., Clinton, S., Crowe, G., Burke, M., Ng, N., Colclough, K., Byrne, M.M., 2020. Phenotype, genotype and glycaemic variability in people with activating mutations in the ABCC8 gene: response to appropriate therapy. Diabet. Med. 37, 876–884.

Reis, A.F., Ye, W.Z., Dubois-Laforgue, D., Bellanne-Chantelot, C., Timsit, J., Velho, G., 2000. Association of a variant in exon 31 of the sulfonylurea receptor 1 (SUR1) gene with type 2 diabetes mellitus in French Caucasians. Hum. Genet. 107, 138–144.

Rezende, L.F., Stoppiglia, L.F., Souza, K.L., Negro, A., Langone, F., Boschero, A.C., 2007. Ciliary neurotrophic factor promotes survival of neonatal rat islets via the BCL-2 anti-apoptotic pathway. J. Endocrinol. 195, 157–165.

Reznik, Y., Dao, T., Coutant, R., Chiche, L., Jeannot, E., Clauin, S., Rousselot, P., Fabre, M., Oberti, F., Fatome, A., Zucman-Rossi, J., Bellanne-Chantelot, C., 2004. Hepatocyte nuclear factor-1 alpha gene inactivation: cosegregation between liver adenomatosis and diabetes phenotypes in two maturity-onset diabetes of the young (MODY)3 families. J. Clin. Endocrinol. Metab. 89, 1476–1480.

Rhee, J., Inoue, Y., Yoon, J.C., Puigserver, P., Fan, M., Gonzalez, F.J., Spiegelman, B.M., 2003. Regulation of hepatic fasting response by PPARgamma coactivator-1alpha (PGC-1): requirement for hepatocyte nuclear factor 4alpha in gluconeogenesis. Proc. Natl. Acad. Sci. U. S. A. 100, 4012–4017.

Richardson, D.R., Ponka, P., 1997. The molecular mechanisms of the metabolism and transport of iron in normal and neoplastic cells. Biochim. Biophys. Acta 1331, 1–40.

Richardson, C.C., Hussain, K., Jones, P.M., Persaud, S., Lobner, K., Boehm, A., Clark, A., Christie, M.R., 2007. Low levels of glucose transporters and K + ATP channels in human pancreatic beta cells early in development. Diabetologia 50, 1000–1005.

Riggs, A.C., Bernal-Mizrachi, E., Ohsugi, M., Wasson, J., Fatrai, S., Welling, C., Murray, J., Schmidt, R.E., Herrera, P.L., Permutt, M.A., 2005. Mice conditionally lacking the Wolfram gene in pancreatic islet beta cells exhibit diabetes as a result of enhanced endoplasmic reticulum stress and apoptosis. Diabetologia 48, 2313–2321.

Ritz-Laser, B., Estreicher, A., Gauthier, B.R., Mamin, A., Edlund, H., Philippe, J., 2002. The pancreatic beta-cell-specific transcription factor Pax-4 inhibits glucagon gene expression through Pax-6. Diabetologia 45, 97–107.

Riva, S., Ghisalberti, C., Parini, R., Furlan, F., Bettinelli, A., Somaschini, M., 2004. The Fanconi-Bickel syndrome: a case of neonatal onset. J. Perinatol. 24, 322–323.

Robson, E.J., He, S.J., Eccles, M.R., 2006. A PANorama of PAX genes in cancer and development. Nat. Rev. Cancer 6, 52–62.

Rodolosse, A., Chalaux, E., Adell, T., Hagege, H., Skoudy, A., Real, F.X., 2004. PTF1alpha/p48 transcription factor couples proliferation and differentiation in the exocrine pancreas [corrected]. Gastroenterology 127, 937–949.

Ron, D., Walter, P., 2007. Signal integration in the endoplasmic reticulum unfolded protein response. Nat. Rev. Mol. Cell Biol. 8, 519–529.

Rondinelli, M., Novara, F., Calcaterra, V., Zuffardi, O., Genovese, S., 2015. Wolfram syndrome 2: a novel CISD2 mutation identified in Italian siblings. Acta Diabetol. 52, 175–178.

Rose, S.D., Swift, G.H., Peyton, M.J., Hammer, R.E., Macdonald, R.J., 2001. The role of PTF1-P48 in pancreatic acinar gene expression. J. Biol. Chem. 276, 44018–44026.

Ross, A.J., Ruiz-Perez, V., Wang, Y., Hagan, D.M., Scherer, S., Lynch, S.A., Lindsay, S., Custard, E., Belloni, E., Wilson, D.I., Wadey, R., Goodman, F., Orstavik, K.H., Monclair, T., Robson, S., Reardon, W., Burn, J., Scambler, P., Strachan, T., 1998. A homeobox gene, HLXB9, is the major locus for dominantly inherited sacral agenesis. Nat. Genet. 20, 358–361.

Rotig, A., De Lonlay, P., Chretien, D., Foury, F., Koenig, M., Sidi, D., Munnich, A., Rustin, P., 1997. Aconitase and mitochondrial iron-sulphur protein deficiency in Friedreich ataxia. Nat. Genet. 17, 215–217.

Rubio-Cabezas, O., Ellard, S., 2013. Diabetes mellitus in neonates and infants: genetic heterogeneity, clinical approach to diagnosis, and therapeutic options. Horm. Res. Paediatr. 80, 137–146.

Rubio-Cabezas, O., Diaz Gonzalez, F., Aragones, A., Argente, J., Campos-Barros, A., 2008. Permanent neonatal diabetes caused by a homozygous nonsense mutation in the glucokinase gene. Pediatr. Diabetes 9, 245–249.

Rubio-Cabezas, O., Patch, A.M., Minton, J.A., Flanagan, S.E., Edghill, E.L., Hussain, K., Balafrej, A., Deeb, A., Buchanan, C.R., Jefferson, I.G., Mutair, A., Neonatal Diabetes International Collaborative Group, Hattersley, A.T., Ellard, S., 2009. Wolcott-Rallison syndrome is the most common genetic cause of permanent neonatal diabetes in consanguineous families. J. Clin. Endocrinol. Metab. 94, 4162–4170.

Rubio-Cabezas, O., Minton, J.A., Kantor, I., Williams, D., Ellard, S., Hattersley, A.T., 2010. Homozygous mutations in NEUROD1 are responsible for a novel syndrome of permanent neonatal diabetes and neurological abnormalities. Diabetes 59, 2326–2331.

Rubio-Cabezas, O., Jensen, J.N., Hodgson, M.I., Codner, E., Ellard, S., Serup, P., Hattersley, A.T., 2011. Permanent neonatal diabetes and enteric anendocrinosis associated with biallelic mutations in NEUROG3. Diabetes 60, 1349–1353.

Rubio-Cabezas, O., Codner, E., Flanagan, S.E., Gomez, J.L., Ellard, S., Hattersley, A.T., 2014a. Neurogenin 3 is important but not essential for pancreatic islet development in humans. Diabetologia 57, 2421–2424.

Rubio-Cabezas, O., Hattersley, A.T., Njolstad, P.R., Mlynarski, W., Ellard, S., White, N., Chi, D.V., Craig, M.E., International Society for Pediatric and Adolescent Diabetes, Adolescent, D., 2014b. ISPAD Clinical Practice Consensus Guidelines 2014. The diagnosis and management of monogenic diabetes in children and adolescents. Pediatr. Diabetes 15 (Suppl. 20), 47–64.

Ruggero, K., Guffanti, A., Corradin, A., Sharma, V.K., De Bellis, G., Corti, G., Grassi, A., Zanovello, P., Bronte, V., Ciminale, V., D'agostino, D.M., 2014. Small noncoding RNAs in cells transformed by human T-cell leukemia virus type 1: a role for a tRNA fragment as a primer for reverse transcriptase. J. Virol. 88, 3612–3622.

Ryan, F., Devaney, D., Joyce, C., Nestorowicz, A., Permutt, M.A., Glaser, B., Barton, D.E., Thornton, P.S., 1998. Hyperinsulinism: molecular aetiology of focal disease. Arch. Dis. Child. 79, 445–447.

Ryffel, G.U., 2001. Mutations in the human genes encoding the transcription factors of the hepatocyte nuclear factor (HNF)1 and HNF4 families: functional and pathological consequences. J. Mol. Endocrinol. 27, 11–29.

Sagen, J.V., Raeder, H., Hathout, E., Shehadeh, N., Gudmundsson, K., Baevre, H., Abuelo, D., Phornphutkul, C., Molnes, J., Bell, G.I., Gloyn, A.L., Hattersley, A.T., Molven, A., Sovik, O., Njolstad, P.R., 2004. Permanent neonatal diabetes due to mutations in KCNJ11 encoding Kir6.2: patient characteristics and initial response to sulfonylurea therapy. Diabetes 53, 2713–2718.

Sakamoto, O., Ogawa, E., Ohura, T., Igarashi, Y., Matsubara, Y., Narisawa, K., Iinuma, K., 2000. Mutation analysis of the GLUT2 gene in patients with Fanconi-Bickel syndrome. Pediatr. Res. 48, 586–589.

Sakikubo, M., Furuyama, K., Horiguchi, M., Hosokawa, S., Aoyama, Y., Tsuboi, K., Goto, T., Hirata, K., Masui, T., Dor, Y., Fujiyama, T., Hoshino, M., Uemoto, S., Kawaguchi, Y., 2018. Ptf1a inactivation in adult pancreatic acinar cells causes apoptosis through activation of the endoplasmic reticulum stress pathway. Sci. Rep. 8, 15812.

Salehi, M.H., Kamalidehghan, B., Houshmand, M., Yong Meng, G., Sadeghizadeh, M., Aryani, O., Nafissi, S., 2014. Gene expression profiling of mitochondrial oxidative phosphorylation (OXPHOS) complex I in Friedreich ataxia (FRDA) patients. PLoS One 9, e94069.

Sanchez-Lechuga, B., Saqlain, M., Ng, N., Colclough, K., Woods, C., Byrne, M., 2020. Case report: adult onset diabetes with partial pancreatic agenesis and congenital heart disease due to a de novo GATA6 mutation. BMC Med. Genet. 21, 70.

Sander, M., Neubuser, A., Kalamaras, J., Ee, H.C., Martin, G.R., German, M.S., 1997. Genetic analysis reveals that PAX6 is required for normal transcription of pancreatic hormone genes and islet development. Genes Dev. 11, 1662–1673.

Sansbury, F.H., Flanagan, S.E., Houghton, J.A., Shuixian Shen, F.L., Al-Senani, A.M., Habeb, A.M., Abdullah, M., Kariminejad, A., Ellard, S., Hattersley, A.T., 2012. SLC2A2 mutations can cause neonatal diabetes, suggesting GLUT2 may have a role in human insulin secretion. Diabetologia 55, 2381–2385.

Sansbury, F.H., Kirel, B., Caswell, R., Allen, H.L., Flanagan, S.E., Hattersley, A.T., Ellard, S., Shaw-Smith, C.J., 2015. Biallelic RFX6 mutations can cause childhood as well as neonatal onset diabetes mellitus. Eur. J. Hum. Genet. 23, 1744–1748.

Saveliev, A., Everett, C., Sharpe, T., Webster, Z., Festenstein, R., 2003. DNA triplet repeats mediate heterochromatin-protein-1-sensitive variegated gene silencing. Nature 422, 909–913.

Sayar, E., Islek, A., Yilmaz, A., Akcam, M., Flanagan, S.E., Artan, R., 2013. Extremely rare cause of congenital diarrhea: enteric anendocrinosis. Pediatr. Int. 55, 661–663.
Sayed, S., Langdon, D.R., Odili, S., Chen, P., Buettger, C., Schiffman, A.B., Suchi, M., Taub, R., Grimsby, J., Matschinsky, F.M., Stanley, C.A., 2009. Extremes of clinical and enzymatic phenotypes in children with hyperinsulinism caused by glucokinase activating mutations. Diabetes 58, 1419–1427.
Schisler, J.C., Fueger, P.T., Babu, D.A., Hohmeier, H.E., Tessem, J.S., Lu, D., Becker, T.C., Naziruddin, B., Levy, M., Mirmira, R.G., Newgard, C.B., 2008. Stimulation of human and rat islet beta-cell proliferation with retention of function by the homeodomain transcription factor Nkx6.1. Mol. Cell. Biol. 28, 3465–3476.
Schulz, J.B., Dehmer, T., Schols, L., Mende, H., Hardt, C., Vorgerd, M., Burk, K., Matson, W., Dichgans, J., Beal, M.F., Bogdanov, M.B., 2000. Oxidative stress in patients with Friedreich ataxia. Neurology 55, 1719–1721.
Schwab, M.H., Bartholomae, A., Heimrich, B., Feldmeyer, D., Druffel-Augustin, S., Goebbels, S., Naya, F.J., Zhao, S., Frotscher, M., Tsai, M.J., Nave, K.A., 2000. Neuronal basic helix-loop-helix proteins (NEX and BETA2/Neuro D) regulate terminal granule cell differentiation in the hippocampus. J. Neurosci. 20, 3714–3724.
Scolding, N.J., Kellar-Wood, H.F., Shaw, C., Shneerson, J.M., Antoun, N., 1996. Wolfram syndrome: hereditary diabetes mellitus with brainstem and optic atrophy. Ann. Neurol. 39, 352–360.
Sellick, G.S., Barker, K.T., Stolte-Dijkstra, I., Fleischmann, C., Coleman, R.J., Garrett, C., Gloyn, A.L., Edghill, E.L., Hattersley, A.T., Wellauer, P.K., Goodwin, G., Houlston, R.S., 2004. Mutations in PTF1A cause pancreatic and cerebellar agenesis. Nat. Genet. 36, 1301–1305.
Selvadurai, L.P., Harding, I.H., Corben, L.A., Georgiou-Karistianis, N., 2018. Cerebral abnormalities in Friedreich ataxia: a review. Neurosci. Biobehav. Rev. 84, 394–406.
Senee, V., Vattem, K.M., Delepine, M., Rainbow, L.A., Haton, C., Lecoq, A., Shaw, N.J., Robert, J.J., Rooman, R., Diatloff-Zito, C., Michaud, J.L., Bin-Abbas, B., Taha, D., Zabel, B., Franceschini, P., Topaloglu, A.K., Lathrop, G.M., Barrett, T.G., Nicolino, M., Wek, R.C., Julier, C., 2004. Wolcott-Rallison syndrome: clinical, genetic, and functional study of EIF2AK3 mutations and suggestion of genetic heterogeneity. Diabetes 53, 1876–1883.
Senee, V., Chelala, C., Duchatelet, S., Feng, D., Blanc, H., Cossec, J.C., Charon, C., Nicolino, M., Boileau, P., Cavener, D.R., Bougneres, P., Taha, D., Julier, C., 2006. Mutations in GLIS3 are responsible for a rare syndrome with neonatal diabetes mellitus and congenital hypothyroidism. Nat. Genet. 38, 682–687.
Servitja, J.M., Pignatelli, M., Maestro, M.A., Cardalda, C., Boj, S.F., Lozano, J., Blanco, E., Lafuente, A., Mccarthy, M.I., Sumoy, L., Guigo, R., Ferrer, J., 2009. Hnf1alpha (MODY3) controls tissue-specific transcriptional programs and exerts opposed effects on cell growth in pancreatic islets and liver. Mol. Cell. Biol. 29, 2945–2959.
Shalev, S.A., Tenenbaum-Rakover, Y., Horovitz, Y., Paz, V.P., Ye, H., Carmody, D., Highland, H.M., Boerwinkle, E., Hanis, C.L., Muzny, D.M., Gibbs, R.A., Bell, G.I., Philipson, L.H., Greeley, S.A., 2014. Microcephaly, epilepsy, and neonatal diabetes due to compound heterozygous mutations in IER3IP1: insights into the natural history of a rare disorder. Pediatr. Diabetes 15, 252–256.
Shan, Y., Schoenfeld, R.A., Hayashi, G., Napoli, E., Akiyama, T., Iodi Carstens, M., Carstens, E.E., Pook, M.A., Cortopassi, G.A., 2013. Frataxin deficiency leads to defects in expression of antioxidants and Nrf2 expression in dorsal root ganglia of the Friedreich's ataxia YG8R mouse model. Antioxid. Redox Signal. 19, 1481–1493.
Sharari, S., Abou-Alloul, M., Hussain, K., Ahmad Khan, F., 2020. Fanconi-Bickel syndrome: a review of the mechanisms that lead to dysglycaemia. Int. J. Mol. Sci. 21, 6286.

Shaw-Smith, C., De Franco, E., Lango Allen, H., Batlle, M., Flanagan, S.E., Borowiec, M., Taplin, C.E., Van Alfen-Van Der Velden, J., Cruz-Rojo, J., Perez De Nanclares, G., Miedzybrodzka, Z., Deja, G., Wlodarska, I., Mlynarski, W., Ferrer, J., Hattersley, A.T., Ellard, S., 2014. GATA4 mutations are a cause of neonatal and childhood-onset diabetes. Diabetes 63, 2888–2894.

Sheets, T.P., Park, K.E., Park, C.H., Swift, S.M., Powell, A., Donovan, D.M., Telugu, B.P., 2018. Targeted mutation of NGN3 gene disrupts pancreatic endocrine cell development in pigs. Sci. Rep. 8, 3582.

Shih, D.Q., Screenan, S., Munoz, K.N., Philipson, L., Pontoglio, M., Yaniv, M., Polonsky, K.S., Stoffel, M., 2001. Loss of HNF-1alpha function in mice leads to abnormal expression of genes involved in pancreatic islet development and metabolism. Diabetes 50, 2472–2480.

Shima, K.R., Usuda, R., Futatani, T., Akahori, H., Kaneko, S., Yorifuji, T., Takamura, T., 2018. Heterogeneous nature of diabetes in a family with a gain-of-function mutation in the ATP-binding cassette subfamily C member 8 (ABCC8) gene. Endocr. J. 65, 1055–1059.

Shimajiri, Y., Sanke, T., Furuta, H., Hanabusa, T., Nakagawa, T., Fujitani, Y., Kajimoto, Y., Takasu, N., Nanjo, K., 2001. A missense mutation of Pax4 gene (R121W) is associated with type 2 diabetes in Japanese. Diabetes 50, 2864–2869.

Shoffner, J.M., Lott, M.T., Lezza, A.M., Seibel, P., Ballinger, S.W., Wallace, D.C., 1990. Myoclonic epilepsy and ragged-red fiber disease (MERRF) is associated with a mitochondrial DNA tRNA(Lys) mutation. Cell 61, 931–937.

Silander, K., Mohlke, K.L., Scott, L.J., Peck, E.C., Hollstein, P., Skol, A.D., Jackson, A.U., Deloukas, P., Hunt, S., Stavrides, G., Chines, P.S., Erdos, M.R., Narisu, N., Conneely, K.N., Li, C., Fingerlin, T.E., Dhanjal, S.K., Valle, T.T., Bergman, R.N., Tuomilehto, J., Watanabe, R.M., Boehnke, M., Collins, F.S., 2004. Genetic variation near the hepatocyte nuclear factor-4 alpha gene predicts susceptibility to type 2 diabetes. Diabetes 53, 1141–1149.

Silva, A.M., Brown, J.M., Buckle, V.J., Wade-Martins, R., Lufino, M.M., 2015. Expanded GAA repeats impair FXN gene expression and reposition the FXN locus to the nuclear lamina in single cells. Hum. Mol. Genet. 24, 3457–3471.

Simpson, C.L., Lemmens, R., Miskiewicz, K., Broom, W.J., Hansen, V.K., Van Vught, P.W., Landers, J.E., Sapp, P., Van Den Bosch, L., Knight, J., Neale, B.M., Turner, M.R., Veldink, J.H., Ophoff, R.A., Tripathi, V.B., Beleza, A., Shah, M.N., Proitsi, P., Van Hoecke, A., Carmeliet, P., Horvitz, H.R., Leigh, P.N., Shaw, C.E., Van Den Berg, L.H., Sham, P.C., Powell, J.F., Verstreken, P., Brown Jr., R.H., Robberecht, W., Al-Chalabi, A., 2009. Variants of the elongator protein 3 (ELP3) gene are associated with motor neuron degeneration. Hum. Mol. Genet. 18, 472–481.

Skopkova, M., Hennig, F., Shin, B.S., Turner, C.E., Stanikova, D., Brennerova, K., Stanik, J., Fischer, U., Henden, L., Muller, U., Steinberger, D., Leshinsky-Silver, E., Bottani, A., Kurdiova, T., Ukropec, J., Nyitrayova, O., Kolnikova, M., Klimes, I., Borck, G., Bahlo, M., Haas, S.A., Kim, J.R., Lotspeich-Cole, L.E., Gasperikova, D., Dever, T.E., Kalscheuer, V.M., 2017. EIF2S3 mutations associated with severe X-linked intellectual disability syndrome MEHMO. Hum. Mutat. 38, 409–425.

Skoric-Milosavljevic, D., Tjong, F.V.Y., Barc, J., Backx, A., Clur, S.B., Van Spaendonck-Zwarts, K., Oostra, R.J., Lahrouchi, N., Beekman, L., Bokenkamp, R., Barge-Schaapveld, D., Mulder, B.J., Lodder, E.M., Bezzina, C.R., Postma, A.V., 2019. GATA6 mutations: characterization of two novel patients and a comprehensive overview of the GATA6 genotypic and phenotypic spectrum. Am. J. Med. Genet. A 179, 1836–1845.

Smith, S.B., Ee, H.C., Conners, J.R., German, M.S., 1999. Paired-homeodomain transcription factor PAX4 acts as a transcriptional repressor in early pancreatic development. Mol. Cell. Biol. 19, 8272–8280.

Smith, S.B., Qu, H.Q., Taleb, N., Kishimoto, N.Y., Scheel, D.W., Lu, Y., Patch, A.M., Grabs, R., Wang, J., Lynn, F.C., Miyatsuka, T., Mitchell, J., Seerke, R., Desir, J., Vanden Eijnden, S., Abramowicz, M., Kacet, N., Weill, J., Renard, M.E., Gentile, M., Hansen, I., Dewar, K., Hattersley, A.T., Wang, R., Wilson, M.E., Johnson, J.D., Polychronakos, C., German, M.S., 2010. Rfx6 directs islet formation and insulin production in mice and humans. Nature 463, 775–780.

Soares, A.R., Santos, M., 2017. Discovery and function of transfer RNA-derived fragments and their role in disease. Wiley Interdiscip. Rev. RNA 8, e1423.

Sohn, Y.S., Tamir, S., Song, L., Michaeli, D., Matouk, I., Conlan, A.R., Harir, Y., Holt, S.H., Shulaev, V., Paddock, M.L., Hochberg, A., Cabanchick, I.Z., Onuchic, J.N., Jennings, P.A., Nechushtai, R., Mittler, R., 2013. NAF-1 and mitoNEET are central to human breast cancer proliferation by maintaining mitochondrial homeostasis and promoting tumor growth. Proc. Natl. Acad. Sci. U. S. A. 110, 14676–14681.

Solomon, B.D., Pineda-Alvarez, D.E., Balog, J.Z., Hadley, D., Gropman, A.L., Nandagopal, R., Han, J.C., Hahn, J.S., Blain, D., Brooks, B., Muenke, M., 2009. Compound heterozygosity for mutations in PAX6 in a patient with complex brain anomaly, neonatal diabetes mellitus, and microophthalmia. Am. J. Med. Genet. A 149A, 2543–2546.

Solorzano-Vargas, R.S., Bjerknes, M., Wang, J., Wu, S.V., Garcia-Careaga, M.G., Pitukcheewanont, P., Cheng, H., German, M.S., Georgia, S., Martin, M.G., 2020. Null mutations of NEUROG3 are associated with delayed onset diabetes mellitus. JCI Insight 5, e127657.

Soragni, E., Gottesfeld, J.M., 2016. Translating HDAC inhibitors in Friedreich's ataxia. Expert Opin. Orphan Drugs 4, 961–970.

Soragni, E., Miao, W., Iudicello, M., Jacoby, D., De Mercanti, S., Clerico, M., Longo, F., Piga, A., Ku, S., Campau, E., Du, J., Penalver, P., Rai, M., Madara, J.C., Nazor, K., O'connor, M., Maximov, A., Loring, J.F., Pandolfo, M., Durelli, L., Gottesfeld, J.M., Rusche, J.R., 2014. Epigenetic therapy for Friedreich ataxia. Ann. Neurol. 76, 489–508.

Sosa-Pineda, B., 2004. The gene Pax4 is an essential regulator of pancreatic beta-cell development. Mol. Cells 18, 289–294.

Sosa-Pineda, B., Chowdhury, K., Torres, M., Oliver, G., Gruss, P., 1997. The Pax4 gene is essential for differentiation of insulin-producing beta cells in the mammalian pancreas. Nature 386, 399–402.

Soyer, J., Flasse, L., Raffelsberger, W., Beucher, A., Orvain, C., Peers, B., Ravassard, P., Vermot, J., Voz, M.L., Mellitzer, G., Gradwohl, G., 2010. Rfx6 is an Ngn3-dependent winged helix transcription factor required for pancreatic islet cell development. Development 137, 203–212.

Spiegel, R., Dobbie, A., Hartman, C., De Vries, L., Ellard, S., Shalev, S.A., 2011. Clinical characterization of a newly described neonatal diabetes syndrome caused by RFX6 mutations. Am. J. Med. Genet. A 155A, 2821–2825.

Stanescu, D.E., Hughes, N., Kaplan, B., Stanley, C.A., De Leon, D.D., 2012. Novel presentations of congenital hyperinsulinism due to mutations in the MODY genes: HNF1A and HNF4A. J. Clin. Endocrinol. Metab. 97, E2026–E2030.

Steele, A.M., Shields, B.M., Shepherd, M., Ellard, S., Hattersley, A.T., Pearson, E.R., 2010. Increased all-cause and cardiovascular mortality in monogenic diabetes as a result of mutations in the HNF1A gene. Diabet. Med. 27, 157–161.

Steele, A.M., Shields, B.M., Wensley, K.J., Colclough, K., Ellard, S., Hattersley, A.T., 2014. Prevalence of vascular complications among patients with glucokinase mutations and prolonged, mild hyperglycemia. JAMA 311, 279–286.

Steiner, D.F., Oyer, P.E., 1967. The biosynthesis of insulin and a probable precursor of insulin by a human islet cell adenoma. Proc. Natl. Acad. Sci. U. S. A. 57, 473–480.

Steinmuller, R., Steinberger, D., Muller, U., 1998. MEHMO (mental retardation, epileptic seizures, hypogonadism and -genitalism, microcephaly, obesity), a novel syndrome: assignment of disease locus to xp21.1-p22.13. Eur. J. Hum. Genet. 6, 201–206.

Steinthorsdottir, V., Thorleifsson, G., Reynisdottir, I., Benediktsson, R., Jonsdottir, T., Walters, G.B., Styrkarsdottir, U., Gretarsdottir, S., Emilsson, V., Ghosh, S., Baker, A., Snorradottir, S., Bjarnason, H., Ng, M.C., Hansen, T., Bagger, Y., Wilensky, R.L., Reilly, M.P., Adeyemo, A., Chen, Y., Zhou, J., Gudnason, V., Chen, G., Huang, H., Lashley, K., Doumatey, A., So, W.Y., Ma, R.C., Andersen, G., Borch-Johnsen, K., Jorgensen, T., Van Vliet-Ostaptchouk, J.V., Hofker, M.H., Wijmenga, C., Christiansen, C., Rader, D.J., Rotimi, C., Gurney, M., Chan, J.C., Pedersen, O., Sigurdsson, G., Gulcher, J.R., Thorsteinsdottir, U., Kong, A., Stefansson, K., 2007. A variant in CDKAL1 influences insulin response and risk of type 2 diabetes. Nat. Genet. 39, 770–775.

Stoffel, M., Duncan, S.A., 1997. The maturity-onset diabetes of the young (MODY1) transcription factor HNF4alpha regulates expression of genes required for glucose transport and metabolism. Proc. Natl. Acad. Sci. U. S. A. 94, 13209–13214.

Stoffers, D.A., Ferrer, J., Clarke, W.L., Habener, J.F., 1997a. Early-onset type-II diabetes mellitus (MODY4) linked to IPF1. Nat. Genet. 17, 138–139.

Stoffers, D.A., Zinkin, N.T., Stanojevic, V., Clarke, W.L., Habener, J.F., 1997b. Pancreatic agenesis attributable to a single nucleotide deletion in the human IPF1 gene coding sequence. Nat. Genet. 15, 106–110.

Stone, S.I., Abreu, D., Mcgill, J.B., Urano, F., 2021. Monogenic and syndromic diabetes due to endoplasmic reticulum stress. J. Diabetes Complications 35, 107618.

St-Onge, L., Sosa-Pineda, B., Chowdhury, K., Mansouri, A., Gruss, P., 1997. Pax6 is required for differentiation of glucagon-producing alpha-cells in mouse pancreas. Nature 387, 406–409.

Stoy, J., Edghill, E.L., Flanagan, S.E., Ye, H., Paz, V.P., Pluzhnikov, A., Below, J.E., Hayes, M.G., Cox, N.J., Lipkind, G.M., Lipton, R.B., Greeley, S.A., Patch, A.M., Ellard, S., Steiner, D.F., Hattersley, A.T., Philipson, L.H., Bell, G.I., Neonatal Diabetes International Collaborative Group, 2007. Insulin gene mutations as a cause of permanent neonatal diabetes. Proc. Natl. Acad. Sci. U. S. A. 104, 15040–15044.

Stoy, J., Steiner, D.F., Park, S.-Y., Ye, H., Philipson, L.H., Bell, G.I, 2010. Clinical and molecular genetics of neonatal diabetes due to mutations in the insulin gene. Rev. Endocr. Metab. Disord. 11, 205–215.

Su, H.C., Wu, S.C., Yen, L.C., Chiao, L.K., Wang, J.K., Chiu, Y.L., Ho, C.L., Huang, S.M., 2020. Gene expression profiling identifies the role of Zac1 in cervical cancer metastasis. Sci. Rep. 10, 11837.

Sudhof, T.C., Rizo, J., 2011. Synaptic vesicle exocytosis. Cold Spring Harb. Perspect. Biol. 3, a005637.

Sujjitjoon, J., Kooptiwut, S., Chongjaroen, N., Tangjittipokin, W., Plengvidhya, N., Yenchitsomanus, P.T., 2016. Aberrant mRNA splicing of paired box 4 (PAX4) IVS7-1G>A mutation causing maturity-onset diabetes of the young, type 9. Acta Diabetol. 53, 205–216.

Sun, J., Ren, D., 2017. IER3IP1 deficiency leads to increased beta-cell death and decreased beta-cell proliferation. Oncotarget 8, 56768–56779.

Sun, J., Have, C.T., Hollensted, M., Grarup, N., Linneberg, A., Pedersen, O., Nielsen, J.S., Rungby, J., Christensen, C., Brandslund, I., Kristiansen, K., Jun, W., Hansen, T., Gjesing, A.P., 2019a. Sequencing reveals protective and pathogenic effects on development of diabetes of rare GLIS3 variants. PLoS One 14, e0220805.

Sun, Z.Q., Cui, Y.H., Yan, B., 2019b. Identification of two novel GATA6 mutations in an adult with acute myocardial infarction, diabetes, and atrial fibrillation: a case report. J. Geriatr. Cardiol. 16, 785–788.

Suzuki, K., Harada, N., Yamane, S., Nakamura, Y., Sasaki, K., Nasteska, D., Joo, E., Shibue, K., Harada, T., Hamasaki, A., Toyoda, K., Nagashima, K., Inagaki, N., 2013. Transcriptional regulatory factor X6 (Rfx6) increases gastric inhibitory polypeptide (GIP) expression in enteroendocrine K-cells and is involved in GIP hypersecretion in high fat diet-induced obesity. J. Biol. Chem. 288, 1929–1938.

Szopa, M., Ludwig-Galezowska, A.H., Radkowski, P., Skupien, J., Machlowska, J., Klupa, T., Wolkow, P., Borowiec, M., Mlynarski, W., Malecki, M.T., 2016. A family with the Arg103Pro mutation in the NEUROD1 gene detected by next-generation sequencing—clinical characteristics of mutation carriers. Eur. J. Med. Genet. 59, 75–79.

Tager, H., Given, B., Baldwin, D., Mako, M., Markese, J., Rubenstein, A., Olefsky, J., Kobayashi, M., Kolterman, O., Poucher, R., 1979. A structurally abnormal insulin causing human diabetes. Nature 281, 122–125.

Taha, D., Barbar, M., Kanaan, H., Williamson Balfe, J., 2003. Neonatal diabetes mellitus, congenital hypothyroidism, hepatic fibrosis, polycystic kidneys, and congenital glaucoma: a new autosomal recessive syndrome? Am. J. Med. Genet. A 122A, 269–273.

Takei, K., Slepnev, V.I., Haucke, V., De Camilli, P., 1999. Functional partnership between amphiphysin and dynamin in clathrin-mediated endocytosis. Nat. Cell Biol. 1, 33–39.

Tanabe, Y., William, C., Jessell, T.M., 1998. Specification of motor neuron identity by the MNR2 homeodomain protein. Cell 95, 67–80.

Teitelman, G., 2019. Heterogeneous expression of proinsulin processing enzymes in beta cells of non-diabetic and type 2 diabetic humans. J. Histochem. Cytochem. 67, 385–400.

Temple, I.K., Gardner, R.J., Mackay, D.J., Barber, J.C., Robinson, D.O., Shield, J.P., 2000. Transient neonatal diabetes: widening the understanding of the etiopathogenesis of diabetes. Diabetes 49, 1359–1366.

Terauchi, Y., Sakura, H., Yasuda, K., Iwamoto, K., Takahashi, N., Ito, K., Kasai, H., Suzuki, H., Ueda, O., Kamada, N., et al., 1995. Pancreatic beta-cell-specific targeted disruption of glucokinase gene. Diabetes mellitus due to defective insulin secretion to glucose. J. Biol. Chem. 270, 30253–30256.

Texido, G., Su, I.H., Mecklenbrauker, I., Saijo, K., Malek, S.N., Desiderio, S., Rajewsky, K., Tarakhovsky, A., 2000. The B-cell-specific Src-family kinase Blk is dispensable for B-cell development and activation. Mol. Cell. Biol. 20, 1227–1233.

Thomas, P.M., Cote, G.J., Hallman, D.M., Mathew, P.M., 1995. Homozygosity mapping, to chromosome 11p, of the gene for familial persistent hyperinsulinemic hypoglycemia of infancy. Am. J. Hum. Genet. 56, 416–421.

Thomas, H., Jaschkowitz, K., Bulman, M., Frayling, T.M., Mitchell, S.M., Roosen, S., Lingott-Frieg, A., Tack, C.J., Ellard, S., Ryffel, G.U., Hattersley, A.T., 2001. A distant upstream promoter of the HNF-4alpha gene connects the transcription factors involved in maturity-onset diabetes of the young. Hum. Mol. Genet. 10, 2089–2097.

Thorens, B., 2015. GLUT2, glucose sensing and glucose homeostasis. Diabetologia 58, 221–232.

Thornton, P.S., Satin-Smith, M.S., Herold, K., Glaser, B., Chiu, K.C., Nestorowicz, A., Permutt, M.A., Baker, L., Stanley, C.A., 1998. Familial hyperinsulinism with apparent autosomal dominant inheritance: clinical and genetic differences from the autosomal recessive variant. J. Pediatr. 132, 9–14.

Thornton, P.S., Macmullen, C., Ganguly, A., Ruchelli, E., Steinkrauss, L., Crane, A., Aguilar-Bryan, L., Stanley, C.A., 2003. Clinical and molecular characterization of a dominant form of congenital hyperinsulinism caused by a mutation in the high-affinity sulfonylurea receptor. Diabetes 52, 2403–2410.

Tokuyama, Y., Matsui, K., Ishizuka, T., Egashira, T., Kanatsuka, A., 2006. The Arg121Trp variant in PAX4 gene is associated with beta-cell dysfunction in Japanese subjects with type 2 diabetes mellitus. Metabolism 55, 213–216.

Tornovsky, S., Crane, A., Cosgrove, K.E., Hussain, K., Lavie, J., Heyman, M., Nesher, Y., Kuchinski, N., Ben-Shushan, E., Shatz, O., Nahari, E., Potikha, T., Zangen, D., Tenenbaum-Rakover, Y., De Vries, L., Argente, J., Gracia, R., Landau, H., Eliakim, A., Lindley, K., Dunne, M.J., Aguilar-Bryan, L., Glaser, B., 2004. Hyperinsulinism of infancy: novel ABCC8 and KCNJ11 mutations and evidence for additional locus heterogeneity. J. Clin. Endocrinol. Metab. 89, 6224–6234.

Torres, A.G., Batlle, E., Ribas De Pouplana, L., 2014. Role of tRNA modifications in human diseases. Trends Mol. Med. 20, 306–314.

Touati, A., Errea-Dorronsoro, J., Nouri, S., Halleb, Y., Pereda, A., Mahdhaoui, N., Ghith, A., Saad, A., Perez De Nanclares, G., H'mida Ben Brahim, D., 2019. Transient neonatal diabetes mellitus and hypomethylation at additional imprinted loci: novel ZFP57 mutation and review on the literature. Acta Diabetol. 56, 301–307.

Tsai, C.L., Barondeau, D.P., 2010. Human frataxin is an allosteric switch that activates the Fe-S cluster biosynthetic complex. Biochemistry 49, 9132–9139.

Tsaytler, P., Harding, H.P., Ron, D., Bertolotti, A., 2011. Selective inhibition of a regulatory subunit of protein phosphatase 1 restores proteostasis. Science 332, 91–94.

Tsukuda, K., Suzuki, Y., Kameoka, K., Osawa, N., Goto, Y., Katagiri, H., Asano, T., Yazaki, Y., Oka, Y., 1997. Screening of patients with maternally transmitted diabetes for mitochondrial gene mutations in the tRNA[Leu(UUR)] region. Diabet. Med. 14, 1032–1037.

Turque, N., Plaza, S., Radvanyi, F., Carriere, C., Saule, S., 1994. Pax-QNR/Pax-6, a paired box- and homeobox-containing gene expressed in neurons, is also expressed in pancreatic endocrine cells. Mol. Endocrinol. 8, 929–938.

Tutak, E., Satar, M., Yapicioglu, H., Altintas, A., Narli, N., Herguner, O., Bayram, Y., 2009. A Turkish newborn infant with cerebellar agenesis/neonatal diabetes mellitus and PTF1A mutation. Genet. Couns. 20, 147–152.

Urakami, T., 2019. Maturity-onset diabetes of the young (MODY): current perspectives on diagnosis and treatment. Diabetes Metab. Syndr. Obes. 12, 1047–1056.

Valenzuela, I., Boronat, S., Martinez-Saez, E., Clemente, M., Sanchez-Montanez, A., Munell, F., Carrascosa, A., Macaya, A., 2017. Microcephaly with simplified gyral pattern, epilepsy and permanent neonatal diabetes syndrome (MEDS). A new patient and review of the literature. Eur. J. Med. Genet. 60, 517–520.

Valkovicova, T., Skopkova, M., Stanik, J., Gasperikova, D., 2019. Novel insights into genetics and clinics of the HNF1A-MODY. Endocr. Regul. 53, 110–134.

Van De Bunt, M., Lako, M., Barrett, A., Gloyn, A.L., Hansson, M., Mccarthy, M.I., Beer, N.L., Honore, C., 2016. Insights into islet development and biology through characterization of a human iPSC-derived endocrine pancreas model. Islets 8, 83–95.

Van Den Ouweland, J.M., Lemkes, H.H., Ruitenbeek, W., Sandkuijl, L.A., De Vijlder, M.F., Struyvenberg, P.A., Van De Kamp, J.J., Maassen, J.A., 1992. Mutation in mitochondrial tRNA(Leu)(UUR) gene in a large pedigree with maternally transmitted type II diabetes mellitus and deafness. Nat. Genet. 1, 368–371.

Vankan, P., 2013. Prevalence gradients of Friedreich's ataxia and R1b haplotype in Europe co-localize, suggesting a common Palaeolithic origin in the Franco-Cantabrian ice age refuge. J. Neurochem. 126 (Suppl. 1), 11–20.

Vaxillaire, M., Populaire, C., Busiah, K., Cave, H., Gloyn, A.L., Hattersley, A.T., Czernichow, P., Froguel, P., Polak, M., 2004. Kir6.2 mutations are a common cause of permanent neonatal diabetes in a large cohort of French patients. Diabetes 53, 2719–2722.

Verkarre, V., Fournet, J.C., De Lonlay, P., Gross-Morand, M.S., Devillers, M., Rahier, J., Brunelle, F., Robert, J.J., Nihoul-Fekete, C., Saudubray, J.M., Junien, C., 1998. Paternal mutation of the sulfonylurea receptor (SUR1) gene and maternal loss of 11p15 imprinted genes lead to persistent hyperinsulinism in focal adenomatous hyperplasia. J. Clin. Invest. 102, 1286–1291.

Vilardo, E., Amman, F., Toth, U., Kotter, A., Helm, M., Rossmanith, W., 2020. Functional characterization of the human tRNA methyltransferases TRMT10A and TRMT10B. Nucleic Acids Res. 48, 6157–6169.

Villamayor, L., Rodriguez-Seguel, E., Araujo, R., Carrasco, M., Bru-Tari, E., Mellado-Gil, J.M., Gauthier, B.R., Martinelli, P., Quesada, I., Soria, B., Martin, F., Cano, D.A., Rojas, A., 2018. GATA6 controls insulin biosynthesis and secretion in adult beta-cells. Diabetes 67, 448–460.

Vionnet, N., Stoffel, M., Takeda, J., Yasuda, K., Bell, G.I., Zouali, H., Lesage, S., Velho, G., Iris, F., Passa, P., et al., 1992. Nonsense mutation in the glucokinase gene causes early-onset non-insulin-dependent diabetes mellitus. Nature 356, 721–722.

Vissers, L., Kalvakuri, S., De Boer, E., Geuer, S., Oud, M., Van Outersterp, I., Kwint, M., Witmond, M., Kersten, S., Polla, D.L., Weijers, D., Begtrup, A., Mcwalter, K., Ruiz, A., Gabau, E., Morton, J.E.V., Griffith, C., Weiss, K., Gamble, C., Bartley, J., Vernon, H.J., Brunet, K., Ruivenkamp, C., Kant, S.G., Kruszka, P., Larson, A., Afenjar, A., Billette De Villemeur, T., Nugent, K., Study, D.D.D., Raymond, F.L., Venselaar, H., Demurger, F., Soler-Alfonso, C., Li, D., Bhoj, E., Hayes, I., Hamilton, N.P., Ahmad, A., Fisher, R., Van Den Born, M., Willems, M., Sorlin, A., Delanne, J., Moutton, S., Christophe, P., Mau-Them, F.T., Vitobello, A., Goel, H., Massingham, L., Phornphutkul, C., Schwab, J., Keren, B., Charles, P., Vreeburg, M., De Simone, L., Hoganson, G., Iascone, M., Milani, D., Evenepoel, L., Revencu, N., Ward, D.I., Burns, K., Krantz, I., Raible, S.E., Murrell, J.R., Wood, K., Cho, M.T., Van Bokhoven, H., Muenke, M., Kleefstra, T., Bodmer, R., De Brouwer, A.P.M., 2020. De novo variants in CNOT1, a central component of the CCR4-NOT complex involved in gene expression and RNA and protein stability, cause neurodevelopmental delay. Am. J. Hum. Genet. 107, 164–172.

Voevoda, M.I., Ivanova, A.A., Shakhtshneider, E.V., Ovsyannikova, A.K., Mikhailova, S.V., Astrakova, K.S., Voevoda, S.M., Rymar, O.D., 2016. Molecular genetics of maturity-onset diabetes of the young. Ter. Arkh. 88, 117–124.

Wabitsch, M., Lahr, G., Van De Bunt, M., Marchant, C., Lindner, M., Von Puttkamer, J., Fenneberg, A., Debatin, K.M., Klein, R., Ellard, S., Clark, A., Gloyn, A.L., 2007. Heterogeneity in disease severity in a family with a novel G68V GCK activating mutation causing persistent hyperinsulinaemic hypoglycaemia of infancy. Diabet. Med. 24, 1393–1399.

Walther, C., Gruss, P., 1991. Pax-6, a murine paired box gene, is expressed in the developing CNS. Development 113, 1435–1449.

Wang, J., Takeuchi, T., Tanaka, S., Kubo, S.K., Kayo, T., Lu, D., Takata, K., Koizumi, A., Izumi, T., 1999. A mutation in the insulin 2 gene induces diabetes with severe pancreatic beta-cell dysfunction in the Mody mouse. J. Clin. Invest. 103, 27–37.

Wang, H., Maechler, P., Antinozzi, P.A., Hagenfeldt, K.A., Wollheim, C.B., 2000. Hepatocyte nuclear factor 4alpha regulates the expression of pancreatic beta -cell genes implicated in glucose metabolism and nutrient-induced insulin secretion. J. Biol. Chem. 275, 35953–35959.

Wang, J., Elghazi, L., Parker, S.E., Kizilocak, H., Asano, M., Sussel, L., Sosa-Pineda, B., 2004a. The concerted activities of Pax4 and Nkx2.2 are essential to initiate pancreatic beta-cell differentiation. Dev. Biol. 266, 178–189.

Wang, L., Coffinier, C., Thomas, M.K., Gresh, L., Eddu, G., Manor, T., Levitsky, L.L., Yaniv, M., Rhoads, D.B., 2004b. Selective deletion of the Hnf1beta (MODY5) gene in beta-cells leads to altered gene expression and defective insulin release. Endocrinology 145, 3941–3949.

Wang, H., Brun, T., Kataoka, K., Sharma, A.J., Wollheim, C.B., 2007. MAFA controls genes implicated in insulin biosynthesis and secretion. Diabetologia 50, 348–358.

Wang, Q., Fang, W.H., Krupinski, J., Kumar, S., Slevin, M., Kumar, P., 2008. Pax genes in embryogenesis and oncogenesis. J. Cell. Mol. Med. 12, 2281–2294.

Wang, C.H., Chen, Y.F., Wu, C.Y., Wu, P.C., Huang, Y.L., Kao, C.H., Lin, C.H., Kao, L.S., Tsai, T.F., Wei, Y.H., 2014a. Cisd2 modulates the differentiation and functioning of adipocytes by regulating intracellular Ca2+ homeostasis. Hum. Mol. Genet. 23, 4770–4785.

Wang, C.H., Kao, C.H., Chen, Y.F., Wei, Y.H., Tsai, T.F., 2014b. Cisd2 mediates lifespan: is there an interconnection among Ca(2)(+) homeostasis, autophagy, and lifespan? Free Radic. Res. 48, 1109–1114.

Wang, X., Lu, Z., Gomez, A., Hon, G.C., Yue, Y., Han, D., Fu, Y., Parisien, M., Dai, Q., Jia, G., Ren, B., Pan, T., He, C., 2014c. N6-methyladenosine-dependent regulation of messenger RNA stability. Nature 505, 117–120.

Wang, M., Liu, H., Zheng, J., Chen, B., Zhou, M., Fan, W., Wang, H., Liang, X., Zhou, X., Eriani, G., Jiang, P., Guan, M.X., 2016. A deafness- and diabetes-associated tRNA mutation causes deficient pseudouridinylation at position 55 in tRNAGlu and mitochondrial dysfunction. J. Biol. Chem. 291, 21029–21041.

Wang, Y., Branicky, R., Noe, A., Hekimi, S., 2018. Superoxide dismutases: dual roles in controlling ROS damage and regulating ROS signaling. J. Cell Biol. 217, 1915–1928.

Wang, X., Sterr, M., Ansarullah, Burtscher, I., Bottcher, A., Beckenbauer, J., Siehler, J., Meitinger, T., Haring, H.U., Staiger, H., Cernilogar, F.M., Schotta, G., Irmler, M., Beckers, J., Wright, C.V.E., Bakhti, M., Lickert, H., 2019. Point mutations in the PDX1 transactivation domain impair human beta-cell development and function. Mol. Metab. 24, 80–97.

Wat, M.J., Shchelochkov, O.A., Holder, A.M., Breman, A.M., Dagli, A., Bacino, C., Scaglia, F., Zori, R.T., Cheung, S.W., Scott, D.A., Kang, S.H., 2009. Chromosome 8p23.1 deletions as a cause of complex congenital heart defects and diaphragmatic hernia. Am. J. Med. Genet. A 149A, 1661–1677.

Watt, A.J., Zhao, R., Li, J., Duncan, S.A., 2007. Development of the mammalian liver and ventral pancreas is dependent on GATA4. BMC Dev. Biol. 7, 37.

Weedon, M.N., Owen, K.R., Shields, B., Hitman, G., Walker, M., Mccarthy, M.I., Love-Gregory, L.D., Permutt, M.A., Hattersley, A.T., Frayling, T.M., 2004. Common variants of the hepatocyte nuclear factor-4alpha P2 promoter are associated with type 2 diabetes in the U.K. population. Diabetes 53, 3002–3006.

Weedon, M.N., Cebola, I., Patch, A.M., Flanagan, S.E., De Franco, E., Caswell, R., Rodriguez-Segui, S.A., Shaw-Smith, C., Cho, C.H., Allen, H.L., Houghton, J.A., Roth, C.L., Chen, R., Hussain, K., Marsh, P., Vallier, L., Murray, A., International Pancreatic Agenesis Consortium, Ellard, S., Ferrer, J., Hattersley, A.T., 2014. Recessive mutations in a distal PTF1A enhancer cause isolated pancreatic agenesis. Nat. Genet. 46, 61–64.

Wen, J.H., Chen, Y.Y., Song, S.J., Ding, J., Gao, Y., Hu, Q.K., Feng, R.P., Liu, Y.Z., Ren, G.C., Zhang, C.Y., Hong, T.P., Gao, X., Li, L.S., 2009. Paired box 6 (PAX6) regulates glucose metabolism via proinsulin processing mediated by prohormone convertase 1/3 (PC1/3). Diabetologia 52, 504–513.

Whitnall, M., Suryo Rahmanto, Y., Sutak, R., Xu, X., Becker, E.M., Mikhael, M.R., Ponka, P., Richardson, D.R., 2008. The MCK mouse heart model of Friedreich's ataxia: alterations in iron-regulated proteins and cardiac hypertrophy are limited by iron chelation. Proc. Natl. Acad. Sci. U. S. A. 105, 9757–9762.

Wiley, S.E., Andreyev, A.Y., Divakaruni, A.S., Karisch, R., Perkins, G., Wall, E.A., Van Der Geer, P., Chen, Y.F., Tsai, T.F., Simon, M.I., Neel, B.G., Dixon, J.E., Murphy, A.N., 2013. Wolfram syndrome protein, Miner1, regulates sulphydryl redox status, the unfolded protein response, and Ca2+ homeostasis. EMBO Mol. Med. 5, 904–918.

Winkler, G.S., Mulder, K.W., Bardwell, V.J., Kalkhoven, E., Timmers, H.T., 2006. Human Ccr4-Not complex is a ligand-dependent repressor of nuclear receptor-mediated transcription. EMBO J. 25, 3089–3099.

Wortham, N.C., Proud, C.G., 2015. eIF2B: recent structural and functional insights into a key regulator of translation. Biochem. Soc. Trans. 43, 1234–1240.
Wright, N.M., Metzger, D.L., Borowitz, S.M., Clarke, W.L., 1993. Permanent neonatal diabetes mellitus and pancreatic exocrine insufficiency resulting from congenital pancreatic agenesis. Am. J. Dis. Child. 147, 607–609.
Xiao, D., Liu, X., Zhang, M., Zou, M., Deng, Q., Sun, D., Bian, X., Cai, Y., Guo, Y., Liu, S., Li, S., Shiang, E., Zhong, H., Cheng, L., Xu, H., Jin, K., Xiang, M., 2018. Direct reprogramming of fibroblasts into neural stem cells by single non-neural progenitor transcription factor Ptf1a. Nat. Commun. 9, 2865.
Yamagata, K., 2003. Regulation of pancreatic beta-cell function by the HNF transcription network: lessons from maturity-onset diabetes of the young (MODY). Endocr. J. 50, 491–499.
Yamagata, K., 2014. Roles of HNF1alpha and HNF4alpha in pancreatic beta-cells: lessons from a monogenic form of diabetes (MODY). Vitam. Horm. 95, 407–423.
Yamagata, K., Oda, N., Kaisaki, P.J., Menzel, S., Furuta, H., Vaxillaire, M., Southam, L., Cox, R.D., Lathrop, G.M., Boriraj, V.V., Chen, X., Cox, N.J., Oda, Y., Yano, H., Le Beau, M.M., Yamada, S., Nishigori, H., Takeda, J., Fajans, S.S., Hattersley, A.T., Iwasaki, N., Hansen, T., Pedersen, O., Polonsky, K.S., Bell, G.I., et al., 1996. Mutations in the hepatocyte nuclear factor-1alpha gene in maturity-onset diabetes of the young (MODY3). Nature 384, 455–458.
Yang, Y., Chang, B.H., Yechoor, V., Chen, W., Li, L., Tsai, M.J., Chan, L., 2011. The Kruppel-like zinc finger protein GLIS3 transactivates neurogenin 3 for proper fetal pancreatic islet differentiation in mice. Diabetologia 54, 2595–2605.
Yang, Y., Chang, B.H., Chan, L., 2013. Sustained expression of the transcription factor GLIS3 is required for normal beta cell function in adults. EMBO Mol. Med. 5, 92–104.
Yano, H., Kitano, N., Morimoto, M., Polonsky, K.S., Imura, H., Seino, Y., 1992. A novel point mutation in the human insulin gene giving rise to hyperproinsulinemia (proinsulin Kyoto). J. Clin. Invest. 89, 1902–1907.
Yasuda, T., Kajimoto, Y., Fujitani, Y., Watada, H., Yamamoto, S., Watarai, T., Umayahara, Y., Matsuhisa, M., Gorogawa, S., Kuwayama, Y., Tano, Y., Yamasaki, Y., Hori, M., 2002. PAX6 mutation as a genetic factor common to aniridia and glucose intolerance. Diabetes 51, 224–230.
Yasukawa, T., Suzuki, T., Ueda, T., Ohta, S., Watanabe, K., 2000. Modification defect at anticodon wobble nucleotide of mitochondrial tRNAs(Leu)(UUR) with pathogenic mutations of mitochondrial myopathy, encephalopathy, lactic acidosis, and stroke-like episodes. J. Biol. Chem. 275, 4251–4257.
Yau, D., De Franco, E., Flanagan, S.E., Ellard, S., Blumenkrantz, M., Mitchell, J.J., 2017. Case report: maternal mosaicism resulting in inheritance of a novel GATA6 mutation causing pancreatic agenesis and neonatal diabetes mellitus. Diagn. Pathol. 12, 1.
Yew, T.W., Mccreight, L., Colclough, K., Ellard, S., Pearson, E.R., 2016. tRNA methyltransferase homologue gene TRMT10A mutation in young adult-onset diabetes with intellectual disability, microcephaly and epilepsy. Diabet. Med. 33, e21–e25.
Yoon, T., Cowan, J.A., 2003. Iron-sulfur cluster biosynthesis. Characterization of frataxin as an iron donor for assembly of [2Fe-2S] clusters in ISU-type proteins. J. Am. Chem. Soc. 125, 6078–6084.
Yoon, T., Cowan, J.A., 2004. Frataxin-mediated iron delivery to ferrochelatase in the final step of heme biosynthesis. J. Biol. Chem. 279, 25943–25946.
Yorifuji, T., Kurokawa, K., Mamada, M., Imai, T., Kawai, M., Nishi, Y., Shishido, S., Hasegawa, Y., Nakahata, T., 2004. Neonatal diabetes mellitus and neonatal polycystic, dysplastic kidneys: phenotypically discordant recurrence of a mutation in the hepatocyte nuclear factor-1beta gene due to germline mosaicism. J. Clin. Endocrinol. Metab. 89, 2905–2908.

Yorifuji, T., Nagashima, K., Kurokawa, K., Kawai, M., Oishi, M., Akazawa, Y., Hosokawa, M., Yamada, Y., Inagaki, N., Nakahata, T., 2005. The C42R mutation in the Kir6.2 (KCNJ11) gene as a cause of transient neonatal diabetes, childhood diabetes, or later-onset, apparently type 2 diabetes mellitus. J. Clin. Endocrinol. Metab. 90, 3174–3178.

Yorifuji, T., Kawakita, R., Hosokawa, Y., Fujimaru, R., Yamaguchi, E., Tamagawa, N., 2012. Dominantly inherited diabetes mellitus caused by GATA6 haploinsufficiency: variable intrafamilial presentation. J. Med. Genet. 49, 642–643.

Yoshida, Y., Suzuki, K., Yamamoto, A., Sakai, N., Bando, M., Tanimoto, K., Yamaguchi, Y., Sakaguchi, T., Akhter, H., Fujii, G., Yoshimura, S., Ogata, S., Sohda, M., Misumi, Y., Nakamura, N., 2008. YIPF5 and YIF1A recycle between the ER and the Golgi apparatus and are involved in the maintenance of the Golgi structure. Exp. Cell Res. 314, 3427–3443.

Young-Baird, S.K., Shin, B.S., Dever, T.E., 2019. MEHMO syndrome mutation EIF2S3-I259M impairs initiator Met-tRNAiMet binding to eukaryotic translation initiation factor eIF2. Nucleic Acids Res. 47, 855–867.

Yourshaw, M., Solorzano-Vargas, R.S., Pickett, L.A., Lindberg, I., Wang, J., Cortina, G., Pawlikowska-Haddal, A., Baron, H., Venick, R.S., Nelson, S.F., Martin, M.G., 2013. Exome sequencing finds a novel PCSK1 mutation in a child with generalized malabsorptive diarrhea and diabetes insipidus. J. Pediatr. Gastroenterol. Nutr. 57, 759–767.

Zatyka, M., Ricketts, C., Da Silva Xavier, G., Minton, J., Fenton, S., Hofmann-Thiel, S., Rutter, G.A., Barrett, T.G., 2008. Sodium-potassium ATPase 1 subunit is a molecular partner of Wolframin, an endoplasmic reticulum protein involved in ER stress. Hum. Mol. Genet. 17, 190–200.

Zeruth, G.T., Takeda, Y., Jetten, A.M., 2013. The Kruppel-like protein Gli-similar 3 (Glis3) functions as a key regulator of insulin transcription. Mol. Endocrinol. 27, 1692–1705.

Zhang, J.S., Moncrieffe, M.C., Kaczynski, J., Ellenrieder, V., Prendergast, F.G., Urrutia, R., 2001. A conserved alpha-helical motif mediates the interaction of Sp1-like transcriptional repressors with the corepressor mSin3A. Mol. Cell. Biol. 21, 5041–5049.

Zhang, P., Mcgrath, B., Li, S., Frank, A., Zambito, F., Reinert, J., Gannon, M., Ma, K., Mcnaughton, K., Cavener, D.R., 2002. The PERK eukaryotic initiation factor 2 alpha kinase is required for the development of the skeletal system, postnatal growth, and the function and viability of the pancreas. Mol. Cell. Biol. 22, 3864–3874.

Zhang, W., Feng, D., Li, Y., Iida, K., Mcgrath, B., Cavener, D.R., 2006. PERK EIF2AK3 control of pancreatic beta cell differentiation and proliferation is required for postnatal glucose homeostasis. Cell Metab. 4, 491–497.

Zhao, L., Guo, M., Matsuoka, T.A., Hagman, D.K., Parazzoli, S.D., Poitout, V., Stein, R., 2005. The islet beta cell-enriched MafA activator is a key regulator of insulin gene transcription. J. Biol. Chem. 280, 11887–11894.

Zheng, X., Dumitru, R., Lackford, B.L., Freudenberg, J.M., Singh, A.P., Archer, T.K., Jothi, R., Hu, G., 2012. Cnot1, Cnot2, and Cnot3 maintain mouse and human ESC identity and inhibit extraembryonic differentiation. Stem Cells 30, 910–922.

Zhou, Z., Sun, B., Huang, S., Jia, W., Yu, D., 2019. The tRNA-associated dysregulation in diabetes mellitus. Metabolism 94, 9–17.

Zhu, X., Orci, L., Carroll, R., Norrbom, C., Ravazzola, M., Steiner, D.F., 2002. Severe block in processing of proinsulin to insulin accompanied by elevation of des-64,65 proinsulin intermediates in islets of mice lacking prohormone convertase 1/3. Proc. Natl. Acad. Sci. U. S. A. 99, 10299–10304.

Zung, A., Kori, M., Burundukov, E., Ben-Yosef, T., Tatoor, Y., Granot, E., 2015. Homozygous deletion of TRMT10A as part of a contiguous gene deletion in a syndrome of failure to thrive, delayed puberty, intellectual disability and diabetes mellitus. Am. J. Med. Genet. A 167A, 3167–3173.

CHAPTER FOUR

Role of microRNA in pancreatic beta cell function

Flora Brozzi[a,b,*]
[a]Department of Fundamental Neurosciences, University of Lausanne, Lausanne, Switzerland
[b]Department of Biomedical Sciences, University of Lausanne, Lausanne, Switzerland
*Corresponding author: e-mail address: flora.brozzi@unil.ch

Contents

Abstract

MicroRNA (miRNAs) are small non-coding RNA involved in gene expression regulation. Emerging evidences identify miRNAs as key regulators of beta cell physiology. Their role in fine-tuned gene expression regulation is crucial in the differentiation of insulin-producing cells and contributes to the acquisition and management of their unique phenotype. Dysregulation of miRNA expression causes beta cell dysfunction and promotes the development of different forms of diabetes mellitus.

1. Introduction

MicroRNAs (miRNAs) are small non-coding RNA that fine tune the gene expression at the posttranscriptional level, by binding to the 3′ untranslated regions (UTR) of target messenger RNAs (mRNAs) and inhibiting their expression (Bartel, 2018).

In general, the miRNA genes are located within protein-coding genes and are transcribed together with their host-gene. Otherwise, they can be

International Review of Cell and Molecular Biology, Volume 359
ISSN 1937-6448
https://doi.org/10.1016/bs.ircmb.2021.02.012

localized in intergenic regions and can have their own promoter (Bartel, 2018). Sometimes several miRNA genes form clusters and are transcribed in a single long transcript (pri-miRNA) that is processed by the microprocessor complex Drosha–DGCR8 in the nucleus (Lee et al., 2002) (Fig. 1).

In any of the above cases, miRNAs originate from a longer hairpin-folded RNA transcript of 70 nucleotides (pre-miRNA) that is cleaved by the endonuclease Dicer to produce the mature 21–23 nucleotide double-stranded miRNA structure (Fig. 1).

Mature miRNAs interact with the RNA-induced silencing complex (RISC) containing Dicer, Argonaute (Ago) protein and many other associated proteins (Rana, 2007). Only a single guide strand of the mature miRNA binds to Ago protein and is loaded into the RISC, permitting the recognition of complementary sites within target mRNAs (Bartel, 2018).

The paring among miRNAs and their target mRNA is essential for the miRNA function. This interaction occurs mainly between the seed region of the miRNAs (nucleotides at position 2–7) and the 3′ UTR of the mRNA. The consequence of this specific interaction is the post-transcriptional repression through mRNA degradation, translational repression and/or

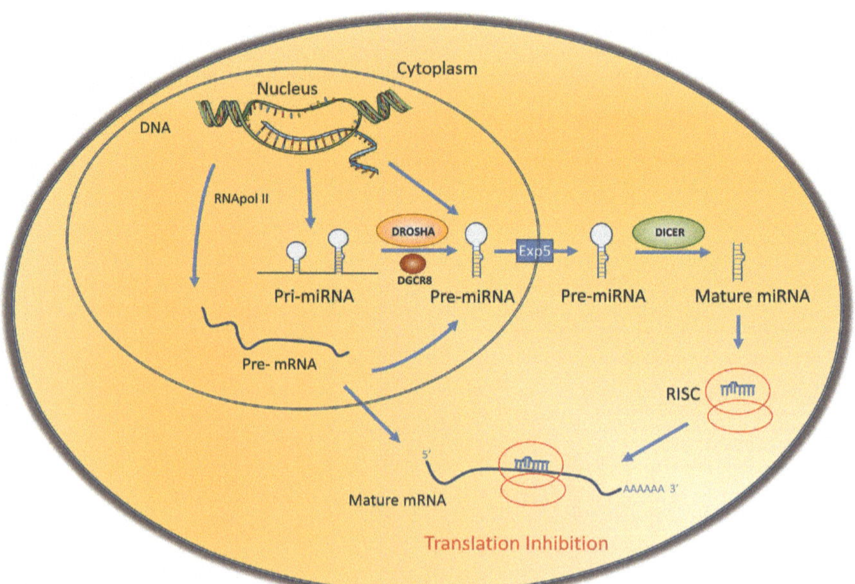

Fig. 1 Biogenesis of miRNAs. See text for details. DGCR8, DiGeorge Syndrome Critical Region 8; EXP5, Exportin 5; RNA pol II, RNA polymerase II; RISC, RNA-induced silencing complex.

mRNA destabilization. The seed region of the miRNAs is relatively small so one miRNA can interact with different target sequences in different mRNAs. Moreover, full complementary pairing of the miRNA with the target sequence is not required, so each miRNA can control the expression of many target genes and each mRNA can be regulated by several miRNAs (Bartel, 2009).

Since their discovery in 1993, where miRNAs have been described in *Caenorhabditis elegans* gene regulation (Lee et al., 1993; Wightman et al., 1993), they have been extensively studied in vertebrates and plants, and have emerged as crucial mediators of a wide range of physiological and pathological processes.

2. miRNAs in pancreatic beta cell physiology

The first reported pancreatic islet-abounded miRNA identified is miR-375 (Poy et al., 2004). The overexpression of miR-375 suppresses glucose-induced insulin secretion, and conversely, the inhibition of endogenous miR-375 function enhances insulin secretion. MiR-375 expression has a direct effect on insulin exocytosis and Myotrophin (Mtpn) was initially identified as one of its targets. Since then, many mRNA targets and functions of miR-375 in beta cells have been identified and numerous other miRNAs have been demonstrated to play a significant role in beta cell activity regulation (Dumortier and Van Obberghen, 2012; Eliasson and Esguerra, 2014; Guay et al., 2011; Jacovetti et al., 2015; Lynn et al., 2007). This chapter will describe the role of miRNAs in the differentiation and functional maturation of beta cells, in beta cell proliferation, and in the regulation of insulin secretion. Deregulation of miRNAs during Type 1 and Type 2 diabetes pathogenesis will also be described.

2.1 miRNAs in pancreatic beta cell development

The differentiation and maturation of insulin-producing pancreatic beta cells is the results of a complex developmental program. During embryonic development, cells of the posterior foregut assume a pancreatic identity, then cells within the expanding pancreatic primordia adopt an endocrine fate, and a subset of these precursors becomes competent to generate beta cells (Murtaugh, 2007).

Pancreatic progenitor cells are defined by the expression of specific transcriptional factors which will lead the differentiation into three different cell types composing the pancreas: endocrine, exocrine, and ductal cells (Cano et al., 2014; Maestro et al., 2003). The transient activation of the transcription factor neurogenin3 (*Ngn3*), is crucial for the differentiation of the pancreatic endocrine lineage. Indeed, even if its expression is gradually reduced by the embryonic age E15.5, its downstream transcriptional factors direct the differentiation into the distinct islet cell subtypes and the activation of the mature islet transcription factors MafA, Paired box gene 6 (Pax6) and Insulin gene enhancer protein Isl-1. Among Ngn3-induced transcription factors, the Regulatory factor X-box 6 (Rfx6) plays a key role in islet cell differentiation, both in mouse and human. Rfx6 KO mice failed to generate any islet cell types except for the pancreatic-polypeptide-producing cells. Of note, human infants with a similar autosomal recessive syndrome of neonatal diabetes, have mutations in the RFX6 gene. These studies demonstrate a unique position for Ngn3 and Rfx6 in the hierarchy of factors that coordinate pancreatic islet development in both mice and humans (Smith et al., 2010).

MiRNAs have an important role in the specification of progenitors into the endocrine lineage of the pancreas (Martinez-Sanchez et al., 2016). This is corroborated by Dicer lost-of-function studies in different mouse models at different ages, and it is confirmed by miRNA profiling studies in the developing pancreas. The Knock-out (KO) of *Dicer1* in mouse developing pancreas (E9.5) leads to reduction in the Ngn3-expressing endocrine cell progenitors, agenesis of the pancreas, and death a few days after birth (Lynn et al., 2007). Deletion of *Dicer1* specifically in pancreatic endocrine progenitor cells (E12.5) causes altered hormone levels after birth leading to chronic hyperglycemia and diabetes onset by the age of 2 weeks (Kanji et al., 2013). Postnatal cell-specific *Dicer1* conditional deletion in pancreatic beta cells, under the control of the rat insulin promoter 1 (RIP1), reduces beta cell mass and number of islets, decreases insulin production and release, and triggers diabetes by 25 weeks (Kalis et al., 2011). Taken together, the above loss-of-function studies demonstrate a role for *Dicer1* and miRNAs in the early stages of pancreatic cell lineage differentiation, even though they do not identify specific miRNAs involved.

MiRNA profiling studies performed in the developing pancreas identified a set of miRNAs whose expression was modify during the differentiation of pancreatic endocrine cells (Martinez-Sanchez et al., 2016; Wong et al., 2018). Some of these miRNAs important in humans are miR-7, miR-9, miR-15a/b, miR-16, miR-195, miR-124a, miR-218, miR-375, miR-376a, miR-503, and miR-541 (Correa-Medina et al., 2009; Joglekar et al., 2009b; Sun and Lai, 2013). MiR-7 and miR-375 are highly expressed in islet cells and their expression increases during pancreatic endocrine cell differentiation, both in human and mouse (Correa-Medina et al., 2009; Joglekar et al., 2009a; Kredo-Russo et al., 2012a; Nieto et al., 2012). Moreover, miR-7 and miR-375 expression is decreased in Ngn3-null mice, a condition that abrogates endocrine pancreatic differentiation (Kredo-Russo et al., 2012a). MiR-7, miR-375 and miR-373 are highly upregulated also in the early stages of human embryonic stem cell (hESC) differentiation (Lopez-Beas et al., 2018). Increased levels of miR-375 and miR-7 promote hESC differentiation into insulin-expressing cells and modulate insulin secretion in response to glucose (Conrad et al., 2014; Francis et al., 2014). More in detail, the increased expression level of miR-7 is concomitant to the increased expression of pancreatic markers like Forkhead box protein A2 (FOXA2), Pancreatic and duodenal homeobox 1 (PDX1), and insulin, in early hESC differentiation.

Studies in animal models, aimed to better identify the molecular pathways controlled by miR-7 during endocrine cell differentiation, highlight the complexity and the time dependent-effect of the expression of miR-7 (Kredo-Russo et al., 2012b; Latreille et al., 2014; Nieto et al., 2012). MiR-7 overexpression at both E12.5 and E15.5 leads to Pax6 downregulation and inhibition of alpha- and beta-cell differentiation, favoring the appearance of ghrelin-positive cells. Accordingly, miR-7 knockdown results in Pax6 upregulation and promotes alpha- and beta-cell differentiation (Kredo-Russo et al., 2012b). Beta cell-specific miR-7 overexpression at adult age does not affect islet architecture or beta cell mass, but decreases *Ins1* and *Ins2* levels in the islets, with a consequent impact on beta cell function (Latreille et al., 2014). Globally, the dynamics of miR-7 expression during islet cell differentiation is crucial for the proper development of functional endocrine cells (Martinez-Sanchez et al., 2016).

MiR-375 follows a similar expression pattern: the generation of pancreatic islet cells is associated with an increase in miR-375 and the maturation of beta cells is associated with its decrease (Wei et al., 2013). In fact, whole-body miR-375 null mice did not show any major defect in beta cell number or insulin expression in the first 2 weeks of life, and altered islet architecture could only be observed after 3 weeks of age, suggesting that miR-375 is not essential for beta cell formation, but is required for maturation and function (Poy et al., 2009). Indeed, overexpression of miR-375 recovered beta cell phenotype in de-differentiated human islet cells, suggesting the important role of miR-375 in regulation of human beta cell phenotype (Nathan et al., 2015). Some of the miR-375 targets are hepatic nuclear factor 6 (HNF6), insulinoma-associated 1 transcription factor (INSM1), and PDX1, which are active in the development and function of pancreatic islets (Avnit-Sagi et al., 2009; Keller et al., 2007). Moreover, other predicted mRNA targets of miR-375 are transcription factors (such as Gata6, Hnf1α, and Pax6) that play important roles during the specification of pancreatic progenitors and terminal maturation of endocrine cells. Taken together these discoveries suggest that a miR-375 circuit directs the appropriate beta cell development through temporal controls of beta cell transcription factor expression.

MiR-7 and miR-375 are only two example to highlight the complex interplay between miRNAs and transcription factors involved in beta cells formation. Many others are involved and more detailed illustrated elsewhere (Francis et al., 2014; Kaviani et al., 2016; Martinez-Sanchez et al., 2016; Wong et al., 2018).

Generating insulin-producing cells (IPCs) from human-induced pluripotent stem cells (hIPSC) is an emerging and promising method for studying the molecular mechanism underlying pancreas development. A further understanding of the role of miRNAs in pancreatic development and function may be important in the on-going quest to generate insulin-secreting cells from pluripotent stem cells. Pioneer studies on the importance of miRNAs in the generation of IPCs demonstrated that overexpression of miR-375 (Lahmy et al., 2014) and miR-7 (Shaer et al., 2016) in hiPSC induced their differentiation into insulin-secreting cells and into functional islet-like cell clusters respectively.

Moreover, microRNA expression profiles during in vitro differentiation of hiPSCs into insulin-producing cells showed a progressive upregulation of 11 microRNAs and downregulation of 7 microRNAs. Among progressively upregulated microRNAs the authors confirmed miR-10a, miR-99a-3p, miR-149 and miR-342-3p, while among those downregulated miR-302c-5p, miR-373, miR-518b and miR-520c-3p were described. Target gene analysis showed that microRNAs synergistically regulate cohorts of genes involved in pancreas organogenesis (Sebastiani et al., 2017).

More recently, miR-181c-5p was shown to promote the formation of insulin-producing cells from hiPSCs, by directly targeting smad7 and TGIF2 mRNAs. hiPSCs were transduced with lentiviral vectors containing human miR-181c-5p precursor, which significantly increased the endodermal markers SOX17, FOXA2, CXCR4 and GATA4 and pancreatic endocrine-specific gene expression, including PDX1, MAFA and Insulin. When transplanted, these cells exhibit glucose-stimulated C-peptide secretion in vivo and protect mice from chemically induced diabetes. These results suggest that miR-181c-5p is critically involved in pancreatic lineage commitment through direct repression of smad7 and TGIF2 and that it modulates TGF-β-smad2/3 signaling activation. This represents a new strategy for generating IPCs in vitro by miR-181c-5p overexpression and a potential source of IPCs for transplantation therapy of type 1 diabetes mellitus (Li et al., 2020) (Fig. 2).

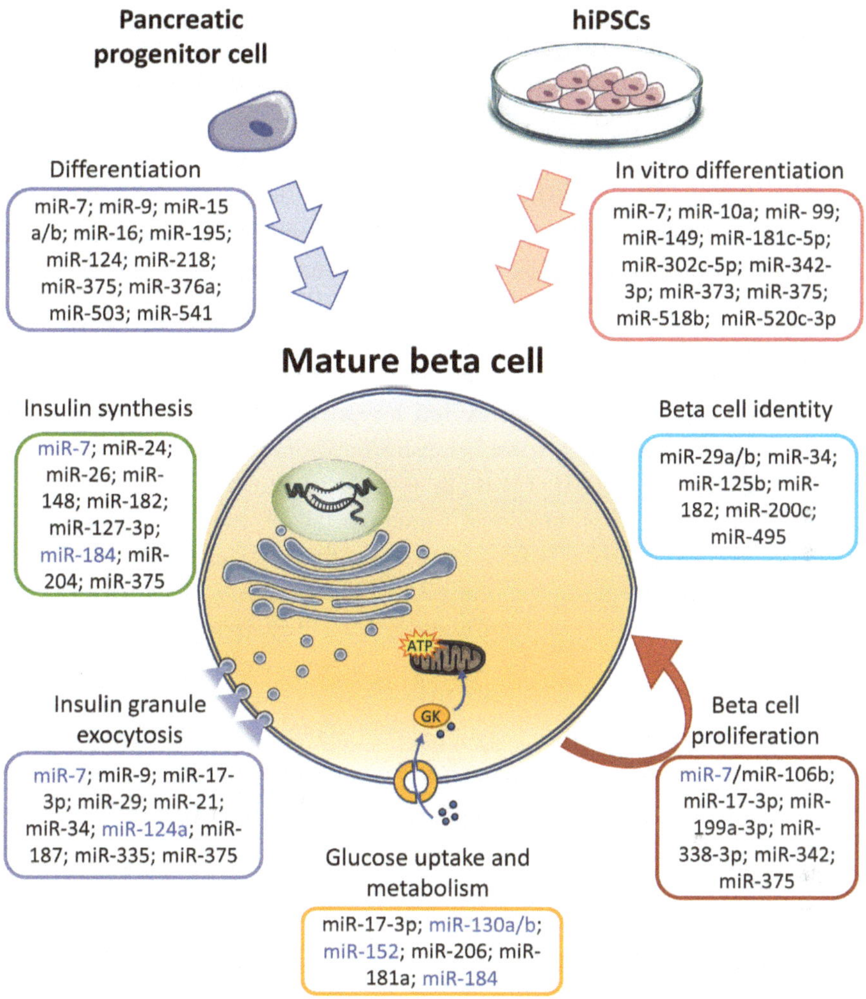

Fig. 2 MicroRNAs (miRNAs) involved in pancreatic beta cell differentiation and in the main beta cell functions. Listed are the miRNAs implicated in beta cell differentiation from either a pancreatic progenitor cell or from human-induced pluripotent stem cells (hiPSCs). In the mature beta cell, are indicated miRNAs involved in insulin expression, insulin granule exocytosis, glucose uptake, proliferation, and beta cell identity. miR-7a, miR-124a, miR-130a/b, miR-152 and miR-184 (highlighted in blue) are differentially expressed in islets of T2D donors compare to the healthy controls. *Latreille et al., 2014; Ofori et al., 2017; Sebastiani et al., 2015; Tattikota et al., 2014.*

2.2 miRNAs in pancreatic beta cell phenotype

Changes in the level of several miRNAs contribute to postnatal beta cell maturation and to the acquisition and maintenance of mature and functional beta cells (Jacovetti et al., 2015). Indeed, the achievement of a fully functional beta cell phenotype requires the reprogramming of gene networks expressed in newly generated islet cells that occurs during the neonatal period. Around 60 "disallowed" genes, which are highly expressed and play housekeeping roles in most other mammalian tissues, are unusually low in beta cells. MiRNAs are important players in this process and specifically inhibit the expression of genes that negatively affect beta cell identity and function (Kaspi et al., 2014; Lemaire et al., 2016; Martinez-Sanchez et al., 2016). Beta cell-specific inactivation of *Dicer1* at adulthood, in a tamoxifen-inducible Pdx1CreER transgene model, leads to the up-regulation of several beta cell "prohibited" genes, causing reduced insulin content, decreased beta cell mass and glucose intolerance (Martinez-Sanchez et al., 2015). One more specific example is represented by miR-29a and b, expressed at elevated levels in mature beta cells, which directly target the lactate/pyruvate transporter monocarboxylate transporter 1 (*Mct1*) (Pullen et al., 2011). MCTs 1–4 play essential metabolic roles in most tissues and their expression is fundamental for energy metabolism in the brain, skeletal muscle, heart, tumor cells, for T-lymphocyte activation, for gluconeogenesis, spermatogenesis, bowel metabolism and drug transport (Halestrap and Wilson, 2012). However, *Mct1* downregulation avoids the induction of insulin secretion in response to a rise in lactate generated during physical activities (Pullen et al., 2012) and deregulated *Mct1* expression causes pancreatic beta cell failure (Esguerra et al., 2018). Mct1/Slc16a1 expression is also inhibited by miR-495 in human pancreatic endoderm embryonic stem cell (hESC) and their transplant in a murine model for T2D improve diabetic symptoms (Liang et al., 2015).

The inhibition of glucagon expression is also fundamental for the beta cell identity maintenance. In alpha cells glucagon expression is promoted by Maf B (Artner et al., 2006; Katoh et al., 2018) and cMaf. Interestingly, cMaf is repressed by miR-125b, miR-182 and miR-200c in the mouse insulinoma MIN6 cells (Klein et al., 2013).

Platelet-derived growth factor receptor, alpha (Pdgfra) is another disallowed gene repressed by miRNAs in rat and human beta cells (Tugay et al., 2016). Pdgfra is a regulator of beta cell proliferation and its inhibition contributes to limit the proliferative capacity of mature beta cells (Chen et al., 2011).

MiR-34 targets Pdgfra mRNA and inhibits its expression. Interestingly, miR-34 resulted upregulated in rat and human beta cells with age (Tugay et al., 2016).

One potential trigger of the miRNA expression profile change, during the post-natal life, is the nutritional switch occurring at weaning from fat-rich maternal milk to a carbohydrate-rich diet (Jacovetti et al., 2015; Stolovich-Rain et al., 2015). Nutrition-driven miRNA changes in islets enable the coupling of glucose metabolism with insulin secretion (Jacovetti et al., 2015) and partially control the circadian rhythms during postnatal beta cell maturation (Jacovetti et al., 2017). Among the miRNAs differentially expressed during neonatal beta cell maturation, miR-17-5p and miR-181a regulates the expression of several genes involved in glucose metabolism, insulin secretion and proliferation (Jacovetti et al., 2015). Moreover, miR-17-3p and miR-29b-3p directly regulate the expression of clock and period circadian protein homolog 3 (per3), respectively, suggesting a role for miRNAs driving beta cell maturation in the regulation of adult circadian gene expression (Jacovetti et al., 2017). However, the mechanisms involved in the initiation of the circadian clock in the developing islets is still mainly unknown and is probably due to the action of multiple miRNAs together with other factors.

2.3 miRNAs in pancreatic beta cell proliferation

Pancreatic beta cells have a high proliferative rate during the first postnatal period, before maturing in adult and glucose-sensitive beta cells (Finegood et al., 1995; Gregg et al., 2012). During this phase, the expansion of the beta cells is critical for the acquisition and the maintenance of a fully functional beta cell mass in adulthood. Beta cell proliferation strongly declines thereafter and is almost undetectable in adults, particularly in human pancreas (Wang et al., 2015). Different studies suggest that defects in the expansion and the functional maturation of the beta cells occurring early in life can potentially affect the ability to respond to increased insulin demand occurring later in life, such as pregnancy or obesity, predisposing to the development of gestational diabetes and T2D.

The availability of nutrients in the intrauterine environment can alter the epigenome during critical periods of the development of an individual (Martin-Gronert and Ozanne, 2012; Sandovici et al., 2011) modifying the structure of chromatin, DNA methylation, and the expression of specific miRNAs, which in turn results in changes in the transcriptome (Neu et al., 2007). More specifically, a low availability of proteins in the maternal diet can produce alterations in offspring that include changes in insulin and glucose metabolism, a decrease in the size and number of cells of pancreatic islets, and alteration in pancreatic beta cell secretion (Sosa-Larios et al., 2015). Perturbation on the expression of key miRNAs in the endocrine pancreas of progeny contributes to the effects of early malnutrition on the establishment of beta cell mass and, consequently, on the long-term health of the organism.

Indeed, a change in the pancreatic miRNA expression profile in the progeny of rat dams fed a low-protein (LP) diet during gestation and lactation (Dumortier et al., 2014) increases the probability of the offspring to develop diabetes, compared to normal diet-fed dams (Fernandez-Twinn et al., 2019; Martin-Gronert and Ozanne, 2012). In particular, the persistent upregulation of miR-375 expression in the islets of adult rats born from mothers maintained on LP diet, inhibits postnatal beta cell mass expansion (Dumortier et al., 2014). In another independent study, miR-199a-3p, miR-342 and miR-7 were found upregulated in offspring mouse islets of LP-fed mothers, and their inhibition ameliorated insulin secretion and insulin content, restoring the mTOR signaling pathway. mTOR is a serine/threonine kinase that is essential for cell growth and proliferation and exposure to a low-protein diet during pregnancy decreases mTOR signaling in neonatal and adult beta cells, caused by the direct bind of miR-199a-3p to mTOR 3′UTR. Moreover, transient induction of mTORC1 signaling in vivo during embryonic life until birth, improved glucose tolerance in adult progeny of mothers fed a LP diet (Alejandro et al., 2014).

During obesity and pregnancy, reduced insulin sensitivity of peripheral tissues results in increased insulin demand. The beta cells respond to this metabolic challenge by increasing their proliferation and their secretory capacity. While in humans the obesity-induced beta cell expansion is still under debate (Linnemann et al., 2014), in rodents there is a dramatic increase in beta cell mass with obesity (Hull et al., 2005). Particularly in mice, beta cell mass expansion occurs in both diet-induced obesity models and genetic models of obesity. MiRNAs are important contributors to successful beta cell adaptation during insulin resistance. High glucose exposure induces expression of miR-17 in MIN-6 cells and the consequent downregulation of Menin, a well-established negative regulator of beta cell proliferation (Lu et al., 2015). MiR-17 overexpression positive correlate with the expansion of the beta cell mass occurring during the neonatal period

(Jacovetti et al., 2015) and a conditional KO mouse model, in which the miR-17-92/miR-106b-25 cluster is deleted, displays reduced beta cell mass and up to 50% reduction in total pancreatic insulin content (Mandelbaum et al., 2019).

Among many miRNAs, miR-338-3p has a major role in in compensatory beta cell mass expansion during insulin resistance. Activated G protein–coupled estrogen receptor GPR30 and the glucagon-like peptide 1 (GLP1) receptor decrease the expression level of miR-338-3p in gestation and obesity. Blockade of miR-338-3p in beta cells using specific anti-miR molecules mimicked gene expression changes occurring during beta cell mass expansion and resulted in increased proliferation and improved survival both in vitro and in vivo (Jacovetti et al., 2012).

MiR-7 is also involved in beta cell compensation mechanisms during insulin resistance. Mir-7 expression is strongly reduced in islets of diet-induced obesity (DIO) mice, in obese non-diabetic *ob/ob* mice and in human islets transplant under the kidney capsule of DIO mice (Latreille et al., 2014).

Both leptin-deficient (*ob/ob*) C57BL/6 and BTBR (Black and Tan Brachyury) mice become obese, but only the BTBR strain becomes diabetic (Keller et al., 2008; Stoehr et al., 2000). The C57BL/6 mice display increased beta cell proliferation in response to obesity and can meet the increased demand for insulin, whereas the BTBR mice cannot. MiRNA profile analysis comparison in islets from lean heterozygous diabetes resistant C57BL/6 (B6) and diabetes-susceptible BTBR mice, and their obese peers, B6-*ob/ob* and BTBR-*ob/ob*, permitted to discern the contribution of genetics and obesity on miRNA regulation (Zhao et al., 2009). The authors identified changes in a subset of obesity-dependent miRNAs in islets of both, B6-*ob/ob* and BTBR-*ob/ob* mice compared with their respective lean controls, but the magnitude of the upregulation of some miRNAs was dependent on the strain. The strain differences in the pancreatic miRNA profile in response to obesity may help understanding the relationship between miRNA regulation and diabetes resistance (B6) or susceptibility (BTBR) (Zhao et al., 2009).

2.4 miRNAs in insulin production and secretion

Pancreatic beta cells are the only cells in the human body able to produce insulin and release it in response to glucose. Their ability of coupling glucose metabolism with insulin secretion is unique and is permitted by the expression of a complex molecular machinery able to uptake glucose, metabolize it and transduce it into insulin production and secretion. MiRNAs are involved in the fine-tuned regulation of all those steps, from the glucose uptake to insulin granule exocytosis.

Insulin gene expression and insulin mRNA stability are mainly regulated by glucose (Poitout et al., 2006; Welsh et al., 1985). Insulin mRNA is translated in pre-proinsulin, cleaved in pro-insulin and finally in insulin and C-peptide during granules packaging and maturation. The reduction of insulin expression after *Dicer* KO confirms the involvement of miRNAs in insulin production (Kalis et al., 2011). Indeed, the downregulation of miR-24, miR-26, miR-182 and miR-148 in beta cells reduces insulin promoter activity and consequently insulin content (Melkman-Zehavi et al., 2011). Moreover, a positive correlation between miR-127-3p, miR-184 and miR-375 and insulin mRNA level has been described in human islet cells (Bolmeson et al., 2011). MafA is one of the transcription factor that regulates insulin gene expression and it is a target of miR-204, upregulated during diabetes development (Xu et al., 2013). Interestingly, both miR-204 and miR-184 expression is regulated by glucose. Indeed, glucose-mediated regulation of thioredoxin-interacting protein (TXNIP) increases miR-204 expression (Xu et al., 2013), while glucose regulation of AMPK activity reduces the expression of miR-184 (Martinez-Sanchez et al., 2018). As mention in Section 2.2, other changes in miRNA expression regulated by glucose happens during weaning, passing from a fat-rich diet to a carbohydrate-rich diet. All together, these discoveries suggest that glucose-regulated miRNAs are essential for a functional correlation between glucose, insulin production and insulin secretion.

Glucose uptake occurs through the glucose transporters GLUT1/2, at the membrane of the beta cell. Once inside, glucose is phosphorylated in glucose-6-phosphate by glucokinase (GK) and enters glycolysis and KREBs cycle, followed by ATP production through oxidative phosphorylation. The resulting ATP increase triggers ATP-dependent potassium channel closure (Ashcroft and Rorsman, 2013), which induces beta cell membrane depolarization and consequent opening of the voltage-dependent calcium channels. Cytosolic Ca^{++} increase triggers insulin granules exocytosis.

Glucose to glucose-6-phosphate conversion is a rate-limiting step and GK mutations causes reduced glycolysis and altered intracellular ADP/ATP ratio that affect potassium channel and thus decreased insulin secretion (Gloyn, 2003). Heterozygous mutations in *GK* gene has been reported to cause maturity onset diabetes of young type 2 (MODY2) (Froguel et al., 1992), while homozygous mutations in *GK* leads to permanent neonatal diabetes mellitus (PNDM) (Njolstad et al., 2001). GK expression and ATP/ADP level are regulated by miR130a/b and miR-152, through the inhibition of GK and Pdha1 mRNAs, the last one coding for a mitochondrial enzyme complex subunit linking glycolysis and KREBs cycle (Ofori et al., 2017). Interestingly, the expression of miR130a/b and miR-152 is regulated by glucose in human pancreatic islets. GCK mRNA is also a direct target of miR-206 and islets isolated from miR206-KO mice display higher Gk mRNA levels and activity. Moreover, miR-206 was reported to regulate glucose tolerance in normal and high-fat diet-fed mice, suggesting that miR-206 is involved in GK-dependent beta cell compensation in the context of obesity (Vinod et al., 2016). However, unlike mouse islets, human islets do not exhibit any correlation between miR-206 expression level and GCK transcripts (Joglekar et al., 2018), suggesting the species-specificity of some miRNA functions (Fig. 2).

Insulin granules exocytosis is a highly regulated process, essential for beta cell functionality. MiRNAs mainly act on insulin exocytosis modulating the expression of different components of the secretory machinery, for example, SNAP Receptor (SNARE) proteins, Granuphilin, Syntaxin1a and the vesicle-associated membrane protein (VAMP).

MiR-9 and miR-29 modulate insulin exocytosis through direct targeting the Granuphilin transcriptional inhibitor Onecut-2 (OC-2) and decreasing its expression. OC-2 directly binds to the granuphilin promoter and stops its transcription (Plaisance et al., 2006; Roggli et al., 2012), therefore the inhibition of OC-2 increases Granuphilin expression. Since Granuphilin mediates the docking of secretory granules onto the plasma membrane, but it inhibits the release of insulin (Kato et al., 2006), the increased expression of miR-9 and miR-29 further enhance granuphilin-dependent inhibition of insulin secretion. Interestingly, miR-29 upregulation is observed in the islets of non-obese diabetic (NOD) mice, the murine model of type 1 diabetes, and it is induced by pro-inflammatory cytokines (Roggli et al., 2012) and high glucose (Bagge et al., 2012), in both murine and human islets. Moreover, in high glucose-treated rodent beta cells, the upregulation of miR-29 reduces Syntaxin1A expression and insulin secretion (Bagge et al., 2013). Multiple experimental evidences demonstrate the role of miR-29 in the regulation of beta cell insulin secretion and suggest its potential role in diabetes pathogenesis.

MiR-7 is also a negative regulator of glucose-stimulated insulin secretion (GSIS) in beta cells by directly regulating genes that control late stages of insulin granule fusion and ternary SNARE complex activity (Latreille et al., 2014). Transgenic mice overexpressing miR-7a in beta cells developed diabetes due to impaired insulin secretion and beta cell dedifferentiation (Latreille et al., 2014).

MiR-124a directly regulates the expression of genes involved in insulin granule exocytosis or in glucose sensing, such as Foxa2, in insulin-secreting MIN6 cells. Moreover, miR-124 inhibits insulin biosynthesis and insulin release by downregulating Foxa2 levels, which in turn resulted in diminished of Pdx1, Kir6.2 and Sur1, and by modulating the levels of other regulators of insulin exocytosis such as SNAP25, Rab3A, synapsin-1A, Rab27A and Noc2 (Baroukh et al., 2007; Lovis et al., 2008a; Sebastiani et al., 2015). Remarkably, the expression of miR-124a is elevated in human pancreatic islets of T2D donor, and it could contribute to the insulin secretion inhibition during the disease (Sebastiani et al., 2015).

MiR-335 is another key regulator of the exocytosis in pancreatic beta cells. Indeed miR-335 negatively correlated with secretion index in human islets of individuals with prediabetes. Overexpression of miR-335 in human EndoC-βH1 and in rat INS-1 cells results in decreased glucose-stimulated insulin secretion, and downregulation of three exocytotic proteins: SNAP25, Syntaxin-binding protein 1 (STXBP1), and synaptotagmin 11 (SYT11) (Salunkhe et al., 2017) (Fig. 2).

3. Deregulation of beta cell miRNAs during diabetes

Diabetes mellitus is the most common metabolic disorder worldwide and is a major public health concern. Pancreatic beta cells play a central role in the pathophysiology of the disease. Indeed, the release of appropriate amounts of insulin in response to a rise in circulating levels of glucose and other nutrients is essential to maintain blood glucose homeostasis. Insulin binds to the insulin receptor of the cells in many target tissues, including liver, skeletal muscles and adipose tissue, and stimulates intracellular signaling pathways that ultimately increase glucose uptake. Insufficient or absent insulin supply leads to chronic hyperglycemia and to the occurrence of different forms of diabetes mellitus. Type 1 diabetes (T1D) is characterized by the complete lack of insulin production due to the autoimmune destruction of pancreatic beta cells. Without supplemental insulin treatment, type 1 diabetes is lethal. In contrast Type 2 diabetes (T2D), representing about 90% of the cases, occurs when the amount of insulin released by beta-cells is insufficient to maintain blood glucose concentration within normal physiological ranges. The pathogenesis of T2D involves a combination of genetic and environmental factors and is frequently associated with obesity (Zimmet, 1999). Obesity diminishes insulin sensitivity of target tissues and is usually compensated by an expansion of the functional beta cells mass, at least in rodent models. However, in about one third of obese individuals this adaptive process fails, resulting in relative insulin deficiency and in the development of T2D (Butler and Cone, 2003).

MicroRNAs make an important contribution to alterations in gene expression observed in dysfunctional beta cells and in insulin resistant tissues, playing a crucial role in diabetes pathogenesis.

3.1 Type 1 diabetes

During T1D pathogenesis, the activated immune cells progressively infiltrate the pancreas and invade the islets, in a phenomenon called insulitis. The exact mechanisms causing the specific autoimmune attack against the beta cells remains unknown. During insulitis (which means inflammation of the islets), the immune cells and the beta cells exchange pro-inflammatory cytokines, chemokines and exosomes with the result of beta cell death (Eizirik et al., 2009; Guay et al., 2019). The symptoms of the disease appear only after the destruction of the beta cells, making the study of the human beta cells during the progression of the disease quite challenging. The non-obese diabetic (NOD) mice spontaneously develop autoimmune diabetes similarly to what observed in human (Pearson et al., 2016), with insulitis starting around 6–8 weeks of age and a total invasion of the islets by 12–14 weeks of age. The diabetes onset occurs in 80% of female NOD mice by the age of 16–20 weeks.

Global miRNA profiling in islets of 4, 8 and 12 weeks-old female NOD mice led to the identification several miRNAs dysregulated during T1D development (Roggli et al., 2012). As previously discussed (Section 2.3), miR-29 family members were found to be highly upregulated along disease progression, due to the effect of pro-inflammatory cytokines. Beside the inhibition on GSIS (see Section 2.3), miR-29 upregulation increased beta cell apoptosis by increasing *Mcl1* expression. Masking miR-29 binding site on *Mcl1* mRNA with a protector oligonucleotide prevented beta cell death induced by miR-29 or pro-inflammatory cytokines (Roggli et al., 2012).

In human islet cells, the pro-inflammatory cytokine IL-1β alone induce the upregulation of miR-21a-5p, miR-34a-5p and miR-146a-5p, that are also upregulated in pre-diabetic NOD mice (Roggli et al., 2010). Interestingly, miR-146a-5p expression is induced by NF-kB, a well-known mediator of cytokine-induced beta cell death, and the inhibition of miR-146a-5p in MIN6 cells prevents IL-1β-induced apoptosis. As expected, overexpression of this miRNA increases apoptosis (Roggli et al., 2010). Augmented levels of miR-34a-5p also promotes apoptosis via p53 (Lovis et al., 2008b) and affects insulin expression, content and secretion in response to glucose. The effect of miR-34a on insulin release is in part mediated by inhibition of VAMP2 and Rab3a, two proteins involved in insulin exocytosis (Roggli et al., 2010). Those proteins are also a target of the IL-1β-induced miR-21a-5p (Bravo-Egana et al., 2012; Roggli et al., 2010).

Several other miRNAs were found dysregulated in both human and mouse islets challenged with pro-inflammatory cytokines (Bravo-Egana et al., 2012; Grieco et al., 2017, 2019; Roggli et al., 2010). Among them, miR-23a/b-3p, miR-149-5p, miR-204-5p and miR-211-5p were described to be downregulated in human islets and human EndoC-βH1 cells treated with IL-1β and IFNγ. Cytokine-induced decreased of those miRNAs in human beta cells augmented the expression of the pro-apoptotic genes *PUMA* and *DP5*, contributing to cytokine-induced beta cell death (Grieco et al., 2017, 2019). Moreover, the inhibition of both miR-211-5p and miR-204-5p in human EndoC-βH1 cells induced ER stress in a PERK-dependent manner (Grieco et al., 2019). ER stress markers are upregulated in islets of T1D patients (Marhfour et al., 2012) and display an age-dependent increase in NOD mouse islets during diabetes pathogenesis (Tersey et al., 2012). In vivo inhibition of ER stress pathways before or during insulitis strongly reduce diabetes onset in NOD mice (Engin et al., 2013; Lee et al., 2020), remarking the importance of ER stress on T1D pathogenesis. Interestingly, miR-204-5p and miR-211-5p were reported to directly target, respectively, PERK in beta cells (Xu et al., 2016) and CHOP in fibroblast NIH-3T3 cells, key factors in beta cells ER stress response (Chitnis et al., 2012).

Beside miRNA changes induced by pro-inflammatory cytokines, beta cell dysfunction and death can also be caused by miRNA transfer from exosomes produced by other cells. Indeed, the beta cell miRNA profile during insulitis is dynamic and is the results not only of pro-inflammatory cytokine modulation, but also of the direct transfer of specific miRNA via exosomes. The specific transfer of miR-142-3p, miR-142-5p and miR-155 from T-cells to pancreatic beta cells induced inflammation and NF-kB-dependent beta cell apoptosis in NOD mice. In vivo inhibition of the action of those specific three miRNAs inside the beta cells decreased by 50% diabetes incidence (Guay et al., 2019). Moreover, miRNAs are horizontally transferred between murine beta cells in vitro through exosomes (Guay et al., 2015) and exposure of MIN6 to pro-inflammatory cytokines, affect the composition of exosomal miRNAs and induce apoptosis in the recipient beta cells (Guay et al., 2015).

MiRNAs islet cell exosomes can also actively contribute to the crosstalk with immune cells and stimulate their activation during T1D pathogenesis. Different studies show that beta cell-released exosomes can activate T cells (Sheng et al., 2011) and B cells (Bashratyan et al., 2013) in pre-diabetic NOD mice. Moreover, human islet exosomes can be internalized by monocytes and B cells, and induce T-helper cytokine production and T-/B-cell proliferation (Rutman et al., 2018). However, the cargo of these exosomes was not analyzed, thus the potential role of beta cell miRNAs in the observed outcomes remains to be demonstrated.

3.2 Type 2 diabetes

T2D develops when pancreatic beta cells are unable to release enough insulin to compensate for the diminished sensitivity of peripheral tissues (See Section 2.3). Different mouse models exist to study the etiology of T2D. Ob/ob mice (lacking leptin) and db/db mice (lacking leptin receptor), display massive obesity and hyperinsulinemia, with or without chronic hyperglycemia according to the genetic background (Kjorholt et al., 2005; Lindstrom, 2007). Diet-induced obesity (DIO) mice are also a useful model, where mice are fed on a high-fat diet for 2–3 months leading to progressive weight gain, hyperglycemia and hyperinsulinemia (Peyot et al., 2010). Contrary to the ob/ob and db/db mice, these animals do not develop obesity and model the largest fraction of T2D patients. The only T2D rat model is the Goto–Kakizaki (GK) rats, that spontaneously developed diabetes caused by a defective glucose-stimulated insulin secretion (Portha et al., 2009).

Differently from T1D, intact pancreatic islets form T2D donor are relatively accessible, allowing studies on global miRNA profile to identify dysregulated miRNAs in human islets of T2D donors compare to healthy donors. Using small RNA-seq and RT-qPCR-based array, these studies reported pattern of miRNAs that are upregulated (miR-187-5p, miR-187-3p, miR-224 and miR-589) and others that are downregulated (miR-7, miR-369, miR-487a, miR-655 and miR-65) in diabetic islets compare to islets from non-diabetic donors (Kameswaran et al., 2014; Locke et al., 2014). Interestingly, a miRNA cluster in the *DLK1-MEG3* locus, which is one of the largest miRNA-generating clusters in humans containing 54 miRNAs, results repressed in islets from T2D donors. The downregulation of miR-376a and miR-432, two miRNAs generated from this locus, was proposed to be responsible of the increased expression of the pro-apoptotic genes islet amyloid polypeptide (*IAPP*) and p53-induced nuclear protein 1 (TP53INP1), and the consequent beta cell apoptosis, in T2D patients (Kameswaran et al., 2014). Moreover, the identification of epigenetic modification at this locus that correlates with its expression in human diabetic donors' islets, support a role for microRNAs and their epigenetic control in the pathogenesis of T2DM.

In another independent study, a miRNA array identified the increased expression of miR-187-3p in T2D islets (Locke et al., 2014). MiR-187-3p upregulation in T2D islets negatively affect glucose-stimulated insulin secretion and overexpression of miR-187 in rat beta cells inhibited insulin secretion accordingly. In this contest, the direct target of miR-187-3p was identify on the homeodomain-interacting protein kinase-3 (HIPK3), a known regulator of insulin secretion, which is reduced in islets from individuals with T2D (Locke et al., 2014). Besides miRNA profiling studies, specific miRNAs of interest were reported to be differentially expressed in the islets of T2D patients. Significant increased expression of miR-21, and a mild increased of miR-375 and miR-127-3p, were found in islets of glucose-intolerant individuals, compare to the islets of healthy controls (Bolmeson et al., 2011). However miR-375 expression does not change in T2D islets, were miR-124a was instead identify as highly expressed and impairing insulin secretion (Sebastiani et al., 2015) (see Section 2.3). While miR-184 levels are not different in the cohort previously described (Kameswaran et al., 2014), it was reported reduced in the islets of T2D human subjects and in different animal models of obesity and diabetes (Nesca et al., 2013; Tattikota et al., 2014). Indeed, miR-184 appears to positively contribute to glucose-induced insulin production (see Section 2.3) and compensatory beta cell mass expansion in response to obesity in rodents (Nesca et al., 2013; Tattikota et al., 2015) (Fig. 2).

Numerous studies conducted in islets from T2D animal models describe changes in miRNA expression occurring during disease progression (Guay and Regazzi, 2015a, b; Lapierre and Stoffel, 2017; Martinez-Sanchez et al., 2016; Osmai et al., 2016; Wendt et al., 2018). However, only few of them were consistently modulated in both human and animal models. Upregulation of miR-130 a/b expression in islets form T2D donor and in the GK rats is one of those positive examples. Indeed, islets from hyperglycemic human donors contain elevated levels of miR-130a, miR-130b and miR-152 (Ofori et al., 2017), and the over-expression of the same miRNAs are validated in the pancreatic islets of the T2D GK rat model (Esguerra et al., 2011). This crossover between rodent and human to these miRNAs, suggests that they play an important role in beta cell function or maintenance. Actually, miR-130a, miR-130b and miR-152 have been shown to influence the metabolic control of GSIS via modulation of ATP levels, partially through targeting of pyruvate dehydrogenase E1alpha 1 subunit (PDHA1) and GCK in the pancreatic beta cell (Ofori et al., 2017) (see Section 2.3).

MiR-7 downregulation was described in T2D islets and in islets of obese mouse models (HFD and *ob/ob*), in the attempt of compensating insulin resistance. However, miR-7 upregulation was reported in islets of *db/db* mice which develop hyperglycemia and display reduced insulinemia. Furthermore, beta-cell specific transgenic mice overexpressing miR-7 developed diabetes due to impaired insulin secretion and beta cell dedifferentiation (Latreille et al., 2014). Consistently, reduced miR-7 levels are present in human islets transplanted in HFD mice and therefore exposed to insulin resistance inputs, but increases significantly 10 weeks after transplantation (Latreille et al., 2014). The authors of this study observed lower levels of miR-7a in obese and nondiabetic subjects, but also in islets of moderate T2D human samples. This result was explained by the fact that this cohort includes patients with beta cells still able to compensate for increased insulin needs in response to insulin resistance. Altogether, these observations indicate that with low levels of miR-7 the beta cell are still able to cope with insulin resistance and preserve the secretory function, but increased miR-7 expression levels positively correlate with diabetes and insulin secretion dysfunction.

The discrepancy between the miRNA profile of T2D islets from human donors and the miRNA profile in the islets of diabetic animal models can be explained by multiple factors. In fact, differently from rodent models, the number of human samples are usually relatively low, making the confounding effects such as genetic variability, differences linked to diet, T2D treatment, lifestyle, age, gender and others difficult to control. In addition, part of the differences in miRNA expression described in diabetes animal models are the consequence of chronically elevated blood glucose levels, situation that is avoided in T2D patients by pharmacological treatment. Additionally, the analysis of global expression profiles is influenced by methodological variability, such as the data normalization. All these limitations should be considered for a critical interpretation of the published results and for future studies.

As has been demonstrated in T1D, miRNAs imported into beta cells through extracellular vesicles (EV) can modulate gene expression and take part in T2D pathogenesis. Indeed, skeletal muscle cells of HFD-fed mice transfer specific miRNAs to the beta cells via exosome-like vesicles (ELVs) that induce beta cell proliferation (Jalabert et al., 2016). Moreover, hepatocytes from HFD-fed mice, transfer specific miRNAs to the beta cells via exosomes and stimulate beta cell proliferation by miR-7218-5p transfer (Fu et al., 2019). These discoveries suggest the existence of a vesicle-mediated cross-talk between peripheral organ/tissues and pancreatic beta cells, that partially regulates beta mass compensation during insulin resistance (Fu et al., 2019; Jalabert et al., 2016). A complex signaling network between pancreas, liver, adipose tissue and muscles, is essential to regulate glucose and lipid metabolism, to achieve metabolic homeostasis and energy balance and to avoid obesity and T2D. Accumulating evidence support the importance of extracellular miRNAs as mediators of this crosstalk, complementing the well-known role of hormones and metabolites (Guay and Regazzi, 2017; Thomou et al., 2017; Ying et al., 2017).

4. Conclusions and perspectives

During the last 20 years, a large amount of data has been produced to obtain an exhaustive picture of the miRNAs expressed in beta cells and to identify changes in the miRNA profile occurring in association with obesity, inflammation and diabetes.

It has been largely demonstrated that key miRNAs regulate the secretory activity of β-cells and/or control the β-cell mass (Guay et al., 2020). Genetic and environmental factors involved in the pathogenesis of Type 1 and T2D influence the expression of these miRNAs that are associated with alteration in insulin secretion (e.g., miR-130a/b and miR-152), in compensatory adaptations of beta cells (e.g., miR-184 and miR-212/miR-132), in inflammation and apoptosis (e.g., miR-146-3p, miR-142-3p/-5p and miR-155). Strategies to modulate the level of miRNAs are developing with the aim of design new therapeutic approach to prevent and treat diabetes.

Several miRNA drug candidates are in phase 1 and phase 2 clinical trials for other diseases. These studies utilized Locked Nucleic Acid (LNA) based drugs to inhibit the function of specific miRNAs (Hanna et al., 2019). The few studies already published in the attempt to treat T2D in animal models utilize adeno-associated virus (AAV) construct to overexpress specific miRNAs usually under the control of the insulin promoter to ensure beta-cell specificity. Other approaches employ the inhibition of specific miRNAs through in vivo injection of cholesterol-conjugated antagomirs (Bijkerk et al., 2019). The use of antagomirs might hold a therapeutic potential, especially if coupled with targeted locked nucleic acid molecules, which are more stable in blood (Eliasson and Regazzi, 2020).

However, for diabetes treatment, the main challenge is to target one or more miRNAs specifically in beta cells, since none of the miRNAs studied so far is exclusively expressed in insulin-secreting cells. Therefore, without a beta cell-specific delivery of these miRNA-targeting tools, any of these therapeutic strategies will most likely cause unacceptable side effects. Theranostic nanomedicine is a novel interdisciplinary approach that integrates nanotechnology, biomedicine and imaging (Panwar et al., 2019; Wagner et al., 2006). Theranostic nanodrugs can deliver therapeutics including small molecules, peptides, antibodies and oligonucleotides to targeted tissues, while imaging allows the non-invasive assessment of their biodistribution and pharmacokinetics. Using theranostic nanodrug consisting of magnetic nanoparticles it is also possible to deliver miRNA mimics or inhibiting LNA antisense oligonucleotides to beta cells. In vivo delivery of a specific miR-216a mimic nanodrug to pancreatic islets, monitored by MRI, resulted in increased insulin secretion and beta cell proliferation (Wang et al., 2020). This strategy could potentially be used to deliver multiple RNA molecules to alter various aspect of diabetogenesis. The ability to monitor this delivery in vivo by MRI is an added value for the future clinical implementation of this theranostic approach.

In the meantime, the discovery of the presence of miRNAs in plasma and other body fluids addressed many groups to evaluate these circulating non-coding RNAs as biomarkers of beta cell function and/or loss. However, until now there are no indications on how discriminate the miRNAs coming from the beta cells from those secreting by other cell types, and the measurements of the total pool of blood miRNAs does not reflect specific changes occurring in beta cells.

Moreover, miRNAs are not the only non-coding RNA express in pancreatic beta cells. PiRNAs, tRNA-derived fragments, long non-coding

RNAs and circular RNAs are also expressed in beta cells and we are only beginning to understand their functions and the potential interactions existing between these different classes of non-coding RNAs (Guay et al., 2020). Unveiled these questions will provide a bigger picture of the complex regulatory networks governing islet cell functions under normal and disease conditions.

References

Alejandro, E.U., Gregg, B., Wallen, T., Kumusoglu, D., Meister, D., Chen, A., Merrins, M.J., Satin, L.S., Liu, M., Arvan, P., Bernal-Mizrachi, E., 2014. Maternal diet-induced microRNAs and mTOR underlie beta cell dysfunction in offspring. J. Clin. Invest. 124, 4395–4410.

Artner, I., Le Lay, J., Hang, Y., Elghazi, L., Schisler, J.C., Henderson, E., Sosa-Pineda, B., Stein, R., 2006. MafB: an activator of the glucagon gene expressed in developing islet alpha- and beta-cells. Diabetes 55, 297–304.

Ashcroft, F.M., Rorsman, P., 2013. K(ATP) channels and islet hormone secretion: new insights and controversies. Nat. Rev. Endocrinol. 9, 660–669.

Avnit-Sagi, T., Kantorovich, L., Kredo-Russo, S., Hornstein, E., Walker, M.D., 2009. The promoter of the pri-miR-375 gene directs expression selectively to the endocrine pancreas. Plos One 4, E5033.

Bagge, A., Clausen, T.R., Larsen, S., Ladefoged, M., Rosenstierne, M.W., Larsen, L., Vang, O., Nielsen, J.H., Dalgaard, L.T., 2012. MicroRNA-29a is up-regulated in beta-cells by glucose and decreases glucose-stimulated insulin secretion. Biochem. Biophys. Res. Commun. 426, 266–272.

Bagge, A., Dahmcke, C.M., Dalgaard, L.T., 2013. Syntaxin-1a is a direct target of miR-29a in insulin-producing beta-cells. Horm. Metab. Res. 45, 463–466.

Baroukh, N., Ravier, M.A., Loder, M.K., Hill, E.V., Bounacer, A., Scharfmann, R., Rutter, G.A., Van Obberghen, E., 2007. MicroRNA-124a regulates Foxa2 expression and intracellular signaling in pancreatic beta-cell lines. J. Biol. Chem. 282, 19575–19588.

Bartel, D.P., 2009. MicroRNAs: target recognition and regulatory functions. Cell 136, 215–233.

Bartel, D.P., 2018. Metazoan microRNAs. Cell 173, 20–51.

Bashratyan, R., Sheng, H., Regn, D., Rahman, M.J., Dai, Y.D., 2013. Insulinoma-released exosomes activate autoreactive marginal zone-like B cells that expand endogenously in prediabetic NOD mice. Eur. J. Immunol. 43, 2588–2597.

Bijkerk, R., Esguerra, J.L.S., Ellenbroek, J.H., Au, Y.W., Hanegraaf, M.A.J., De Koning, E.J., Eliasson, L., Van Zonneveld, A.J., 2019. In vivo silencing of MicroRNA-132 reduces blood glucose and improves insulin secretion. Nucleic Acid Ther. 29, 67–72.

Bolmeson, C., Esguerra, J.L., Salehi, A., Speidel, D., Eliasson, L., Cilio, C.M., 2011. Differences in islet-enriched miRNAs in healthy and glucose intolerant human subjects. Biochem. Biophys. Res. Commun. 404, 16–22.

Bravo-Egana, V., Rosero, S., Klein, D., Jiang, Z., Vargas, N., Tsinoremas, N., Doni, M., Podetta, M., Ricordi, C., Molano, R.D., Pileggi, A., Pastori, R.L., 2012. Inflammation-mediated regulation of MicroRNA expression in transplanted pancreatic islets. J. Transplant 2012, 723614.

Butler, A.A., Cone, R.D., 2003. Knockout studies defining different roles for melanocortin receptors in energy homeostasis. Ann. N. Y. Acad. Sci. 994, 240–245.

Cano, D.A., Soria, B., Martin, F., Rojas, A., 2014. Transcriptional control of mammalian pancreas organogenesis. Cell. Mol. Life Sci. 71, 2383–2402.

Chen, H., Gu, X., Liu, Y., Wang, J., Wirt, S.E., Bottino, R., Schorle, H., Sage, J., Kim, S.K., 2011. PDGF signalling controls age-dependent proliferation in pancreatic beta-cells. Nature 478, 349–355.

Chitnis, N.S., Pytel, D., Bobrovnikova-Marjon, E., Pant, D., Zheng, H., Maas, N.L., Frederick, B., Kushner, J.A., Chodosh, L.A., Koumenis, C., Fuchs, S.Y., Diehl, J.A., 2012. miR-211 is a prosurvival microRNA that regulates chop expression in a PERK-dependent manner. Mol. Cell 48, 353–364.

Conrad, E., Stein, R., Hunter, C.S., 2014. Revealing transcription factors during human pancreatic beta cell development. Trends Endocrinol. Metab. 25, 407–414.

Correa-Medina, M., Bravo-Egana, V., Rosero, S., Ricordi, C., Edlund, H., Diez, J., Pastori, R.L., 2009. MicroRNA miR-7 is preferentially expressed in endocrine cells of the developing and adult human pancreas. Gene Expr. Patterns 9, 193–199.

Dumortier, O., Van Obberghen, E., 2012. MicroRNAs in pancreas development. Diabetes Obes. Metab. 14 (Suppl. 3), 22–28.

Dumortier, O., Hinault, C., Gautier, N., Patouraux, S., Casamento, V., Van Obberghen, E., 2014. Maternal protein restriction leads to pancreatic failure in offspring: role of mis-expressed microRNA-375. Diabetes 63, 3416–3427.

Eizirik, D.L., Colli, M.L., Ortis, F., 2009. The role of inflammation in insulitis and beta-cell loss in type 1 diabetes. Nat. Rev. Endocrinol. 5, 219–226.

Eliasson, L., Esguerra, J.L., 2014. Role of non-coding RNAs in pancreatic beta-cell development and physiology. Acta Physiol (Oxf.) 211, 273–284.

Eliasson, L., Regazzi, R., 2020. Micro(RNA) management and mismanagement of the islet. J. Mol. Biol. 432, 1419–1428.

Engin, F., Yermalovich, A., Nguyen, T., Hummasti, S., Fu, W., Eizirik, D.L., Mathis, D., Hotamisligil, G.S., 2013. Restoration of the unfolded protein response in pancreatic beta cells protects mice against type 1 diabetes. Sci. Transl. Med. 5, 211ra156.

Esguerra, J.L., Bolmeson, C., Cilio, C.M., Eliasson, L., 2011. Differential glucose-regulation of microRNAs in pancreatic islets of non-obese type 2 diabetes model Goto-Kakizaki rat. Plos One 6, E18613.

Esguerra, J.L.S., Nagao, M., Ofori, J.K., Wendt, A., Eliasson, L., 2018. MicroRNAs in islet hormone secretion. Diabetes Obes. Metab. 20 (Suppl. 2), 11–19.

Fernandez-Twinn, D.S., Hjort, L., Novakovic, B., Ozanne, S.E., Saffery, R., 2019. Intrauterine programming of obesity and type 2 diabetes. Diabetologia 62, 1789–1801.

Finegood, D.T., Scaglia, L., Bonner-Weir, S., 1995. Dynamics of beta-cell mass in the growing rat pancreas. Estimation with a simple mathematical model. Diabetes 44, 249–256.

Francis, N., Moore, M., Rutter, G.A., Burns, C., 2014. The role of microRNAs in the pancreatic differentiation of pluripotent stem cells. Microrna 3, 54–63.

Froguel, P., Vaxillaire, M., Sun, F., Velho, G., Zouali, H., Butel, M.O., Lesage, S., Vionnet, N., Clement, K., Fougerousse, F., et al., 1992. Close linkage of glucokinase locus on chromosome 7p to early-onset non-insulin-dependent diabetes mellitus. Nature 356, 162–164.

Fu, Q., Li, Y., Jiang, H., Shen, Z., Gao, R., He, Y., Liu, Y., Xu, K., Yang, T., 2019. Hepatocytes derived extracellular vesicles from high-fat diet induced obese mice modulate genes expression and proliferation of islet beta cells. Biochem. Biophys. Res. Commun. 516, 1159–1166.

Gloyn, A.L., 2003. Glucokinase (GCK) mutations in hyper- and hypoglycemia: maturity-onset diabetes of the young, permanent neonatal diabetes, and hyperinsulinemia of infancy. Hum. Mutat. 22, 353–362.

Gregg, B.E., Moore, P.C., Demozay, D., Hall, B.A., Li, M., Husain, A., Wright, A.J., Atkinson, M.A., Rhodes, C.J., 2012. Formation of a human beta-cell population within pancreatic islets is set early in life. J. Clin. Endocrinol. Metab. 97, 3197–3206.

Grieco, F.A., Sebastiani, G., Juan-Mateu, J., Villate, O., Marroqui, L., Ladriere, L., Tugay, K., Regazzi, R., Bugliani, M., Marchetti, P., Dotta, F., Eizirik, D.L., 2017.

MicroRNAs miR-23a-3p, miR-23b-3p, and miR-149-5p regulate the expression of proapoptotic BH3-only proteins DP5 and PUMA in human pancreatic beta-cells. Diabetes 66, 100–112.

Grieco, F.A., Schiavo, A.A., Brozzi, F., Juan-Mateu, J., Bugliani, M., Marchetti, P., Eizirik, D., 2019. The microRNAs miR-211-5p and miR-204-5p modulate ER stress in human beta cells. J. Mol. Endocrinol. 63, 139–149.

Guay, C., Regazzi, R., 2015a. MicroRNAs and the functional beta cell mass: for better or worse. Diabetes Metab. 41, 369–377.

Guay, C., Regazzi, R., 2015b. Role of islet microRNAs in diabetes: which model for which question? Diabetologia 58, 456–463.

Guay, C., Regazzi, R., 2017. Exosomes as new players in metabolic organ cross-talk. Diabetes Obes. Metab. 19 (Suppl. 1), 137–146.

Guay, C., Roggli, E., Nesca, V., Jacovetti, C., Regazzi, R., 2011. Diabetes mellitus, a microRNA-related disease? Transl. Res. 157, 253–264.

Guay, C., Menoud, V., Rome, S., Regazzi, R., 2015. Horizontal transfer of exosomal microRNAs transduce apoptotic signals between pancreatic beta-cells. Cell Commun. Signal 13, 17.

Guay, C., Kruit, J.K., Rome, S., Menoud, V., Mulder, N.L., Jurdzinski, A., Mancarella, F., Sebastiani, G., Donda, A., Gonzalez, B.J., Jandus, C., Bouzakri, K., Pinget, M., Boitard, C., Romero, P., Dotta, F., Regazzi, R., 2019. Lymphocyte-derived exosomal microRNAs promote pancreatic beta cell death and may contribute to type 1 diabetes development. Cell Metab. 29, 348–361.e6.

Guay, C., Jacovetti, C., Bayazit, M.B., Brozzi, F., Rodriguez-Trejo, A., Wu, K., Regazzi, R., 2020. Roles of noncoding RNAs in islet biology. Compr. Physiol. 10, 893–932.

Halestrap, A.P., Wilson, M.C., 2012. The monocarboxylate transporter family—role and regulation. IUBMB Life 64, 109–119.

Hanna, J., Hossain, G.S., Kocerha, J., 2019. The potential for microRNA therapeutics and clinical research. Front. Genet. 10, 478.

Hull, R.L., Kodama, K., Utzschneider, K.M., Carr, D.B., Prigeon, R.L., Kahn, S.E., 2005. Dietary-fat-induced obesity in mice results in beta cell hyperplasia but not increased insulin release: evidence for specificity of impaired beta cell adaptation. Diabetologia 48, 1350–1358.

Jacovetti, C., Abderrahmani, A., Parnaud, G., Jonas, J.C., Peyot, M.L., Cornu, M., Laybutt, R., Meugnier, E., Rome, S., Thorens, B., Prentki, M., Bosco, D., Regazzi, R., 2012. MicroRNAs contribute to compensatory beta cell expansion during pregnancy and obesity. J. Clin. Invest. 122, 3541–3551.

Jacovetti, C., Matkovich, S.J., Rodriguez-Trejo, A., Guay, C., Regazzi, R., 2015. Postnatal beta-cell maturation is associated with islet-specific microRNA changes induced by nutrient shifts at weaning. Nat. Commun. 6, 8084.

Jacovetti, C., Rodriguez-Trejo, A., Guay, C., Sobel, J., Gattesco, S., Petrenko, V., Saini, C., Dibner, C., Regazzi, R., 2017. MicroRNAs modulate core-clock gene expression in pancreatic islets during early postnatal life in rats. Diabetologia 60, 2011–2020.

Jalabert, A., Vial, G., Guay, C., Wiklander, O.P., Nordin, J.Z., Aswad, H., Forterre, A., Meugnier, E., Pesenti, S., Regazzi, R., Danty-Berger, E., Ducreux, S., Vidal, H., El-Andaloussi, S., Rieusset, J., Rome, S., 2016. Exosome-like vesicles released from lipid-induced insulin-resistant muscles modulate gene expression and proliferation of beta recipient cells in mice. Diabetologia 59, 1049–1058.

Joglekar, M.V., Joglekar, V.M., Hardikar, A.A., 2009a. Expression of islet-specific microRNAs during human pancreatic development. Gene Expr. Patterns 9, 109–113.

Joglekar, M.V., Patil, D., Joglekar, V.M., Rao, G.V., Reddy, D.N., Mitnala, S., Shouche, Y., Hardikar, A.A., 2009b. The miR-30 family microRNAs confer epithelial phenotype to human pancreatic cells. Islets 1, 137–147.

Joglekar, M.V., Wong, W.K.M., Maynard, C.L., Umrani, M.R., Martin, D., Loudovaris, T., Thomas, H.E., Dalgaard, L.T., Hardikar, A.A., 2018. Expression of miR-206 in human islets and its role in glucokinase regulation. Am. J. Physiol. Endocrinol. Metab. 315, E634–E637.

Kalis, M., Bolmeson, C., Esguerra, J.L., Gupta, S., Edlund, A., Tormo-Badia, N., Speidel, D., Holmberg, D., Mayans, S., Khoo, N.K., Wendt, A., Eliasson, L., Cilio, C.M., 2011. Beta-cell specific deletion of Dicer1 leads to defective insulin secretion and diabetes mellitus. Plos One 6, E29166.

Kameswaran, V., Bramswig, N.C., Mckenna, L.B., Penn, M., Schug, J., Hand, N.J., Chen, Y., Choi, I., Vourekas, A., Won, K.J., Liu, C., Vivek, K., Naji, A., Friedman, J.R., Kaestner, K.H., 2014. Epigenetic regulation of the DLK1-MEG3 microRNA cluster in human type 2 diabetic islets. Cell Metab. 19, 135–145.

Kanji, M.S., Martin, M.G., Bhushan, A., 2013. Dicer1 is required to repress neuronal fate during endocrine cell maturation. Diabetes 62, 1602–1611.

Kaspi, H., Pasvolsky, R., Hornstein, E., 2014. Could microRNAs contribute to the maintenance of beta cell identity? Trends Endocrinol. Metab. 25, 285–292.

Kato, T., Shimano, H., Yamamoto, T., Yokoo, T., Endo, Y., Ishikawa, M., Matsuzaka, T., Nakagawa, Y., Kumadaki, S., Yahagi, N., Takahashi, A., Sone, H., Suzuki, H., Toyoshima, H., Hasty, A.H., Takahashi, S., Gomi, H., Izumi, T., Yamada, N., 2006. Granuphilin is activated by SREBP-1c and involved in impaired insulin secretion in diabetic mice. Cell Metab. 4, 143–154.

Katoh, M.C., Jung, Y., Ugboma, C.M., Shimbo, M., Kuno, A., Basha, W.A., Kudo, T., Oishi, H., Takahashi, S., 2018. MafB is critical for glucagon production and secretion in mouse pancreatic alpha cells in vivo. Mol. Cell. Biol. 38, e00504–17.

Kaviani, M., Azarpira, N., Karimi, M.H., Al-Abdullah, I., 2016. The role of microRNAs in islet beta-cell development. Cell Biol. Int. 40, 1248–1255.

Keller, D.M., Mcweeney, S., Arsenlis, A., Drouin, J., Wright, C.V., Wang, H., Wollheim, C.B., White, P., Kaestner, K.H., Goodman, R.H., 2007. Characterization of pancreatic transcription factor Pdx-1 binding sites using promoter microarray and serial analysis of chromatin occupancy. J. Biol. Chem. 282, 32084–32092.

Keller, M.P., Choi, Y., Wang, P., Davis, D.B., Rabaglia, M.E., Oler, A.T., Stapleton, D.S., Argmann, C., Schueler, K.L., Edwards, S., Steinberg, H.A., Chaibub Neto, E., Kleinhanz, R., Turner, S., Hellerstein, M.K., Schadt, E.E., Yandell, B.S., Kendziorski, C., Attie, A.D., 2008. A gene expression network model of type 2 diabetes links cell cycle regulation in islets with diabetes susceptibility. Genome Res. 18, 706–716.

Kjorholt, C., Akerfeldt, M.C., Biden, T.J., Laybutt, D.R., 2005. Chronic hyperglycemia, independent of plasma lipid levels, is sufficient for the loss of beta-cell differentiation and secretory function in the db/db mouse model of diabetes. Diabetes 54, 2755–2763.

Klein, D., Misawa, R., Bravo-Egana, V., Vargas, N., Rosero, S., Piroso, J., Ichii, H., Umland, O., Zhijie, J., Tsinoremas, N., Ricordi, C., Inverardi, L., Dominguez-Bendala, J., Pastori, R.L., 2013. MicroRNA expression in alpha and beta cells of human pancreatic islets. Plos One 8, E55064.

Kredo-Russo, S., Mandelbaum, A.D., Ness, A., Alon, I., Lennox, K.A., Behlke, M.A., Hornstein, E., 2012a. Pancreas-enriched miRNA refines endocrine cell differentiation. Development 139, 3021–3031.

Kredo-Russo, S., Ness, A., Mandelbaum, A.D., Walker, M.D., Hornstein, E., 2012b. Regulation of pancreatic microRNA-7 expression. Exp. Diabetes Res. 2012, 695214.

Lahmy, R., Soleimani, M., Sanati, M.H., Behmanesh, M., Kouhkan, F., Mobarra, N., 2014. MiRNA-375 promotes beta pancreatic differentiation in human induced pluripotent stem (hiPS) cells. Mol. Biol. Rep. 41, 2055–2066.

Lapierre, M.P., Stoffel, M., 2017. MicroRNAs as stress regulators in pancreatic beta cells and diabetes. Mol. Metab. 6, 1010–1023.

Latreille, M., Hausser, J., Stutzer, I., Zhang, Q., Hastoy, B., Gargani, S., Kerr-Conte, J., Pattou, F., Zavolan, M., Esguerra, J.L., Eliasson, L., Rulicke, T., Rorsman, P., Stoffel, M., 2014. MicroRNA-7a regulates pancreatic beta cell function. J. Clin. Invest. 124, 2722–2735.

Lee, R.C., Feinbaum, R.L., Ambros, V., 1993. The C. elegans heterochronic gene lin-4 encodes small RNAs with antisense complementarity to lin-14. Cell 75, 843–854.

Lee, Y., Jeon, K., Lee, J.T., Kim, S., Kim, V.N., 2002. MicroRNA maturation: stepwise processing and subcellular localization. EMBO J. 21, 4663–4670.

Lee, H., Lee, Y.S., Harenda, Q., Pietrzak, S., Oktay, H.Z., Schreiber, S., Liao, Y., Sonthalia, S., Ciecko, A.E., Chen, Y.G., Keles, S., Sridharan, R., Engin, F., 2020. Beta cell dedifferentiation induced by IRE1alpha deletion prevents type 1 diabetes. Cell Metab. 31, 822–836.e5.

Lemaire, K., Thorrez, L., Schuit, F., 2016. Disallowed and allowed gene expression: two faces of mature islet beta cells. Annu. Rev. Nutr. 36, 45–71.

Li, N., Jiang, D., He, Q., He, F., Li, Y., Deng, C., Li, F., 2020. MicroRNA-181c-5p promotes the formation of insulin-producing cells from human induced pluripotent stem cells by targeting smad7 and TGIF2. Cell Death Dis. 11, 462.

Liang, D., Zhang, Y., Han, J., Wang, W., Liu, Y., Li, J., Jiang, X., 2015. Embryonic stem cell-derived pancreatic endoderm transplant with MCT1-suppressing miR-495 attenuates type II diabetes in mice. Endocr. J. 62, 907–920.

Lindstrom, P., 2007. The physiology of obese-hyperglycemic mice [ob/ob mice]. ScientificWorldJournal 7, 666–685.

Linnemann, A.K., Baan, M., Davis, D.B., 2014. Pancreatic beta-cell proliferation in obesity. Adv. Nutr. 5, 278–288.

Locke, J.M., Da Silva Xavier, G., Dawe, H.R., Rutter, G.A., Harries, L.W., 2014. Increased expression of miR-187 in human islets from individuals with type 2 diabetes is associated with reduced glucose-stimulated insulin secretion. Diabetologia 57, 122–128.

Lopez-Beas, J., Capilla-Gonzalez, V., Aguilera, Y., Mellado, N., Lachaud, C.C., Martin, F., Smani, T., Soria, B., Hmadcha, A., 2018. miR-7 modulates hESC differentiation into insulin-producing beta-like cells and contributes to cell maturation. Mol. Ther.–Nucleic Acids 12, 463–477.

Lovis, P., Gattesco, S., Regazzi, R., 2008a. Regulation of the expression of components of the exocytotic machinery of insulin-secreting cells by microRNAs. Biol. Chem. 389, 305–312.

Lovis, P., Roggli, E., Laybutt, D.R., Gattesco, S., Yang, J.Y., Widmann, C., Abderrahmani, A., Regazzi, R., 2008b. Alterations in microRNA expression contribute to fatty acid-induced pancreatic beta-cell dysfunction. Diabetes 57, 2728–2736.

Lu, Y., Fei, X.Q., Yang, S.F., Xu, B.K., Li, Y.Y., 2015. Glucose-induced microRNA-17 promotes pancreatic beta cell proliferation through down-regulation of Menin. Eur. Rev. Med. Pharmacol. Sci. 19, 624–629.

Lynn, F.C., Skewes-Cox, P., Kosaka, Y., Mcmanus, M.T., Harfe, B.D., German, M.S., 2007. MicroRNA expression is required for pancreatic islet cell genesis in the mouse. Diabetes 56, 2938–2945.

Maestro, M.A., Boj, S.F., Luco, R.F., Pierreux, C.E., Cabedo, J., Servitja, J.M., German, M.S., Rousseau, G.G., Lemaigre, F.P., Ferrer, J., 2003. Hnf6 and Tcf2 (MODY5) are linked in a gene network operating in a precursor cell domain of the embryonic pancreas. Hum. Mol. Genet. 12, 3307–3314.

Mandelbaum, A.D., Kredo-Russo, S., Aronowitz, D., Myers, N., Yanowski, E., Klochendler, A., Swisa, A., Dor, Y., Hornstein, E., 2019. miR-17-92 and miR-106b-25 clusters regulate beta cell mitotic checkpoint and insulin secretion in mice. Diabetologia 62, 1653–1666.

Marhfour, I., Lopez, X.M., Lefkaditis, D., Salmon, I., Allagnat, F., Richardson, S.J., Morgan, N.G., Eizirik, D.L., 2012. Expression of endoplasmic reticulum stress markers in the islets of patients with type 1 diabetes. Diabetologia 55, 2417–2420.

Martinez-Sanchez, A., Nguyen-Tu, M.S., Rutter, G.A., 2015. DICER inactivation identifies pancreatic beta-cell "disallowed" genes targeted by microRNAs. Mol. Endocrinol. 29, 1067–1079.

Martinez-Sanchez, A., Rutter, G.A., Latreille, M., 2016. MiRNAs in beta-cell development, identity, and disease. Front. Genet. 7, 226.

Martinez-Sanchez, A., Nguyen-Tu, M.S., Cebola, I., Yavari, A., Marchetti, P., Piemonti, L., De Koning, E., Shapiro, A.M.J., Johnson, P., Sakamoto, K., Smith, D.M., Leclerc, I., Ashrafian, H., Ferrer, J., Rutter, G.A., 2018. MiR-184 expression is regulated by AMPK in pancreatic islets. FASEB J. 32, 2587–2600.

Martin-Gronert, M.S., Ozanne, S.E., 2012. Metabolic programming of insulin action and secretion. Diabetes Obes. Metab. 14 (Suppl. 3), 29–39.

Melkman-Zehavi, T., Oren, R., Kredo-Russo, S., Shapira, T., Mandelbaum, A.D., Rivkin, N., Nir, T., Lennox, K.A., Behlke, M.A., Dor, Y., Hornstein, E., 2011. miRNAs control insulin content in pancreatic beta-cells via downregulation of transcriptional repressors. EMBO J. 30, 835–845.

Murtaugh, L.C., 2007. Pancreas and beta-cell development: from the actual to the possible. Development 134, 427–438.

Nathan, G., Kredo-Russo, S., Geiger, T., Lenz, A., Kaspi, H., Hornstein, E., Efrat, S., 2015. MiR-375 promotes redifferentiation of adult human beta cells expanded in vitro. Plos One 10, E0122108.

Nesca, V., Guay, C., Jacovetti, C., Menoud, V., Peyot, M.L., Laybutt, D.R., Prentki, M., Regazzi, R., 2013. Identification of particular groups of microRNAs that positively or negatively impact on beta cell function in obese models of type 2 diabetes. Diabetologia 56, 2203–2212.

Neu, J., Hauser, N., Douglas-Escobar, M., 2007. Postnatal nutrition and adult health programming. Semin. Fetal Neonatal Med. 12, 78–86.

Nieto, M., Hevia, P., Garcia, E., Klein, D., Alvarez-Cubela, S., Bravo-Egana, V., Rosero, S., Damaris Molano, R., Vargas, N., Ricordi, C., Pileggi, A., Diez, J., Dominguez-Bendala, J., Pastori, R.L., 2012. Antisense miR-7 impairs insulin expression in developing pancreas and in cultured pancreatic buds. Cell Transplant. 21, 1761–1774.

Njolstad, P.R., Sovik, O., Cuesta-Munoz, A., Bjorkhaug, L., Massa, O., Barbetti, F., Undlien, D.E., Shiota, C., Magnuson, M.A., Molven, A., Matschinsky, F.M., Bell, G.I., 2001. Neonatal diabetes mellitus due to complete glucokinase deficiency. N. Engl. J. Med. 344, 1588–1592.

Ofori, J.K., Salunkhe, V.A., Bagge, A., Vishnu, N., Nagao, M., Mulder, H., Wollheim, C.B., Eliasson, L., Esguerra, J.L., 2017. Elevated miR-130a/miR130b/miR-152 expression reduces intracellular ATP levels in the pancreatic beta cell. Sci. Rep. 7, 44986.

Osmai, M., Osmai, Y., Bang-Berthelsen, C.H., Pallesen, E.M., Vestergaard, A.L., Novotny, G.W., Pociot, F., Mandrup-Poulsen, T., 2016. MicroRNAs as regulators of beta-cell function and dysfunction. Diabetes Metab. Res. Rev. 32, 334–349.

Panwar, N., Soehartono, A.M., Chan, K.K., Zeng, S., Xu, G., Qu, J., Coquet, P., Yong, K.T., Chen, X., 2019. Nanocarbons for biology and medicine: sensing, imaging, and drug delivery. Chem. Rev. 119, 9559–9656.

Pearson, J.A., Wong, F.S., Wen, L., 2016. The importance of the non obese diabetic (NOD) mouse model in autoimmune diabetes. J. Autoimmun. 66, 76–88.

Peyot, M.L., Pepin, E., Lamontagne, J., Latour, M.G., Zarrouki, B., Lussier, R., Pineda, M., Jetton, T.L., Madiraju, S.R., Joly, E., Prentki, M., 2010. Beta-cell failure in diet-induced obese mice stratified according to body weight gain: secretory dysfunction and altered islet lipid metabolism without steatosis or reduced beta-cell mass. Diabetes 59, 2178–2187.

Plaisance, V., Abderrahmani, A., Perret-Menoud, V., Jacquemin, P., Lemaigre, F., Regazzi, R., 2006. MicroRNA-9 controls the expression of granuphilin/Slp4 and the secretory response of insulin-producing cells. J. Biol. Chem. 281, 26932–26942.

Poitout, V., Hagman, D., Stein, R., Artner, I., Robertson, R.P., Harmon, J.S., 2006. Regulation of the insulin gene by glucose and fatty acids. J. Nutr. 136, 873–876.

Portha, B., Lacraz, G., Kergoat, M., Homo-Delarche, F., Giroix, M.H., Bailbe, D., Gangnerau, M.N., Dolz, M., Tourrel-Cuzin, C., Movassat, J., 2009. The GK rat beta-cell: a prototype for the diseased human beta-cell in type 2 diabetes? Mol. Cell. Endocrinol. 297, 73–85.

Poy, M.N., Eliasson, L., Krutzfeldt, J., Kuwajima, S., Ma, X., Macdonald, P.E., Pfeffer, S., Tuschl, T., Rajewsky, N., Rorsman, P., Stoffel, M., 2004. A pancreatic islet-specific microRNA regulates insulin secretion. Nature 432, 226–230.

Poy, M.N., Hausser, J., Trajkovski, M., Braun, M., Collins, S., Rorsman, P., Zavolan, M., Stoffel, M., 2009. miR-375 maintains normal pancreatic alpha- and beta-cell mass. Proc. Natl. Acad. Sci. U. S. A. 106, 5813–5818.

Pullen, T.J., Da Silva Xavier, G., Kelsey, G., Rutter, G.A., 2011. miR-29a and miR-29b contribute to pancreatic beta-cell-specific silencing of monocarboxylate transporter 1 (Mct1). Mol. Cell. Biol. 31, 3182–3194.

Pullen, T.J., Sylow, L., Sun, G., Halestrap, A.P., Richter, E.A., Rutter, G.A., 2012. Overexpression of monocarboxylate transporter-1 (SLC16a1) in mouse pancreatic beta-cells leads to relative hyperinsulinism during exercise. Diabetes 61, 1719–1725.

Rana, T.M., 2007. Illuminating the silence: understanding the structure and function of small RNAs. Nat. Rev. Mol. Cell Biol. 8, 23–36.

Roggli, E., Britan, A., Gattesco, S., Lin-Marq, N., Abderrahmani, A., Meda, P., Regazzi, R., 2010. Involvement of microRNAs in the cytotoxic effects exerted by proinflammatory cytokines on pancreatic beta-cells. Diabetes 59, 978–986.

Roggli, E., Gattesco, S., Caille, D., Briet, C., Boitard, C., Meda, P., Regazzi, R., 2012. Changes in microRNA expression contribute to pancreatic beta-cell dysfunction in prediabetic NOD mice. Diabetes 61, 1742–1751.

Rutman, A.K., Negi, S., Gasparrini, M., Hasilo, C.P., Tchervenkov, J., Paraskevas, S., 2018. Immune response to extracellular vesicles from human islets of Langerhans in patients with type 1 diabetes. Endocrinology 159, 3834–3847.

Salunkhe, V.A., Ofori, J.K., Gandasi, N.R., Salo, S.A., Hansson, S., Andersson, M.E., Wendt, A., Barg, S., Esguerra, J.L.S., Eliasson, L., 2017. MiR-335 overexpression impairs insulin secretion through defective priming of insulin vesicles. Physiol. Rep. 5, e13493.

Sandovici, I., Smith, N.H., Nitert, M.D., Ackers-Johnson, M., Uribe-Lewis, S., Ito, Y., Jones, R.H., Marquez, V.E., Cairns, W., Tadayyon, M., O'neill, L.P., Murrell, A., Ling, C., Constancia, M., Ozanne, S.E., 2011. Maternal diet and aging alter the epigenetic control of a promoter-enhancer interaction at the Hnf4a gene in rat pancreatic islets. Proc. Natl. Acad. Sci. U. S. A. 108, 5449–5454.

Sebastiani, G., Po, A., Miele, E., Ventriglia, G., Ceccarelli, E., Bugliani, M., Marselli, L., Marchetti, P., Gulino, A., Ferretti, E., Dotta, F., 2015. MicroRNA-124a is hyperexpressed in type 2 diabetic human pancreatic islets and negatively regulates insulin secretion. Acta Diabetol. 52, 523–530.

Sebastiani, G., Valentini, M., Grieco, G.E., Ventriglia, G., Nigi, L., Mancarella, F., Pellegrini, S., Martino, G., Sordi, V., Piemonti, L., Dotta, F., 2017. MicroRNA expression profiles of human iPSCs differentiation into insulin-producing cells. Acta Diabetol. 54, 265–281.

Shaer, A., Azarpira, N., Karimi, M.H., Soleimani, M., Dehghan, S., 2016. Differentiation of human-induced pluripotent stem cells into insulin-producing clusters by microRNA-7. Exp. Clin. Transplant. 14, 555–563.

Sheng, H., Hassanali, S., Nugent, C., Wen, L., Hamilton-Williams, E., Dias, P., Dai, Y.D., 2011. Insulinoma-released exosomes or microparticles are immunostimulatory and can activate autoreactive T cells spontaneously developed in nonobese diabetic mice. J. Immunol. 187, 1591–1600.

Smith, S.B., Qu, H.Q., Taleb, N., Kishimoto, N.Y., Scheel, D.W., Lu, Y., Patch, A.M., Grabs, R., Wang, J., Lynn, F.C., Miyatsuka, T., Mitchell, J., Seerke, R., Desir, J., Vanden Eijnden, S., Abramowicz, M., Kacet, N., Weill, J., Renard, M.E., Gentile, M., Hansen, I., Dewar, K., Hattersley, A.T., Wang, R., Wilson, M.E., Johnson, J.D., Polychronakos, C., German, M.S., 2010. Rfx6 directs islet formation and insulin production in mice and humans. Nature 463, 775–780.

Sosa-Larios, T.C., et al., 2015. Epigenetic alterations caused by nutritional stress during fetal programming of the endocrine pancreas. Arch. Med. Res. 46 (2), 93–100.

Stoehr, J.P., Nadler, S.T., Schueler, K.L., Rabaglia, M.E., Yandell, B.S., Metz, S.A., Attie, A.D., 2000. Genetic obesity unmasks nonlinear interactions between murine type 2 diabetes susceptibility loci. Diabetes 49, 1946–1954.

Stolovich-Rain, M., Enk, J., Vikesa, J., Nielsen, F.C., Saada, A., Glaser, B., Dor, Y., 2015. Weaning triggers a maturation step of pancreatic beta cells. Dev. Cell 32, 535–545.

Sun, K., Lai, E.C., 2013. Adult-specific functions of animal microRNAs. Nat. Rev. Genet. 14, 535–548.

Tattikota, S.G., Rathjen, T., Mcanulty, S.J., Wessels, H.H., Akerman, I., Van De Bunt, M., Hausser, J., Esguerra, J.L., Musahl, A., Pandey, A.K., You, X., Chen, W., Herrera, P.L., Johnson, P.R., O'carroll, D., Eliasson, L., Zavolan, M., Gloyn, A.L., Ferrer, J., Shalom-Feuerstein, R., Aberdam, D., Poy, M.N., 2014. Argonaute2 mediates compensatory expansion of the pancreatic beta cell. Cell Metab. 19, 122–134.

Tattikota, S.G., Rathjen, T., Hausser, J., Khedkar, A., Kabra, U.D., Pandey, V., Sury, M., Wessels, H.H., Mollet, I.G., Eliasson, L., Selbach, M., Zinzen, R.P., Zavolan, M., Kadener, S., Tschop, M.H., Jastroch, M., Friedlander, M.R., Poy, M.N., 2015. miR-184 regulates pancreatic beta-cell function according to glucose metabolism. J. Biol. Chem. 290, 20284–20294.

Tersey, S.A., Nishiki, Y., Templin, A.T., Cabrera, S.M., Stull, N.D., Colvin, S.C., Evans-Molina, C., Rickus, J.L., Maier, B., Mirmira, R.G., 2012. Islet beta-cell endoplasmic reticulum stress precedes the onset of type 1 diabetes in the nonobese diabetic mouse model. Diabetes 61, 818–827.

Thomou, T., Mori, M.A., Dreyfuss, J.M., Konishi, M., Sakaguchi, M., Wolfrum, C., Rao, T.N., Winnay, J.N., Garcia-Martin, R., Grinspoon, S.K., Gorden, P., Kahn, C.R., 2017. Adipose-derived circulating miRNAs regulate gene expression in other tissues. Nature 542, 450–455.

Tugay, K., Guay, C., Marques, A.C., Allagnat, F., Locke, J.M., Harries, L.W., Rutter, G.A., Regazzi, R., 2016. Role of microRNAs in the age-associated decline of pancreatic beta cell function in rat islets. Diabetologia 59, 161–169.

Vinod, M., Patankar, J.V., Sachdev, V., Frank, S., Graier, W.F., Kratky, D., Kostner, G.M., 2016. MiR-206 is expressed in pancreatic islets and regulates glucokinase activity. Am. J. Physiol. Endocrinol. Metab. 311, E175–E185.

Wagner, V., Dullaart, A., Bock, A.K., Zweck, A., 2006. The emerging nanomedicine landscape. Nat. Biotechnol. 24, 1211–1217.

Wang, P., Fiaschi-Taesch, N.M., Vasavada, R.C., Scott, D.K., Garcia-Ocana, A., Stewart, A.F., 2015. Diabetes mellitus—advances and challenges in human beta-cell proliferation. Nat. Rev. Endocrinol. 11, 201–212.

Wang, P., Liu, Q., Zhao, H., Bishop, J.O., Zhou, G., Olson, L.K., Moore, A., 2020. miR-216a-targeting theranostic nanoparticles promote proliferation of insulin-secreting cells in type 1 diabetes animal model. Sci. Rep. 10, 5302.

Wei, R., Yang, J., Liu, G.Q., Gao, M.J., Hou, W.F., Zhang, L., Gao, H.W., Liu, Y., Chen, G.A., Hong, T.P., 2013. Dynamic expression of microRNAs during the differentiation of human embryonic stem cells into insulin-producing cells. Gene 518, 246–255.

Welsh, M., Nielsen, D.A., Mackrell, A.J., Steiner, D.F., 1985. Control of insulin gene expression in pancreatic beta-cells and in an insulin-producing cell line, RIN-5F cells. II. Regulation of insulin mRNA stability. J. Biol. Chem. 260, 13590–13594.

Wendt, A., Esguerra, J.L., Eliasson, L., 2018. Islet microRNAs in health and type-2 diabetes. Curr. Opin. Pharmacol. 43, 46–52.

Wightman, B., Ha, I., Ruvkun, G., 1993. Posttranscriptional regulation of the heterochronic gene lin-14 by lin-4 mediates temporal pattern formation in C. elegans. Cell 75, 855–862.

Wong, W.K.M., Sorensen, A.E., Joglekar, M.V., Hardikar, A.A., Dalgaard, L.T., 2018. Non-coding RNA in pancreas and beta-cell development. Noncoding RNA 4.

Xu, G., Chen, J., Jing, G., Shalev, A., 2013. Thioredoxin-interacting protein regulates insulin transcription through microRNA-204. Nat. Med. 19, 1141–1146.

Xu, G., Chen, J., Jing, G., Grayson, T.B., Shalev, A., 2016. miR-204 targets PERK and regulates UPR signaling and beta-cell apoptosis. Mol. Endocrinol. 30, 917–924.

Ying, W., Riopel, M., Bandyopadhyay, G., Dong, Y., Birmingham, A., Seo, J.B., Ofrecio, J.M., Wollam, J., Hernandez-Carretero, A., Fu, W., Li, P., Olefsky, J.M., 2017. Adipose tissue macrophage-derived exosomal miRNAs can modulate in vivo and in vitro insulin sensitivity. Cell 171, 372–384.e12.

Zhao, E., Keller, M.P., Rabaglia, M.E., Oler, A.T., Stapleton, D.S., Schueler, K.L., Neto, E.C., Moon, J.Y., Wang, P., Wang, I.M., Lum, P.Y., Ivanovska, I., Cleary, M., Greenawalt, D., Tsang, J., Choi, Y.J., Kleinhanz, R., Shang, J., Zhou, Y.P., Howard, A.D., Zhang, B.B., Kendziorski, C., Thornberry, N.A., Yandell, B.S., Schadt, E.E., Attie, A.D., 2009. Obesity and genetics regulate microRNAs in islets, liver, and adipose of diabetic mice. Mamm. Genome 20, 476–485.

Zimmet, P.Z., 1999. Diabetes epidemiology as a tool to trigger diabetes research and care. Diabetologia 42, 499–518.

CHAPTER FIVE

Physiology of pancreatic β-cells: Ion channels and molecular mechanisms implicated in stimulus-secretion coupling

Minerva Gil-Rivera[a,*], Regla M. Medina-Gali[b,c], Juan Martínez-Pinna[a,c], and Sergi Soriano[a,c,*]
[a]Departamento de Fisiología, Genética y Microbiología, Universidad de Alicante, Alicante, Spain
[b]Centro de Investigación Biomédica en Red de Diabetes y Enfermedades Metabólicas Asociadas (CIBERDEM), Madrid, Spain
[c]Instituto de Investigación, Desarrollo e Innovación en Biotecnología Sanitaria de Elche (IDiBE), Universidad Miguel Hernández, Elche, Spain
*Corresponding authors: e-mail address: mgr72@alu.ua.es; sergi.soriano@ua.es

Contents

Abstract

The human and mouse islet of Langerhans is an endocrine organ composed of five different cells types; insulin-secreting β-cells, glucagon-producing α-cells, somatostatin-producing δ-cells, pancreatic polypeptide-secreting PP cells and ε-cells that secretes ghrelin. The most important cells are the pancreatic β-cells that comprise around 45–50% of human islets and 75–80% in the mouse. Pancreatic β-cells secrete insulin at high glucose concentration, thereby finely regulating glycaemia by the hypoglycaemic effects of this hormone. Different ion channels are implicated in the stimulus-secretion coupling of insulin. An increase in the intracellular ATP concentration leads to closure K_{ATP} channels, depolarizing the cell and opening voltage-gated calcium channels. The increase of intracellular calcium concentration induced by calcium entry

International Review of Cell and Molecular Biology, Volume 359
ISSN 1937-6448
https://doi.org/10.1016/bs.ircmb.2021.02.006

through voltage-gated calcium channels promotes insulin secretion. Here, we briefly describe the diversity of ion channels present in pancreatic β-cells and the different mechanisms that are responsible to induce insulin secretion in human and mouse cells. Moreover, we described the pathophysiology due to alterations in the physiology of the main ion channels present in pancreatic β-cell and its implication to predispose metabolic disorders as type 2 diabetes mellitus.

1. Introduction

The pancreas is a mixed glandular organ located in the retroperitoneal cavity. From histological and physiological point of view, the pancreas is divided in two parts. The exocrine part is made up of pancreatic acini, which secrete digestive enzymes and bicarbonate (pancreatic juice) to the small intestine during food's digestion. The endocrine part is made up of the islets of Langerhans, responsible for releasing the hormones that maintain glucose homeostasis.

The islets of Langerhans were described by the German pathologist and physiologist Paul Langerhans in 1869. These spherical structures are randomly distributed among the exocrine parenchyma of the pancreas (Rorsman and Ashcroft, 2018). Pancreatic islets are composed of five different cell types. The majority cell type within the islet is the β-cell, which secretes insulin. Glucagon-secreting α-cells are the second most abundant cell type within the islet. Three other cell types are found with smaller percentage of representation: somatostatin-secreting δ-cells, pancreatic polypeptide-secreting PP-cells, and ghrelin-secreting ε-cells (Cabrera et al., 2006; Da Silva Xavier, 2018).

The distribution of different cell types varies when comparing mice islets with human islets. In mice, pancreatic β-cells are available in the central part, occupying most of the islet (75%). The rest of the cells are located on the periphery, with α-cells being more abundant (≈20%) than δ-cells (≈5%) (Fig. 1). In humans, the proportion of β-cells is lower than in mice (54%), but it remains the predominant cell type in the pancreatic islet. β-cells are intermingled with α-cells and δ-cells in a 35% and 10% proportion, respectively (Fig. 1) (Brissova et al., 2005; Cabrera et al., 2006).

The maintenance of glycaemic homeostasis is essential for the proper function of the organism. To this purpose, there are two antagonistic hormones that are involved in this process: insulin and glucagon, secreted by β-cells and α-cells, respectively. Insulin is the only hormone with hypoglycaemic effect. After food intake, pancreatic β-cells secrete insulin into

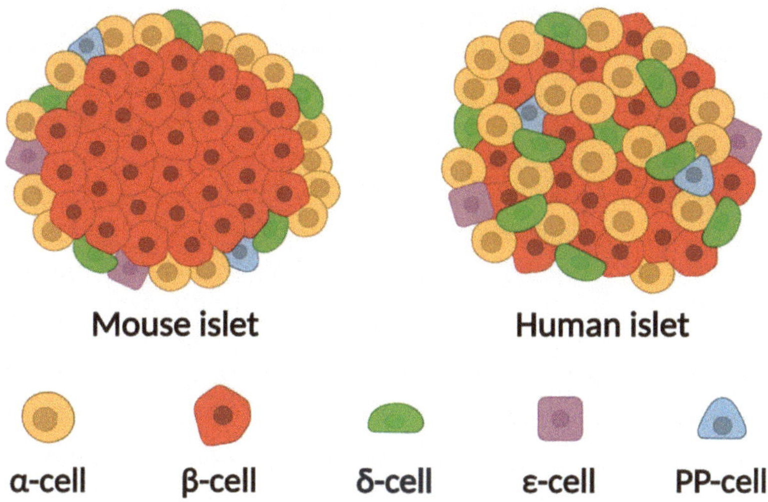

Fig. 1 Architecture of the islets of Langerhans in mouse (left) and human (right). Each cell type is represented by a different color: α-cell (yellow), β-cell (red), δ-cell (green), ε-cell (purple) and PP-cell (blue).

the bloodstream which exerts its action on three peripheral tissues mainly: liver, skeletal muscle, and adipose tissue. Insulin promotes glucose uptake in these tissues, leading to processes such as glycogenesis (glycogen synthesis) in the liver or lipogenesis (fatty acid synthesis) in adipose tissue. In muscle, it promotes translocation of glucose transporters to the cell membrane, which increases glucose entry into these cells. On the other hand, when blood glucose levels drop, pancreatic α-cells secrete glucagon, which exerts anti-insulin effects on peripheral tissues. In this case, glucagon induces the release of glucose from target tissues into the bloodstream. To do this, the processes of glycogenolysis and gluconeogenesis in the liver and lipolysis in adipose tissue are initiated. In addition, skeletal muscle decreases glucose uptake (Roder et al., 2016).

Hormonal secretion of the pancreatic islet is highly regulated, as there are paracrine relationships between the different cell types of the islet that establish a negative feedback model. Insulin secreted by the β-cell has an inhibitory effect on glucagon secretion by α-cells. Glucagon, in turn, has a stimulating effect on insulin secretion. In addition, somatostatin secreted by δ-cells has an inhibitory effect on α-cells and β-cells (Briant et al., 2018).

Although PP cells and ε cells are in a very small percentage within the islet and do not directly regulate glycaemic homeostasis, they perform important functions that contribute to the proper functioning of the organism.

Pancreatic polypeptide produced by PP cells is involved in regulating pancreatic exocrine secretion. In addition, it produces an inhibitory effect in the intestine, reducing intestinal motility. On the other hand, ghrelin secreted by ε cells is a powerful appetite stimulator and is commonly known as the "hunger hormone" (Da Silva Xavier, 2018). Ghrelin released by pancreatic ε cells complements the action of ghrelin released by stomach cells, the main release site of this hormone in the body.

Pancreatic islets are highly vascularized structures and receive innervation of the autonomous nervous system, both sympathetic and parasympathetic (Rodriguez-Diaz and Caicedo, 2014; Thorens, 2014). In hyperglycaemia situations, parasympathetic branches that innervate pancreatic islets enhance insulin secretion through the release of acetylcholine and its binding to muscarinic receptors (m3AchR) present in the β-cell. Certain peptides, such as pituitary adenylate cyclase activator polypeptide (PACAP), vasoactive intestinal peptide (VIP), or gastrin-releasing peptide (GRP), can contribute to parasympathetic control in the β-cell. Parasympathetic innervation is also involved in controlling the cephalic phase of insulin secretion, an anticipatory response that is triggered in the organism by a psychic, visual or olfactory stimulus or during food mastication.

The sympathetic branch of the nervous system affects the function of both α-cell and β-cell (Rodriguez-Diaz and Caicedo, 2014; Thorens, 2014). In hypoglycaemia situations, adrenaline and noradrenaline, both secreted by adrenal glands, stimulate glucagon secretion through binding to adrenergic β2 receptors present in the α-cell. In turn, these catecholamines inhibit insulin secretion in the β-cell by binding their adrenergic α2 receptors.

Human islets have less parasympathetic innervation than mouse islets. In mice, parasympathetic nerve fibers make direct contact with β-cells. As for the sympathetic innervation, human islets establish connections to the smooth muscle of the vessels that supply the pancreas, allowing control of blood flow within the islet. In mice, sympathetic nerve fibers make direct contact with the cells (Rodriguez-Diaz and Caicedo, 2014; Thorens, 2014).

The complementary action of hormones secreted by the endocrine pancreas allows blood glucose values to be maintained around 70–100 mg/dL in fasting conditions and 140 mg/dL after food intake. However, if this negative feedback system fails, metabolic diseases such as diabetes may appear, in which blood glucose values exceed 126 mg/dL in fasting conditions or 200 mg/dL after food intake (Fig. 2). Diabetes occurs when the β-cell is unable to produce and release insulin (type 1 diabetes) or when insulin is

	Fasting plasma glucose	Two-hour plasma glucose*	Glycosylated hemoglobin (HbA1c)
Normoglycemia	70-100 mg/dL	< 140 mg/dL	< 5.7 %
Impaired fasting glucose	110-125 mg/dL	< 140 mg/dL	5.7-6.4 %
Impaired glucose tolerance	≥ 126 mg/dL	140-200 mg/dL	5.7-6.4 %
Diabetes	≥ 126 mg/dL	≥ 200 mg/dL	> 6.5%

*Oral glucose tolerance test (OGTT): 75 g glucose dissolved in water.

Fig. 2 Glycaemic values for the diagnosis of diabetes mellitus and other categories of glycaemic disorders that occur with hyperglycaemia by performing an oral glucose tolerance test.

unable to function in peripheral tissues, as they have resistance to the hormone's action (type 2 diabetes) (Gomis and Aguilar Diosdado, 2007).

There are certain nutrients, in addition to glucose, that promote insulin secretion (Rorsman and Ashcroft, 2018). These include some amino acids (arginine, lysine, alanine, glutamate, leucine), fatty acids, hormones (glucagon, estrogen) and neurotransmitters (acetylcholine). In patients with diabetes, medicines that increase insulin secretion, such as sulfonylureas or glinides, are used.

2. Ion channels in pancreatic β-cells

All cells are surrounded by a lipid bilayer that is impermeable to charged and/or excessively large molecules, called plasma membrane. This membrane helps define and shape the cell, maintaining the balance between the intracellular and extracellular environment. In addition to its static functions, it contains proteins specialized in the exchange of various ions and molecules between the cytoplasm and the extracellular medium.

Traditionally, two types of proteins responsible for transport through the plasma membrane have been described: transporters and channels.

Transporters, including the Ca^{2+} pump, the Na^+/Ca^{2+} exchanger, or the Na^+/K^+ pump, transport ions against their concentration gradient at the expense of metabolic energy consumption. The channels, on the other hand, are proteins that form aqueous pores in the phospholipid plasma membrane through which ions can passively flow down their electrochemical gradient. Depending on the stimulus that leads to the opening of the ion channel, two main types of ion channels can be distinguished: (a) ligand-gated ion channels, require the specific binding of a molecule directly to the channel or to an associated protein; and (b) voltage-gated ion channels, the opening is coupled to a change in the transmembrane potential. The presence of negatively charged non-diffusible proteins in the cell cytoplasm, together with the selective ionic permeability of the channels and the action of transporters, makes the plasma membrane polarized with an excess of negative charge in the inside with respect to the outside at rest, stablishing the resting membrane potential. Excitable cells respond to the appropriate stimulus with a transient and rapid change of the membrane potential (usually reaching positive values) known as the action potential. The frequency of action potential firing carries encoded information to control a physiological process. The different pancreatic cell types, α, β and δ are excitable cells. The expression and differential regulation of the ion channels, transporters and pumps present in their plasma membrane allows them to maintain the membrane potential polarized and respond to a certain stimulus (increase in glycaemia for β-cells) with an alteration of the underlying ionic conductance, causing a characteristic pattern of electrical activity.

In the case of pancreatic cells, the information encoded in their pattern of action potential triggering is related to the control of hormonal secretion. Hormone secretion in the islet of Langerhans is a Ca^{2+}-dependent process, which is activated after the increase in the intracellular Ca^{2+} concentration ($[Ca^{2+}]_i$) caused by the entry of this ion through voltage-gated calcium channels during action potential firing (Drews et al., 2010; Kanno et al., 2002a; Rorsman, 1997). In this way, the regulation that glucose exerts on the excitability of pancreatic cells allows a precise control of hormonal secretion in response to the functional demands that characterize the periods of hypo and hyperglycemia (stimulus-secretion coupling). The application of the patch-clamp technique (Hamill et al., 1981) developed by Erwin Neher and Bert Sakmann in the 1970s (Neher and Sakmann, 1976), allowed the knowledge of the ionic conductances involved in the generation of electrical activity in small cells such as pancreatic cells. The introduction of this technique to the study of pancreatic cells improved the understanding of the mechanisms that control the secretion of different hormones present in the islet of Langerhans.

The establishment and maintenance of transmembrane voltage at rest and the characteristic shape of the action potential and the dynamics in the relationship of the active and quiescent phases in the pancreatic β-cells, are determined by a complex orchestrated relationship of ionic movements through various types of ion channels. The application of the patch-clamp technique has shown that each β-cell has multiple copies (100–5000) of 10–20 different ion channels (Ashcroft and Rorsman, 1989).

The objective of the following sections is providing information on some basic aspects of the structure and function of the ion channels, which have special relevance in generating the electrical activity controlling β-cell insulin secretion.

2.1 Potassium channels

Potassium (K^+) channels represent the largest and most diverse group of all ion channels. In humans, there are approximately 78 genes that encode for α-subunits and 13 genes that encode for auxiliary subunits. These channels are divided into three groups depending on their structure (Kuang et al., 2015): voltage-gated and calcium-gated potassium channels (K_V and K_{Ca}), with 6 or 7 transmembrane segments; inwardly rectifying potassium channels (K_{ir}), with 2 transmembrane segments; and tandem pore domain potassium channels (K2P), with 4 transmembrane segments. Both mouse and human β-cells express these three groups of K^+ channels.

2.1.1 Inwardly rectifying K^+ channels

Inwardly rectifying K^+ channels (Kir) are opened at rest, at very negative potentials, and are closed when the cell depolarizes. These channels contribute to the maintenance of the resting membrane potential (close to the K^+ equilibrium potential). In addition, they modulate signaling processes of G protein-coupled receptors, which establishes a link between cellular metabolism and membrane excitability (Hibino et al., 2010). Kir channels can be modulated by ions, phospholipids, or binding proteins (Kuang et al., 2015). These molecules interact directly with the channel to regulate properties such as ionic flow and channel opening and closing kinetics (Hibino et al., 2010).

The primary structure of Kir channels is composed of 2 transmembrane segments joined by a loop (P) that forms the pore-forming region of the channel. The N-terminal and C-terminal domains are found in the cytoplasmic region and regulate the gating of the channel. These channels have an ion selectivity filter with a highly conserved sequence made up of the amino acids threonine, valine, glycine, tyrosine and glycine (TVGYG). This sequence is maintained in other K^+ channels. The structure of Kir channels lacks the characteristic voltage sensor of voltage-gated channels, making it insensitive to changes in membrane potential. The formation of a functional ion channel needs the union of 4 subunits, being the final structure a tetrameric complex (Hibino et al., 2010; Kuang et al., 2015).

There are 15 genes that encode for 7 families of Kir channel subunits (Kir1.x-Kir7.x). These families are divided into 4 functional groups (Hibino et al., 2010): classical Kir channels (Kir2.x), G protein-gated Kir channels (Kir3.x), ATP-sensitive K^+ channels (Kir6.x) and K^+-transport channels (Kir1.x, Kir4.x, Kir5.x and Kir7.x). In the pancreatic β cell, the predominant Kir channel is the ATP-sensitive K^+ channel.

2.1.1.1 ATP-sensitive K^+ channels

The ATP-sensitive K^+ channel (K_{ATP}) is essential in glucose-induced insulin secretion in the pancreatic β-cell. In the absence of glucose, the K_{ATP} channel is opened and permeates K^+ outside the cell, maintaining the membrane potential at −70mV. When blood glucose concentration increases, it enters the β-cell and is metabolized in the mitochondria. Consequently, ATP is generated, which blocks the K_{ATP} channel. The closure of the channel causes an increase in positive charges inside the cell, which favors the depolarization of the cell, an increase in Ca^{2+} influx through voltage-gated Ca^{2+} channels and insulin secretion (Ashcroft and Rorsman, 2013; Rorsman and Ashcroft, 2018).

The K_{ATP} channel is a hetero-octameric complex made up of four Kir6.2 subunits (encoded by the KCNJ11 gene, on chromosome 7) and four regulatory subunits of the sulfonylurea receptor SUR1 (encoded by the ABCC8 gene, on chromosome 11) (Ashcroft and Rorsman, 2013, Rorsman and Ashcroft, 2018). SUR1 belongs to subfamily C of the family of proteins called ATP-binding cassettes or ABC proteins. This protein is composed of two domains (TMD1 and TMD2) with 6 transmembrane segments each harboring two folds involved in the binding and hydrolysis of adenine nucleotides (NBF1 and NBF2). SUR1 also contains a N-terminal domain with 5 transmembrane segments (TMD0) whose function is related to the regulation of the traffic of Kir6.2 subunits to the plasma membrane and the gating of the channel (Campbell et al., 2003; Nichols, 2006).

The SUR1 subunits surround the Kir6.2 subunits, which form the central structure of the channel and contain the channel pore. The N-terminal domain of each SUR1 subunit is anchored to the first transmembrane segment of each Kir6.2 subunit. The metabolic regulation of this channel is mediated by changes in the intracellular concentrations of ATP, Mg-ADP and PIP_2. Inhibition of the K_{ATP} channel occurs by binding of ATP to Kir6.2, while activation of the channel occurs by binding of Mg-ADP or PIP_2 to SUR1 (Rorsman and Ashcroft, 2018).

The opening kinetics of the K_{ATP} channel are complex. There are several studies that point to the existence of two independent gates in the selective entry pathway, one of fast voltage-dependent kinetics, but ligand-independent that is found in the region that confers K^+ selectivity and gives rise to the opening and closing events within a burst (Proks et al., 2001); and another one of slow kinetics that is regulated by the binding of adenine nucleotides and PIP_2, which gives rise to the events between bursts (Drain et al., 2004; Enkvetchakul et al., 2000). Considering only the ligand-dependent mechanism, it is known that binding to ATP (in the presence or absence of Mg^{2+}) shifts the conformational equilibrium toward the closed state, while PIP_2 shifts it toward the opened state, both through its binding to the four specific sites that are formed at the interface between the cytoplasmic tails of each Kir6.2 subunit (Enkvetchakul and Nichols, 2003).

There are a wide variety of pharmacological agents that can activate or inhibit the K_{ATP} channel by binding to the SUR1 subunit. Some of the inhibitory agents are sulfonylureas, such as chlorpropamide, tolbutamide or glibenclamide, which are used to treat patients with type 2 diabetes. On the other hand, activating agents include pinacidil, nicorandil and diazoxide. These substances are known as K^+ channel openers (KCOs). KCOs can be used for the treatment of hypoglycaemic and/or hyperinsulinemic states not associated with defects in the K_{ATP} channel (Ashcroft and Rorsman, 2013; Rorsman and Ashcroft, 2018).

2.1.2 Voltage-gated K^+ channels

Generally, voltage-gated K^+ channels (K_V) open when the membrane is depolarized and close when the membrane is hyperpolarized (Kuang et al., 2015). In the β-cell, K_V channels are responsible for the repolarization of the action potential (Rorsman and Ashcroft, 2018). These currents were described in 1986 in murine models (Rorsman and Trube, 1986) and later in humans, in 1991 (Kelly et al., 1991). These channels are composed of 4 α-subunits that conform a tetramer. Each of these subunits contains 6 transmembrane segments with α-helix structure. The S1–S4 segments act as a voltage sensor, due to the amino acid positive charges that form their structure (arginine and lysine). The S5-S6 segments, joined by a loop (P), form the pore-forming domain of the channel. The P-loop, which penetrates the plasma membrane without passing through it, contributes to the permeability of the pore. The ion selectivity filter consists of a highly conserved sequence of amino acids (TVGYG) that is present in other K^+ channels. Typically, α-subunits are associated with auxiliary β-subunits that modulate the channel's activity (Kuang et al., 2015; MacDonald and Wheeler, 2003).

K_V channels are divided into 12 families (K_V1–K_V12). Depending on their electrophysiological properties, there are 2 types of K_V channels: low-voltage-activated (LVA) K^+ channels and high-voltage-activated (HVA) K^+ channels. LVA channels are activated at potentials between −70 and −40 mV. This group includes members of the K_V1, K_V4, K_V7 families. Moreover, the current evoked by HVA channels is activated at potentials between −30 and −10 mV. In this group, the K_V2 and K_V3 channels are remarkable (Johnston et al., 2010).

Based on the activation and inactivation kinetics of K_V channels, delayed rectifying K^+ channels and A-type current K^+ channels can be distinguished (MacDonald and Wheeler, 2003). Delayed rectifying K^+ channels are activated with a small delay after the application of a depolarizing pulse and once activated, their inactivation kinetics are very slow. In contrast, A-type current K^+ channels have very fast activation and inactivation kinetics. In both mouse and human β-cells, delayed rectifying K^+ channels have higher expression than A-type current K^+ channels. $K_V2.1$ and $K_V2.2$ are predominantly expressed in mice and humans, respectively. Both channels are activated at a potential of −20 mV and are sensitive to tetraethylammonium (TEA) and stromatoxin. A-type currents mediated by $K_V3.4$ can also be detected in insulin secreting cells.

K_V channel antagonists act as insulinotropic agents, opening the possibility of developing drugs that block the channel and reverse hypoglycaemia situations.

The use of TEA, a wide-spectrum antagonist of K^+ channels, abolish the K^+ current in the β-cell, increasing amplitude and prolonging the duration of the action potential. As a result, intracellular Ca^{2+} availability increases and insulin secretion is increased. This effect only occurs in the presence of glucose or when cell depolarization occurs. K_V channels are closed at rest and TEA is unable to modify the resting membrane potential of the cell.

Therefore, inhibition of K_V channels with TEA in the absence of glucose would have no effect on the channel and, in turn, would have no effect on insulin secretion (MacDonald and Wheeler, 2003).

There are other antagonists that inhibit K_V channels and alter their function (MacDonald and Wheeler, 2003). Stromatoxin, a K_V2 channel inhibitor, has the same effect as TEA. The specific $K_V2.1$ channel antagonist C-1 is capable of blocking approximately 80% of the K^+ current in mouse β-cell. This causes an increase in insulin release in both the first and second phases of secretion. Hanatoxin, an antagonist of voltage-gated potassium channels $K_v2.1$ and $K_v4.2$, increases Ca^{2+} oscillations and insulin secretion in mice.

The use of these antagonists shows that the voltage-gated K^+ current mediated by K_V2 channels plays a key role in the repolarization of the action potential and, therefore, in regulating glucose-induced insulin secretion in pancreatic β-cell in mice and humans (MacDonald and Wheeler, 2003).

Alterations in the function of K_V channels could indirectly contribute to the onset of diabetes. Increased activity of K_V channels could alter insulin secretion due to an early repolarization of the action potential (MacDonald and Wheeler, 2003). In contrast, a decrease in the activity of K_V channels could have a hyperexcitatory effect on the β-cell and increase intracellular Ca^{2+} levels, causing cell apoptosis by the activation of the caspases' pathway (Chandra et al., 2001).

2.1.3 Ca^{2+}-activated K^+ channels

The repolarization phase of the action potential in the β-cell is not only represented by K_V channels. There is another K^+ current that is involved in the repolarization of the membrane: the current evoked by Ca^{2+}-activated K^+ channels (K_{Ca}). K_{Ca} channels are divided into 5 families (KCa1-KCa5) (Berkefeld et al., 2010). There are three types of K_{Ca} channels that are classified according to their conductance: low-conductance Ca^{2+}-activated K^+ channels (SK), which include $K_{Ca}2.1$, $K_{Ca}2.2$, $K_{Ca}2.3$ channels (Stocker, 2004); intermediate conductance Ca^{2+}-activated K^+ channels (IK), which include $K_{Ca}3.1$ channels; and high conductance Ca^{2+}-activated K^+ channels (BK or maxi K), which include $K_{Ca}1.1$ channels (Magleby, 2003). SK and IK channels are voltage-independent, while BK channels can be activated by changes in membrane potential and increases in intracellular Ca^{2+} concentration. Of the three types, BK channels are the most abundant in the β-cell, in both mouse and human, with higher expression levels in humans.

BK channels, like K_V channels, have a tetrameric structure composed of 4 subunits. However, these subunits are made up of 7 transmembrane segments, instead of 6. The S1–S4 segments act as a voltage sensor and the S5–S6 segments form the pore-forming domain of the channel. The additional segment (S0), located before segment S1, causes the N-terminal domain to be in the extracellular part. The C-terminal domain joins two RCK domains (RCK1 and RCK2) that have the regions responsible for channel's Ca^{2+} sensitivity. Each subunit has an auxiliary subunit that contributes to the modulation of the channel's activity (Berkefeld et al., 2010; Magleby, 2003).

BK channels are inactive at the resting potential of the cell. The K^+ current can be activated when the membrane potential reaches 20 mV and when the intracellular Ca^{2+} concentration reaches values around 10 μM (Rorsman and Ashcroft, 2018). Ca^{2+} joins the C-terminal domain of the channel and allows it to be opened. This channel can be blocked with TEA or iberiotoxin. The blocking of BK current increases the amplitude and the duration of the action potential and insulin secretion in mouse and human β-cell. Activation of this channel may also be altered by decreases in intracellular pH (Cook et al., 1984).

Both K_V and K_{Ca} channels play an important role in glucose-induced insulin secretion in the β-cell. Both currents participate in the repolarization of the β-cell, limiting the entry of Ca^{2+} into the β-cell and decreasing insulin secretion. However, 80–85% of the total K^+ current is represented by K_V channels, while the remaining 10–15% corresponds to the current produced by K_{Ca} channels (MacDonald and Wheeler, 2003).

2.2 Calcium channels

The ability of calcium (Ca^{2+}) ions to alter local electrostatic fields and protein conformations is a key element in cellular signal transduction. The intracellular Ca^{2+} concentration ($[Ca^{2+}]_i$) of pancreatic beta cells increase to trigger glucose-induced insulin release (Wollheim and Sharp, 1981). This rise in $[Ca^{2+}]_i$ is mainly due to influx of extracellular Ca^{2+}, with intracellular Ca^{2+} stores contributing only marginally. This influx of extracellular Ca^{2+} in the cytoplasm of β-cells occurs primary through voltage-gated Ca^{2+} channels. The opening of these channels plays a crucial role in the upstroke of the β-cell action potential (Rorsman et al., 2011).

Voltage-gated Ca^{2+} channels are composed of a pore-forming α_1-subunit and two auxiliary subunits that influence protein function and intracellular distribution (Dolphin, 2016). Mouse β-cells express L-type Ca^{2+} channels (Cacna1c and Cacna1d), contributing to 50% of the whole

Ca^{2+} current, and smaller amounts of N-type (Cacna1b), P/Q-type (Cacna1a) and R-type (Cacna1e) Ca^{2+} channels. There is no experimental evidence of low-threshold T-type Ca^{2+} currents in mouse β-cells. The molecular composition of Ca^{2+} currents in the human β-cell differs from that of the mouse. Human β-cells express L-type (CACNA1C and CACNA1D), P/Q-type (CACNA1A), and T-type (CACNA1H) Ca^{2+} channels, with N-type (CACNA1B) Ca^{2+} channels expressed at lower levels than in mouse β-cells, and R-type (CACNA1E) Ca^{2+} channels being undetectable (Rorsman and Ashcroft, 2018). These differences in the expression of subtypes of Ca^{2+} channels between mouse and human β-cells reinforces the concept that the results obtained in mice are not always directly extrapolated to humans.

Biophysical measurement of exocytosis (using patch-clamp capacitance measurements) and secretion studies on β-cell-specific knockouts of L-type Ca^{2+} channel subunits suggest that *Cacna1c* ($Ca_V1.2$) is important for the first phase of insulin secretion (Schulla et al., 2003). R-type Ca^{2+} channels (*Cacna1e*) have been proposed to underlie the second phase (sustained) of insulin secretion in mice (Jing et al., 2005).

The biophysical properties of voltage-gated Ca^{2+} currents in the mouse β-cell permit that membrane depolarization to −40 mV or above activates Ca^{2+} currents (Bokvist et al., 1991; Trube and Rorsman, 1986) with maximal activation at 0 mV and with current peak amplitude of 160 pA when measured at a physiological extracellular Ca^{2+} concentration of 2.6 mM (Villar-Pazos et al., 2018). At positive voltages, the current amplitude is smaller, induced by the decreased electrochemical gradient for Ca^{2+} entry. The Ca^{2+} current magnitude may be reduced by islet dissociation and subsequent culture of single β-cells. During membrane depolarization, pancreatic β-cells Ca^{2+} currents undergo both Ca^{2+} and voltage-dependent inactivation (Plant, 1988). The rapid inactivation of the Ca^{2+} current during depolarization to 10 mV suggests that the current may undergo significant inactivation during a single action potential, which typically peaks at 10 mV and lasts 30 ms (Hoppa et al., 2009).

The recovery from rapid inactivation has a time constant of 140 ms. With an action potential frequency of 5 Hz, the interval between successive action potentials is 170 ms. This is insufficient for the Ca^{2+} current to recover fully and leads to progressive inactivation (Kinard and Satin, 1996). Single Ca^{2+} channel currents have been recorded from cell-attached patches in the presence of the L-type Ca^{2+} channel activator BAY K 8644, hence they are likely to be L-type Ca^{2+} channels. With 100 mM Ba^{2+} as the charge carrier, the single-channel conductance is 24 pS (Rorsman et al., 1988), while at 10 mM extracellular Ca^{2+} the conductance is 6 pS and the single-channel current is 0.15 pA (Smith et al., 1993). The number of voltage-gated Ca^{2+} channels per β-cell has been estimated as ~500 by nonstationary fluctuation analysis of whole-cell Ca^{2+} currents in isolated β-cells in culture (Barg et al., 2001).

The T-type Ca^{2+} channel in human β-cells activates at much more negative membrane potentials than high-voltage-activated Ca^{2+} channel and, hence, it is referred to as low-voltage-activated Ca^{2+} channel. T-type Ca^{2+} channels give rise to a transient, rapidly inactivating current that can be detected at membrane potentials as negative as −60 mV with a whole-cell current maximal at −40 to −30 mV. T-type Ca^{2+} channels undergo voltage-dependent inactivation. At −50 mV, the current inactivates with a time constant of 40 ms but becomes faster at more depolarized membrane potentials. Probably, T-type Ca^{2+} channels play a role in pacemaking, facilitating initiation of action potential firing, as it happens in other excitable cells (Perez-Reyes, 2003). T-type Ca^{2+} channels expressed in human β-cells are blocked by micromolar concentrations of zinc (Traboulsie et al., 2007). Given the high intragranular concentration of Zn^{2+} in β-cells (30 mM) it seems possible that Zn^{2+} released during exocytosis may exert an inhibitory autocrine effect on electrical activity and insulin secretion in human β-cells (Braun et al., 2012; Khan et al., 2014). At potentials positive to about −40 mV, other Ca^{2+} channels, high-voltage-activated, contribute to the whole-cell Ca^{2+} current in human cells as it happens in mice, although with different subtype composition.

The Ca^{2+} channel activity in pancreatic β-cells is regulated by glucose metabolism. In mouse and rat β-cells, glucose increases the whole-cell Ca^{2+} current by increasing Ca^{2+} channel activity (Kato et al., 1996; Smith et al., 1989). Furthermore, the glucose metabolite glyceraldehyde increases Ca^{2+} channel activity by a protein kinase C-dependent mechanism in RINm5F cells (Velasco et al., 1988). These effects are reversed by mannoheptulose (which blocks glucose phosphorylation) and inhibitors of mitochondrial ATP production (Smith et al., 1989). The effects of glucose on Ca^{2+} channels are not always observed, probably reflecting metabolic heterogeneity between β-cells. The underlying mechanism of glucose-induced Ca^{2+} current potentiation remains to be elucidated. The Ca^{2+} channel activity in pancreatic β-cells is also regulated by protein phosphorylation. L-type Ca^{2+} channels are regulated by protein phosphorylation in mouse β-cells. In fact, forskolin, which elevates cAMP, increases the Ca^{2+} current by 80%

(Ammala et al., 1993), suggesting the channel is activated by protein kinase A. Furthermore, the PKC activator PMA also increases the β-cell Ca^{2+} current (Ammala et al., 1994). In addition, it has been reported that glucose-induced activation of the AMPK-related kinase salt-inducible kinase 2 leads to increased Ca^{2+} entry and insulin secretion in mouse β-cells (Sakamaki et al., 2014).

3. Stimulus-secretion coupling in pancreatic β-cells

As we showed before, pancreatic β-cells are excitable cells, and they are composed of multiple ion channels. Dean and Matthews were the first to record electrical activity in pancreatic β-cells (Dean and Matthews, 1968). Later, many laboratories have demonstrated that this cell type responds to multiple stimulators that promotes insulin secretion. However, in 1984, Ashcroft and colleges demonstrated at the first time that glucose metabolism was the responsible on the electrical activity of the pancreatic β-cells (Ashcroft et al., 1984). At low glucose concentration, K_{ATP} channels were active and pancreatic β-cells remained in a hyperpolarized state. High glucose concentration (>5 mM glucose) closed K_{ATP} channels, leading to depolarization of β-cells resulting in calcium influx and insulin secretion. In this section, we will describe two models of glucose stimulus-secretion coupling (SCC) in pancreatic β-cells and how different metabolites can regulate insulin secretion. Moreover, we will explain paracrine interactions with neighboring cells (pancreatic α- and δ-cells).

To date, insulin is the unique hormone that is responsible to lower high glucose levels in blood. For this reason, failures in pancreatic β-cells can produce severe disorders in glucose homeostasis, as Type 1 and Type 2 Diabetes Mellitus, as we will describe in last section.

Secretagogues that initiate electrical activity and promote insulin secretion in pancreatic β-cells can be classified into three groups: on one hand, those that are considered "initiators." These include glucose, leucine, glyceraldehyde, ketoisocaproate or antidiabetic drugs as sulfonylureas. These substances are capable to induce insulin secretion on their own. On the other hand, many other nutrients are capable to enhanced insulin secretion in presence of sub-stimulatory concentration of "initiators." They are referred as "potentiators." Hormones as glucagon, oestrogens, vasoactive intestinal peptide (VIP) or neurotransmitters as acetylcholine are included in this group. In this regard, recently it has been demonstrated that exposure pancreatic β-cells to carbamoylcholine (a muscarinic agonist receptor) increases the size of glucose-induced depolarizing waves (slow wave activity, see below) and this potentiation requires the simultaneous action of carbachol and glucose (Sanchez-Andres et al., 2020). A third group is classified as "inhibitors" and they decreased or inhibit insulin secretion. "Inhibitors" decrease intracellular calcium concentration (via membrane repolarization) or reduce exocytosis (Sharp, 1996). Hormones as somatostatin secreted by pancreatic δ-cells, or galanin and calcitonin gen related peptide (CGPR) are contained in this third group (Ashcroft and Rorsman, 1989; Henquin, 2004; Rorsman and Ashcroft, 2018).

Pancreatic β-cells act as a fuel sensor and the main initiator of insulin secretion is glucose. Glucose is transported into β-cells by facilitated diffusion thought glucose transporters that are present in plasma membrane. Glucose metabolism is initiated by glucokinase (Hexokinase IV) and after that mitochondrial metabolism increase ATP/ADP ratio, that it is crucial for glucose-stimulated electrical activity and insulin secretion. It has been demonstrated that there is a good correlation between this ratio and insulin secretion (Detimary et al., 1996, 1998). This phenomenon is essential to close K_{ATP} channels, calcium influx and insulin secretion. As we will describe later, calcium influx is important to insulin secretion in mice and men. However, there are some examples where insulin secretion does not correlate with increases in intracellular calcium. For example, the presence of some "potentiators" as acetylcholine with low glucose concentration causes a much greater increase in insulin secretion than glucose alone with any change in intracellular calcium concentration (Ashcroft et al., 1994; Berggren et al., 1993; Gembal et al., 1992; Rorsman et al., 1994).

3.1 Stimulus-secretion coupling in mouse pancreatic β-cell

Stimulus-secretion coupling has been well described in the mouse pancreatic β-cell. This animal model has been widely used to describe this mechanism because islets of Langerhans are easy to isolate compared to human islets. In both mice and humans, K_{ATP} channel is the main channel that links increases in blood glucose concentration to insulin secretion. This channel is an octameric complex of four-pore forming subunits Kir6.2 (Kcnj11) and four regulatory subunits SUR1 (abcc8), as we discussed above (Aguilar-Bryan et al., 1998; Sakura et al., 1995). At low glucose concentrations (<6 mM glucose) the K_{ATP} channel is in an open state, maintaining this cell type close to the K^+ equilibrium potential (−70 mV). In fact, it is the only channel that remains active in the silent phase. However, stimulatory glucose concentrations (6–20 mM Glucose) close K_{ATP} channel, which depolarize β-cell between −60 and − 50 mV. This depolarization causes the opening of the L-type voltage-dependent Ca^{2+} channels (Cacna1c and Cacna1d), increasing intracellular calcium concentration and triggering exocytosis of insulin granules (Fig. 3).

Other voltage-gated Ca2+ channels may contribute to Ca^{2+} influx (see above). Increasing extracellular glucose concentration from non-stimulatory concentrations (< 6 mM glucose) to high concentrations (6–20 mM) promotes a slow depolarization which duration is glucose-dependent. This electrical activity consists of two phases:

1. Initial phase: In this first step, glucose metabolism hyperpolarizes mitochondrial membrane potential and thus increase ATP generation closing

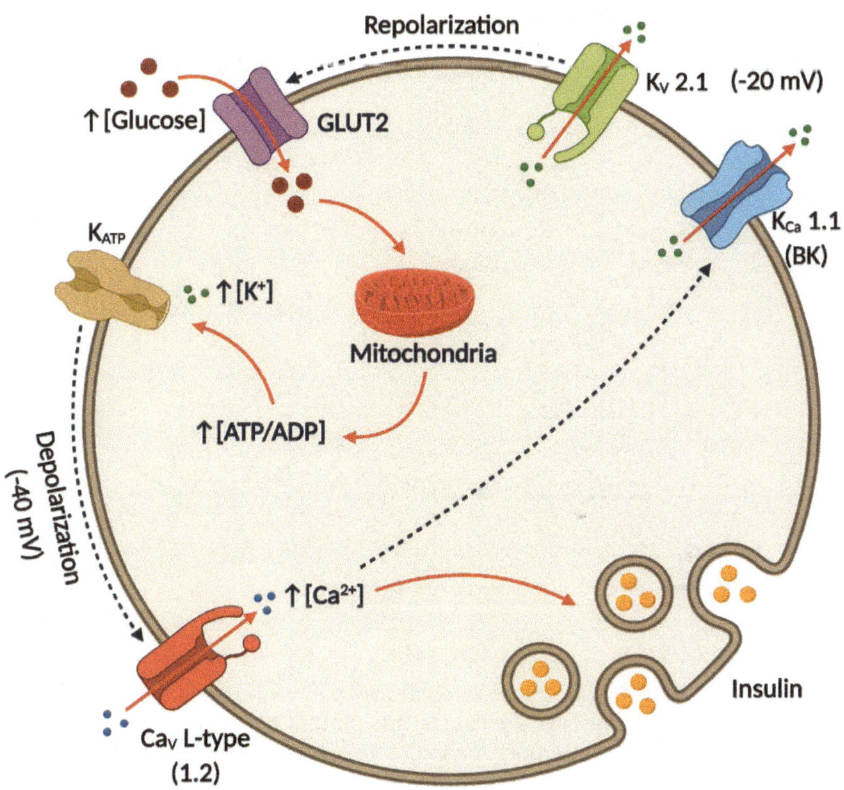

Fig. 3 Stimulus-secretion coupling in mouse pancreatic β-cell. Schematic representation of the ion channels and cellular processes involved in the initiation and generation of the characteristic electrical activity of the mouse β-cell.

of K_{ATP} channels and promoting membrane depolarization. In this phase a single burst of action potentials is generated for 20–60 s. Afterwards cell repolarizes again to the resting membrane potential, activating the second phase, where membrane potential (V_m) oscillates between two states: active and silent phase.

2. Slow wave activity: The V_m presents a characteristic pattern of slow oscillations between two states, a silent state (close to resting membrane potential) and a depolarized plateau (between −60 and −50 mV) where action potentials are generated (Fig. 4).

Oscillations in membrane potential in presence of stimulatory glucose concentration are critical to insulin secretion. The consensus model describes

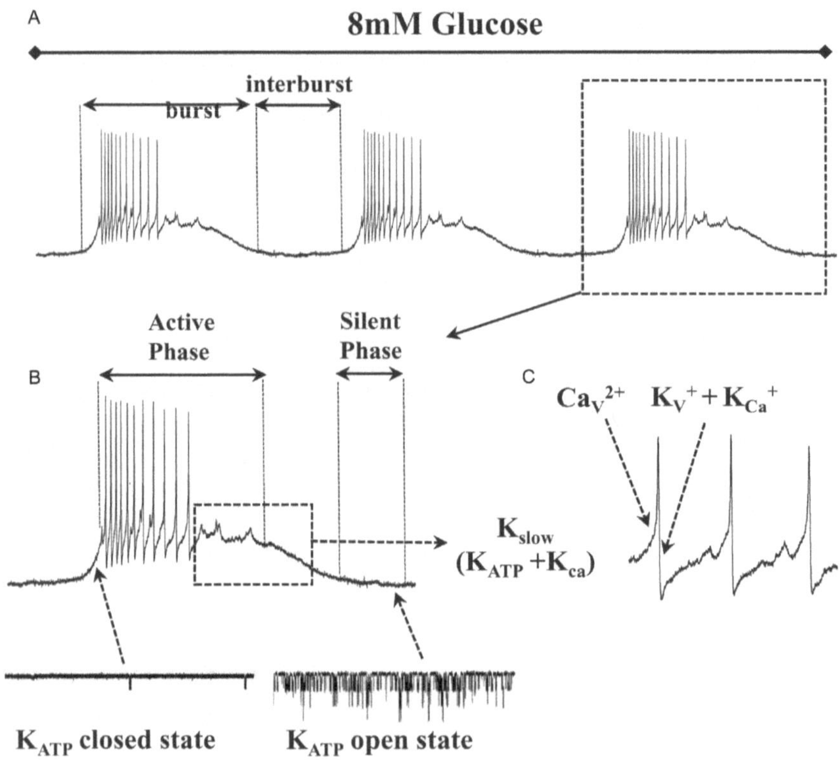

Fig. 4 Electrical activity in freshly isolated mouse pancreatic β-cell. (A) In presence of 8 mM glucose the V_m presents a characteristic pattern of slow oscillations between two states, a silent state (interburst period) and a depolarized plateau (burst) where action potentials are generated. (B) Closure of K_{ATP} channels (B_2) promotes membrane depolarization and it generates action potentials. Activation of K_{slow} current repolarizes membrane potential. (C) Different contribution of ion channels in action potential generation.

that the increase in $[Ca^{2+}]_i$ in the initial phase hyperpolarizes the mitochondrial membrane potential, reducing ATP production and opening K_{ATP} channels. At the same time, Ca^{2+} influx activates K_{Ca} channels. Therefore, pancreatic β-cell repolarized to resting membrane potential reducing the intracellular calcium concentration and activating again the mitochondrial ATP production. Finally, a new cycle of oscillation is generated. This model implies that burst repolarization is due to the activation of the K_{slow} current, composed of K_{ATP} and K_{Ca} (Fig. 4) (Gopel et al., 1999; Kanno et al., 2002b; Krippeit-Drews et al., 2000; Rolland et al., 2002).

However, it has been proposed an additional mechanism that is independent on K_{ATP} channel activity. Ravier et al. demonstrated that pancreatic β-cells lacking pore-forming subunit Kir6.2 exhibited an oscillation pattern in presence of stimulatory glucose concentration (Ravier et al., 2009). In this case, the K_{Ca} component of K_{slow} may gain importance in repolarizing membrane potential. (Gopel et al., 1999). Another possibility is an activation of the Na^+-K^+-ATPase by glucose or metabolism inducing a hyperpolarizing current that repolarized pancreatic β-cells.

Increasing glucose concentration generate an increase in the burst duration and the period between them decreases (interburst period). Firing become continuous at glucose concentrations above 20 mM (Drews et al., 2010; Henquin and Meissner, 1984).

This oscillatory pattern recorded in islets of Langerhans in vitro has also been demonstrated in vivo under physiological conditions. In both studies, it has been observed that the action potentials are calcium dependent. Moreover, it has been demonstrated that in vivo pancreatic β-cells are more sensitivity to different glucose concentrations. This effect remarks the importance of the "potentiators" in stimulus-secretion coupling mechanisms (Sanchez-Andres et al., 1995).

It is important to highlight the following aspects for glucose concentration beyond 8 mM:

(a) Initial phase is usually longer than the active phases in slow wave activity.

(b) In slow wave activity, the resting membrane potential is not maintained; a slow depolarization takes place until reaching depolarized plateau where action potentials are generated (Meissner, 1976; Meissner et al., 1978).

(c) Action potentials into the active phase are not homogeneous; several populations can be distinguished with different amplitudes (Soria and Ferrer, 1986).

3.2 Stimulus-secretion coupling in human pancreatic β-cell

As in mouse, the human body uses the pancreatic islets to maintain the systemic regulation of metabolism. They secrete mainly two hormones: insulin and glucagon. Both hormones have opposite effects on the glucose concentration. On one hand, insulin is secreted at high glucose concentration. On the other hand, in presence of low glucose concentration glucagon is secreted by pancreatic α-cells. As in mice, several voltage-dependent and independent channels can be found in human pancreatic β-cell membrane (Ashcroft and Rorsman, 1989; Gromada et al., 2007).

Studies using human pancreatic β-cells are very scarce. Thus, stable human cell line Endoc-βH1 (pancreatic β-cell), is a necessary tool for detailed studies of the biology of human β-cells (Andersson et al., 2015). However, today studies on human cells have not been as elaborate as the mouse β-cell. However, everything seems to indicate that the "stimulus-secretion coupling model" described before in murine model is very similar and it applies to human β-cells.

First, the human pancreatic β-cell is capable to detect an extracellular glucose elevation as a stimulus. Glucose is transported into the cell by the GLUT1 transporter (SLC2A1), which constitutes the primary glucose transporter in human β-cells (Heimberg et al., 1995; McCulloch et al., 2011). This transporter has a much lower K_m than GLUT2 (SLCC2A2), which could explain why insulin secretion is initiated at low glucose concentrations in human islets, as opposed to mouse islets. Other results obtained from RNA_{seq} suggest that human β-cells also express the Na-dependent glucose transporters SGLT1, SLC5A1 and SLC5A2 but at levels considerably lower than those found for GLUT1 (Blodgett et al., 2015).

Once glucose is internalized into the cell, its metabolism begins and it is immediately phosphorylated to glucose-6-phosphate by the enzyme glucokinase (GCK) also known as hexokinase IV (Bonner et al., 2015), which also has a low affinity for glucose (4 mM) in human β-cells (Doliba et al., 2012). Some results indicate that glucose phosphorylation is the rate-limiting step in insulin secretion, and loss-of-function mutations in GCK impair glucose-induced insulin secretion (GIIS) (Gloyn et al., 2005). Similarly, the critical role of GCK in GIIS is also demonstrated by the effects of enzyme inhibitors and activators, such is the case of mannoheptulose, a GCK inhibitor, strongly suppresses GIIS (Randle, 1993) and, in contrast, the glucokinase activators stimulate GIIS (Johnson et al., 2007). One of these activators, pyragliatin, restores GIIS in islets isolated from T2DM donors to levels comparable to those observed in islets from non-diabetic donors (Doliba et al., 2012).

Most of all the glucose that enters the cell, passes directly to the Krebs cycle (Prentki et al., 2013; Schuit et al., 1997), being redistributed in the synthesis of glycogen (influencing the uptake of new glucose), the conversion of glucose into lipids, the synthesis of amino acids and the sorbitol and pentose phosphate pathways, the latter two being of less importance. Instead, a relatively small amount is metabolized to lactate by the enzyme lactate dehydrogenase (LDH1) (Thorrez et al., 2011). In this case, the scarcity of both genes is necessary to prevent the activation of insulin secretion during exercise, due to the circulation of pyruvate or lactate. There are studies that reveal that increased expression of MCT1 in human cells can lead to hypoglycaemia induced by exercise (Otonkoski et al., 2003).

In parallel, ATP production occurs in the mitochondria, and this facilitated by efficient transfer of glucose-derived NADH from the cytosol, via glycerol-3 phosphate and malate/aspartate transports. This glucose metabolism produces ATP, and it is the responsible for closing the K_{ATP} channels, resulting influx of Ca^{2+} that promotes the exocytosis of the insulin granules.

It is important to bear in mind that mitochondrial metabolism, which includes the processes of anaplerosis and cataplerosis, is essential for a better activity of the Krebs cycle and the production of ATP, as well as for the generation of other additional factors that participate in the response of insulin secretion generated by glucose stimuli (Islam, 2020). Anaplerosis reactions are chemical processes that form intermediates of a metabolic pathway. When certain intermediates of the Krebs cycle, such as citrate, are elevated, it improves the catalytic activity of the cycle, and in addition this intermediate participates in additional metabolic pathways that lead to the production of other components used in the cytoplasm (for example, malonyl-CoA, glutamate, NADPH). This process in turn is complemented by cataplerosis, which refers to the exit of intermediate products of the Krebs cycle from the mitochondrial matrix to the cytoplasm. In this step, there are several transporters located in the inner mitochondrial membrane that facilitate the trans-mitochondrial movement of the intermediates of the Krebs cycle, such as di- and tricarboxylate. On the other hand, some of the intermediate products of the Krebs cycle (citrate, isocitrate, α-ketoglutarate and malate) participate in the pyruvate recycling processes promoting the generation of cytoplasmic NADPH (Kleiber, 1947).

Several studies have been carried out on the profile of metabolites at basal and stimulating glucose levels in the metabolic responses stimulated by glucose. The profile of these metabolites is very similar to those generated in murine docking (Andersson et al., 2015). Although there are small differences, especially in the basal levels, these could be species-specific, since the metabolic rate decreases with increasing body size (Kleiber, 1947; Prentki et al., 2013).

The differences in metabolic responses could be due to alterations in mitochondrial metabolism or the coupling of cytosolic and mitochondrial metabolism. In the case of studies in human pancreatic β-cells, they indicate a difference in mitochondrial metabolic flexibility since there is less relative stimulation of pyruvate-induced respiration. Therefore, in both glycolysis and respiration the human β-cell has a less response (Andersson et al., 2015). Although the glucose-driven respiration process is associated with a lower relative leakage of protons in human pancreatic β-cells, the overall coupling efficiency is the same when compared to the model described in mice (Martin and Soria, 1996; Misler et al., 1992; Rolland et al., 2002).

There are few ion channels recordings on human pancreatic β-cells because obtaining and maintaining human islets is very difficult. As in mouse islets, human β-cells are hyperpolarized (−70 mV) and electrically silent at low glucose levels (Gilon et al., 1993; Misler et al., 1992). In human islets insulin secretion is stimulated at 3 mM glucose. On one hand, ATP production decreases K_{ATP} channel activity that depolarizes membrane potential. On the other hand, in human pancreatic β-cells low-threshold voltage dependent Ca^{2+} channels (T-Type Ca^{2+} channels) are expressed. The combination of two effects initiates electrical activity and it promotes insulin secretion. The presence of low-threshold voltage dependent Ca^{2+} channels are the responsible that the action potentials of human cells activate from a more negative membrane potential than in mouse cells (−60 mV instead of −50 to −40 mV) reaching a maximum at potentials greater than 0 mV (Rorsman and Braun, 2013). Upstroke of action potential in human pancreatic β-cells is due to opening of TTX-sensitive sodium channels. At peak of action potential, activation of P/Q-type Ca^{2+} channels trigger insulin secretion. Finally, elevation of $[Ca^{2+}]_i$ activates BK channels (SK3 channels) and action potential repolarizes (Fig. 5). Human pancreatic β-cells also present an oscillatory pattern on membrane potential, but it is not as well-defined as in mouse pancreatic β-cells.

High glucose concentration promotes that the amplitude of the action potential gradually decreases. This is because TTX-sensitive sodium channels and T-Type Ca^{2+} channels undergo progressive inactivated. A hallmark of electrical activity in human cells is that, while it starts rapidly, it is only

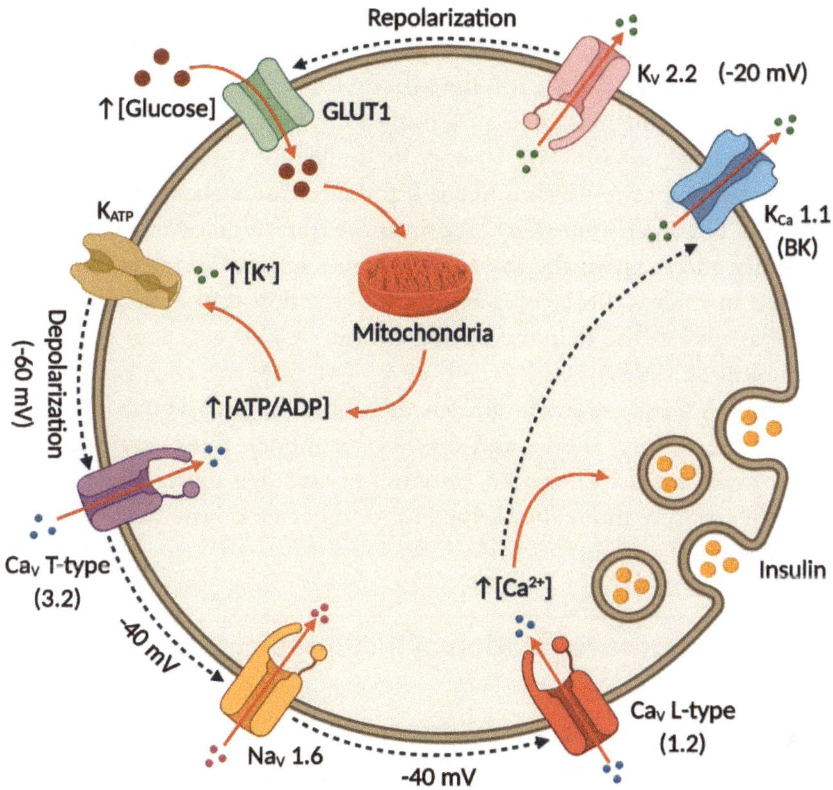

Fig. 5 Stimulus-secretion coupling in human pancreatic β-cell. Schematic representation of the ion channels and cellular processes involved in the initiation and generation of the characteristic electrical activity of the human β-cell.

slowly reversible upon returning to low glucose levels and takes up to 30 min after exposure to 22 mM glucose to reactivate sodium and calcium channels. As the cell begins to repolarize, the amplitude of the action potential also increases, and insulin begins to secrete after exposure to high concentrations of glucose (20 mM) (Henquin et al., 2015).

However, glucose-induced $[Ca^{2+}]_i$ oscillations are observed in human cells, and these are driven by oscillatory electrical activity. Furthermore, there is evidence that the electrical responses of the cell in human islets are less synchronized than in mouse islets (Cabrera et al., 2006). This suggests that the electrical coupling between individual human cells may be weaker, perhaps due to the different architectures of the mouse and human islets. Whereas the islets of mice contain a central nucleus of electrically coupled β-cells that functions as a syncytium; β-cells in human islets are organized into several subdomains, with β-cells in one domain being electrically isolated from the other cell types that make up the islet (Bosco et al., 2010; Cabrera et al., 2006). It is here in these subdomains where records of synchronized oscillations of $[Ca^{2+}]_i$ induced by glucose have been obtained (Quesada et al., 2006). The absence of oscillations in EndoC-βH1 cells indicates that more research is needed, to better understand the underlying processes. This decrease in Ca^{2+} supports the notion of lower NADH shuttle activity and explains the lower TCA cycle activity and insulin response observed in EndoC-βH1 cells. Thus, the activation of the malate-aspartate shuttle, as well as the dehydrogenases of the TCA cycle, depend on Ca^{2+} (Marmol et al., 2009).

Although there are small differences in the stimulus-secretion coupling between the murine model and the human model, the overall metabolic function was similar. Therefore, previous knowledge about β-cell metabolism in the murine model has helped us to understand how behavior can be in the human β-cell model.

4. Paracrine regulation of insulin secretion in pancreatic β-cells

It is important to highlight that rodent and human pancreatic β-cells are connected by connexin channels (Meda, 1996). This communication is relevant to regulate insulin secretion in response to different glucose concentration. In fact, single β-cells, which are not coupled by connexion channels, release low amounts of insulin (Meda et al., 1990). Recently it has been described that the coordinated response to insulin release depends on a privileged role for individual β-cell. Johnston et al. demonstrated the presence of hub cells with pacemaker characteristics and silencing these cells abolished coordinated islet responses to stimulatory glucose concentration (Johnston et al., 2016).

Another important mechanism that regulates insulin secretion is the paracrine interactions with α and δ-cells. Somatostatin is a hormone that is released by pancreatic δ-cells, that comprise about 5% of the cells of islet of Langerhans. Recently, Briant et al. described that β and δ-cells are electrically coupled via gap-junctions (Briant et al., 2018). Somatostatin is secreted in presence of high glucose concentration in a similar manner that pancreatic β-cells secrete insulin. The effects of somatostatin on insulin producing cells are well characterized. Somatostatin activates somatostatin receptor 3 (SSTR3) and opens G protein-activated inward rectifier K^+

channels (GIRK channels) which cause membrane repolarization decreasing insulin secretion. Moreover, somatostatin inhibits voltage-dependent calcium channels decreasing $[Ca^{2+}]_i$ and inhibiting insulin secretion (Rorsman and Huising, 2018). On the other hand, glucagon is secreted by pancreatic α-cells (comprise about 15–20% of the cells of Islet of Langerhans) at low glucose concentration and this hormone modulates insulin secretion. Classically, glucagon secretion stimulates insulin secretion (Samols et al., 1965). Svendsen et al. described that this is a direct effect mediated by GLP1 receptors, although with a lower potency compared with its interactions with the glucagon receptor (Svendsen et al., 2018).

In summary, and as we have described in previous sections, it is important to note that the electrical oscillatory pattern in the pancreatic β-cells depends on the different ion channels present in this cell type. Moreover, insulin secretion requires paracrine interaction with the different cell types that are present in islets of Langerhans and the neural control. In the last section, we will describe different insulin secretory disorders related to malfunction in ion channels present in pancreatic β-cells.

5. Ion channels and insulin secretion: Implications in Type 2 diabetes mellitus

K_{ATP} channels are the main ion channels that are present in human and mice pancreatic β-cells. These channels regulate insulin secretion. When blood glucose concentration increases, mitochondrial ATP production decreases K_{ATP} channel activity, depolarizing membrane potential and promoting insulin secretion. Consequently, these channels have a critical role regulating glucose-dependent insulin secretion. It is not surprising that malfunction or disruption of these channels induces to disease. Gain-of-function mutations in genes that encoded Kir6.2 or SUR1 cause neonatal diabetes mellitus (NDM) which is characterized by hyperglycaemia within the first months of life. NDM can be either transient or permanent. It has been estimated to occur in 1 over 20.000–500.000 live births (Ashcroft and Rorsman, 2013; Nansseu et al., 2016). Around 20% patients also have mental and motor delay, muscle hypotonia, and epilepsy (Ashcroft, 2005). In NDM the increased K_{ATP} channel activity prevents β-cells depolarization in presence of high glucose concentration and impairs insulin secretion. Patients suffering NDM were treated with insulin. Nowadays, sulfonylurea therapy, which closes K_{ATP} channel, is a more effective treatment to NDM (Ashcroft, 2010; Pearson et al., 2006). On the other hand, loss-of-function mutations in genes that encoded Kir6.2 or SUR1 results in congenital hyperinsulinism (HI). HI is characterized by inappropriate insulin secretion at low blood glucose concentration. HI can be diffuse (all pancreatic β-cells are affected) or focal (affected islets are in a region of the pancreas) (Ashcroft and Rorsman, 2013). The incidence is estimated at 1/50.000 live births, but it may be as 1/2.500 in countries with elevated consanguinity (Arnoux et al., 2011). HI is frequently treated by agonist of K_{ATP} channels, such diazoxide, but in some cases surgical pancreatectomy is required.

Moreover, in recent years it has been demonstrated that oestrogens have an important role regulating K_{ATP} channel activity in pancreatic β-cells. Soriano et al. demonstrated that 17β-oestradiol can regulate these channels through the estrogen receptor β (ERβ) and type A guanylate cyclase receptor increasing cGMP levels (Soriano et al., 2009, 2011). In synergy with glucose, application of 1 nM of 17β-oestradiol decreased K_{ATP} channel activity rapidly and promoted insulin secretion in mice pancreatic β-cells. Oestradiol has an important role in the adaptation of pancreatic β-cells to obesity. In obesity, adipose tissue is one of the most source of estrogen synthesis. Oestradiol secreted by adipocytes may be a signal for pancreatic β-cells to adapt to the insulin resistant produced by obesity via K_{ATP} channels.

Over last years, it has been demonstrated that chemical pollutants can predispose to suffer metabolic diseases. In this regard, endocrine disrupting chemicals (EDCs) can directly alter ion channel activity in pancreatic β-cells. EDCs are pollutants defined as "an exogenous chemical, or mixture of chemicals, that interferes with any aspect of hormone action" (Zoeller et al., 2012). One of the most EDCs used is bisphenol A (BPA). BPA is produced in large quantities of polycarbonate and epoxy resins and mimics oestradiol effects. For this reason, it is considered as a xenoestrogen. BPA exposed has been associated with diabetes. Soriano et al. showed that 1 nM BPA decreased K_{ATP} channel activity in mice and human pancreatic β-cells and impaired insulin secretion (Soriano et al., 2012). Moreover, recent paper has been demonstrated that BPA treatment can alter function and expression of several ion channels in pancreatic β-cells. Specifically, BPA decreased both sodium (Na_v 1.7) and potassium currents (K_v 2.1/2.2 channels and K_{Ca} 1.1 channels) via ERβ receptor. On the other hand, BPA increased burst duration and it altered action potential (Martinez-Pinna et al., 2019).

Moreover, other ion channels dysfunction can promote T2DM. Several studies indicate that the alteration of Ca^{2+} channels in pancreatic β-cells may contribute to diabetes risk in some patients. For example, the expression of CACNA1D is reduced in islets from donors with type 2 diabetes and a polymorphism (rs312480) in the gene, causing a reduced glucose-induced insulin secretion (Reinbothe et al., 2013). Pancreatic β-cells can secrete frizzled-related protein 4 (SFRP4) in response to interleukin1. SFRP4 has been shown to decrease β-cell L- and P/Q-type Ca^{2+} channel activity. Interestingly, SFRP expression is increased in type 2 diabetes patients, and can be detected years before the diagnosis and could be used as a predictor of later development of type 2 diabetes (Mahdi et al., 2012). A recent study shows that enhanced expression of rat β-cell T-type Ca^{2+} channels impairs insulin release and glucose homeostasis by means of initial excessive Ca^{2+} influx, subsequent activation of calcineurin, consequent dephosphorylation and nuclear retention of FoxO1, and eventual FoxO1-mediated down-regulation of β cell exocytotic proteins (Yu et al., 2020).

Finally, ethology of type 2 diabetes mellitus is multifactorial where both genes and environment have an important role. Understanding which ion channels are altered by mutations or by EDCs is a special interest to treatment of some metabolic disorders, as type 2 diabetes mellitus. Moreover, the lasts findings support that endocrine disruptors, as BPA, may be considered as a risk factor for metabolic diseases.

6. Concluding remarks

High glucose concentration promotes insulin secretion in mice and human pancreatic β-cells. In stimulus-secretion coupling ATP production induces closure of K_{ATP} channels and it depolarizes pancreatic β-cells. Therefore, voltage dependent calcium channels are activated increasing intracellular calcium concentration and promoting insulin secretion. Glucose regulation is mainly bihormonal mechanism, where insulin and glucagon secreted by pancreatic α-cells interact to counteract high or low glucose levels, respectively. However, and as we describe before, insulin is the unique hormone that is responsible to lower high glucose levels in blood. Therefore, failures in insulin secretion can promote metabolic disorders as type 2 diabetes mellitus (T2DM). Although it is a multifactorial disease, dysfunction in pancreatic β-cells are the main reason to develop T2DM. Here we have described briefly how gain or loss-of-function of K_{ATP} channels can affect to insulin secretion, promoting diabetes neonatal (NDM) or congenital hyperinsulinism (HI) respectively. Also, presence of hormones as oestrogens or endocrine disrupting chemicals (EDCs) can alter insulin secretion, predisposing to T2DM.

In the last decade, diabetes research has allowed the development of different treatments for this disease. Therapies can be personalized for each patient depending on the symptoms and severity of diabetes. In the early stages, changes in the patient's lifestyle, such as exercise and following a proper diet, can help in blood glucose control without requiring medication. When diet and exercise are not enough, drug treatment is necessary.

In patients who do not have altered insulin secretion, but their tissues show resistance to the action of this hormone, treatments with non-insulin drugs are chosen. There is a wide variety of drugs that increase insulin secretion and counteract resistance to its action in peripheral tissues. Among these drugs, the most widely used are insulin secretagogues, such as sulfonylureas and metiglinides, or biguanides, such as metformin. Insulin secretagogues act by increasing insulin secretion through binding to the sulfonylurea receptor (SUR) on the K_{ATP} channel of the pancreatic β-cell promoting insulin secretion. Within this group are drugs such as tolbutamide, chlorpropamide or glibenclamide. In the case of biguanides, such as metformin, they do not directly affect insulin secretion, they rather promote the body's response to the action of this hormone. To do this, they decrease glucose absorption in the intestine and reduce glucose production in the liver, stimulating glycolysis and inhibiting gluconeogenesis. Other drugs used include peroxisome proliferator-activated receptor agonists (PPARs), alpha-glucosidase inhibitors, amylin agonists, incretin mimetics (GLP-1 agonists and DPP-IV inhibitors) and SGLT2 antagonists/inhibitors (Padhi et al., 2020).

In advanced stages of the disease in which the β-cell is not capable of secreting insulin, the main treatment is the administration of insulin, which may or may not be combined with non-insulin drugs depending on the patient's profile. For example, in some patients with persistent hyperglycaemia dual therapy is required, where GLP-1 receptor agonist combined with metformin. In overweight patients triple therapy is used, combining metformin, GLP-1 receptor agonist and inhibitors of sodium-glucose transporter 2 (SGLT-2). It has been demonstrated that these therapies decrease postprandial hyperglycaemia (DeFronzo, 2017; Goncalves and Bell, 2018).

An aspect to consider when establishing a treatment is the side effects of medications and the possible complications that the disease can lead to. Patients must be able to recognize and treat possible hypoglycaemia that may be caused by using these medications. Hypertension is a major risk factor for the appearance of diabetes complications. It is also important to carry out ophthalmological controls to rule out vascular alterations in the eye that could lead to diabetic retinopathy, as well as to carry out periodic measurements of the albumin-creatinine ratio in the urine for the early detection of possible diabetic nephropathy (Padhi et al., 2020).

Another possible treatment for the most advanced stages of T2DM and for type 1 diabetes patients, who, due to the loss of β-cells, are not able to produce insulin, would be the generation of insulin-producing β-cells. This work has been encountering different difficulties, because of obtaining cells from healthy donors and the problem of avoiding immune rejection because they come from another person.

In recent years, another very promising alternative has been the generation of "own" β-cells from stem cells. Also, more specifically from pluripotent cells that are obtained by modifying other cell types. A possible solution for type 1 diabetes could be linked to the re-education of the immune system and promote "trans-regeneration" mechanisms to restore the population of insulin-secreting β-cells. This would facilitate the "dialogue" between the immune system and pancreatic islets, making it possible to achieve both objectives: successfully recovering immune tolerance and restoring a functional mass of pancreatic β-cells (Bruin et al., 2015).

An example of this is the work carried out by Trott et al. where human multipotent pancreatic cells achieve self-renewal, which in development are more proximal to the specialized cells of the adult pancreas, improving culture conditions (Trott et al., 2017). These cultured pancreatic progenitor cells manage to express key transcription factors such as PDX1 and SOX9, and exhibit transcriptomes closely related to their counterparts in vivo. Upon exposure to differentiation signals, these cells give rise to endocrine, acinar and ductal pancreatic lineages, indicating multi-lineage potency. Furthermore, the cells were able to generate cells with characteristics like β-cells, suggesting that they offer a convenient alternative to pluripotent cells as a source of adult cell types to model pancreatic development and diabetes.

Finally, in the last 30 years, development of new techniques has allowed us to better understand ionic mechanism in stimulus-secretion coupling in mice and human pancreatic β-cells (Rorsman and Ashcroft, 2018) and it helped to develop new therapies. However, certain aspects remain to be determined such the role on voltage-dependent sodium channels in mice pancreatic β-cells or why dispersed pancreatic β-cells are not able to generate an oscillatory pattern that is present in freshly isolated mice islets. Although diabetes is a disease that primarily affects pancreatic β-cell secretion, research on other islet cell types is also important for the future to understand their role in the etiology of T2DM.

References

Aguilar-Bryan, L., Clement, J.P.T., Gonzalez, G., Kunjilwar, K., Babenko, A., Bryan, J., 1998. Toward understanding the assembly and structure of KATP channels. Physiol. Rev. 78, 227–245.

Ammala, C., Ashcroft, F.M., Rorsman, P., 1993. Calcium-independent potentiation of insulin release by cyclic AMP in single beta-cells. Nature 363, 356–358.

Ammala, C., Eliasson, L., Bokvist, K., Berggren, P.O., Honkanen, R.E., Sjoholm, A., Rorsman, P., 1994. Activation of protein kinases and inhibition of protein phosphatases play a central role in the regulation of exocytosis in mouse pancreatic beta cells. Proc. Natl. Acad. Sci. U. S. A. 91, 4343–4347.

Andersson, L.E., Valtat, B., Bagge, A., Sharoyko, V.V., Nicholls, D.G., Ravassard, P., Scharfmann, R., Spegel, P., Mulder, H., 2015. Characterization of stimulus-secretion coupling in the human pancreatic EndoC-betaH1 beta cell line. PLoS One 10, e0120879.

Arnoux, J.B., Verkarre, V., Saint-Martin, C., Montravers, F., Brassier, A., Valayannopoulos, V., Brunelle, F., Fournet, J.C., Robert, J.J., Aigrain, Y., Bellanne-Chantelot, C., De Lonlay, P., 2011. Congenital hyperinsulinism: current trends in diagnosis and therapy. Orphanet J. Rare Dis. 6, 63.

Ashcroft, F.M., 2005. ATP-sensitive potassium channelopathies: focus on insulin secretion. J. Clin. Invest. 115, 2047–2058.

Ashcroft, F.M., 2010. New uses for old drugs: neonatal diabetes and sulphonylureas. Cell Metab. 11, 179–181.

Ashcroft, F.M., Rorsman, P., 1989. Electrophysiology of the pancreatic beta-cell. Prog. Biophys. Mol. Biol. 54, 87–143.

Ashcroft, F.M., Rorsman, P., 2013. K(ATP) channels and islet hormone secretion: new insights and controversies. Nat. Rev. Endocrinol. 9, 660–669.

Ashcroft, F.M., Harrison, D.E., Ashcroft, S.J., 1984. Glucose induces closure of single potassium channels in isolated rat pancreatic beta-cells. Nature 312, 446–448.

Ashcroft, F.M., Proks, P., Smith, P.A., Ammala, C., Bokvist, K., Rorsman, P., 1994. Stimulus-secretion coupling in pancreatic beta cells. J. Cell. Biochem. 55 (Suppl), 54–65.

Barg, S., Ma, X., Eliasson, L., Galvanovskis, J., Gopel, S.O., Obermuller, S., Platzer, J., Renstrom, E., Trus, M., Atlas, D., Striessnig, J., Rorsman, P., 2001. Fast exocytosis with few Ca(2+) channels in insulin-secreting mouse pancreatic B cells. Biophys. J. 81, 3308–3323.

Berggren, P.O., Arkhammar, P., Islam, M.S., Juntti-Berggren, L., Khan, A., Kindmark, H., Kohler, M., Larsson, K., Larsson, O., Nilsson, T., et al., 1993. Regulation of cytoplasmic free Ca2+ in insulin-secreting cells. Adv. Exp. Med. Biol. 334, 25–45.

Berkefeld, H., Fakler, B., Schulte, U., 2010. Ca2+–activated K+ channels: from protein complexes to function. Physiol. Rev. 90, 1437–1459.

Blodgett, D.M., Nowosielska, A., Afik, S., Pechhold, S., Cura, A.J., Kennedy, N.J., Kim, S., Kucukural, A., Davis, R.J., Kent, S.C., Greiner, D.L., Garber, M.G., Harlan, D.M., Diiorio, P., 2015. Novel observations from next-generation RNA sequencing of highly purified human adult and fetal islet cell subsets. Diabetes 64, 3172–3181.

Bokvist, K., Ammala, C., Berggren, P.O., Rorsman, P., Wahlander, K., 1991. Alpha 2-adrenoreceptor stimulation does not inhibit L-type calcium channels in mouse pancreatic beta-cells. Biosci. Rep. 11, 147–157.
Bonner, C., Kerr-Conte, J., Gmyr, V., Queniat, G., Moerman, E., Thevenet, J., Beaucamps, C., Delalleau, N., Popescu, I., Malaisse, W.J., Sener, A., Deprez, B., Abderrahmani, A., Staels, B., Pattou, F., 2015. Inhibition of the glucose transporter SGLT2 with dapagliflozin in pancreatic alpha cells triggers glucagon secretion. Nat. Med. 21, 512–517.
Bosco, D., Armanet, M., Morel, P., Niclauss, N., Sgroi, A., Muller, Y.D., Giovannoni, L., Parnaud, G., Berney, T., 2010. Unique arrangement of alpha- and beta-cells in human islets of Langerhans. Diabetes 59, 1202–1210.
Braun, M., Ramracheya, R., Rorsman, P., 2012. Autocrine regulation of insulin secretion. Diabetes Obes. Metab. 14 (Suppl. 3), 143–151.
Briant, L.J.B., Reinbothe, T.M., Spiliotis, I., Miranda, C., Rodriguez, B., Rorsman, P., 2018. Delta-cells and beta-cells are electrically coupled and regulate alpha-cell activity via somatostatin. J. Physiol. 596, 197–215.
Brissova, M., Fowler, M.J., Nicholson, W.E., Chu, A., Hirshberg, B., Harlan, D.M., Powers, A.C., 2005. Assessment of human pancreatic islet architecture and composition by laser scanning confocal microscopy. J. Histochem. Cytochem. 53, 1087–1097.
Bruin, J.E., Saber, N., Braun, N., Fox, J.K., Mojibian, M., Asadi, A., Drohan, C., O'dwyer, S., Rosman-Balzer, D.S., Swiss, V.A., Rezania, A., Kieffer, T.J., 2015. Treating diet-induced diabetes and obesity with human embryonic stem cell-derived pancreatic progenitor cells and antidiabetic drugs. Stem Cell Rep. 4, 605–620.
Cabrera, O., Berman, D.M., Kenyon, N.S., Ricordi, C., Berggren, P.O., Caicedo, A., 2006. The unique cytoarchitecture of human pancreatic islets has implications for islet cell function. Proc. Natl. Acad. Sci. U. S. A. 103, 2334–2339.
Campbell, J.D., Sansom, M.S., Ashcroft, F.M., 2003. Potassium channel regulation. EMBO Rep. 4, 1038–1042.
Chandra, J., Zhivotovsky, B., Zaitsev, S., Juntti-Berggren, L., Berggren, P.O., Orrenius, S., 2001. Role of apoptosis in pancreatic beta-cell death in diabetes. Diabetes 50 (Suppl. 1), S44–S47.
Cook, D.L., Ikeuchi, M., Fujimoto, W.Y., 1984. Lowering of pHi inhibits Ca2+–activated K+ channels in pancreatic B-cells. Nature 311, 269–271.
Da Silva Xavier, G., 2018. The cells of the islets of Langerhans. J. Clin. Med. 7 (54), 1–17.
Dean, P.M., Matthews, E.K., 1968. Electrical activity in pancreatic islet cells. Nature 219, 389–390.
Defronzo, R.A., 2017. Combination therapy with GLP-1 receptor agonist and SGLT2 inhibitor. Diabetes Obes. Metab. 19, 1353–1362.
Detimary, P., Van Den Berghe, G., Henquin, J.C., 1996. Concentration dependence and time course of the effects of glucose on adenine and guanine nucleotides in mouse pancreatic islets. J. Biol. Chem. 271, 20559–20565.
Detimary, P., Gilon, P., Henquin, J.C., 1998. Interplay between cytoplasmic Ca2+ and the ATP/ADP ratio: a feedback control mechanism in mouse pancreatic islets. Biochem. J. 333 (Pt. 2), 269–274.
Doliba, N.M., Qin, W., Najafi, H., Liu, C., Buettger, C.W., Sotiris, J., Collins, H.W., Li, C., Stanley, C.A., Wilson, D.F., Grimsby, J., Sarabu, R., Naji, A., Matschinsky, F.M., 2012. Glucokinase activation repairs defective bioenergetics of islets of Langerhans isolated from type 2 diabetics. Am. J. Physiol. Endocrinol. Metab. 302, E87–E102.
Dolphin, A.C., 2016. Voltage-gated calcium channels and their auxiliary subunits: physiology and pathophysiology and pharmacology. J. Physiol. 594, 5369–5390.

Drain, P., Geng, X., Li, L., 2004. Concerted gating mechanism underlying KATP channel inhibition by ATP. Biophys. J. 86, 2101–2112.

Drews, G., Krippeit-Drews, P., Dufer, M., 2010. Electrophysiology of islet cells. Adv. Exp. Med. Biol. 654, 115–163.

Enkvetchakul, D., Nichols, C.G., 2003. Gating mechanism of KATP channels: function fits form. J. Gen. Physiol. 122, 471–480.

Enkvetchakul, D., Loussouarn, G., Makhina, E., Shyng, S.L., Nichols, C.G., 2000. The kinetic and physical basis of K(ATP) channel gating: toward a unified molecular understanding. Biophys. J. 78, 2334–2348.

Gembal, M., Gilon, P., Henquin, J.C., 1992. Evidence that glucose can control insulin release independently from its action on ATP-sensitive K+ channels in mouse B cells. J. Clin. Invest. 89, 1288–1295.

Gilon, P., Shepherd, R.M., Henquin, J.C., 1993. Oscillations of secretion driven by oscillations of cytoplasmic Ca2+ as evidences in single pancreatic islets. J. Biol. Chem. 268, 22265–22268.

Gloyn, A.L., Odili, S., Zelent, D., Buettger, C., Castleden, H.A., Steele, A.M., Stride, A., Shiota, C., Magnuson, M.A., Lorini, R., D'annunzio, G., Stanley, C.A., Kwagh, J., Van Schaftingen, E., Veiga-Da-Cunha, M., Barbetti, F., Dunten, P., Han, Y., Grimsby, J., Taub, R., Ellard, S., Hattersley, A.T., Matschinsky, F.M., 2005. Insights into the structure and regulation of glucokinase from a novel mutation (V62M), which causes maturity-onset diabetes of the young. J. Biol. Chem. 280, 14105–14113.

Gomis, R., Aguilar Diosdado, M., 2007. Tratado SED de diabetes mellitus. Editorial Médica Panamericana, Buenos Aires.

Goncalves, E., Bell, D.S.H., 2018. Combination treatment of SGLT2 inhibitors and GLP-1 receptor agonists: symbiotic effects on metabolism and cardiorenal risk. Diabetes Ther. 9, 919–926.

Gopel, S.O., Kanno, T., Barg, S., Eliasson, L., Galvanovskis, J., Renstrom, E., Rorsman, P., 1999. Activation of Ca(2+)-dependent K(+) channels contributes to rhythmic firing of action potentials in mouse pancreatic beta cells. J. Gen. Physiol. 114, 759–770.

Gromada, J., Franklin, I., Wollheim, C.B., 2007. Alpha-cells of the endocrine pancreas: 35 years of research but the enigma remains. Endocr. Rev. 28, 84–116.

Hamill, O.P., Marty, A., Neher, E., Sakmann, B., Sigworth, F.J., 1981. Improved patch-clamp techniques for high-resolution current recording from cells and cell-free membrane patches. Pflugers Arch. 391, 85–100.

Heimberg, H., De Vos, A., Pipeleers, D., Thorens, B., Schuit, F., 1995. Differences in glucose transporter gene expression between rat pancreatic alpha- and beta-cells are correlated to differences in glucose transport but not in glucose utilization. J. Biol. Chem. 270, 8971–8975.

Henquin, J.C., 2004. Pathways in beta-cell stimulus-secretion coupling as targets for therapeutic insulin secretagogues. Diabetes 53 (Suppl. 3), S48–S58.

Henquin, J.C., Meissner, H.P., 1984. Significance of ionic fluxes and changes in membrane potential for stimulus-secretion coupling in pancreatic B-cells. Experientia 40, 1043–1052.

Henquin, J.C., Dufrane, D., Kerr-Conte, J., Nenquin, M., 2015. Dynamics of glucose-induced insulin secretion in normal human islets. Am. J. Physiol. Endocrinol. Metab. 309, E640–E650.

Hibino, H., Inanobe, A., Furutani, K., Murakami, S., Findlay, I., Kurachi, Y., 2010. Inwardly rectifying potassium channels: their structure, function, and physiological roles. Physiol. Rev. 90, 291–366.

Hoppa, M.B., Collins, S., Ramracheya, R., Hodson, L., Amisten, S., Zhang, Q., Johnson, P., Ashcroft, F.M., Rorsman, P., 2009. Chronic palmitate exposure inhibits insulin secretion by dissociation of Ca(2+) channels from secretory granules. Cell Metab. 10, 455–465.

Islam, S., 2020. Advances in Experimental Medicine and Biology, second ed. vol. 1131, Springer, ISBN: 978-3-030-12456-4.
Jing, X., Li, D.Q., Olofsson, C.S., Salehi, A., Surve, V.V., Caballero, J., Ivarsson, R., Lundquist, I., Pereverzev, A., Schneider, T., Rorsman, P., Renstrom, E., 2005. CaV2.3 calcium channels control second-phase insulin release. J. Clin. Invest. 115, 146–154.
Johnson, D., Shepherd, R.M., Gill, D., Gorman, T., Smith, D.M., Dunne, M.J., 2007. Glucose-dependent modulation of insulin secretion and intracellular calcium ions by Gka50, a glucokinase activator. Diabetes 56, 1694–1702.
Johnston, J., Forsythe, I.D., Kopp-Scheinpflug, C., 2010. Going native: voltage-gated potassium channels controlling neuronal excitability. J. Physiol. 588, 3187–3200.
Johnston, N.R., Mitchell, R.K., Haythorne, E., Pessoa, M.P., Semplici, F., Ferrer, J., Piemonti, L., Marchetti, P., Bugliani, M., Bosco, D., Berishvili, E., Duncanson, P., Watkinson, M., Broichhagen, J., Trauner, D., Rutter, G.A., Hodson, D.J., 2016. Beta cell hubs dictate pancreatic islet responses to glucose. Cell Metab. 24, 389–401.
Kanno, T., Gopel, S.O., Rorsman, P., Wakui, M., 2002a. Cellular function in multicellular system for hormone-secretion: electrophysiological aspect of studies on alpha-, beta- and delta-cells of the pancreatic islet. Neurosci. Res. 42, 79–90.
Kanno, T., Rorsman, P., Gopel, S.O., 2002b. Glucose-dependent regulation of rhythmic action potential firing in pancreatic beta-cells by K(ATP)-channel modulation. J. Physiol. 545, 501–507.
Kato, S., Ishida, H., Tsuura, Y., Tsuji, K., Nishimura, M., Horie, M., Taminato, T., Ikehara, S., Odaka, H., Ikeda, I., Okada, Y., Seino, Y., 1996. Alterations in basal and glucose-stimulated voltage-dependent Ca2+ channel activities in pancreatic beta cells of non-insulin-dependent diabetes mellitus Gk rats. J. Clin. Invest. 97, 2417–2425.
Kelly, R.P., Sutton, R., Ashcroft, F.M., 1991. Voltage-activated calcium and potassium currents in human pancreatic beta-cells. J. Physiol. 443, 175–192.
Khan, S., Yan-Do, R., Duong, E., Wu, X., Bautista, A., Cheley, S., Macdonald, P.E., Braun, M., 2014. Autocrine activation of P2Y1 receptors couples ca (2+) influx to ca (2+) release in human pancreatic beta cells. Diabetologia 57, 2535–2545.
Kinard, T.A., Satin, L.S., 1996. Temperature modulates the Ca2+ current of HIT-T15 and mouse pancreatic beta-cells. Cell Calcium 20, 475–482.
Kleiber, M., 1947. Body size and metabolic rate. Physiol. Rev. 27, 511–541.
Krippeit-Drews, P., Dufer, M., Drews, G., 2000. Parallel oscillations of intracellular calcium activity and mitochondrial membrane potential in mouse pancreatic B-cells. Biochem. Biophys. Res. Commun. 267, 179–183.
Kuang, Q., Purhonen, P., Hebert, H., 2015. Structure of potassium channels. Cell. Mol. Life Sci. 72, 3677–3693.
Macdonald, P.E., Wheeler, M.B., 2003. Voltage-dependent K(+) channels in pancreatic beta cells: role, regulation and potential as therapeutic targets. Diabetologia 46, 1046–1062.
Magleby, K.L., 2003. Gating mechanism of Bk (Slo1) channels: so near, yet so far. J. Gen. Physiol. 121, 81–96.
Mahdi, T., Hanzelmann, S., Salehi, A., Muhammed, S.J., Reinbothe, T.M., Tang, Y., Axelsson, A.S., Zhou, Y., Jing, X., Almgren, P., Krus, U., Taneera, J., Blom, A.M., Lyssenko, V., Esguerra, J.L., Hansson, O., Eliasson, L., Derry, J., Zhang, E., Wollheim, C.B., Groop, L., Renstrom, E., Rosengren, A.H., 2012. Secreted frizzled-related protein 4 reduces insulin secretion and is overexpressed in type 2 diabetes. Cell Metab. 16, 625–633.
Marmol, P., Pardo, B., Wiederkehr, A., Del Arco, A., Wollheim, C.B., Satrustegui, J., 2009. Requirement for aralar and its Ca2+—binding sites in Ca2+ signal transduction in mitochondria from INS-1 clonal beta-cells. J. Biol. Chem. 284, 515–524.
Martin, F., Soria, B., 1996. Glucose-induced [Ca2+]i oscillations in single human pancreatic islets. Cell Calcium 20, 409–414.

Martinez-Pinna, J., Marroqui, L., Hmadcha, A., Lopez-Beas, J., Soriano, S., Villar-Pazos, S., Alonso-Magdalena, P., Dos Santos, R.S., Quesada, I., Martin, F., Soria, B., Gustafsson, J.A., Nadal, A., 2019. Oestrogen receptor beta mediates the actions of bisphenol-A on ion channel expression in mouse pancreatic beta cells. Diabetologia 62, 1667–1680.

McCulloch, L.J., Van De Bunt, M., Braun, M., Frayn, K.N., Clark, A., Gloyn, A.L., 2011. GLUT2 (SLC2A2) is not the principal glucose transporter in human pancreatic beta cells: implications for understanding genetic association signals at this locus. Mol. Genet. Metab. 104, 648–653.

Meda, P., 1996. The role of gap junction membrane channels in secretion and hormonal action. J. Bioenerg. Biomembr. 28, 369–377.

Meda, P., Bosco, D., Chanson, M., Giordano, E., Vallar, L., Wollheim, C., Orci, L., 1990. Rapid and reversible secretion changes during uncoupling of rat insulin-producing cells. J. Clin. Invest. 86, 759–768.

Meissner, H.P., 1976. Electrophysiological evidence for coupling between beta cells of pancreatic islets. Nature 262, 502–504.

Meissner, H.P., Henquin, J.C., Preissler, M., 1978. Potassium dependence of the membrane potential of pancreatic B-cells. FEBS Lett. 94, 87–89.

Misler, S., Barnett, D.W., Gillis, K.D., Pressel, D.M., 1992. Electrophysiology of stimulus-secretion coupling in human beta-cells. Diabetes 41, 1221–1228.

Nansseu, J.R., Ngo-Um, S.S., Balti, E.V., 2016. Incidence, prevalence and genetic determinants of neonatal diabetes mellitus: a systematic review and meta-analysis protocol. Syst. Rev. 5, 188.

Neher, E., Sakmann, B., 1976. Single-channel currents recorded from membrane of denervated frog muscle fibres. Nature 260, 799–802.

Nichols, C.G., 2006. KATP channels as molecular sensors of cellular metabolism. Nature 440, 470–476.

Otonkoski, T., Kaminen, N., Ustinov, J., Lapatto, R., Meissner, T., Mayatepek, E., Kere, J., Sipila, I., 2003. Physical exercise-induced hyperinsulinemic hypoglycemia is an autosomal-dominant trait characterized by abnormal pyruvate-induced insulin release. Diabetes 52, 199–204.

Padhi, S., Nayak, A.K., Behera, A., 2020. Type ii diabetes mellitus: a review on recent drug based therapeutics. Biomed. Pharmacother. 131, 110708.

Pearson, E.R., Flechtner, I., Njolstad, P.R., Malecki, M.T., Flanagan, S.E., Larkin, B., Ashcroft, F.M., Klimes, I., Codner, E., Iotova, V., Slingerland, A.S., Shield, J., Robert, J.J., Holst, J.J., Clark, P.M., Ellard, S., Sovik, O., Polak, M., Hattersley, A.T., Neonatal Diabetes International Collaborative Group, 2006. Switching from insulin to oral sulfonylureas in patients with diabetes due to Kir6.2 mutations. N. Engl. J. Med. 355, 467–477.

Perez-Reyes, E., 2003. Molecular physiology of low-voltage-activated t-type calcium channels. Physiol. Rev. 83, 117–161.

Plant, T.D., 1988. Properties and calcium-dependent inactivation of calcium currents in cultured mouse pancreatic B-cells. J. Physiol. 404, 731–747.

Prentki, M., Matschinsky, F.M., Madiraju, S.R., 2013. Metabolic signaling in fuel-induced insulin secretion. Cell Metab. 18, 162–185.

Proks, P., Capener, C.E., Jones, P., Ashcroft, F.M., 2001. Mutations within the P-loop of Kir6.2 modulate the intraburst kinetics of the ATP-sensitive potassium channel. J. Gen. Physiol. 118, 341–353.

Quesada, I., Todorova, M.G., Alonso-Magdalena, P., Beltra, M., Carneiro, E.M., Martin, F., Nadal, A., Soria, B., 2006. Glucose induces opposite intracellular Ca2+ concentration oscillatory patterns in identified alpha- and beta-cells within intact human islets of Langerhans. Diabetes 55, 2463–2469.

Randle, P.J., 1993. Glucokinase and candidate genes for type 2 (non-insulin-dependent) diabetes mellitus. Diabetologia 36, 269–275.

Ravier, M.A., Nenquin, M., Miki, T., Seino, S., Henquin, J.C., 2009. Glucose controls cytosolic Ca2+ and insulin secretion in mouse islets lacking adenosine triphosphate-sensitive K+ channels owing to a knockout of the pore-forming subunit Kir6.2. Endocrinology 150, 33–45.

Reinbothe, T.M., Alkayyali, S., Ahlqvist, E., Tuomi, T., Isomaa, B., Lyssenko, V., Renstrom, E., 2013. The human L-type calcium channel Cav1.3 regulates insulin release and polymorphisms in Cacna1D associate with type 2 diabetes. Diabetologia 56, 340–349.

Roder, P.V., Wu, B., Liu, Y., Han, W., 2016. Pancreatic regulation of glucose homeostasis. Exp. Mol. Med. 48, e219.

Rodriguez-Diaz, R., Caicedo, A., 2014. Neural control of the endocrine pancreas. Best Pract. Res. Clin. Endocrinol. Metab. 28, 745–756.

Rolland, J.F., Henquin, J.C., Gilon, P., 2002. Feedback control of the ATP-sensitive K(+) current by cytosolic Ca(2+) contributes to oscillations of the membrane potential in pancreatic beta-cells. Diabetes 51, 376–384.

Rorsman, P., 1997. The pancreatic beta-cell as a fuel sensor: an electrophysiologist's viewpoint. Diabetologia 40, 487–495.

Rorsman, P., Ashcroft, F.M., 2018. Pancreatic beta-cell electrical activity and insulin secretion: of mice and men. Physiol. Rev. 98, 117–214.

Rorsman, P., Braun, M., 2013. Regulation of insulin secretion in human pancreatic islets. Annu. Rev. Physiol. 75, 155–179.

Rorsman, P., Huising, M.O., 2018. The somatostatin-secreting pancreatic delta-cell in health and disease. Nat. Rev. Endocrinol. 14, 404–414.

Rorsman, P., Trube, G., 1986. Calcium and delayed potassium currents in mouse pancreatic beta-cells under voltage-clamp conditions. J. Physiol. 374, 531–550.

Rorsman, P., Ashcroft, F.M., Trube, G., 1988. Single ca channel currents in mouse pancreatic B-cells. Pflugers Arch. 412, 597–603.

Rorsman, P., Bokvist, K., Ammala, C., Eliasson, L., Renstrom, E., Gabel, J., 1994. Ion channels, electrical activity and insulin secretion. Diabetes Metab. 20, 138–145.

Rorsman, P., Eliasson, L., Kanno, T., Zhang, Q., Gopel, S., 2011. Electrophysiology of pancreatic beta-cells in intact mouse islets of Langerhans. Prog. Biophys. Mol. Biol. 107, 224–235.

Sakamaki, J., Fu, A., Reeks, C., Baird, S., Depatie, C., Al Azzabi, M., Bardeesy, N., Gingras, A.C., Yee, S.P., Screaton, R.A., 2014. Role of the SIK2-p35-PJA2 complex in pancreatic beta-cell functional compensation. Nat. Cell Biol. 16, 234–244.

Sakura, H., Ammala, C., Smith, P.A., Gribble, F.M., Ashcroft, F.M., 1995. Cloning and functional expression of the CDNA encoding a novel ATP-sensitive potassium channel subunit expressed in pancreatic beta-cells, brain, heart and skeletal muscle. FEBS Lett. 377, 338–344.

Samols, E., Marri, G., Marks, V., 1965. Promotion of insulin secretion by glucagon. Lancet 2, 415–416.

Sanchez-Andres, J.V., Gomis, A., Valdeolmillos, M., 1995. The electrical activity of mouse pancreatic beta-cells recorded in vivo shows glucose-dependent oscillations. J. Physiol. 486 (Pt. 1), 223–228.

Sanchez-Andres, J.V., Pomares, R., Malaisse, W.J., 2020. Adaptive short-term associative conditioning in the pancreatic beta-cell. Physiol. Rep. 8, e14403.

Schuit, F., De Vos, A., Farfari, S., Moens, K., Pipeleers, D., Brun, T., Prentki, M., 1997. Metabolic fate of glucose in purified islet cells. Glucose-regulated anaplerosis in beta cells. J. Biol. Chem. 272, 18572–18579.

Schulla, V., Renstrom, E., Feil, R., Feil, S., Franklin, I., Gjinovci, A., Jing, X.J., Laux, D., Lundquist, I., Magnuson, M.A., Obermuller, S., Olofsson, C.S., Salehi, A., Wendt, A.,

Klugbauer, N., Wollheim, C.B., Rorsman, P., Hofmann, F., 2003. Impaired insulin secretion and glucose tolerance in beta cell-selective Ca(v)1.2 Ca2+ channel null mice. EMBO J. 22, 3844–3854.

Sharp, G.W., 1996. Mechanisms of inhibition of insulin release. Am. J. Phys. 271, C1781–C1799.

Smith, P.A., Rorsman, P., Ashcroft, F.M., 1989. Modulation of dihydropyridine-sensitive Ca2+ channels by glucose metabolism in mouse pancreatic beta-cells. Nature 342, 550–553.

Smith, P.A., Aschroft, F.M., Fewtrell, C.M., 1993. Permeation and gating properties of the L-type calcium channel in mouse pancreatic beta cells. J. Gen. Physiol. 101, 767–797.

Soria, B., Ferrer, R., 1986. Graded spike electrogenesis in mouse pancreatic beta-cells. Adv. Exp. Med. Biol. 211, 235–246.

Soriano, S., Ropero, A.B., Alonso-Magdalena, P., Ripoll, C., Quesada, I., Gassner, B., Kuhn, M., Gustafsson, J.A., Nadal, A., 2009. Rapid regulation of K(ATP) channel activity by 17{beta}-estradiol in pancreatic {beta}-cells involves the estrogen receptor {beta} and the atrial natriuretic peptide receptor. Mol. Endocrinol. 23, 1973–1982.

Soriano, S., Ripoll, C., Fuentes, E., Gonzalez, A., Alonso-Magdalena, P., Ropero, A.B., Quesada, I., Nadal, A., 2011. Regulation of K(ATP) channel by 17beta-estradiol in pancreatic beta-cells. Steroids 76, 856–860.

Soriano, S., Alonso-Magdalena, P., Garcia-Arevalo, M., Novials, A., Muhammed, S.J., Salehi, A., Gustafsson, J.A., Quesada, I., Nadal, A., 2012. Rapid insulinotropic action of low doses of bisphenol-A on mouse and human islets of Langerhans: role of estrogen receptor beta. PLoS One 7, e31109.

Stocker, M., 2004. Ca(2+)-activated K+ channels: molecular determinants and function of the SK family. Nat. Rev. Neurosci. 5, 758–770.

Svendsen, B., Larsen, O., Gabe, M.B.N., Christiansen, C.B., Rosenkilde, M.M., Drucker, D.J., Holst, J.J., 2018. Insulin secretion depends on intra-islet glucagon signaling. Cell Rep. 25 (1127–1134), e2.

Thorens, B., 2014. Neural regulation of pancreatic islet cell mass and function. Diabetes Obes. Metab. 16 (Suppl. 1), 87–95.

Thorrez, L., Laudadio, I., Van Deun, K., Quintens, R., Hendrickx, N., Granvik, M., Lemaire, K., Schraenen, A., Van Lommel, L., Lehnert, S., Aguayo-Mazzucato, C., Cheng-Xue, R., Gilon, P., Van Mechelen, I., Bonner-Weir, S., Lemaigre, F., Schuit, F., 2011. Tissue-specific disallowance of housekeeping genes: the other face of cell differentiation. Genome Res. 21, 95–105.

Traboulsie, A., Chemin, J., Chevalier, M., Quignard, J.F., Nargeot, J., Lory, P., 2007. Subunit-specific modulation of T-type calcium channels by zinc. J. Physiol. 578, 159–171.

Trott, J., Tan, E.K., Ong, S., Titmarsh, D.M., Denil, S., Giam, M., Wong, C.K., Wang, J., Shboul, M., Eio, M., Cooper-White, J., Cool, S.M., Rancati, G., Stanton, L.W., Reversade, B., Dunn, N.R., 2017. Long-term culture of self-renewing pancreatic progenitors derived from human pluripotent stem cells. Stem Cell Rep. 8, 1675–1688.

Trube, G., Rorsman, P., 1986. Calcium and potassium currents recorded from pancreatic beta-cells under voltage clamp control. Adv. Exp. Med. Biol. 211, 167–175.

Velasco, J.M., Petersen, J.U., Petersen, O.H., 1988. Single-channel Ba2+ currents in insulin-secreting cells are activated by glyceraldehyde stimulation. FEBS Lett. 231, 366–370.

Villar-Pazos, S., Martinez-Pinna, J., Castellano-Munoz, M., Alonso-Magdalena, P., Marroqui, L., Quesada, I., Gustafsson, J.A., Nadal, A., 2018. Author correction: molecular mechanisms involved in the non-monotonic effect of bisphenol-a on Ca(2+) entry in mouse pancreatic beta-cells. Sci. Rep. 8, 4262.

Wollheim, C.B., Sharp, G.W., 1981. Regulation of insulin release by calcium. Physiol. Rev. 61, 914–973.
Yu, J., Shi, Y., Zhao, K., Yang, G., Yu, L., Li, Y., Andersson, E.M., Ammala, C., Yang, S.N., Berggren, P.O., 2020. Enhanced expression of beta cell CaV3.1 channels impairs insulin release and glucose homeostasis. Proc. Natl. Acad. Sci. U. S. A. 117, 448–453.
Zoeller, R.T., Brown, T.R., Doan, L.L., Gore, A.C., Skakkebaek, N.E., Soto, A.M., Woodruff, T.J., Vom Saal, F.S., 2012. Endocrine-disrupting chemicals and public health protection: a statement of principles from the Endocrine Society. Endocrinology 153, 4097–4110.

CHAPTER SIX

Long non-coding RNA-regulated pathways in pancreatic β cells: Their role in diabetes

Itziar González-Moro[a,b] and Izortze Santin[a,b,c,*]
[a]Department of Biochemistry and Molecular biology, University of the Basque Country, Leioa, Spain
[b]Biocruces Bizkaia Health Research Institute, Barakaldo, Spain
[c]CIBER (Centro de Investigación Biomédica en Red) de Diabetes y Enfermedades Metabólicas Asociadas (CIBERDEM), Instituto de Salud Carlos III, Madrid, Spain
*Corresponding author: e-mail address: izortze.santin@ehu.eus

Contents

Abstract

Long non-coding RNAs (lncRNAs) are transcripts of more than 200 nucleotides that have not coding potential, but act as gene expression regulators through several molecular mechanisms. Several studies have identified tons of lncRNAs that are expressed in pancreatic β cells and many of them have been shown to have β cell-specific expression, suggesting a potential role in the regulation of basal β cell functions. Indeed, accumulating evidence based on numerous studies, has highlighted the implication of lncRNAs in the regulation of pancreatic β cell differentiation and proliferation, insulin synthesis and secretion, and apoptosis.

In addition, several lncRNAs have shown to be implicated in pancreatic β cell dysfunction linked to different types of diabetes, including type 1 and type 2 diabetes,

ISSN 1937-6448
https://doi.org/10.1016/bs.ircmb.2021.02.007

and monogenic forms of the disease. Pathogenic conditions linked to diabetes (inflammation or lipoglucotoxicity, for example) dysregulate the expression of several lncRNAs, suggesting that changes in lncRNA may alter potentially important pathways for β cell function, and eventually leading to β cell dysfunction and diabetes development. In this sense, functional characterization of some lncRNAs has demonstrated that these non-coding molecules participate in the regulation of several crucial pathways at the pancreatic β cell level, and dysregulation of these pathways leads to pathogenic phenotypes.

In this review, we provide an overview of the action mechanisms of functionally characterized lncRNAs in healthy β cells and describe the contribution of some diabetes-associated lncRNAs to pancreatic β cell failure.

1. Introduction

According to the central dogma of molecular biology, genetic information flows from DNA to RNA and finally, to protein. However, during the last years, this dogma has turned upside down due to the discovery of tons of genes that transcribe into RNA molecules that do not produce any protein. These non-coding RNA molecules were vastly called "junk DNA" and were supposed not to carry out any relevant biological function, however recent advantages in our understanding of non-coding RNA biology has revealed that these molecules are crucial regulators of several key biological processes, including random X chromosome inactivation, imprinting, cell cycle, organogenesis, differentiation and pluripotency (Guttman et al., 2011; Hung et al., 2011; Klattenhoff et al., 2013; Penny et al., 1996).

Advances in high-throughput sequencing techniques and in the annotation of the human genome have revealed that only about 2% of the genome has coding potential and hence, that most of the genome is constituted by non-coding genes (Dunham et al., 2012) (Fig. 1). Many of these non-coding genes transcribe for long non-coding RNAs (lncRNAs), which are defined as RNA molecules longer than 200 bp without potential to encode any protein. In contrast to protein-coding transcripts, lncRNAs are usually expressed in lower abundance and their expression is generally tissue- or cell-specific (Mercer et al., 2009).

LncRNAs are crucial regulators of gene transcription through different molecular mechanisms. Initially, these transcripts were shown to function locally, near their sites of synthesis, by regulating the expression of neighboring genes. More recently, it has been demonstrated that lncRNAs are able to interact with chromatin at several thousand different locations across

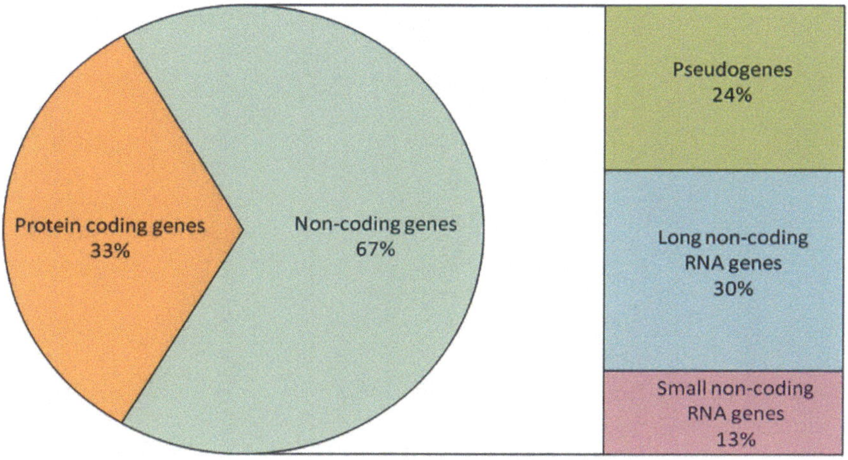

Fig. 1 Human genome sequence distribution. Data retrieved from V34 of the Encyclopedia of DNA Elements Consortium (ENCODE).

multiple chromosomes to modulate large-scale gene expression (Vance and Ponting, 2014). Although the mode of action of most lncRNAs remains to be clarified, several studies have linked lncRNAs with the control of several processes including protein synthesis, RNA maturation, splicing and stability, transport and transcriptional gene silencing/activation through regulation of the chromatin structure (Bhat et al., 2016; Böhmdorfer and Wierzbicki, 2015; Romero-Barrios et al., 2018; Wang and Chang, 2011) (Fig. 2).

As lncRNAs are implicated in several regulatory processes, they emerge as potential mediators in the development of several human pathologies (Delás and Hannon, 2017; DiStefano, 2018). Indeed, many lncRNAs are enriched for disease-associated single nucleotide polymorphisms (SNPs), suggesting that these genetic variants might alter the function of lncRNAs and have an impact in relevant pathways for disease pathogenesis which include cancer, immune-related disorders and cardiometabolic traits (Castellanos-Rubio and Ghosh, 2019; Giral et al., 2018; Kumar et al., 2013; Mirza et al., 2014).

Taking into account that interactions of lncRNAs with other molecules (DNA, RNA or proteins) are governed by their structure, rather than by their sequences, polymorphisms altering the secondary structure of lncRNAs may have an impact in their capacity to interact with other molecules, leading to altered biological pathways. Disease-associated SNPs may also impact in lncRNAs function by disturbing their expression levels or by

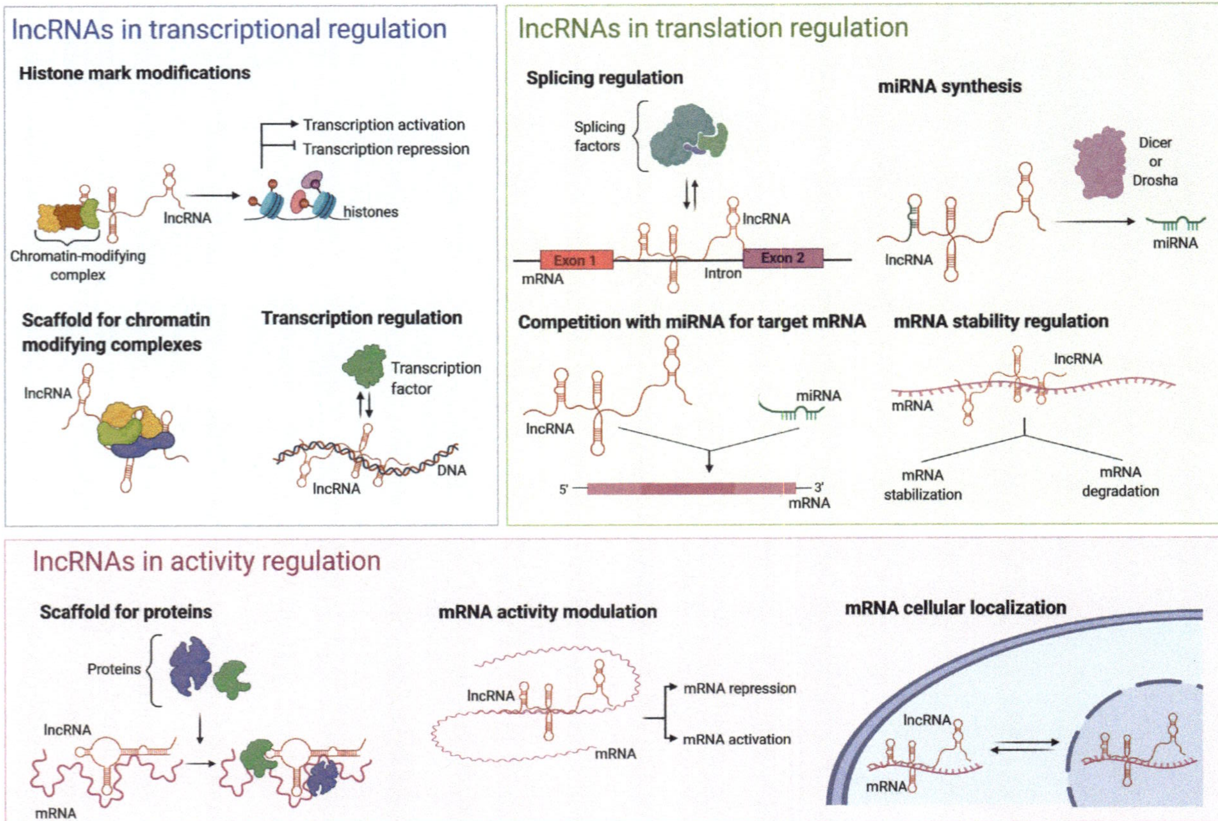

Fig. 2 Mechanisms of action of lncRNAs. The function of lncRNAs is mainly classified in 3 groups; regulating transcription, translation or activity modulation.

modifying the alternative splicing of the transcript, for example (Kumar et al., 2013). Although many SNPs located in lncRNAs have been associated with pathogenic conditions, the impact of these polymorphisms in the function of lncRNAs and their contribution to disease pathogenesis remains to be clarified.

In this chapter, we focus on lncRNAs expressed in pancreatic islets and β cells to describe their function in basal biological processes and their implication in pathogenic conditions linked to pancreatic β cell dysfunction, such as type 1 and type 2 diabetes mellitus. We provide an overview of the action mechanisms of functionally characterized lncRNAs in healthy β cells and describe the contribution of some diabetes-associated lncRNAs to pancreatic β cell failure.

2. The non-coding transcriptome of pancreatic β cells

During the last decade, several whole transcriptome studies of pancreatic islets have revealed the existence of thousands of lncRNAs that are specifically expressed in pancreatic β cells. By integrating sequence-based transcriptome and chromatin maps of human islets and β cells, more than 1100 islet lncRNAs were identified (Morán et al., 2012). The comparison with other 18 human tissues revealed that only a 9.4% of coding genes were specifically expressed in pancreatic islets, whereas more than 50% of identified lncRNAs were islet-specific. Interestingly, islet lncRNAs were dynamically regulated by glucose changes and were shown to be crucial components of β cell-specific maturation and differentiation program (Morán et al., 2012).

Some islet-specific lncRNAs seem to be mediators of lineage-specific differentiation and cellular processes within the different cell types present in pancreatic islets. In this sense, the first comparison of gene expression profiles between β cells, whole islets, and β cell depleted islet preparations, showed that β cells have a specific expression signature of lncRNAs (Nica et al., 2013). Indeed, 132 overexpressed lncRNAs were identified in β cells compared to the whole islet and 148 overexpressed lncRNAs in β cells compared to non-β cells. These results corroborated the cell-type specific expression of lncRNAs previously observed in other human tissues (Cabili et al., 2011).

Similarly to human β cells, mouse β cells also have a specific expression pattern of lncRNAs. Using a bitransgenic reporter mouse that labeled all β and α cells in islets to enable FACS-based purification, transcriptome analysis of both islet cell populations was performed (Benner et al., 2014). The transcriptome data identified 145 novel β cell specific lncRNAs that were enriched for significant transcription factor binding sites of key β cell transcription factors. Comparison of the set of 145 mouse lncRNAs with human lncRNAs revealed evidence for only two human orthologous lncRNAs. These data suggest that the repertoire of islet lncRNAs is dynamic and has changed considerably since the divergence of primates and rodents, in line with the general notion that the lncRNA repertoire across vertebrates is already quite dynamic (Benner et al., 2014). By sequencing poly-adenylated RNAs in FACS-purified pancreatic β cells from MIP-GFP mice, 1359 lncRNA genes were identified (Ku et al., 2012). Interestingly, 160 showed very high β cell specificity (more than 200-fold enrichment in β cells) and 108 lncRNAs were only detected in pancreatic β cells.

As previously mentioned, the majority of lncRNAs are of very low abundance, having a copy number of even lower than one per cell (Derrien et al., 2012; Mercer et al., 2011; Seiler et al., 2017); however, their expression may change upon the presence of different stimuli, such as pro-inflammatory cytokines, glucose or specific drugs (Fadista et al., 2014; Kutty et al., 2018; Perry et al., 2014), or even under specific pathophysiological conditions, such as obesity or cancer (Lu et al., 2018; Sun et al., 2016b). In this sense, in a whole RNA sequencing analysis performed in pancreatic islets of mice fed with a high-fat diet, a model of mild diabetes associated with obesity that corresponds to early diabetes in human; a set of more than 1500 novel lncRNAs was identified (Motterle et al., 2017). In addition, more than 100 lncRNAs were differentially expressed (23 up- and 104 downregulated lncRNAs) in mice fed with a high-fat diet, when compared to animals fed with a regular diet. These data indicated that diet-induced changes in lncRNA expression at the pancreatic β cell level may be important for the development of obesity-driven β cell dysfunction in type 2 diabetes (T2D). Indeed, analogous changes of specific lncRNAs were observed in islets of obese diabetic db/db mice and in the islets of T2D donors (Motterle et al., 2017). Moreover, comparisons of islet transcriptomes from human islets isolated from normoglycemic and hyperglycemic donors uncovered lncRNAs that were differentially expressed in the two groups as well as those associated with hemoglobin A1c (HbA1c) levels (a measure of long-term glycaemia).

Inflammatory mediators, such as pro-inflammatory cytokines, have also been linked to changes in lncRNA expression in rodent pancreatic β cells

(Sun et al., 2016a). Exposure of the MIN-6 rodent β cell line to a cocktail of pro-inflammatory cytokines composed of IL-1β, TNFα and IFNγ, led to the identification of 23 differently regulated lncRNAs in the cytokine-stimulated group compared to the control group. In another study following the same approach, the authors observed that pro-inflammatory cytokine treatment in MIN-6 cells induced a sixfold increase in lncRNA expression when compared to cells under basal condition (Motterle et al., 2015). Moreover, they observed that some lncRNAs were also upregulated with the development of insulitis in non-diabetic obese (NOD) mice and their modulation increased β cell apoptosis. Altogether, these data suggested that lncRNAs might play a role in inflammation-induced pancreatic β cell demise in the context of type 1 diabetes.

In summary, human and mouse pancreatic β cells exhibit a portfolio of more than 1000 islet-specific lncRNAs. The expression of these specific lncRNAs is modified by different diabetogenic stimuli linked to the dysfunction of pancreatic β cells, and thus, lncRNAs have emerged as potential key regulators of diabetes-associated pathways at the pancreatic β cell level.

3. The implication of LncRNAs in pancreas development and β cell differentiation

Pancreatic development is a complex process in which two morphologically distinct tissue types (exocrine and endocrine tissue) must derive from one simple epithelium. These two tissues, exocrine (including acinar cells, centro-acinar cells, and ducts) and endocrine cells (including α, β, δ, pancreatic polypeptide and ghrelin cells), have different functions and an entirely different morphology (Gittes, 2009). Pancreas development is an orchestrated process that involves highly regulated expression of several transcription factors (PDX1, GATA6, HNF1α and GLIS3, among others) (Wong et al., 2018).

Pancreas development and differentiation has been widely studied in mouse models. However, the limited availability of human fetal pancreatic tissue has limited the study of this process in humans and few works have addressed the role of lncRNAs in the development of human pancreatic endocrine tissue and β cell differentiation. In this sense, Morán et al. (2012) studied the temporal expression of islet lncRNAs in human embryonic pancreases and in embryonic stem (ES) cell- based *in vitro* models of β cell differentiation (Morán et al., 2012). Among their observations, it is very interesting the fact that islet-specific lncRNAs were preferentially located in genomic regions where the closest protein-coding genes were linked to β cell function, development, and transcription (for example, islet protein-coding genes paired with intergenic lncRNAs included *MAFB*, *FOXA2*, *PC*SK1, and *ISL1*) (Morán et al., 2012). By analyzing the expression of 13 β cell selective lncRNAs in dissected Carnegie stage human embryonic pancreas, they observed that all except one were silent or expressed at low levels in pancreatic progenitors and were subsequently active in adult islets (Morán et al., 2012). In the same line, the expression profiling of islet lncRNAs in a human embryonic stem-cell differentiation model, revealed that some lncRNAs were expressed at very low or undetectable levels throughout all *in vitro* differentiation steps and were only activated during the *in vivo* maturation step. Overall, these data suggested that islet-specific lncRNA expression is highly linked to the pancreatic endocrine differentiation stage, and therefore, lncRNAs are candidate markers and/or regulators of pancreatic β cell differentiation and maturation (Morán et al., 2012).

Interestingly, islet-specific lncRNAs are not limited to developmental roles, as inhibition of a β cell-specific lncRNAs has been shown to influence gene expression in mature β cells. Indeed, suppression of *HI-LNC25*, a lncRNA that shares a regulatory domain with the transcription factor MAFB, led to a decreased in *GLIS3* gene expression (Morán et al., 2012). Interestingly, GLIS3 is a transcription factor that participates in pancreatic β cell development and insulin gene expression (Kang et al., 2009), and has been associated with T1D (Barrett et al., 2009) and T2D (Cho et al., 2012).

Another study focused in the functional characterization of pancreatic β cell-specific lncRNAs showed that β cell-specific lncRNAs frequently regulate genes associated with clusters of islet enhancers, which have previously been shown to be the primary functional targets of islet-specific transcription factors that play crucial roles in pancreas development and β cell function (e.g., *HNF1A*, *MAFB*, and *PDX1*, among others) (Akerman et al., 2017). Moreover, they showed that the lncRNA *PLUTO* affected local 3D chromatin structure and transcription of *PDX1*, encoding a key β cell transcription factor. Given the importance of β cell related transcription factors in β cell differentiation, and taking into account that some β cell specific lncRNAs play analogous roles, it is conceivable to think that lncRNAs can be harnessed to promote β cell differentiation, function, or cellular mass (Akerman et al., 2017).

4. LncRNAs and pancreatic β cell function

The main function of a pancreatic β cells is to produce and secrete insulin, the hormone responsible for regulating levels of glucose in the blood by inducing glucose intake by cells. Indeed, transcriptome analysis of pancreatic islets showed that around 20% of exonic reads in islets and 45% in β cells were originated from the proinsulin mRNA (Morán et al., 2012). Although it has been widely described that the expression of the insulin gene is tightly regulated by several islet enriched transcription factors (Melloul et al., 2002), during the last years, islet lncRNAs have emerged as novel regulators of insulin production and release (Singer et al., 2015) (Fig. 3).

Meg3 (*Maternally expressed gene 3*) is an imprinted non-coding gene transcribed from the maternal allele that is located on mouse chromosome 12. In humans, its homolog resides on chromosome 14, and although it is widely expressed in many normal tissues, its expression is lost in many human tumors and tumor derived cell lines (Miyoshi et al., 2000; Zhang et al., 2010). Within the islet of Langerhans, *Meg3* is more expressed (20-fold higher) in human β cells than in glucagon-producing α cells (Dorrell et al., 2011), suggesting its potential implication in a β cell-specific function. Indeed, *Meg3* expression is dynamically modulated by glucose and its inhibition by specific siRNAs in MIN6 mouse β cells led to a decreased in insulin synthesis and secretion, and increased β cell apoptosis (You et al., 2016). Moreover, administration of intravenous *Meg3*-targeting siRNAs in normal Balb/c female mice induced impaired insulin synthesis and secretion. Consistent with these findings, islets from these mice showed decreased mRNA levels of *Ins2*, *PDX1*, and *MAFA*, when compared to control mice. Although the exact molecular mechanisms by which *Meg3* regulates insulin production remains to be clarified, its nuclear localization in pancreatic β cells suggests a potential role in transcriptional regulation. Indeed, studies in other tissues have demonstrated that *Meg3* regulates gene expression by recruiting the histone methyltransferase PRC2 (polycomb repressive complex 2) (Mondal et al., 2015).

The *HI-LNC78* lncRNA is a glucose-regulated islet transcript (Morán et al., 2012) that is orthologous to mouse *Tunar* non-coding transcript. Knockdown of *HI-LNC78* in T antigen-excised EndoC-βH3 cells resulted in reduced insulin content and impaired glucose-stimulated insulin secretion (Akerman et al., 2017). Moreover, inhibition of *HI-LNC78* correlated with alterations in the expression of genes known to play crucial roles in

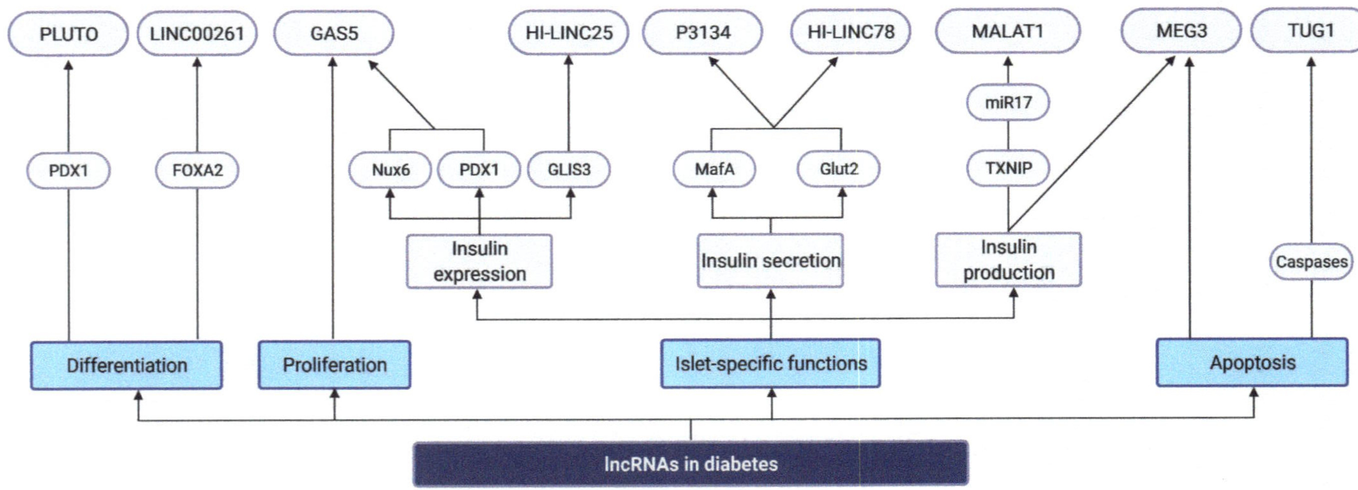

Fig. 3 lncRNAs and their function in pancreatic β cells. Several studies have implicated lncRNAs in different pancreatic β cell functions, including β cell differentiation and proliferation, insulin expression and release, and β cell apoptosis, among others.

pancreatic β cell function. Interestingly, the transcriptional changes that occurred after knocking down *HI-LNC78* significantly correlated with those observed after inhibition of key transcription factors in β cells, such as HNF1A and MAFB, suggesting that this lncRNA and islet-specific transcription factors can regulate common gene expression programs tightly related to pancreatic β cell function.

The lncRNA *taurine upregulated gene 1* (*TUG1*) is a spliced and polyadenylated RNA that is highly expressed in mouse pancreas tissues and islets of Langerhans compared with other organ tissues (liver, lung, colon and brain, among others) (Yin et al., 2015). Knockdown of *TUG1* in mouse pancreatic β cells decreased insulin content and secretion, suggesting its potential implication in insulin production. Indeed, *TUG1* knockdown in normal BALB/c mice led to reduced insulin level and increased blood glucose levels when compared to control mice. The diabetogenic phenotype in *TUG1*-inhibited mice correlated with decreased expression of PDX1, NeuroD1 and MAFA, suggesting that *TUG1* may mediate insulin expression by regulating the expression of these key β cell transcription factors.

The *metastasis-associated lung adenocarcinoma transcript 1* (*MALAT1*) is a highly conserved lncRNA that is dysregulated in many tumors (Ji et al., 2003) and expressed in several cell types, including pancreatic α and β cells (Dorrell et al., 2011). *MALAT1* gene is located in chromosome 11 and encoded within an active enhancer cluster with several binding sites for islet-transcription factors (Pasquali et al., 2014). Interestingly, its expression has recently been shown to predict human islet isolation quality, suggesting that *MALAT1* can be used as an additional biomarker to predict post-isolation islet outcome to further improve the prediction for clinical islet transplantation (Wong et al., 2019). Taking into account that *MALAT1* is encoded inside an active enhancer cluster, it is possible that *MALAT1* could be facilitating endocrine gene expression within human islets (Wong et al., 2019). On the other hand, *MALAT1* is regulated by hypoxia (Brock et al., 2017; Sallé-Lefort et al., 2016; Wenger et al., 2015), a major determinant of post-isolation islet quality (Lau et al., 2009); and thus, *MALAT1* could be influencing islet survival under hypoxic conditions. In a recent study in which the effect of cigarette smoke extract (CSE) in pancreatic β cell function was determined, it was shown that CSE inhibits insulin production by upregulating thioredoxin-interacting protein (TXNIP) *via* *MALAT1*-mediated downregulation of miR-17 (Sun et al., 2018). Taking together, these data suggest a role of *MALAT1* lncRNA in pancreas and β cell function. However, *MALAT1* knockout mouse did not show differences in the histology of pancreases and islets (Eißmann et al., 2012; Nakagawa et al., 2012), neither in proliferation or viability of hepatocellular carcinoma cells (Eißmann et al., 2012).

Growth arrest-specific 5 (*Gas5*) is a lncRNA highly expressed in mouse pancreas and downregulated in db/db mice, a type 2 diabetes (T2D) mouse model (Jin et al., 2018). *Gas5* knockdown in mouse pancreatic β cells inhibited pancreatic β cell proliferation and impaired insulin synthesis and secretion. Moreover, *Gas5* knockdown reduced the expression of islet key transcription factors, such as PDX1 and MAFA, as well as the expression of GLUT2, a glucose transporter essential for the activation of insulin secretion (Jin et al., 2018). These data indicate that lncRNA *Gas5* might perform as a regulator to maintain β cell identity and function through regulation of insulin synthesis and secretion. Indeed, in human islets and in the human β cell line EndoC-β H1, upregulation of lncRNA *Gas5* using short oligonucleotides with GAS5 HREM (hormone response element motif), alleviated glucocorticoid-induced insulin secretion defect, evidencing a role of this lncRNA in the regulation of insulin production and secretion (Esguerra et al., 2020). Interestingly, in line with these results, *Gas5* knockdown in pancreatic β cells downregulated several transcription factors implicated in the regulation of insulin expression, including PDX1 and Nkx6.1 (Esguerra et al., 2020).

p3134 is a lncRNA highly expressed in adipose tissue and in circulating exosomes secreted by islet β cells, that has been shown to enhance insulin synthesis and release *via* regulation of islet-specific transcription factors (Ruan et al., 2018). In addition, overexpression of lncRNA-p3134 decreased apoptosis ratio and partially reversed the effect of glucotoxicity on glucose-stimulated insulin secretion in mouse pancreatic β cells. This effect was confirmed *in vivo* as upregulation of lncRNA-p3134 in db/db mice (T2D mouse model) ameliorated insulin synthesis and release in response to glucose. Interestingly, the mass of insulin-positive cells detected by insulin immunostaining analysis in db/db mice was significantly increased after p3134 upregulation, and consistent with these findings, islets from p3134-overexpressing mice showed increased mRNA and protein levels of PDX1, MAFA, GLUT2 and TCF7L2 when compared to db/db control mice (Ruan et al., 2018).

5. Micropeptides encoded by LncRNAs and their potential role in pancreatic β cells

In the recent years, translation events of noncoding genes have been frequently described in different cell types and tissues. The resulted micropeptides have been implicated in several functions of importance in biological and pathological processes in different cell types and tissues (Anderson et al., 2015, 2016; Bhatta et al., 2020; Cai et al., 2017; Guo et al., 2020; Makarewich et al., 2018; Wu et al., 2020; Yeasmin et al., 2018). Accumulating evidence support that small open reading frames (smORFs) present in the nucleotide sequence of lncRNAs have the potential to encode micropeptides with biological functions (Wang et al., 2019) (Fig. 4). However, up to now the existence of these micropeptides at the pancreatic β cell level has been poorly documented. In order to identify protein coding potential for lncRNAs expressed in human pancreas, Cao et al. used a bioinformatics approach and identified 30 lncRNAs that were likely to encode micropeptides (Cao et al., 2019). In the same study, the authors identified one conserved micropeptide named Beta cell Glucose Regulated Micropeptide 1(BGRM1) that was implicated in the regulation of calcium homeostasis and insulin secretion in beta cells. This study was the first to identify that some lncRNAs expressed in human pancreas might actually be translated as micropeptides and play important roles for pancreatic functions.

More recently, using two computational algorithms named PhyloCSF and RNAcode, to comprehensively survey the protein-coding potential of lncRNAs expressed in the human pancreatic islet, a putative micropeptide encoded from the lncRNA *TUNAR* (also known as *TUNA*, *HI-LNC78* or *LINC00617*) was identified (Li et al., 2020). *TUNAR* is a lncRNA highly expressed in pancreatic β and neuronal cells that has been shown to participate in the regulation of insulin secretion in human pancreatic β cells (Akerman et al., 2017). Interestingly, *TUNAR* was predicted to harbor a smORF at the beginning of the last exon (which is shared among the three transcript isoforms) which, additionally, was highly conserved among 100 vertebrate genomes (Li et al., 2020). Moreover, ribosomal profiling suggested that this smORF was actively translated into a 48 amino acid micropeptide; and the conservation analysis of the resulted micropeptide revealed that its sequence was highly conserved in 88 vertebrate species. The micropeptide, named by the authors as Beta cell- and Neural cell-regulin (BNLN), was enriched in the insulin producing β cells of human pancreatic islets. BNLN was primarily localized at the endoplasmic reticulum (ER) of pancreatic β cells and its overexpression lowered ER calcium levels, increased cytosolic calcium levels, and maintained ER homeostasis in response to high glucose challenge. Interestingly, in mice fed with a high fat diet, BNLN expression was downregulated when compared to animals fed with normal diet. In line with these results, BNLN overexpression

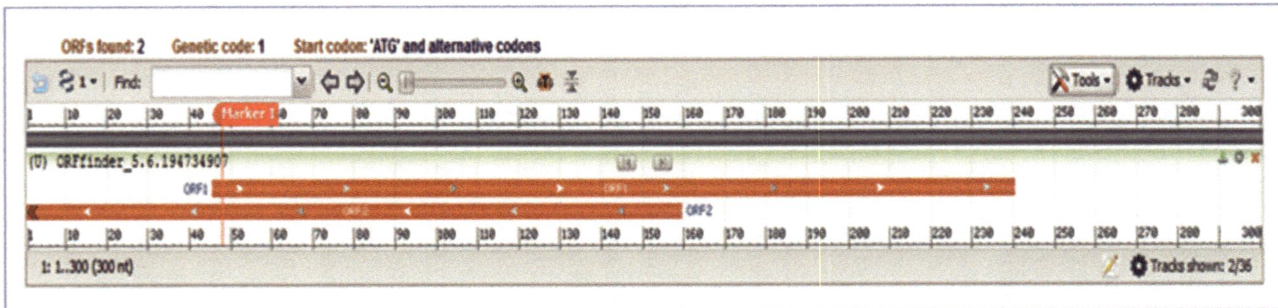

ORF number	Nucleotide number	Micropeptide length
1	159 nucleotides	53 aa
2	196 nucleotides	64 aa

Fig. 4 *In silico* prediction of Open Reading Frames (ORFs) in a lncRNA. In red, the schematic representation of two ORFs identified in a given lncRNA. ORF1 has 159 nucleotides and potentially codifies a micropeptide of 53 aminoacids, and ORF2 has 196 nucleotides and potentially codifies a micropeptide of 64 aminoacids. Prediction was performed using the software ORF finder.

enhanced insulin secretion in islets from lean and obese mice as well as from humans. Taken together the results of this study suggested that dysregulation of BNLN in the pancreas might contribute to the obesity-associated malfunction of insulin secretion (Li et al., 2020).

Another study using a human embryonic stem cell (hESC)-based pancreatic differentiation system has recently shown that many lncRNAs in direct vicinity of lineage-determining transcription factors (TFs) are dynamically regulated, predominantly cytosolic, and highly translated (Gaertner et al., 2020). Genetic ablation of 10 lncRNAs with coding potential showed that 9 were essential for pancreatic endocrine cell development. The deletion of one of them, *LINC00261*, diminished insulin positive cells, in a manner independent of the nearby islet transcription factor FOXA2. However, one-by-one removal of smORFs within *LINC00261* did not show any impact in endocrine cell development, and thus, the authors concluded that the RNA, rather than the produced microproteins, was required for endocrine cell formation. According to the authors, *LINC00261*'s microproteins and smORFs may have redundant functions, which could explain why deletion of individual smORFs produced no apparent phenotype. Thus, despite limited sequence similarity and stark differences in translation rates between the identified translated smORFs in *LINC00261*, they could not rule out that different microproteins produced by *LINC00261* compensate when one smORF was deleted.

Multiple micropeptides encoded from smORFs in lncRNAs have been implicated in biological and pathological processes (63–70). However this field of research has just started to flourish and there are several unknowns that to remain to be clarified. In the context of β cell function and diabetes development, beyond the potential implication of these micropeptides in basal pancreatic β functions, the hypothesis that these micropeptides could also act as new autoantigens under specific stress stimuli should be considered when studying autoimmune diabetes. Indeed, a micropeptide encoded by lncRNA *MIR155HG* has been shown to regulate antigen presentation and suppress autoimmune inflammation (Niu et al., 2020).

In summary, these findings are all suggestive of a role for lncRNAs (both, as RNA or by encoding functional micropepides) in islet regulome and β cell biology and function. Nevertheless, the implication of lncRNAs in pancreatic β cell dysfunction and diabetes development has been poorly studied. The implication of lncRNAs in the regulation of basal β cell functions, such as insulin production and release, and the fact that several diabetes-associated genetic variants and mutations are located in lncRNAs, suggest a role of these non-coding RNA molecules in the pathogenesis of diabetes. In the following sections of this chapter, we provide an overview of the implication of lncRNAs in the pathogenesis of different types of diabetes, including monogenic and polygenic forms of the disease.

6. LncRNAs and diabetes

Diabetes mellitus is a group of metabolic diseases characterized by hyperglycemia resulting from defects in insulin secretion, insulin action, or both. The vast majority of cases of diabetes fall into two broad etiopathogenetic categories: type 1 and type 2 diabetes (American Diabetes Association, 2009). While in type 1 diabetes (T1D) there is an absolute deficiency of insulin secretion due to the autoimmune destruction of pancreatic β cells; in type 2 diabetes (T2D), the main cause is a combination of resistance to insulin action and an inadequate compensatory insulin secretory response (American Diabetes Association, 2009). Apart from T1D and T2D, there are other forms of diabetes that include gestational diabetes mellitus (any degree of glucose intolerance with onset or first recognition during pregnancy) (Coustan, 2013) and monogenic forms of diabetes (maturity-onset diabetes of the young (MODY) and neonatal diabetes, for example) (Harris and Naylor, 2019).

During the last years, accumulating evidence have linked ncRNAs to diabetes susceptibility and abnormal expression of ncRNAs in peripheral blood have been observed in patients with different types of diabetes and their complications (Cunnington et al., 2010; Scott et al., 2007; Todd et al., 2007; Wallace et al., 2010; Zhang, 2019) (Fig. 5). In general, few studies have characterized the functional role of ncRNAs in the pathogenesis of diabetes, and most of the studies have centered in the study of microRNAs rather than in the analysis of lncRNAs (Dotta et al., 2018; Filardi et al., 2020; Kaur et al., 2020). This field of research is still barely explored, but most probably, in the following years it will arise to provide new insights into the role of these long non-coding molecules in diabetes pathogenesis and pancreatic β cell dysfunction. The characterization of lncRNAs in pancreatic β cells will open new avenues to identify novel prediction/diagnostic markers and to develop new therapeutic approaches to slow down or arrest the development of diabetes and its complications.

lncRNA	Disease	Interacting gene/protein	Related mechanism	Reference
Lnc13	T1D	*STAT1* gene	Pro-inflammatory signling pathway	89
NONHSAG011351	T1D	*ERBB3* gene	Basal and pro-inflammatory cytokine-induced β cell apoptosis	90
GLIS3	T1D and T2D	Bim protein	Cytokine-induced β cell apoptosis	91, 92, 98
NONHSAG044354	T1D	*BACH2* gene	Cytokine-induced β cell apoptosis	93
LncRNA-1	T1D	*NF-κB* gene	β cell apoptosis	33
MALAT1	T1D T2D	*PDX-1* gene *PTBP1* gene	β cell dysfunction β cell apoptosis and insulin secretion	100 103
MEG3	T1D and T2D	*Dlk1* gene	Sensitivity of β cells to cytokine-induced oxidative stress	106-108
Kcnqlot1	T2D	*Kcnqlot1* gene	Voltage-gated potassium channel	109,110
βlinc3	T2D	-	Body mass index and glycemia levels	31
NONRATT003679.2	T2D	-	Glucolipotoxicity-induced β cell damage	105
HNF4α-AS1	MODY1	HNF4α	β cell dysfunction	114
LINC01339	MODY3	HNF1α	Glycolisis and mitochondrial respiration	116
HYMAI	TNDM	-	Persistent hyperglycaemia	118
H19	Gestational diabetes	IGF2	Insulin resistance in obesity and pregnancy	122-124

Fig. 5 LncRNAs and their role in different types of diabetes.

6.1 LncRNAs in the pathogenesis of type 1 diabetes

6.1.1 T1D-associated lncRNAs and its functional impact in pancreatic β cell dysfunction

Type 1 diabetes is a complex chronic disease characterized by the autoimmune destruction of insulin-producing pancreatic β cells (Ilonen et al., 2019). Several genome wide association studies (GWAS) have identified disease-associated genetic variants in non-coding regions of the genome, and specifically in lncRNAs, suggesting a potential link between these molecules and T1D pathogenesis (Mirza et al., 2014).

The disease-associated SNPs in lncRNAs have the potential to disturb the regulatory function of these molecules, particularly if they disrupt their secondary structure, a feature that is essential for their function (Castellanos-Rubio and Ghosh, 2019). In this sense, a study performed by Mirza et al. (2014) revealed that several SNPs in close proximity of T1D loci-associated candidate genes were predicted to disrupt secondary structure of several lncRNAs, suggesting that the RNA conformation altering polymorphisms might be associated with disease phenotype.

Recently, a study performed in the human pancreatic β cell line EndoC-βH1 showed that a T1D-associated lncRNA named *Lnc13* was implicated in the regulation of the STAT1 pro-inflammatory signaling pathway (Gonzalez-Moro et al., 2020). Interestingly, *Lnc13* regulated STAT1 activation and signaling by stabilizing *STAT1* mRNA molecule in an allele-specific manner. Although overexpression of both *Lnc13* alleles upregulated STAT1 signaling, *Lnc13* harboring the risk allele for T1D induced a higher activation than the protective allele. These results were in line with the observation that individuals harboring the risk genotype for T1D in *Lnc13* presented higher levels of *STAT1* in their pancreatic islets when compared to individuals harboring the heterozygous or homozygous genotype for the protective allele. These findings provided for the first time information on the molecular mechanisms by which one T1D-associated SNP in a lncRNA influence disease pathogenesis at the pancreatic β cell level.

Another antisense lncRNA (NONHSAG011351) located at 5 kb from the T1D candidate *loci ERBB3* has been shown to regulate the expression of its neighboring gene (Kaur et al., 2016). The authors showed that a T1D-associated SNP located in *ERBB3* had a cis-eQTL effect on both, *ERBB3* gene and its associated lncRNA. The protective genotype (GG) for T1D was associated with a lower expression of both genes, and interestingly, *ERBB3* downregulation reduced basal and pro-inflammatory cytokine-induced pancreatic β cell apoptosis when compared to control cells. These data suggested a protective role of *ERBB3*-associated lncRNA in pancreatic β cell destruction.

A T1D and T2D candidate gene named *GLIS3* has been described to play a crucial role in cytokine-induced pancreatic β cell apoptosis *via* regulation of a splice variant of the pro-apoptotic protein Bim (Nogueira et al., 2013). Interestingly, *GLIS3* is linked to the activity of different lncRNAs. On one hand, deletion of the human islet specific *Lnc25* in the human EndoC-βH1 cell line induced a decrease in *GLIS3* mRNA expression, although it did not affect glucose-stimulated insulin secretion (Morán et al., 2012). On the other hand, a recent work has identified a novel lncRNA that is regulated by *GLIS3* in mouse pancreatic β cells (Scoville et al., 2020). In a knockout mouse model in which approximately 80% of the lncRNA was deleted using CRISPR-Cas9, the authors observed a loss of expression of *GLIS3*; however, the mice did not develop any phenotypic characteristic linked to β-cell impairment.

Other T1D candidate gene that overlaps with a sense lncRNA (*NONHSAG044354*) is *BACH2*, a transcription factor implicated in cytokine-induced pancreatic β cell apoptosis (Marroquí et al., 2014). The polymorphisms associated with T1D are located within an intron of both, the coding *BACH2* gene and the lncRNA *NONHSAG044354*. Interestingly, these SNPs are predicted to disrupt the secondary structure of the lncRNA, suggesting a potential impact on its function (Mirza et al., 2014). Taking into account that most lncRNAs exert their regulatory effect in *cis*, it is plausible to think that *NONHSAG044354* might be regulating the expression of *BACH2*. Indeed, analysis of the expression profiles of T1D loci-associated lncRNAs using Human Body Map (HBM) data revealed that *NONHSAG044354* positively correlated with its associated candidate gene *BACH2* (Mirza et al., 2014). Whether *NONHSAG044354* is implicated in cytokine-induced β cell apoptosis through regulation of its coding counterpart remains to be clarified.

6.1.2 Dysregulation of lncRNAs in type 1 diabetes

The initial stages of type 1 diabetes are characterized by a pro-inflammatory environment in which immune cells infiltrate the pancreatic islets. The inflammation should be considered in the context of a "dialog" between invading immune cells and the target β cells. This dialog is partially mediated both by cytokines and chemokines that are released by β cells and immune cells, and by putative, immunogenic signals that are delivered by dying β cells (Eizirik et al., 2009). Interestingly, several studies have pointed to the potential role of noncoding RNA-containing exosomes in a wide range of processes that underlie pathogenesis of T1D, including the dialog between immune and β cells (Chang and Wang, 2019; Guay et al., 2019). Few is known about lncRNA-containing exosomes released by β cells and their role in T1D pathogenesis, but some studies have already described several diabetes-related processes in which microRNA-containing exosomes released by β cells are implicated, including processes such as insulin release (Sedgeman et al., 2019; Xu et al., 2020). Recently, total RNA and small RNA sequencing of islet-derived exosomes isolated from human pancreatic islets treated with or without IL-1β and IFN-γ, showed that under cytokine stress the expression of several lncRNAs is modified in islet exosomes (Krishnan et al., 2019). To understand the function of the differentially expressed lncRNAs, the authors performed a correlation analysis of expressed lncRNAs and mRNAs, and identified 362 highly correlated lncRNA-mRNA pairs. Pathway analysis revealed that some differentially expressed lncRNA/mRNA pairs were involved in potentially pathogenic pathways in β cells, including the JAK-STAT and the NF-κB signaling pathway, the cytokine–cytokine receptor interaction or the PI3K-Akt signaling pathway (Krishnan et al., 2019).

Coming back to inflammation and type 1 diabetes pathogenesis, some studies analyzing lncRNAs in mice models of T1D have revealed the implication of these non-coding molecules in the regulation of several pathogenic processes. In this sense, Motterle et al. (2015) showed that the expression of three lncRNAs previously shown to be upregulated by pro-inflammatory cytokines, was upregulated in NOD mice at 8 and 13 weeks (islets with peri-insulitis and insulitis, respectively) when compared to 4-weeks old mice (minor insulitis). Interestingly, the expression of these lncRNAs was not modified in NOD mice that did not develop type 1 diabetes, demonstrating that their upregulation was linked to the autoimmune reaction. Overexpression of one of these lncRNAs (LncRNA-1) in combination with low doses of IL-1β in MIN6 cells (mouse β cell line) induced basal β cell apoptosis through promotion of nuclear factor-kappa B (NF-κB) expression, suggesting that sensitisation of beta cells overexpressing lncRNA-1 to apoptosis may at least in part be caused by NF-κB nuclear translocation.

In a similar study in which mouse MIN6 cells were exposed to pro-inflammatory cytokines, 444 upregulated and 279 downregulated lncRNAs were identified (Sun et al., 2016a). Based on GO analysis, the top five enriched GO terms that were associated with the upregulated lncRNAs were: innate immune response, inflammatory response, apoptotic process, positive regulation of I-kB kinase/nuclear factor NF-κB cascade, and defenced response to virus. These data suggested a link between inflammation-induced lncRNAs and activation of potentially diabetogenic pathways in pancreatic β cells. In contrast, GO terms associated with down-regulated lncRNAs were linked to basal cellular functions, including transport, cell cycle, oxidation-reduction process and mitosis, among others. In summary, these data suggest that differentially expressed lncRNAs may play key or partial roles in cytokine-mediated β cell dysfunction by potentiating the activation of inflammatory and apoptotic pathways, or by shutting down key basal cellular functions.

Recently, lncRNA *MALAT1* has been described to induce β cell dysfunction *via* reduction of the histone acetylation of the PDX1 promoter in the context of type 1 diabetes (Ding et al., 2020). In this study, IL-1β stimulation significantly promoted the expression of *MALAT1* and decreased the expression of PDX1 in NOD mouse islets and Min6 cells. Moreover, overexpressing *MALAT1* suppressed the H3 histone acetylation of the PDX1 promoter, inhibiting both mRNA and protein expressions of PDX1. Genetic silencing of *MALAT1* restored the decrease of the histone acetylation of the PDX1 promoter, as well as the PDX1 expression, which was reduced by IL-1β stimulation. Thus, based on these observations, it seems that *MALAT1* induces β cell dysfunction through reduction of the H3 histone acetylation of the PDX1 promoter, eventually inhibiting PDX1 expression and suppressing insulin release.

6.2 LncRNAs in the pathogenesis of type 2 diabetes

The role of lncRNAs in type 2 diabetes (T2D) has been extensively studied during the last years. In addition to expression analysis of circulating lncRNAs in T2D patients (Carter et al., 2015; Pengyu et al., 2020; Ruan et al., 2018), several functional studies have been performed to elucidate the function of these non-coding molecules at the pancreatic β cell level (Kong et al., 2019; Xiong et al., 2020; Zhang et al., 2020).

We have already mentioned lncRNA *MALAT1* and its role in pancreatic β cell dysfunction in the context of inflammation (Ding et al., 2020). Interestingly, this lncRNA seems to be implicated not only in the pathogenesis of T1D, but also in the development of T2D. Indeed, a recent studied showed that the expression of *MALAT1* was reduced in the islets of db/db mice (*in vivo* model of T2D) and palmitate-treated MIN6 cells (*in vitro* model of T2D) (Xiong et al., 2020). The same study revealed that inhibition of *MALAT1* significantly induced β cell apoptosis and inhibited insulin secretion. Mechanistic experiments showed that *MALAT1* enhanced polypyrimidine tract-binding protein 1 (PTBP1) stability by direct interaction, thereby adjusting the ratio of pyruvate kinase muscle (PKM) isoforms 1 and 2 (PKM1/PKM2). In addition, chromatin immunoprecipitation experiments indicated that *MALAT1* was transcriptionally activated by the islet specific transcription factor PDX1, through which exendin-4 alleviated lipotoxicity-induced pancreatic β cell damage. These data suggest that *MALAT1* may serve as a therapeutic target for the treatment of T2D.

Other lncRNA that has been genetically associated with both, type 1 and 2 diabetes (Kameswaran et al., 2014; Wallace et al., 2010), is *Meg3*. This lncRNA is transcribed downstream *Dlk1* gene, a non-canonical Notch ligand that is highly expressed in pancreatic β cells (Falix et al., 2012). Interestingly, human islets from T2D individuals showed downregulation of *Meg3* and hypermethylation and misregulation of the *Dlk1* locus (Kameswaran et al., 2014). Using targeted epigenetic modifiers, Kameswaran et al. showed that increased methylation of *Meg3* in mouse pancreatic β cells resulted in decreased transcription of the maternal transcripts associated with this locus, resulting in increased sensitivity of β cells to cytokine-induced oxidative stress (Kameswaran et al., 2018). In addition, they demonstrated that an evolutionarily conserved intronic region in *Meg3* locus acted as an enhancer in β cells. In this sense, they showed that the promoter of *Meg3* physically interacted with the intronic enhancer and other putative regulatory elements in *Meg-Dlk1* imprinted region in human islets. Remarkably, the enhancer was bound in an allele-specific manner by the key transcription factors in β cells, FOXA2, PDX1, and NKX2.2.

Another T2D association signal detected by GWAS lies into a paternally expressed antisense lncRNA named *Kcnq1ot1* that is transcribed from an imprinted locus containing the *Kcnq1* gene, a gene that encodes a voltage-gated potassium channel (Monk et al., 2006). Interestingly, lncRNA *Kcnq1ot1* was shown to be upregulated in pancreatic islets of T2D patients (Morán et al., 2012). In contrast, methylation analysis in human fetal samples showed that homozygous genotype for the T2D-associated SNP in *Kcnq1ot1* correlated with increased methylation at the locus, but it was not related to any change in expression levels of *Kcnq1ot1* or *Kcnq1* genes (Travers et al., 2013).

T2D is a chronic metabolic disorder characterized by elevated blood glucose and/or high serum free fatty acids. Chronic hyperlipidemia causes the dysfunction of pancreatic β cells, which is aggravated in the presence of hyperglycemia (glucolipotoxicity). Thus, in order to identify lncRNAs linked to pancreatic β cell dysfunction and T2D development, some studies have compared the transcriptome of mice fed with regular or high fat diet (Motterle et al., 2017). Using this approach, Motterle et al. identified two pancreatic islet-enriched lncRNAs named βlinc2 and βlinc3 (β long intergenic noncoding RNA 2 and 3) that were differentially expressed in islets from mice fed with high fat diet (upregulated and downregulated compared to control group, respectively) (Motterle et al., 2017). The expression of both lncRNAs correlated with body weight gain and glycaemia levels in obese mice and was also modified in diabetic db/db mice. Interestingly, the human orthologue of *βlinc3* was altered in the islets of T2D patients and was associated with the body mass index (BMI) of the donors.

A recent study in which rat INS-1E cells were exposed to high glucose and palmitate to induce glucolipotoxicity identified several differentially expressed lncRNAs (Kong et al., 2019). Among dysregulated lncRNAs, *NONRATT003679.2* (downregulated in treated cells) shown to be implicated in glucolipotoxicity-induced rat pancreatic β cell damage. Indeed, this lncRNA acted as a molecular sponge for a microRNA, sustaining SIRT1-mediated suppression of glucolipotoxicity-induced ROS accumulation and apoptosis in β cells.

6.3 LncRNAs in other types of diabetes

6.3.1 LncRNAs and monogenic diabetes

Monogenic diabetes are rare forms of diabetes that result from mutations or changes in a single gene. The two main forms of monogenic diabetes are Neonatal diabetes mellitus (NDM), which occurs in newborns and young infants, and maturity-onset diabetes of the young (MODY), which usually appears in adolescence or early adulthood (Redondo et al., 2020). Little is known about the implication of lncRNAs in the development of monogenic forms of diabetes, but there are some studies that have already identified lncRNAs potentially linked to the pathogenesis of these disorders.

MODY is characterized by autosomal dominant inheritance, onset before 25 years of age, absence of β cell autoimmunity, and sustained pancreatic β cell function. To date, mutations have been identified in at least 14 different genes, including six genes encoding proteins that, respectively, correspond to MODY subtypes 1 to 6 (Urakami, 2019).

MODY1 is caused by heterozygous loss-of-function mutations in HNF4α (hepatocyte nuclear factor 4α), a transcription factor that plays a crucial role in pancreatic β cell function (Yamagata, 2014). A lncRNA named *HNF4α-AS1* that is located between the P2 and P1 promoters of *HNF4α* gene has been implicated in the regulation of *HNF4α* gene (Guo and Lu, 2019). Interestingly, several mutations related to MODY1 and liver cancer in P1 promoter of *HNF4α* gene affected the expression of lncRNA *HNF4α-AS1*. Although this lncRNA has been widely characterized in several pathogenic conditions linked to cancer (Chen et al., 2020), nothing is known about its potential role in pancreatic β cell dysfunction in the context of MODY1.

MODY3 subtype is caused by mutations in the gene that encodes *HNF1α* (hepatocyte nuclear factor 1α), a key transcription factor that is mainly implicated is β cell function and growth (Yamagata, 2014). Interestingly, it has been recently shown that in a human model of MODY3 based on pluripotent stem cells, a lncRNA named *LINC01139* was dysregulated and participated in a subset of phenotypes seen in *HNF1α mutants*, mainly related with defects in glycolysis, mitochondrial respiration, and downregulation of several genes downstream of *HNF1α* (Leonardo Cardenas and Leonardo, 2018).

Regarding NDM, up to day the potential implication of lncRNAs in its development has been poorly analyzed. NDM is a rare type of diabetes that develops prior to 6 months of age and most of the cases are explained by mutations in genes associated with β cell function, including *KCNJ11* and *ABCC8* (encoding pancreatic β cell K-ATP channel), among others (Lemelman et al., 2018). Taking into account that several lncRNAs have been linked to pancreas and β cell development, it is plausible to think that lncRNAs might play a role in the pathogenesis of this disease. Indeed, *HYMAI* is an imprinted lncRNA that has been linked to transient NDM (TNDM) (Kamiya et al., 2000). TNDM is a subtype in which persistent hyperglycaemia appears within the first 6 months of life, but affected infants go into remission within a few months, with possible relapse to permanent diabetes in adolescence or adulthood (Lemelman et al., 2018). Methylation studies in patients with TNDM have revealed that the region containing lncRNA *HYMAI* in chromosome 6q24 is hypomethylated when compared to control individuals (Kamiya et al., 2000). Further studies are needed to clarify the potential implication of this lncRNA in TNDM development.

6.3.2 LncRNAs and gestational diabetes

Gestational diabetes mellitus (GDM) is characterized by glucose intolerance that first manifests during pregnancy, leading to hyperglycemia of varying severity (Buchanan et al., 2007). Gestational diabetes is mainly develop in the context of insulin resistance that results from a combination of increased maternal adiposity and the insulin-desensitizing effects of hormones made by the placenta (Ryan et al., 1995). However, several studies have pointed to the existence of a large defect in pancreatic β cell function in women with GDM both during and after pregnancy (Ryan et al., 1995). Some works have analyzed the expression and function of lncRNAs in β cells in the context of pregnancy and gestational diabetes, although it is still a barely studied field.

By analyzing the transcriptome of non-pregnant and pregnant mice at gestational day 14.5, a study identified several lncRNAs differentially expressed in mouse islets during pregnancy (Sisino et al., 2017). Interestingly, some of those lncRNAs were implicated in mouse β cell mass expansion and proliferation, as shown by functional studies in both, *in vitro* and *in vivo* experiments (Sisino et al., 2017), suggesting that they might be implicated in the regulation of potentially pathogenic pathways linked to the development of gestational diabetes.

On the other hand, the lncRNA *H19* is transcribed from a conserved imprinted gene cluster, with the *IGF2* (insulin-like growth factor 2) gene located approximately 100 kb upstream of *H19* (Gabory et al., 2010). This lncRNA is maternally expressed and has been shown to be highly expressed in neonatal pancreatic islets, suggesting a potential role in β cell proliferation (Sanchez-Parra et al., 2018). Moreover, *H19* expression is increased in islets of obese mice under conditions of increased insulin demand, pointing to an important role of this lncRNA in the regulation of compensating mechanisms for insulin resistance in obesity (Sanchez-Parra et al., 2018). Interestingly, fetal disruption of *H19* caused significantly higher glucose concentrations during the last trimester in pregnant mice, however human gene variation in maternally transmitted fetal *H19* alleles (rs2071094) was associated with birth weight and other markers of size, but not with significant changes in maternal glucose tolerance in the third trimester of pregnancy (Petry et al., 2011).

Other works studying the link between lncRNAs and the development of GDM have associated polymorphisms located in or close to lncRNAs with the genetic risk to develop GDM (Hu et al., 2019), or have observed differential expression of several lncRNAs in pregnant women with GDM (Zhang et al., 2018). Nevertheless, the role of all these lncRNAs in pancreatic β cells has not been analyzed yet.

7. Concluding remarks

The identification and characterization of lncRNAs expressed in pancreatic β cells have improved our understanding of β cell development and function, and provided insights into many of the cellular dysfunctions related to diabetes. Even if this field of research is still emerging, accumulating scientific data support the crucial role of these non-coding molecules in pancreatic β cell function and dysfunction.

Functional characterization of lncRNAs in pancreatic β cells will provide important information to further understand the physiology of pancreatic β cells, both in health and disease, and to open new avenues for the development of therapeutic strategies to treat and/or cure diabetes.

Author contributions

I.G.M. and I.S. performed the bibliographic research. I.S. wrote the manuscript and is the guarantor of this work. All authors revised the manuscript and approved the final version.

Conflicts of interest

All the authors declare to have no conflict of interest.

Funds

This work was supported by the European Foundation for the Study of Diabetes (EFSD) - EFSD/JDRF/Lilly Programme on Type 1 Diabetes Research and the Spanish Ministry of Science, Innovation and Universities (PID2019-104475GA-I00). I.G.M. is supported by a Predoctoral Fellowship Grant from the UPV/EHU (Universidad del País Vasco / Euskal Herriko Unibertsitatea).

References

Akerman, I., et al., 2017. Human pancreatic β cell lncRNAs control cell-specific regulatory networks. Cell Metab. 25 (2), 400–411.

American Diabetes Association, 2009. Diagnosis and Classification of Diabetes Mellitus. Diabetes Care, p. S62.

Anderson, D.M., et al., 2015. A micropeptide encoded by a putative long noncoding RNA regulates muscle performance. Cell 160 (4), 595–606.

Anderson, D.M., et al., 2016. Widespread control of calcium signaling by a family of SERCA-inhibiting micropeptides. Sci. Signal. 9 (457), ra119.

Barrett, J.C., et al., 2009. Genome-wide association study and meta-analysis find that over 40 loci affect risk of type 1 diabetes. Nat. Genet. 41 (6), 703–707.

Benner, C., et al., 2014. The transcriptional landscape of mouse beta cells compared to human beta cells reveals notable species differences in long non-coding RNA and protein-coding gene expression. BMC Genomics 15 (1), 620.
Bhat, S.A., et al., 2016. Long non-coding RNAs: mechanism of action and functional utility. Non-coding RNA Res. 1 (1), 43–50.
Bhatta, A., et al., 2020. A mitochondrial micropeptide is required for activation of the Nlrp3 inflammasome. J. Immunol. 204 (2), 428–437.
Böhmdorfer, G., Wierzbicki, A.T., 2015. Control of chromatin structure by long noncoding RNA. Trends Cell Biol. 25 (10), 623–632.
Brock, M., et al., 2017. Analysis of hypoxia-induced noncoding RNAs reveals metastasis-associated lung adenocarcinoma transcript 1 as an important regulator of vascular smooth muscle cell proliferation. Exp. Biol. Med. 242 (5), 487–496.
Buchanan, T.A., et al., 2007. What is gestational diabetes? Diabetes Care 30 (2), S105–S111.
Cabili, M., et al., 2011. Integrative annotation of human large intergenic noncoding RNAs reveals global properties and specific subclasses. Genes Dev. 25 (18), 1915–1927.
Cai, B., et al., 2017. LncRNA-Six1 encodes a micropeptide to activate Six1 in Cis and is involved in cell proliferation and muscle growth. Front. Physiol. 20 (8), 230.
Cao, H., et al., 2019. Comprehensive identification of micropeptides encoded by long noncoding RNAs in human tissues. FASEB J. 33 (S1), 714.
Carter, G., et al., 2015. Circulating long noncoding RNA GAS5 levels are correlated to prevalence of type 2 diabetes mellitus. BBA Clin. 4, 102–107.
Castellanos-Rubio, A., Ghosh, S., 2019. Disease associated SNPs in inflammation-related lncRNAs. Front. Immunol. 10, 420.
Chang, W., Wang, J., 2019. Exosomes and their noncoding RNA cargo are emerging as new modulators for diabetes mellitus. Cells 8 (8), 853.
Chen, L., et al., 2020. The roles of long noncoding RNAs HNF1α-AS1 and HNF4α-AS1 in drug metabolism and human diseases. Noncoding RNA 6 (2), 24.
Cho, Y.S., et al., 2012. Meta-analysis of genome-wide association studies identifies eight new loci for type 2 diabetes in east Asians. Nat. Genet. 44 (1), 67–72.
Coustan, D.R., 2013. Gestational diabetes mellitus. Clin. Chem. 59 (9), 1310–1321.
Cunnington, M.S., et al., 2010. Chromosome 9p21 SNPs associated with multiple disease phenotypes correlate with ANRIL expression. PLoS Genet. 6 (4), 1000899.
Delás, M.J., Hannon, G.J., 2017. lncRNAs in development and disease: from functions to mechanisms. Open Biol. 7 (7), 170121.
Derrien, T., et al., 2012. The GENCODE v7 catalog of human long noncoding RNAs: analysis of their gene structure, evolution, and expression. Genome Res. 22 (9), 1775–1789.
Ding, H., et al., 2020. LncRNA MALAT1 induces the dysfunction of β cells via reducing the histone acetylation of the PDX1 promoter in type 1 diabetes. Exp. Mol. Pathol. 114, 104432.
DiStefano, J.K., 2018. The emerging role of long noncoding RNAs in human disease. Methods Mol. Biol. 1076, 91–110.
Dorrell, C., et al., 2011. Transcriptomes of the major human pancreatic cell types. Diabetologia 54 (11), 2832–2844.
Dotta, F., et al., 2018. MicroRNAs: markers of β-cell stress and autoimmunity. Curr. Opin. Endocrinol. Diabetes Obes. 25 (4), 237–245.
Dunham, I., et al., 2012. An integrated encyclopedia of DNA elements in the human genome. Nature 489 (7414), 57–74.
Eißmann, M., et al., 2012. Loss of the abundant nuclear non-coding RNA MALAT1 is compatible with life and development. RNA Biol. 9 (8), 1076–1087.
Eizirik, D.L., Colli, M.L., Ortis, F., 2009. The role of inflammation in insulitis and beta-cell loss in type 1 diabetes. Nat. Rev. Endocrinol. 5 (4), 219–226.

Esguerra, J.L.S., et al., 2020. Glucocorticoid induces human beta cell dysfunction by involving riborepressor GAS5 LincRNA. Mol. Metab. 32, 160–167.
Fadista, J., et al., 2014. Global genomic and transcriptomic analysis of human pancreatic islets reveals novel genes influencing glucose metabolism. Proc. Natl. Acad. Sci. U. S. A. 111 (38), 13924–13929.
Falix, F.A., et al., 2012. Possible roles of DLK1 in the notch pathway during development and disease. Biochim. Biophys. Acta 1822 (6), 988–995.
Filardi, T., et al., 2020. Non-coding RNA: role in gestational diabetes pathophysiology and complications. Int. J. Mol. Sci. 21 (11), 4020.
Gabory, A., Jammes, H., Dandolo, L., 2010. The H19 locus: role of an imprinted non-coding RNA in growth and development. Bioessays 32 (6), 473–480.
Gaertner, B., et al., 2020. A functional screen of translated pancreatic lncRNAs identifies a microprotein-independent role for LINC00261 in endocrine cell differentiation. elife 9, 58659.
Giral, H., Landmesser, U., Kratzer, A., 2018. Into the wild: GWAS exploration of non-coding RNAs. Front. Cardiovasc. Med. 5, 181.
Gittes, G.K., 2009. Developmental biology of the pancreas: a comprehensive review. Dev. Biol. 326 (1), 4–35.
Gonzalez-Moro, I., et al., 2020. The T1D-associated lncRNA Lnc13 modulates human pancreatic β cell inflammation by allele-specific stabilization of STAT1 mRNA. Proc. Natl. Acad. Sci. U. S. A. 117 (16), 9022–9031.
Guay, C., et al., 2019. Lymphocyte-derived Exosomal MicroRNAs promote pancreatic β cell death and may contribute to type 1 diabetes development. Cell Metab. 29 (2), 348–361.
Guo, S., Lu, H., 2019. Novel mechanisms of regulation of the expression and transcriptional activity of hepatocyte nuclear factor 4α. J. Cell. Biochem. 120 (1), 519–532.
Guo, B., et al., 2020. Micropeptide CIP 2A- BP encoded by LINC 00665 inhibits triple-negative breast cancer progression. EMBO J. 39 (1), e102190.
Guttman, M., et al., 2011. LincRNAs act in the circuitry controlling pluripotency and differentiation. Nature 477 (7364), 295–300.
Harris, A., Naylor, R.N., 2019. Pediatric monogenic diabetes: a unique challenge and opportunity. Pediatr. Ann. 48 (8), e319–e325.
Hu, D., et al., 2019. Genetic variants in AC092159.2 and risk of gestational diabetes mellitus in a Chinese population. DNA Cell Biol. 38 (10), 1069–1077.
Hung, T., et al., 2011. Extensive and coordinated transcription of noncoding RNAs within cell-cycle promoters. Nat. Genet. 43 (7), 621–629.
Ilonen, J., Lempainen, J., Veijola, R., 2019. The heterogeneous pathogenesis of type 1 diabetes mellitus. Nat. Rev. Endocrinol. 15 (11), 635–650.
Ji, P., et al., 2003. MALAT-1, a novel noncoding RNA, and thymosin β4 predict metastasis and survival in early-stage non-small cell lung cancer. Oncogene 22 (39), 8031–8041.
Jin, F., et al., 2018. Downregulation of long noncoding RNA Gas5 affects cell cycle and insulin secretion in mouse pancreatic β cells. Cell. Physiol. Biochem. 43 (5), 2062–2073.
Kameswaran, V., et al., 2014. Epigenetic regulation of the DLK1-MEG3 MicroRNA cluster in human type 2 diabetic islets. Cell Metab. 19 (1), 135–145.
Kameswaran, V., et al., 2018. The dysregulation of the DLK1-MEg3 locus in islets from patients with type 2 diabetes is mimicked by targeted epimutation of its promoter with TALE-DNMT constructs. Diabetes 67 (9), 1807–1815.
Kamiya, M., et al., 2000. The cell cycle control gene ZAC/PLAGL1 is imprinted—a strong candidate gene for transient neonatal diabetes. Hum. Mol. Genet. 9 (3), 453–460.
Kang, H.S., et al., 2009. Transcription factor Glis3, a novel critical player in the regulation of pancreatic β-cell development and insulin gene expression. Mol. Cell. Biol. 29 (24), 6366–6379.

Kaur, S., et al., 2016. The genetic and regulatory architecture of ERBB3-type 1 diabetes susceptibility locus. Mol. Cell. Endocrinol. 419, 83–91.

Kaur, P., et al., 2020. Role of miRNAs in the pathogenesis of T2DM, insulin secretion, insulin resistance, and β cell dysfunction: the story so far. J. Physiol. Biochem. 76 (4), 485–502.

Klattenhoff, C.A., et al., 2013. Braveheart, a long noncoding RNA required for cardiovascular lineage commitment. Cell 152 (3), 570–583.

Kong, X., et al., 2019. LncRNA LEGLTBC functions as a ceRNA to antagonize the effects of miR-34a on the downregulation of SIRT1 in glucolipotoxicity-induced INS-1 beta cell oxidative stress and apoptosis. Oxidative Med. Cell. Longev. 2019, 4010764.

Krishnan, P., et al., 2019. Profiling of RNAs from human islet-derived exosomes in a model of type 1 diabetes. Int. J. Mol. Sci. 20 (23), 5903.

Ku, G.M., et al., 2012. Research resource: RNA-Seq reveals unique features of the pancreatic β-cell transcriptome. Mol. Endocrinol. 26 (10), 1783–1792.

Kumar, V., et al., 2013. Human disease-associated genetic variation impacts large intergenic non-coding RNA expression. PLoS Genet. 9 (1), e1003201.

Kutty, R.K., et al., 2018. Proinflammatory cytokine interferon-γ increases the expression of BANCR, a long non-coding RNA, in retinal pigment epithelial cells. Cytokine 104, 147–150.

Lau, J., et al., 2009. Oxygenation of islets and its role in transplantation. Curr. Opin. Organ Transplant. 14 (6), 688–693.

Lemelman, M.B., Letourneau, L., Greeley, S.A.W., 2018. Neonatal diabetes mellitus: an update on diagnosis and management. Clin. Perinatol. 45 (1), 41–59.

Leonardo Cardenas, F., Leonardo, F., 2018. Modeling Monogenic Diabetes Mody3 Using Human Pluripotent Stem Cells. University of Pennsylvania Scholarly Commons.

Li, M., et al., 2020. A putative long noncoding RNA-encoded micropeptide maintains cellular homeostasis in pancreatic β-cells. bioRxiv. 2020.05.12.091728.

Lu, W., et al., 2018. Identification of differential expressed lncRNAs in human thyroid cancer by a genome-wide analyses. Cancer Med. 7 (8), 3935–3944.

Makarewich, C.A., et al., 2018. MOXI is a mitochondrial micropeptide that enhances fatty acid β-oxidation. Cell Rep. 23 (13), 3701–3709.

Marroquí, L., et al., 2014. BACH2, a candidate risk gene for type 1 diabetes, regulates apoptosis in pancreatic β-cells via JNK1 modulation and crosstalk with the candidate gene PTPN2. Diabetes 63 (7), 2516–2527.

Melloul, D., Marshak, S., Cerasi, E., 2002. Regulation of insulin gene transcription. Diabetologia 45 (3), 309–326.

Mercer, T.R., Dinger, M.E., Mattick, J.S., 2009. Long non-coding RNAs: insights into functions. Nat. Rev. Genet. 10 (3), 155–159.

Mercer, T.R., et al., 2011. Targeted RNA sequencing reveals the deep complexity of the human transcriptome. Nat. Biotechnol. 30 (1), 99–104.

Mirza, A.H., et al., 2014. Effects of GWAS-associated genetic variants on lncRNAs within IBD and T1D candidate loci. PLoS One 9 (8), e105723.

Miyoshi, N., et al., 2000. Identification of an imprinted gene, Meg3/Gtl2 and its human homologue MEG3, first mapped on mouse distal chromosome 12 and human chromosome 14q. Genes Cells 5 (3), 211–220.

Mondal, T., et al., 2015. MEG3 long noncoding RNA regulates the TGF-β pathway genes through formation of RNA-DNA triplex structures. Nat. Commun. 6, 7743.

Monk, D., et al., 2006. Limited evolutionary conservation of imprinting in the human placenta. Proc. Natl. Acad. Sci. U. S. A. 103 (17), 6623–6628.

Morán, I., et al., 2012. Human β cell transcriptome analysis uncovers lncRNAs that are tissue-specific, dynamically regulated, and abnormally expressed in type 2 diabetes. Cell Metab. 16 (4), 435–448.

Motterle, A., et al., 2015. Involvement of long non-coding RNAs in beta cell failure at the onset of type 1 diabetes in NOD mice. Diabetologia 58 (8), 1827–1835.
Motterle, A., et al., 2017. Identification of islet-enriched long non-coding RNAs contributing to β-cell failure in type 2 diabetes. Mol. Metab. 6 (11), 1407–1418.
Nakagawa, S., et al., 2012. Malat1 is not an essential component of nuclear speckles in mice. RNA 18 (8), 1487–1499.
Nica, A.C., et al., 2013. Cell-type, allelic, and genetic signatures in the human pancreatic beta cell transcriptome. Genome Res. 23 (9), 1554–1562.
Niu, L., et al., 2020. A micropeptide encoded by lncRNA MIR155HG suppresses autoimmune inflammation via modulating antigen presentation. Sci. Adv. 6 (21), eaaz2059.
Nogueira, T.C., et al., 2013. GLIS3, a susceptibility gene for type 1 and type 2 diabetes, modulates pancreatic Beta cell apoptosis via regulation of a splice variant of the BH3-only protein Bim. PLoS Genet. 9 (5), 1003532.
Pasquali, L., et al., 2014. Pancreatic islet enhancer clusters enriched in type 2 diabetes risk-associated variants. Nat. Genet. 46 (2), 136–143.
Pengyu, Z., et al., 2020. The differential expression of long noncoding RNAs in type 2 diabetes mellitus and latent autoimmune diabetes in adults. Int. J. Endocrinol. 2020, 9235329.
Penny, G.D., et al., 1996. Requirement for Xist in X chromosome inactivation. Nature 379 (6561), 131–137.
Perry, M.M., et al., 2014. Role of non-coding RNAs in maintaining primary airway smooth muscle cells. Respir. Res. 15 (1), 58.
Petry, C.J., et al., 2011. Maternally transmitted foetal H19 variants and associations with birth weight. Hum. Genet. 130 (5), 663–670.
Redondo, M.J., et al., 2020. The clinical consequences of heterogeneity within and between different diabetes types. Diabetologia 63 (10), 2040–2048.
Romero-Barrios, N., et al., 2018. Splicing regulation by long noncoding RNAs. Nucleic Acids Res. 46 (5), 2169–2184.
Ruan, Y., et al., 2018. Circulating LncRNAs analysis in patients with type 2 diabetes reveals novel genes influencing glucose metabolism and islet β-cell function. Cell. Physiol. Biochem. 46 (1), 335–350.
Ryan, E.A., et al., 1995. Defects in insulin secretion and action in women with a history of gestational diabetes. Diabetes 44 (5), 506–512.
Sallé-Lefort, S., et al., 2016. Hypoxia upregulates Malat1 expression through a CaMKK/AMPK/HIF-1α axis. Int. J. Oncol. 49 (4), 1731–1736.
Sanchez-Parra, C., et al., 2018. Contribution of the long noncoding RNA H19 to B-cell mass expansion in neonatal and adult rodents. Diabetes 67 (11), 2254–2267.
Scott, L.J., et al., 2007. A genome-wide association study of type 2 diabetes in finns detects multiple susceptibility variants. Science 316 (5829), 1341–1345.
Scoville, D.W., Gruzdev, A., Jetten, A.M., 2020. Identification of a novel lncrna (G3r1) regulated by glis3 in pancreatic β-cells. J. Mol. Endocrinol. 65 (3), 59–67.
Sedgeman, L.R., et al., 2019. Beta cell secretion of miR-375 to HDL is inversely associated with insulin secretion. Sci. Rep. 9 (1), 1–13.
Seiler, J., Breinig, M., Caudron-Herger, M., Polycarpou-Schwarz, M., Boutros, M., Diederichs, S., 2017. The lncRNA VELUCT strongly regulates viability of lung cancer cells despite its extremely low abundance. Nucleic Acids Res. 45 (9), 5458–5469.
Singer, R.A., Arnes, L., Sussel, L., 2015. Noncoding RNAs in β cell biology. Curr. Opin. Endocrinol. Diabetes Obes. 22 (2), 77–85.
Sisino, G., et al., 2017. Long noncoding RNAs are dynamically regulated during β-cell mass expansion in mouse pregnancy and control β-cell proliferation in vitro. PLoS One 12 (8), e0182371.

Sun, C., et al., 2016a. Insights from lncRNAs profiling of MIN6 Beta cells undergoing inflammation. Mediat. Inflamm. 2016, 9275106.
Sun, J., et al., 2016b. Differentially expressed circulating LncRNAs and mRNA identified by microarray analysis in obese patients. Sci. Rep. 6, 35421.
Sun, Q., et al., 2018. MALAT1 via microRNA-17 regulation of insulin transcription is involved in the dysfunction of pancreatic β-cells induced by cigarette smoke extract. J. Cell. Physiol. 233 (11), 8862–8873.
Todd, J.A., et al., 2007. Robust associations of four new chromosome regions from genome-wide analyses of type 1 diabetes. Nat. Genet. 39 (7), 857–864.
Travers, M.E., et al., 2013. Insights into the molecular mechanism for type 2 diabetes susceptibility at the KCNQ1 locus from temporal changes in imprinting status in human islets. Diabetes 62 (3), 987–992.
Urakami, T., 2019. Maturity-onset diabetes of the young (MODY): current perspectives on diagnosis and treatment. Diabetes Metab. Syndr. Obes. Targets Ther. 12, 1047–1056.
Vance, K.W., Ponting, C.P., 2014. Transcriptional regulatory functions of nuclear long noncoding RNAs. Trends Genet. 30 (8), 348–355.
Wallace, C., et al., 2010. The imprinted DLK1-MEG3 gene region on chromosome 14q32.2 alters susceptibility to type 1 diabetes. Nat. Genet. 42 (1), 68–71.
Wang, K.C., Chang, H.Y., 2011. Molecular mechanisms of long noncoding RNAs. Mol. Cell 43 (6), 904–914.
Wang, S., Mao, C., Liu, S., 2019. Peptides encoded by noncoding genes: challenges and perspectives. Signal Transduct. Target. Ther. 4, 57.
Wenger, R., et al., 2015. Induction of long noncoding RNA MALAT1 in hypoxic mice. Hypoxia 3, 45.
Wong, W.K.M., et al., 2018. Non-coding RNA in pancreas and β-cell development. Noncoding RNA 4 (4), 41.
Wong, W.K.M., et al., 2019. The long noncoding RNA MALAT1 predicts human islet isolation quality. JCI Insight 4 (16), e129299.
Wu, P., et al., 2020. Emerging role of tumor-related functional peptides encoded by lncRNA and circRNA. Mol. Cancer 19 (1), 22.
Xiong, L., et al., 2020. LncRNA-Malat1 is involved in lipotoxicity-induced ß-cell dysfunction and the therapeutic effect of Exendin-4 via Ptbp1. Endocrinology 161 (7), bqaa065.
Xu, H., et al., 2020. Pancreatic β cell microRNA-26a alleviates type 2 diabetes by improving peripheral insulin sensitivity and preserving β cell function. PLoS Biol. 18 (2), e3000603.
Yamagata, K., 2014. Roles of HNF1α and HNF4α in pancreatic β-cells: lessons from a monogenic form of diabetes (MODY). Vitam. Horm. 95, 407–423.
Yeasmin, F., Yada, T., Akimitsu, N., 2018. Micropeptides encoded in transcripts previously identified as long noncoding RNAs: a new chapter in transcriptomics and proteomics. Front. Genet. 9, 144.
Yin, D., et al., 2015. Downregulation of lncRNA TUG1 affects apoptosis and insulin secretion in mouse pancreatic β cells. Cell. Physiol. Biochem. 35 (5), 1892–1904.
You, L., et al., 2016. Downregulation of long noncoding RNA Meg3 affects insulin synthesis and secretion in mouse pancreatic Beta cells. J. Cell. Physiol. 231 (4), 852–862.
Zhang, H., 2019. Mechanism associated with aberrant lncRNA MEG3 expression in gestational diabetes mellitus. Exp. Ther. Med. 18 (5), 3699.
Zhang, X., et al., 2010. Maternally expressed gene 3 (MEG3) noncoding ribonucleic acid: isoform structure, expression, and functions. Endocrinology 151 (3), 939–947.
Zhang, Y., et al., 2018. Long non-coding RNA MALAT1 expression in patients with gestational diabetes mellitus. Int. J. Gynecol. Obstet. 140 (2), 164–169.
Zhang, F.F., et al., 2020. Obesity-induced reduced expression of the lncRNA ROIT impairs insulin transcription by downregulation of Nkx6.1 methylation. Diabetologia 63 (4), 811–824.

CHAPTER SEVEN

Molecular mechanisms of lipotoxicity-induced pancreatic β-cell dysfunction

Asier Benito-Vicente[a,b], Shifa Jebari-Benslaiman[a,b], Unai Galicia-Garcia[a,c], Asier Larrea-Sebal[a,c], Kepa B. Uribe[d], and Cesar Martin[a,b,*]

[a]Department of Molecular Biophysics, Biofisika Institute (University of Basque Country and Consejo Superior de Investigaciones Científicas (UPV/EHU, CSIC)), Leioa, Spain
[b]Department of Biochemistry and Molecular Biology, University of the Basque Country (UPV/EHU), Leioa, Spain
[c]Department of Molecular Biophysics, Fundación Biofisica Bizkaia, Leioa, Spain
[d]Center for Cooperative Research in Biomaterials (CIC biomaGUNE), Basque Research and Technology Alliance (BRTA), Donostia San Sebastián, Spain
*Corresponding author: e-mail address: cesar.martin@ehu.eus

Contents

International Review of Cell and Molecular Biology, Volume 359
ISSN 1937-6448
https://doi.org/10.1016/bs.ircmb.2021.02.013

Abstract

Type 2 diabetes (T2D), a heterogeneous disorder derived from metabolic dysfunctions, leads to a glucose overflow in the circulation due to both defective insulin secretion and peripheral insulin resistance. One of the critical risk factor for T2D is obesity, which represents a global epidemic that has nearly tripled since 1975. Obesity is characterized by chronically elevated free fatty acid (FFA) levels, which cause deleterious effects on glucose homeostasis referred to as lipotoxicity. Here, we review the physiological FFA roles onto glucose-stimulated insulin secretion (GSIS) and the pathological ones affecting many steps of the mechanisms and modulation of GSIS. We also describe *in vitro* and *in vivo* experimental evidences addressing lipotoxicity in β-cells and the role of saturation and chain length of FFA on the potency of GSIS stimulation. The molecular mechanisms underpinning lipotoxic-β-cell dysfunction are also reviewed. Among them, endoplasmic reticulum stress, oxidative stress and mitochondrial dysfunction, inflammation, impaired autophagy and β-cell dedifferentiation. Finally therapeutic strategies for the β-cells dysfunctions such as the use of metformin, glucagon-like peptide 1, thiazolidinediones, anti-inflammatory drugs, chemical chaperones and weight are discussed.

Abbreviations

ACC acetyl-CoA carboxylase
AMPK AMP-activated protein kinase
ATF6 activating transcription factor 6
cAMP cyclic adenosine monophosphate
Cer ceramides
CerS ceramide synthase
CHOP CCAAT-enhancer-binding protein homologous protein
CPT-1 carnitine palmitoyl transferase I
DAG diacylglycerol
DPP-4 dipeptidyl peptidase-4
ER endoplasmic reticulum
FA fatty acid
FAT fatty acid translocase
FFA free fatty acid
Foxo1 forkhead box protein O1
GLP-1 glucagon-like peptide 1
GLUT2 glucose transporter 2

GPCR	G-protein coupled receptor
GSIS	glucose-stimulated insulin secretion
Id1	inhibitor of DNA binding 1
I-kB	inhibitory kB
IL	interleukin
IL-1R	IL-1 receptor
iNOS	nitric oxide synthase
IP3	inositol triphosphate
IR	insulin resistance
IRE1	inositol-requiring enzyme 1
IRS2	insulin receptor substrate 2
JNK	c-Jun N-terminal kinase
KIF12	Kinesin Family Member 12
LC3	microtubule-associates protein 1 light chain 3
MCL1	myeloid cell leukemia sequence 1
mtDNA	mitochondrial DNA
mTORC1	mTOR complex 1
MUFA	monounsaturated fatty acids
NF-kB	nuclear factor kB
NO	nitric oxide
PA	palmitic acid
PBA	4-phenyl butyric acid
PDX-1	duodenal homeobox factor 1
PE	phosphatidylethanolamine
PERK	PKR-like ER kinase
PIP2	phosphatidylinositol 4,5-bisphosphate
PKC	protein kinase C
PKD	protein kinase D
PLC	phospholipase C
PPARγ	peroxisome proliferator-activated receptor gamma
PUFA	polyunsaturated fatty acids
RNS	reactive nitrogen species
ROS	reactive oxygen species
SCD	stearoyl-CoA desaturase
SFA	saturated fatty acid
SM	sphingomyelin
T1D	type 1 diabetes
T2D	type 2 diabetes
TG	triglyceride
TLR	Toll-like receptor
TRP	transient receptor potential
TUDCA	taurine-conjugated derivative of ursodeoxycholic acid
TZN	thiazolidinediones
UPR	unfolded protein response
VLDL	very low density lipoprotein

1. Introduction

Type 2 diabetes (T2D) is a heterogeneous disorder derived from metabolic dysfunctions, which leads to a glucose overflow in the circulation due to both defective insulin secretion and peripheral insulin resistance (IR). One of the critical risk factor for T2D is obesity, which represents a global epidemic that has nearly tripled since 1975 (World Health Organization). Obesity occurs when energy intake exceeds energy expenditure, and is associated with elevated concentrations of circulating free fatty acids (FFAs) due to a diminished esterification in de adipose tissue (Lewis et al., 2002). The oversupply of FFAs eventually leads to systemic pressure on many tissues and cells that have to increase their lipid uptake. Increased FFA uptake, exceeding the cell's ability to oxidize them in mitochondria, leads to intracellular lipid accumulation, which may result in the development of lipotoxicity leading to β-cell damage (Oh, 2015).

Accumulating evidence correlates high levels of FFAs with T2D pathogenesis and, elevated fasting and post-prandial FFA concentrations with increased risk of developing IR through β-cell dysfunction. These evidences point out a mechanistic link between obesity and T2D. A long-term exposure to high lipid levels contributes to the loss of mass and function of β-cell, which constitutes an early event underlying T2D development and IR. Additionally, it has been shown that lipotoxicity underlies gradual atrophy of β-cells in the pancreas (Cnop et al., 2005). Lipotoxicity is driven by biologically active lipids that inhibit or activate enzymes directly affecting the insulin pathway. Among them, long-chain acyl-CoA (Ellis et al., 2000; Hulver et al., 2003), ceramides (Cer) (Adams et al., 2004; Blachnio-Zabielska et al., 2010; Straczkowski et al., 2004), and DAG (Schmitz-Peiffer, 2002; Yu et al., 2002).

2. FFA-induced lipotoxicity

Obesity is accompanied by increased plasma concentrations of FFA, leading to their intensified uptake by tissues involved in the regulation of glucose homeostasis (skeletal muscle, liver, and pancreas). The source of FFA can be both the subcutaneous and visceral adipose tissue (Belfort et al., 2005). Due to the higher lipolytic activity of visceral adipocytes, the release of FFA from visceral adipose tissue is faster compared to subcutaneous fat, which was observed in both obese and lean patients (Bjorntorp, 1991).

Continuous exposure to excess lipids, especially those rich in saturated FA, is proposed to be the major contributing factor to the initiation and progression of β-cell failure in T2D (Poitout and Robertson, 2008). However, there is some controversy regarding lipotoxic effects of FFA through direct effect on β-cells as it has been difficult to find clear correlations between β-cell and FFA, or other circulating lipids. It has been shown that obese individuals without T2D have high rates of insulin secretion (Camastra et al., 2005; Karam et al., 1963) that compensate for IR and as they progress to T2D, their first-phase GSIS deteriorates in lockstep with rising glucose levels (Brunzell et al., 1976). On the contrary, may studies have shown the critical role of environmental lipids in β-cell pathology, which was first proven in Zucker diabetic fatty rats, a genetic model of obesity and T2D, in which increased lipid availability induces both hyperinsulinemia and IR in parallel fashion (Lee et al., 1994). On the other hand, it was demonstrated that diminishing the levels of plasma FFAs prevents β-cell dysfunction (Unger, 1995). Numerous studies have shown that long-tern incubation of β-cell and islet with high concentration of FA in the presence of glucose causes apoptosis and reduces insulin secretion through activation of multiple stress pathways, including ceramide generation, ER stress, oxidative stress, mitochondrial dysfunction, inflammation and autophagy (Bensellam et al., 2012; Doliba et al., 2017; Poitout and Robertson, 2008). Recent studies, systematic reviews and meta-analyses have investigated the association between total amount, type, and origin of dietary fats and T2D (Guasch-Ferre et al., 2017; Mozaffarian, 2016; Rice Bradley, 2018; Riserus et al., 2009). Prospective studies with subjects at risk for T2D demonstrate a correlation between loss of β-cell function and development of abdominal obesity (Cnop et al., 2005). Additionally, the Prospective Metabolism and Islet Cell Evaluation (PROMISE) cohort study showed a strong association of serum FFAs with insulin sensitivity and β-cell function (Johnston et al., 2018).

The effects of elevated FFAs on glucose metabolism in humans have been mostly studied by administration of an intravenous lipid emulsion to experimentally raise FFA concentrations in participants. The lipid emulsions used in these assays are Intralipid (a purified soybean oil emulsion), or Liposyn (a soybean and safflower oil emulsion). Both of them are mainly composed of polyunsaturated FFAs and are usually administered together with heparin, to release lipoprotein lipase into the circulation and promote triglyceride (TG) hydrolysis. According to these assays, different effects on insulin secretion caused by acute and chronic elevations of FFAs have been observed. Acute FFAs exposure enhances GSIS, compensating for lipid induced IR (Carpentier et al., 1999; Paolisso et al., 1995). However, a more sustained elevation of FFAs (24–48 h) causes β-cell function deterioration, impairing the ability of β-cells to compensate for the prevailing IR (Carpentier et al., 1999, 2010; Giacca et al., 2011). Interestingly, co-infusion of lipids with glucose inhibits the stimulatory effect of hyperglycemia on β-cell function (Leung et al., 2004). The effects caused by elevated FFAs may be more pronounced in genetically predisposed individuals. Subjects with a genetic predisposition to develop T2D have increased susceptibility to β-cell dysfunction induced by lipid infusion, compared to subjects without a family history of T2D (Kashyap et al., 2003; Storgaard et al., 2003). In individuals with established T2D, however, lipid infusion does not further worsen β-cell function and prolonged FFA exposure no longer detectably affects insulin secretion (Carpentier et al., 2000). A potential limitation of these studies is that the concentrations close to β-cells can not be determined and is still to be demonstrated that the FFA levels that actually occur in obesity and T2D are in any way damaging to β-cells, which makes it difficult to understand the relevance of these experiments, therefore additional experiments are required to identify the lipotoxic effects of FFA on β-cells.

FA composition has been described as one of the critical factors in the induction of lipotoxicity in β-cells and T2D (Palomer et al., 2018). FAs are classified according to carbon chain length into three categories, short-chain fatty acids that contain less than 6 carbons, medium-chain fatty acids with 6–12 carbons, and long-chain fatty acids, which contain more than 12 carbons. Moreover, given the double bond configuration, FA can be subdivided into saturated fatty acids and unsaturated fatty acids, which can be sub classified as mono or polyunsaturated fatty acids (MUFA and PUFA, respectively) (Poudyal et al., 2013). The effects of a specific FA on insulin secretion and β-cell survival are related to the degree of saturation and carbon chain length of the FA. It has been shown that β-cell cytotoxicity is induced by long-chain FA with a chain length of 16 carbons or greater (palmitate (C16:0) or stearate (C18:0)), whereas a reduction of the carbon chain length to 14 or 12 carbons (myristate and laurate, respectively) are less toxic (Newsholme et al., 2007).

Dietary fats can modulate T2D development, and among them palmitic acid constitutes one of the main FA involved in the lipotoxicity during progression of T2D. It has been shown that prolonged exposure to palmitate inhibits the secretory capacity of β-cells, impairs insulin gene expression, and increases β-cell apoptosis (Hagman et al., 2005). On the other hand, supplementation of PUFAs has been shown to decrease TG and cholesterol levels and enhance insulin secretion (Acosta-Montano and Garcia-Gonzalez, 2018). Administration of unsaturated FA to β-cells, such as arachidonic acid (C20:4), increases glucose-stimulated insulin secretion (GSIS) and β-cell proliferation (Keane et al., 2011). Docosahexaenoic acid (DHA, C22:6) and eicosapentaenoic acid (EPA, C20:5) also prevent cytokine induced cell death in pancreatic islets and enhance insulin secretion (Baynes et al., 2018).

3. Lipid signaling pathways involved in the FFA modulation of GSIS

3.1 Physiological GSIS and potentiation by FFA

Islet β-cell glucose metabolism is essential for the coupling of glucose sensing to insulin release (Meglasson and Matschinsky, 1986; Newgard and McGarry, 1995; Prentki, 1996). At physiological condition, GSIS is triggered by the ATP-sensitive K^+ (K_{ATP}) channel-dependent pathway in which glucose metabolism increases the ATP/ADP ratio, which results in closure of K_{ATP} channels, plasma membrane depolarization followed by the opening of voltage-dependent Ca^{2+} channels, and subsequent Ca^{2+} triggering of insulin granule exocytosis (Ashcroft and Rorsman, 1990; Prentki, 1996) (Fig. 1). In addition, several amplification pathways can modulate insulin secretion by channeling pyruvate *via* pyruvate carboxylase into the anaplerosis/cataplerosis pathway. This results in increased cytosolic levels of oxaloacetate (OAA) and citrate and generates metabolism-derived signaling molecules such as NADPH from the malate-pyruvate shuttle that is important for insulin secretion (Cline et al., 2004; Lu et al., 2002; MacDonald, 1995). Additionally, citrate cataplerosis by the occurrence of pyruvate/citrate shuttle allows the production of malonyl-CoA (Chen et al., 1994; Cline et al., 2004; Farfari et al., 2000; Flamez et al., 2002; Prentki et al., 1992). Malonyl-CoA inhibits carnitine palmitoyl transferase I (CPT-1) and consequently FA catabolism, making fatty acyl-CoA compounds available in the cytosol for the production of diacylglycerol and other lipids. Long-chain acyl-CoA esters may provide substrate for protein acylation or may modulated C-kinase enzyme(s) activity with resulting insulin release (Fig. 1). Another pathway contributing to GSIS amplification is the glutamate-glutamine cataplerotic cycle, which derives α-ketoglutarate from the Krebs cycle to the cytosol to serve as a source of glutamate (Maechler and Wollheim, 1999). Within the cytoplasm, glutamate uptake would render the insulin granules secretion competent.

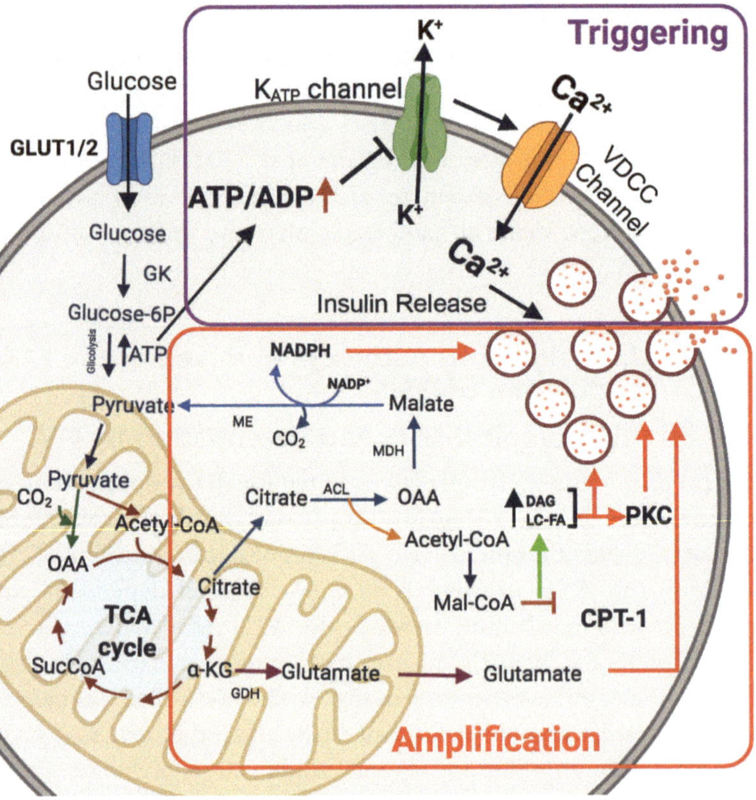

Fig. 1 Glucose-stimulated triggering and amplifying pathways in the β-cell. Glucose enters the β-cell through the GLUT1/2 glucose transporter and is phosphorylated by glucokinase (GK) to yield glucose-6-phosphate. Glucose-6-phosphate enters glycolysis yielding ATP and pyruvate. Pyruvate is transported into the mitochondria where is metabolized by pyruvate carboxylate (PC) to regenerate oxaloacetic acid (OAA) for the tricarboxylic acid cycle (TCA). Pyruvate can also be converted to acetyl-CoA by pyruvate dehydrogenase (PDH) and condenses with OAA yielding citrate. The increase in the ratio of ATP to ADP derived by glycolysis and the TCA cycle lead to closure of the K_{ATP} channel, which activates voltage-dependent calcium channels (VDCC) to allow the influx of calcium. This event triggers insulin secretion. Export of mitochondrial intermediates such as citrate leads to increases in NADPH that is important for insulin secretion. Additionally, citrate allows the production of malonyl-CoA, which inhibits CPT-1 leaving fatty acyl-CoA compounds available for the production of diacylglycerol and other lipids, including long-chain acyl-CoA esters that may act as substrate for protein acylation and modulate PKC activity resulting insulin release. Efflux of α-ketoglutarate to the cytosol serves as a source of glutamate that can render the insulin granules secretion competent. The anaplerosis/cataplerosis pathway also generates lipid-signaling molecules from the malonyl-CoA/long-chain FA-CoA, FA-CoAs accumulate in the cytoplasm to exert stimulation of insulin release.

This could be triggered by the reduction of granular membrane potential (Maycox et al., 1988) or, possibly, through Ca^{2+} uptake into the granules, as Ca^{2+} depletion of this compartment inhibits insulin exocytosis (Scheenen et al., 1998). Finally, the anaplerosis/cataplerosis pathway generates lipid-signaling molecules from the malonyl-CoA/long-chain FA-CoA pathway when citrate leaves the mitochondria to enter the cytoplasm, where it can be transformed to malonyl-CoA *via* two successive reactions catalyzed by citrate-lyase and acetyl-CoA carboxylase (Farfari et al., 2000). Malonyl-CoA plays a pivotal role, as this "signal of plenty" inhibits FA oxidation through its allosteric interaction with CPT-1. Consequently, FA-CoAs accumulate in the cytoplasm to exert stimulation of insulin release (Liu et al., 2002; Prentki and Corkey, 1996; Prentki et al., 2002; Roduit et al., 2004), either directly on exocytosis (Deeney et al., 2000) or indirectly *via* complex lipid formation (Corkey et al., 1989; Farese et al., 1986), protein kinase C activation (Yaney et al., 2000), or protein acylation (Yajima et al., 2000).

Physiological GSIS requires certain level of FFA, and is significantly potentiated by acute FFA rise both *in vivo* and *in vitro* (Nolan et al., 2006; Stein et al., 1996). Potency of FFA-stimulated GSIS is affected by both chain length and saturation of FA (Cen et al., 2016; Stein et al., 1996); in fact PA and oleic acid (18:1, OA), two of the most abundant FA in the circulation and cells, potently stimulate GSIS (Cen et al., 2016; Stein et al., 1997). The augmentation of GSIS by FA is partially mediated by the activation of the cell surface FFA1/GPR40 receptor (FFAR1) (Prentki et al., 2013). In addition, intracellular metabolism of FA in β-cells generates metabolic coupling factors (MCF) that contribute to insulin secretion (Prentki et al., 2013).

At low or normal glucose levels intracellular FFAs are activated into long-chain-CoAs and are oxidized by the β-cell. However, when glucose levels are elevated FFA oxidation is blocked by the generation of malonyl-CoA. The latter inhibits CPT-1, shifting the intracellular FFA metabolism toward esterification and generation of complex lipids, including sphingolipids. As indicated above, the interplay of the Krebs, pyruvate, and glycerolipid/FFA cycles generates essential metabolic coupling factors in the β-cell to amplify signals for insulin secretion but become detrimental under fuel oversupply conditions (Nolan et al., 2006; Prentki and Madiraju, 2012; Prentki et al., 2013).

3.2 Glucolipotoxicity: Synergism of lipotoxicity with hyperglycemia in T2D pathogenesis

The term glucolipotoxicity refers to damaging influences of chronically excessive levels of glucose and lipids on β-cells. In recent years, a number of studies have contributed to a better understanding of the biochemical and molecular mechanisms of glucolipotoxicity in the β-cell.

The deleterious effects of fatty acids on β-cell function that occur in the presence of elevated glucose levels has a biochemical basis (Prentki et al., 2002). As detailed above, at physiological glucose concentrations, fatty acids are transported into the mitochondria and undergo β-oxidation with little, if any, functional consequences. In contrast, when both glucose and fatty acid concentrations are elevated, glucose metabolism in the TCA cycle generates cataplerotic signals, fatty acid oxidation is blocked, long-chain acyl-CoA esters (LC-CoA) accumulate in the cytosol (Prentki and Corkey, 1996) and adversely affects β-cell function (Poitout and Robertson, 2002). In addition to its metabolic effects directing fatty-acid partitioning into esterification, glucose coordinately activates the expression of genes involved in lipogenesis (Roche et al., 1998). AMP-activated protein kinase (AMPK) is a metabolic sensor that detects changes in the cellular energy state and directs the β-cell into a "storage mode" in the face of nutrient oversupply (Ruderman and Prentki, 2004). AMPK activity is inversely correlated with the glucose concentration (Salt et al., 1998) and is stimulated by palmitate (Wang et al., 2007) in β-cells. Once activated, AMPK triggers gene expression involved in fatty acid synthesis through activation of the transcription factor SREBP1c (Foufelle and Ferre, 2002) leading to increased lipogenesis and TG accumulation.

In isolated islets, combined exposure to high levels of glucose and palmitate impairs the expression of the transcription factor MafA and the nuclear translocation of the transcription factor PDX-1 (Hagman et al., 2005) leading to diminished insulin promoter activity (Kelpe et al., 2003). Glucolipotoxic conditions lead to increased expression of PGC1-α, which suppresses the activity of the transcription factor BETA2/NeuroD (Kim et al., 2009). In addition, under these conditions, palmitate induces C/EBPβ expression that reduces insulin promoter activity (Plaisance et al., 2009). These transcriptional changes are mediated, at least in part, *via* intracellular generation of ceramide (Kelpe et al., 2003; Manukyan et al., 2015; Ng et al., 2014) and the activation of stress kinases including c-Jun N-terminal kinase (JNK) (Solinas et al., 2006), extracellular-regulated kinases (ERK) 1/2, and Per-Arnt-Sim (PAS) kinase (Fontes et al., 2009).

As described above, in experimental models, glucolipotoxicity adversely affect insulin production and secretion however, the mechanisms of action remain unclear but likely involve the generation of intracellular lipid metabolites and stress-induced pathways. Nowadays there is some controversy related to the real role of this phenomenon in human T2D. Among the reasons that generate this controversy are the difficulty of mimicking *in vivo* FFA levels in *in vitro* systems, the limitation of simulate the progressive and chronic changes caused by glucolipotoxicity and also the difficulty to accurately quantify β-cell function and β-cell mass in humans.

3.3 FFA receptors involved in FFA-mediated GSIS

FFAs homeostasis is maintained by specific receptors, which triggers a range of downstream effects (Georgiadi and Kersten, 2012) and is affected by long-term exposition of β-cells to FFAs. FFAs bind to several cell-surface receptors of the G-protein coupled receptor (GPCR) family, including GPR40, GPR43, GPR41, GPR120, and GPR84 (Dranse et al., 2013; Nakamura et al., 2014). Another important membrane-bound protein for FFA sensing is the CD36 fatty acid translocase (FAT) protein (Ly et al., 2017). The mechanism by which the two main receptors involved in lipotoxic effects induced by long-term exposure to palmitic acid (PA) are discussed below.

3.4 GPR40

Several reports have shown the involvement of different pathways in GPR40-mediated GSIS potentiation. GPR40 is activated by medium- and long-chain FFAs (especially, eicosatrienoic acid (C20:3)) and facilitates GSIS in pancreatic β-cells (Itoh et al., 2003; Miyauchi et al., 2010) presumably to promote uptake of carbohydrate- and fat-derived-energy-rich molecules after a meal by liver and adipose tissue. Ligand binding to GPR40 activates Gα subunit by GDP/GTP exchange and thus dissociates from the trimeric G-protein. As a result, phospholipase C (PLC) is activated, and thereby the intracellular concentration of inositol triphosphate (IP3) and diacylglycerol (DAG) increases with the consequent hydrolysis of phosphatidylinositol 4,5-bisphosphate (PIP2) (Alquier et al., 2009; Fujiwara et al., 2005; Itoh et al., 2003; Salehi et al., 2005; Shapiro et al., 2005). It has been demonstrated that DAG activates protein kinase D (PKD) 1 and promotes actin depolymerization, which subsequently increases insulin secretion (Ferdaoussi et al., 2012) (Fig. 2). Simultaneously, the increased IP3 levels promote the release of Ca^{2+} from the endoplasmic reticulum (ER) and potentiate GSIS in MIN6 cells (Sakuma et al., 2016). However, the mechanism by which ER Ca^{2+} release leads to GSIS potentiation remains unknown.

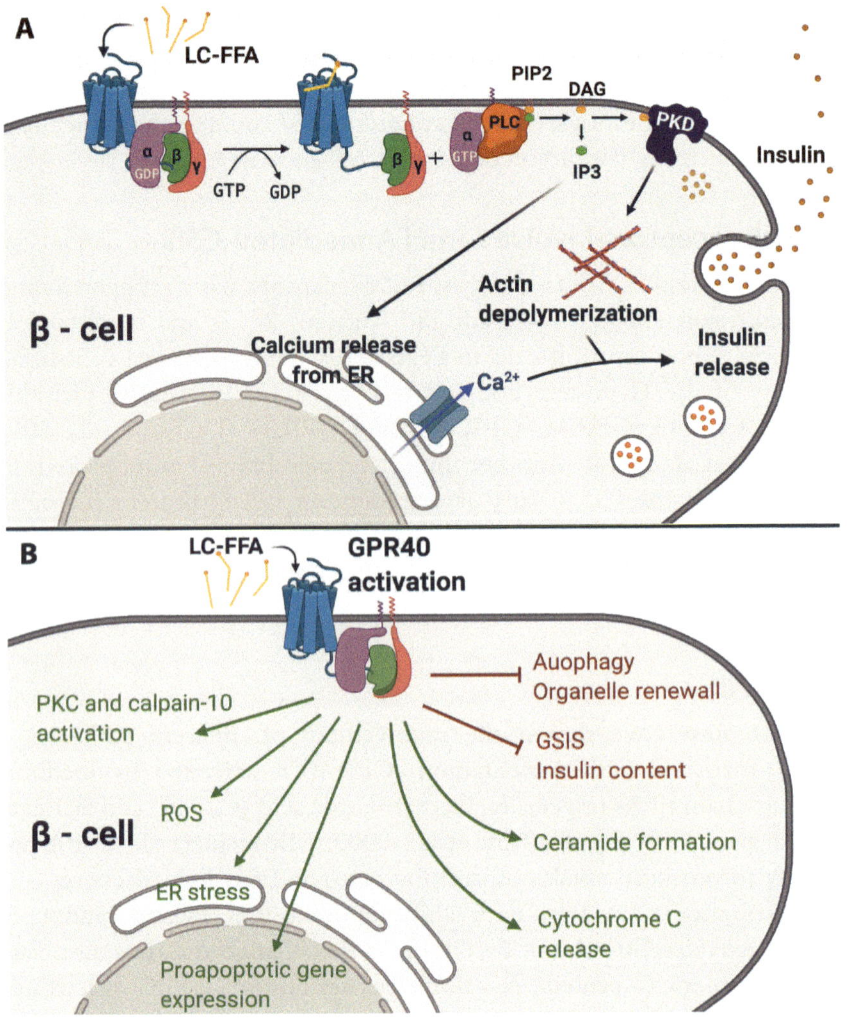

Fig. 2 Involvement of long-chain FA in GPR40-mediated insulin release in physiological and detrimental condition. (A) Physiological conditions: ligand binding to GPR40 activates Gα subunit and dissociates from the trimeric G-protein activating PLC. As a consequence intracellular concentration of IP3 and DAG increases with the consequent hydrolysis of PIP2. DAG activates PKD1 and promotes actin depolymerization leading to insulin secretion. Simultaneously, IP3 promotes Ca^{2+} release from the ER and potentiate GSIS. (B) Detrimental conditions: long-term exposure to LC-FFA inhibits autophagy, GSIS response and insulin content. On the other hand, ceramide formation is increased, Cytochrome *c* release and ER stress are induced, PCK and calpain-10 are activated, ROS are produced and proapoptotic gene expression is upregulated.

The role and mechanisms of GPR40 signaling have been extensively studied. However, conflicting results have been published regarding the contribution of GPR40 to β-cell dysfunction under chronic palmitate exposure (Ferdaoussi et al., 2012; Fujiwara et al., 2005; Latour et al., 2007; Salehi et al., 2005; Steneberg et al., 2005), it has been shown that PA triggers G-protein coupled receptor GPR40 activation (Itoh et al., 2003), which is involved in the long-term effects of chronically elevated palmitate levels (Kristinsson et al., 2013). Continuous PA exposure affects negatively both to GSIS and insulin content by decreasing them 70% and 25%, respectively (Kristinsson et al., 2013). It has been shown that palmitate amplifies both phases of GSIS (first transient phase and the sustained second phase) and that activation of GPR40 affects the sustained phase of GSIS but not the first one, as indicated by the use of specific inhibitors or agonists (Kristinsson et al., 2013). This may indicate that the GPR40 pathway does not regulate the immediate release of granules from the readily releasable pool. Instead, the receptor-mediated pathway seems to stimulate the recruitment of granules, cytoskeleton and the increased second phase of GSIS (Ferdaoussi et al., 2012). Long-term exposure of human islets to palmitate cause low levels of secretion after 7 days and is accompanied by a gradual reduction in insulin content, which suggest that GPR40 signaling is involved in the long-term negative effects of palmitate (Kristinsson et al., 2013). The same authors demonstrated that inhibition of GPR40 by the GPR40 antagonist ANT203 restores the fatty acid-induced impairment of β-cell function and protective effects of ANT203 on GSIS and insulin content has been also observed in palmitate-treated MIN6 cells (Kristinsson et al., 2013).

In particular PA interferes with glucose metabolism by affecting several mechanisms including ceramide formation (Lupi et al., 2002b; Shimabukuro et al., 1998a), cytochrome *c* release (Maestre et al., 2003), activation of protein kinase C (PKC) (Eitel et al., 2003), activation of calpain-10 (Johnson et al., 2004) and ER stress (Karaskov et al., 2006), impairing autophagy and organelle renewal (Jaishy and Abel, 2016), promoting reactive oxygen species (ROS) overproduction (Chen et al., 2019) and proapoptotic gene expression (Oh et al., 2018). GPR40 is the most closely receptor related to lipotoxicity-induced β-cell apoptosis.

3.5 CD36

CD36 is an N-linked glycosylated transmembrane protein, also known as FAT, which is upregulated in response to high glucose in the β-cell leading to increased uptake of FA thus ameliorating GSIS and impairing oxidative metabolism (Wallin et al., 2010). After being transported across the cell membrane *via* CD36, FFAs are activated by fatty acyl-CoA synthetase and derived to β-oxidation as acyl-CoA. In addition, acyl-CoA also enters the glycerolipid/FFA cycle and can participate in sphingolipid synthesis leading to the synthesis of ceramides and sphingosine-1 phosphate (Moulle et al., 2017). The binding of long-chain FFA to CD36 stimulates the tyrosine phosphorylation of downstream proteins, including proinflammatory response associated with diabetes (Silverstein et al., 2010). Studies with the Zucker diabetic fatty rat show that prolonged high levels of FFA induce ceramide accumulation, which triggers β-cells by apoptosis (Shimabukuro et al., 1998b) (Fig. 3). Similarly, high plasma levels of ceramides have been observed in obese patients with T2D, which suggests that these atypical ceramides may serve as biomarkers for the diagnosis and treatment of obesity and diabetes (Haus et al., 2009; Samad et al., 2006; Wigger et al., 2017). CD36 leads

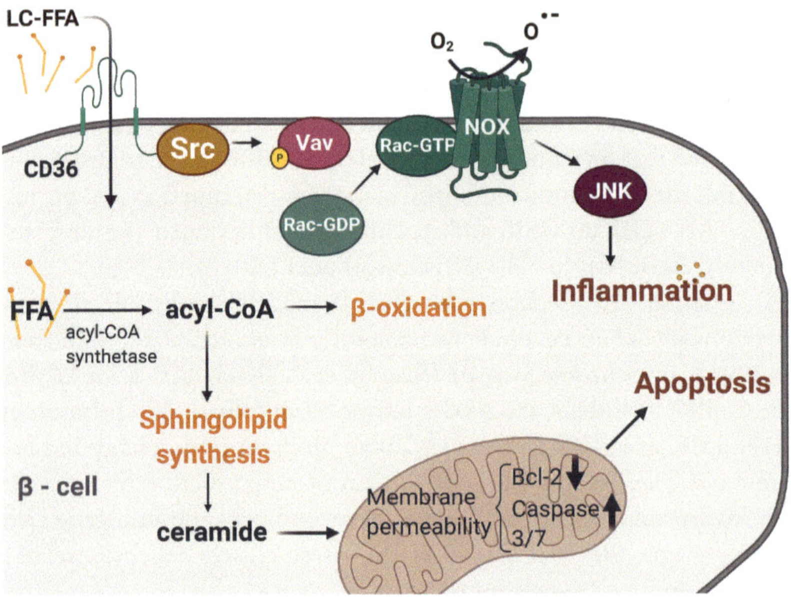

Fig. 3 Involvement CD36 in lipotoxicity induced by long-chain FFA exposure. FFAs are transported by CD36 across the cell membrane and activated by fatty acyl-CoA synthetase. Them activated FA can be derived to β-oxidation as acyl-CoA, can enter the glycerolipid/FFA cycle and can participate in sphingolipid synthesis leading to the synthesis of ceramides and can also stimulate the tyrosine phosphorylation of downstream proteins, including proinflammatory response associated with diabetes.

to ceramide induced β-cell functional defect through Vav family guanine nucleotide exchange factors (GEF) (Karunakaran et al., 2015), which is tightly regulated by Src-mediated tyrosine phosphorylation (Bustelo, 2014). CD36 signaling leads to Src-mediated Vav tyrosine phosphorylation evidences suggest that saturated fatty acids cause abnormal Src signaling which impairs insulin secretion and induces IR (Kominato et al., 2008). In addition, FFAs mediates Src-dependent phosphorylation of Vav and the subsequent engagement of Rac1-NOX signaling coordinates C-Jun N-terminal kinase (JNK) activation contributes to IR and the production of inflammatory cytokines (Hirosumi et al., 2002; Kant et al., 2017). CD36 signaling causes JNK activation as part of a generalized stress response by ceramide (Moon et al., 2020).

3.6 Stearoyl-CoA desaturase

Stearoyl-CoA desaturase (SCD) is a highly expressed enzyme in primary human β-cells (Krizhanovskii et al., 2017; Oshima et al., 2020; Scharfmann et al., 2014) that is involved in PA metabolism and has a protective role against lipotoxicity in β-cells (Flowers et al., 2007; Scharfmann et al., 2014). SCD catalyzes insertion of the double bond at the delta-9 position of 12–19 carbon in saturated fatty acids (SFAs; preferentially palmitate (16:0) and stearate (18:0)), thereby converting them to MUFAs (palmitoleate (16:1n-7) and oleate (18:1n-9)) (Ntambi and Miyazaki, 2004). SCD controls intracellular SFA/MUFA equilibrium, which is physiologically crucial for proper β-cell function given the cytotoxicity of long-chain SFAs higher than long-chain MUFAs (Cnop et al., 2001; Flowers et al., 2007; Morgan, 2009). It has been observed that SCD ablation is associated with β-cell failure and T2D development and results in lower GSIS, proliferation, and apoptosis induction in β-cells (Janikiewicz et al., 2015). Importantly, SCD gene expression is lower in β-cell enriched tissue from individuals with T2D compared with healthy donors (Marselli et al., 2010). It has been proposed that, over time, in the course of T2D progression, SCD expression by β-cells is first induced during compensation in response to IR, and as the duration of diabetes increases, SCD expression decreases leading to a decline in β-cell function (Oshima et al., 2020). Inhibition of SCD activity affects autophagosome-lysosome fusion through perturbations in cellular membrane integrity, leading to an aberrant stress response and β-cell failure and alters DNA methylation levels in 3T3-L1 adipocytes (Dziewulska et al., 2020; Janikiewicz et al., 2015; Malodobra-Mazur et al., 2014).

4. Lipid metabolites involved in FFA-mediated lipotoxicity

4.1 Ceramide

Exposure to excess long-chain saturated FFAs induces ceramide accumulation by serine palmitoyl transferase and ceramide synthase (CerS) (Chavez and Summers, 2003; Summers, 2006). It has been suggested that *de novo* ceramide synthesis induces apoptosis in β-cell by inhibition of Akt by ceramide as demonstrated by inhibition of sphingomyelinase (SMase), an enzyme that hydrolyses sphingomyelin (SM) to generate ceramide, which protects β-cells from apoptosis (Lei et al., 2010). Mitochondrial membrane permeability is also increased by ceramide leading to the activation of intrinsic pathways by lowering anti-apoptotic molecules Bcl-2 and increasing caspase 3/7 (Veret et al., 2011). Ceramide also inactivates Akt, a mechanism that has been observed in human treated with saturated fat who showed ceramide accumulation (Bachmann et al., 2001; Chavez et al., 2003). In addition, ceramide disrupts electron transport at complex I and complex III thus enhancing ROS generation, which leads to cytochrome *c* release and caspase activation (Garcia-Ruiz et al., 2002).

4.2 DAG

DAG is an essential lipid to maintain a proper insulin secretion upon glucose stimulation. Indeed it is well documented that DAG generation is favored upon glucose stimulation and contributes to the proper β-cell function. As aforementioned, DAG activates PKD1 and mediates the activation of other pathways (PKC, Ras guanine nucleotide-releasing protein (Ras-GRP), transient receptor potential (TRP) channels or Munc-13 dependent pathway) that regulate insulin vesicle release (Kaneko and Ishikawa, 2015; Rhee et al., 2002). DAG is metabolized through different mechanisms that include its conversion through DAG-kinase into phosphatidic acid; a terminator of DAG mediated signaling. DAG conversion into phosphatidic acid favors insulin vesicle release and improves GSIS. This process is highly dependent on intracellular Ca^{2+} availability and thus the inability of β-cells to regulate Ca^{2+} homeostasis under lipotoxic environments reduce the DAG/phosphatidic acid conversion and consequently inhibits insulin release and promote DAG accumulation (Thomas and Pek, 1992). On the other hand, high DAG levels also contribute to defective insulin signaling by preventing the phosphorylation of insulin receptor substrates and contribute to overall β-cell dysfunction (Yu et al., 2002).

5. Molecular mechanism leading to lipotoxic β-cell dysfunction

High FFA levels and obesity are considered important risk factors for the development of T2D (Weir, 2020) as they are associated to peripheral insulin resistance, which increases insulin secretory demand in order to compensate the low glucose uptake. However their direct effect onto the β-cell function in humans is still controversial. In the last 20 years many studies have sheet light on the mechanism that lead to β-cell lipotoxicity using *in vitro* and *in vivo* models. In this section we will summarized the most relevant ones.

5.1 FFAs and endoplasmic reticulum stress

Protein synthesis, folding and maturation in β-cells is primarily driven in the ER that mediates this processes through the action of different chaperones and enzymes allowing the formation of completely functional proteins. Moreover, ER is able to identify and degrade incorrectly folded proteins and stabilize posttranscriptional modification. However, several pathological conditions negatively affect ER function and disrupt its protein folding activity causing ER stress that promotes β-cell dysfunction and eventually cell death. These conditions include hypoxia, accumulation of mutated proteins that cannot properly fold, increased secretory load, loss of Ca^{2+} homeostasis that is essential for the proper functioning of different chaperones and, metabolic disturbances (Lin et al., 2008; Mollazadeh et al., 2018; Sicari et al., 2020). In order to re-establish ER homeostasis, β-cells, that are especially susceptible to ER stress conditions, activate unfolded protein response (UPR) system reducing global protein synthesis rate, upregulate chaperone and antioxidant protein synthesis, degrade aberrant mRNA and activate autophagy pathways (Pinto et al., 2019).

The UPR system is initiated by three ER-membrane bound proteins: PKR-like ER kinase (PERK), inositol-requiring enzyme 1 (IRE1), and activating transcription factor 6 (ATF6) (Fribley et al., 2009). The three of them can recognize unfolded proteins by their luminal domain and activate different pathways under ER stress conditions. The activation of the UPR system by ER stress promotes ER homeostasis recovery. If maintained, ER stress activates the effectors downstream of ER stress that promotes mitochondrial permeabilization and subsequent β-cell apoptosis (Biden et al., 2014; Cunha et al., 2008, 2012). Possibly, the most important effect of the activation of UPR pathways involves the upregulation and activation of CCAAT-enhancer-binding protein homologous protein (CHOP). This protein mediates the upregulation of several proapoptotic genes and downregulates the expression of anti-apoptotic proteins tilting the balance toward proapoptotic phenotype (Keane et al., 2015).

Chronic FFA elevation has been shown to induce UPR signaling molecules, such as eIF-2a phosphorylation, ATF4, CHOP, BiP, ATF6, and spliced XBP-1 in β-cells (Cnop et al., 2007; Kharroubi et al., 2004) (Fig. 4). Sustained high levels of FFA in the pancreas has been shown to induce ER stress-dependent β-cell apoptosis; however, this process seems to be more restricted to long chain saturated-FFA while unsaturated-FFA could even have a protective effect (Boslem et al., 2012). PA can induce ER stress compromising the folding capacity of the ER and promoting misfolded protein accumulation that activates the UPR system (Fig. 4). In addition, PA may also produce aberrant palmitoylations, a posttranslational

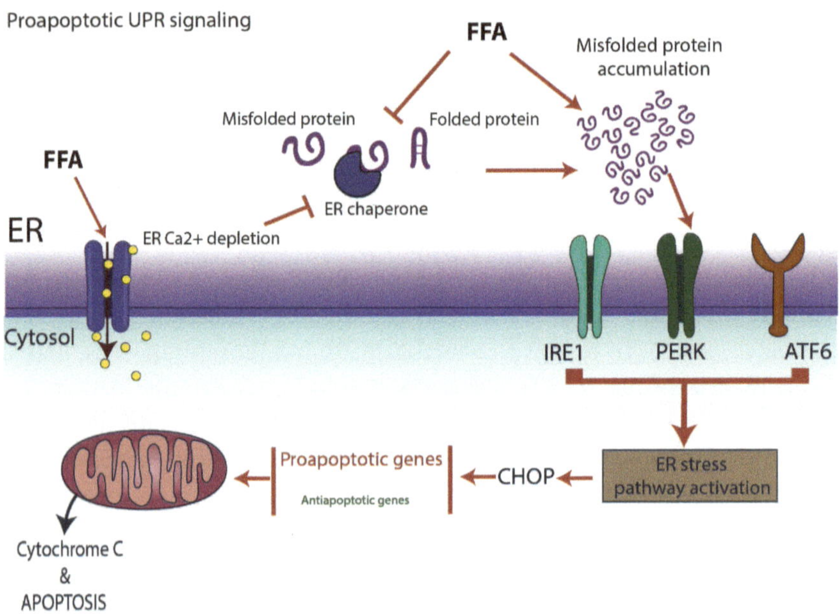

Fig. 4 ER stress induced apoptosis in β-cell exposed to lipotoxic environment. Chronic exposure to FFA induces an excessive misfolded protein accumulation in β-celsl ER. PA, an especially lipotoxic FA, promotes aberrant protein palmitoylations and ER Ca^{2+} depletion, affecting calcium-dependent chaperone activity and contributing to misfolded protein accumulation. Under these circumstances ER stress dependent UPR pathway is activated (IRE1, PERK and ATF6), inducing the activation of nuclear factors (mainly CHOP) that stimulate proapoptotic gene expression and apoptosis.

modification that alter protein stability and trafficking that contributes to ER protein accumulation (Robson-Doucette et al., 2011). PA activates JNK, a downstream metabolite of IRE1 pathways, and reduces the phosphorylation of several anti-apoptotic proteins such as myeloid cell leukemia sequence 1 (MCL1) through the activation of PERK pathway (Supale et al., 2012; Victor et al., 2011). Moreover, PA promotes ER Ca^{2+} depletion affecting calcium-dependent chaperone activity and favors ER stress and β-cell apoptosis (Gwiazda et al., 2009; Okado-Matsumoto and Fridovich, 2001). This process is enhanced under hyperglycemic conditions due PA-induced ER Ca^{2+} depletion (Gwiazda et al., 2009). Long-chain saturated FFAs like PA, stearate, arachidate also induce *de novo* synthesis and accumulation of both ceramides and different sphingolipids that activate the UPR system and trigger ER stress derived β-cell dysfunction (Thomas et al., 2009; Yaribeygi et al., 2020). Indeed, the inhibition of *de novo* ceramide synthesis through serine palmitoyl transferase or ceramide synthase inhibitors attenuates FFAs-induced β-cell dead (Shimabukuro et al., 1998a; Tang et al., 2017) (Fig. 4).

Additionally, when adipose tissue storage capacity is overcome, as it happens in obese people, FFA are released from adipose tissue and stored in extra-adipose tissues inducing lipid accumulation that inhibits their normal functioning and induces IR (Galicia-Garcia et al., 2020). The IR, especially in adipose tissue, muscle and liver, elevates circulating glucose levels. In this condition, pancreatic β-cells increase insulin production over their capacity which may result in the accumulation of unfolded proteins in the ER, since the amount of chaperones necessary for the prevention of protein misfolding does not suffice protein synthesis so they are accumulated triggering ER stress (Newsholme et al., 2016).

5.2 Lipotoxicity and oxidative stress

Accumulation of ROS can result in oxidative stress and irreversible β-cell injury (Drews et al., 2010). Pancreatic β-cells are among the most metabolically active cells and their function is highly dependent on energy availability, mainly ATP coming from mitochondrial oxidation. This process is especially relevant under elevated glucose condition where insulin must be synthesized and secreted in order to maintain glucose homeostasis. The oxidative processes that allow ATP production also generate ROS, a group of molecules that describe highly reactive molecules containing oxygen atoms such as superoxide (O_2^-), hydrogen peroxide (H_2O_2) or hydroxyl radicals ($^{\bullet}OH$) among others (Pacher et al., 2007). ROS are primarily synthesized in the mitochondrial respiratory chain and more specifically in the first and the third complexes located in the inner membrane of the mitochondria (Lenaz, 2001). The mitochondrial respiratory chain promotes superoxide formation that is considered the first stage in the ROS-generated cascade that will generate other ROS and reactive nitrogen species (RNS) (Yuzefovych et al., 2010). In addition to mitochondrial synthesis, ROS can also be produced by cytosolic and plasma membrane oxidoreductases that oxidize NADPH and produce ROS through the reduction of molecular oxygen (Gray and Heart, 2010). These molecules, at physiological conditions, mediate insulin synthesis and secretion and are essential for the correct β-cell function (Leloup et al., 2009). However, when glucose clearance starts to be impaired due to peripheral insulin resistance, glucose influx increases β-cell cells promoting an excessive ROS production that disturbs the balance between ROS activity and antioxidant activity that stabilize ROS and generates oxidative stress.

The deleterious effects of an excessive ROS/RNS production include ribonucleic acid, protein, and lipid damage by nitrilation, nitrosylation, carbonylation and peroxidation. These modifications alter enzyme activities, ion channel transport, or receptor signal transduction and consequently dysregulate gene expression and induce apoptosis (Keane et al., 2015; Newsholme et al., 2012). Moreover, possibly due to the positive effects ROS into insulin metabolism, β-cell possess low antioxidant levels that make them especially vulnerable to oxidative stress (Gehrmann et al., 2010). Oxidative stress targets in β-cells include duodenal homeobox factor 1 (PDX-1), which plays an important role in pancreatic development and maintaining β-cell function. Indeed, the exposure of rat islets to H_2O_2 reduces PDX-1-DNA binding *via* JNK pathway and consequently insulin lowers insulin-related gene expression. Additionally, oxidative stress downregulates the expression of other proteins such as MafA, a transcription factor involved in insulin gene expression and contributes to diabetes development.

FFA and excessive glucose content are potent inducers of ROS production. As in ER stress, long-chain FFA are the most harmful lipids as they are catabolized through both mitochondrial β-oxidation and peroxisomal β-oxidation generating ROS. Long-chain FFAs, such as palmitic and stearic acid, can be metabolized through β-oxidation in the peroxisomes as well as in the mitochondria. However, in contrast with mitochondrial β-oxidation, peroxisomal β-oxidation produces H_2O_2 but not reducing equivalents (Elsner et al., 2011). Moreover, in contrast with other tissues, β-cell peroxisomes lack catalases, which impede the inactivation of peroxisomal H_2O_2 and promote the accumulation of ROS (Tiedge et al., 1997).

Oxidative stress also impairs mitochondrial function and integrity as mitochondrial DNA (mtDNA) and other mitochondrial component are at higher risk of oxidative injury due to the proximity to ROS/RNS primary source (Li et al., 2008). The lack of mtDNA protective histones and repair mechanism present in the nucleus also contribute to increase the risk of oxidative injury (Grishko et al., 2005). FFA have been shown to induce mtDNA damage in a dose dependent manner *in vitro* in INS-1 cell thorough an exacerbated activation of the inducible form of nitric oxide synthase (iNOS) that promotes nitric oxide (NO) overproduction (Grishko et al., 2005). In addition NO compromises ER calcium storages and contributes to ER stress (Oyadomari et al., 2001).

Long-term FFA exposure inhibits the transcription of Kinesin Family Member 12 (KIF12), a microtubule motor protein in β-cells with potent antioxidative capacity. KIF12 promotes the function of peroxisomes by stabilizing the newly synthesized transcription factor Sp1, which in turn induces the expression of Hsc70 (Yang et al., 2014). Oxidative stress can also activate the nuclear factor *k*B (NF-*k*B) pathway thus increasing inflammation (Morgan and Liu, 2011).

Finally, similarly to as happening with ER stress, when adipose tissue exceed its storage capacity and IR is generated, β-cell have to increase the insulin synthesis and secretion rates to reduce plasma glucose levels. The insulin overproduction increases the ATP demand and consequently ROS/RNS are overproduced damaging β-cell metabolism and inducing cell damage and apoptosis (Gerber and Rutter, 2017).

5.3 Lipotoxicity and inflammation

Obesity epidemic has resulted in a dramatic increase in obesity-associated T2D and its known that inflammation, derived from a defective adipose tissue expansion, contributes to many of the obesity-associated diseases such as diabetes. Indeed, proinflammatory cytokines have been detected in obese patients with T2D. In contrast with type 1 diabetes mellitus (T1D), the involvement of immune processes that lead to TD2 development and β-cell dysfunction are still poorly understood.

Several studies have pointed out the importance of inflammatory processes and proinflammatory macrophage recruitment to β-islets as mediators in β-cell dysfunction and T2D development (Boni-Schnetzler et al., 2009;

Eguchi et al., 2012; Richardson et al., 2009). For instance, in histological sections of T2D patients an increase of proinflammatory CD68^{+} macrophages have been observed in contrast with histological sections of non-diabetic patients (Richardson et al., 2009). IL-1β is also upregulated in T2D patients and its inhibition by IL-1β receptor agonist improves β-cell function. IL-1β induces iNOS activity through NF-*k*B activation and promotes the accumulation and overproduction of NO reducing β-islet number (Steer et al., 2006). NF-*k*B is present in its inactive form in the cytoplasm, bound to inhibitory *k*B (I-*k*B) proteins. Following IL-1 receptor (IL-1R) stimulation by IL-1β, I-*k*Bs are phosphorylated allowing NF-*k*B to translocate into the nucleus regulating the expression of several pro and anti-apoptotic protein. Moreover, NF-*k*B induces the synthesis and secretion of different chemokines such as CCE and CXCL1 that contribute to intra-islet macrophage recruitment and cytokines such as IL-6, IL-8, IL-1β or TNF-α that enhance macrophage recruitment and cause deficient intracellular calcium handling and decrease glucose-stimulated insulin secretion (Dula et al., 2010; O'Neill et al., 2013).

Saturated FFA can trigger inflammatory pathways in β-cell, both *via* direct activation, mediating macrophage recruitments or secondary to oxidative and ER stress (Fig. 5). FFAs activate the inflammatory program *via* toll-like receptors (TLRs) in many cell types (Lytrivi et al., 2018). TLR4 are synthesized in the β-cells and recognize several FFAs like PA promoting CD11b^{+}Ly-6C^{+} M1-type macrophage recruitment (Eguchi et al., 2012). Once in the pancreatic islets proinflammatory macrophages exposed to PA express and secrete IL-1β cytokines that are recognized by IL-1R receptors in the β-cells and promote NF-*k*B translocation. These results were confirmed in mouse models depleted of M1-macrophages before ethyl palmitate infusion where mice were protected from β-cell dysfunction (Eguchi et al., 2012). In addition, palmitate and high glucose synergistically favor islet secretion of the S100 calcium-binding protein A8 (S100A8), a damage associated pattern that activate macrophages through TLR4 receptor and induce subsequent cytokine production leading to β-cell apoptosis (Inoue et al., 2018) (Fig. 5).

5.4 Lipotoxicity and autophagy

Autophagy is characterized by lysosomal degradation of cellular components that regulates the balance between synthesis, degradation and recycling of cellular membranes, cytoplasm content, proteins and other cellular

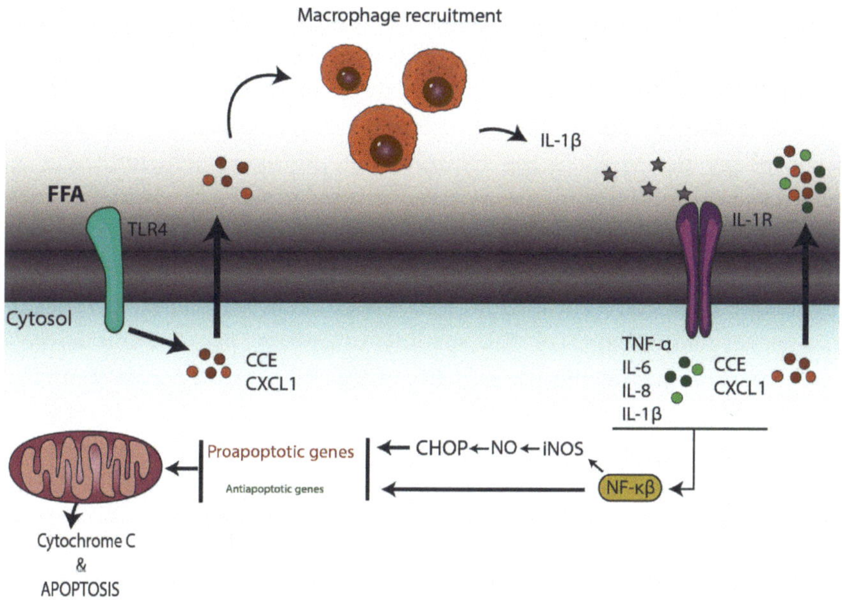

Fig. 5 Inflammatory processes and the crosstalk between immune and β-cells is enhances under lipotoxic conditions. FFAs are recognized by β-cell tool-like receptors (TLRs) inducing the synthesis and secretion of chemoattractant molecules (CCE and CXCL1) that promote macrophage intra-islet recruitment. Proinflammatory macrophages produce IL-1β that is recognized by its receptor in β-cell promoting the activation of different chemokines and cytokines that enhance macrophage recruitment. The binding of IL-1β to its receptor also activates NF-κβ that increases NO production through iNOS and activate proapoptotic signaling pathways that induce mitochondrial cytochrome *c* liberation to the cytosol and caspase-pathway mediated apoptosis.

components and organelles (Asfari et al., 1992). Autophagy is critical for the proper functioning of the β-cells as it has been demonstrated *in vivo*, *ex vivo* and *in vitro* when autophagic processes were compromised (Ebato et al., 2008; Yamamoto et al., 2018).

Autophagy is divided into three stages: initiation/nucleation, expansion and degradation/turnover. In the first stage, autophagy inducers such as rapamycin or nutrient deprivation, inhibit mTOR complex 1 (mTORC1) which is the starting signal for autophagosome formation (Yu et al., 2010). mTORC1 inhibition allows membrane isolation that will constitute the autophagosomes, a double membrane that surrounds organelles and other cell constituents, which is the hallmark of autophagy. The second step of the autophagosome is controlled by the Atg protein family that allows the elongation of initial membranes to form spherical structures. In this process Atg8 (also called microtubule-associated protein 1 light chain 3 (LC3)) ubiquitin-like protein is conjugated with phosphatidylethanolamine (PE) in the membrane forming LC3-II, which is released to the cytosol promoting the fusion of the autophagosome with the lysosome, which constitutes the third phase of autophagy that induces the hydrolysis of the entrapped molecules and organelles through lysosomal enzymes (Lee et al., 2019).

Autophagic machinery can also be restricted to specific organelles through specialized machinery in addition to the common autophagic pathway. Indeed, selective autophagic processes such as mitophagy (mitochondrial specific autophagy) or ER specific autophagy play a central role in β-cell function and in the maintenance of glucose homeostasis (Fumagalli et al., 2016; Grumati et al., 2017; Le Guerroue et al., 2017; Pickrell and Youle, 2015). These processes are especially relevant in β-cells as they are highly susceptible to damage from dysfunctional mitochondrial derived ROS and ER stress. For instance, in autophagy-deficient β-cells (*Atg*7-K0 β-cell), the ER-UPR system is markedly depressed thus reducing their ability to adequately respond to ER stress (Li et al., 1993; Merksamer et al., 2008). Other studies highlighted a deficient respond to stressors of primary islets coming from Atg7 deficient mice when comparing to wild type islets. Moreover, *in vivo* studies in *Atg*7-K0 mice showed an impaired glucose tolerance and when crossed with ob/ob mice, that require UPR system upregulation, developed severe diabetes (Quan et al., 2012). In addition, autophagy-deficient mice showed a reduce expression of antioxidant genes making the mice more vulnerable to ROS accumulation and oxidative stress (Wu et al., 2009).

The deleterious effect of lipotoxicity in autophagy has been confirmed *in vitro* using INS-1E cells exposed to oleate or palmitate. These cells showed an increase LC3-II levels, the hallmark of autophagosome formation; however, the autophagosome induction was followed by the inhibition of autophagosome turnover due to impaired lysosome-autophagosome fusion thus generating subsequent autophagosome accumulation (Choi et al., 2009; Komiya et al., 2010; Las et al., 2011). Furthermore, electron microscopy images confirmed the presence of abnormal accumulation of autophagosomes that likely confirm the *in vitro* results (Masini et al., 2009). All the data together indicate that lipotoxicity, and more specifically PA and oleate, promotes autophagy impairment that contribute to β-cell dysfunction and T2D development.

5.5 Lipotoxicity and β-cell dedifferentiation

In T2D there is a reduction in the β-cell mass of the pancreas triggered by apoptosis through caspase system and through complex cellular processes such as β-cell dedifferentiation or transdifferentiation (Bensellam et al., 2018; Cinti et al., 2016; Jonas et al., 1999). These processes, beside cell death, have been proposed as contributing factors to the loss of β-cell content and the impairment of the correct plasma glucose management. β-cell dedifferentiation involves the downregulation of characteristic genes and pathways that include those related with glucose metabolism or genes involved in protein processing and protein secretion. During dedifferentiation there is also an increased expression of the so-called forbidden genes, genes that under physiological conditions are suppressed or slightly expressed (Lemaire et al., 2016, 2017). Moreover, the presence of progenitor markers have been observed in dedifferentiated β-islets of diabetic animals and in some cases even makers of alfa-cells that suggest transdifferentiation processes (Spijker et al., 2013; Stancill et al., 2017).

Under physiological conditions glucose improves, stimulates and maintains β-cell differentiation. The beneficial effects include increased insulin mRNA level, upregulation of critical genes triggering insulin synthesis and secretion such as glucose transporter 2 (*glut2*) or subunits of K_{ATP} channel *Kir6.2* and *Sur1* and genes that participate in glycolysis and Krebs cycle among others (Accili et al., 2016; Bensellam et al., 2018). On the other hand, glucotoxicity has been shown to promote β-cell dedifferentiation in multiple studies where, under continued high glucose stimulation, β-cells were not able to maintain their phenotype reducing insulin secretion and inducing T2D development (Bensellam et al., 2018). However, while the role of glucotoxicity in β-cell dedifferentiation has been studied since the late 1990s, the role of lipotoxicity in β-cell dedifferentiation process is still under study.

Recent studies pointed out that forkhead box protein O1 (Foxo1) could be implicated in lipotoxic derived β-cell dedifferentiation. Foxo1 translocates to the nucleus upon Akt-dependent phosphorylation as well as NAD^+-dependent acetylation and enables the maintenance of glucose/lipid oxidation ratio toward the former and preserves β-cell identity (Martinez et al., 2008; Talchai et al., 2012). PA induces Foxo1 translocation as adaptive response; however, prolonged Foxo1 activation leads to its degradation and thus β-cell dedifferentiation. Moreover, mice fed a high-fat diet showed ALDH1A3, a β-cell differentiation marker, upregulation and Foxo1 downregulation by week 13 in contrast with mice feeding a low-fat diet (Tersey et al., 2018).

In 2011 Akerfeldt and colleagues published an article in which they studied the role of inhibitor of DNA binding 1 (Id1) protein for the development of glucose intolerance, impaired insulin secretion and β-cell dedifferentiation under conditions of chronic lipid oversupply (Akerfeldt and Laybutt, 2011; Busch et al., 2002). Id1 expression is induced *in vitro* in chronically fatty acid–treated MIN6 β-cells, an *in vitro* model characterized by a deficient insulin secretion and β-cell dedifferentiation. The inhibition of Id1 *in vitro* and *in vivo* reduced islet stress, improved insulin secretion and maintained β-cell phenotype (Akerfeldt and Laybutt, 2011).

5.6 Treatments for lipotoxic β-cell dysfunction

The objective of anti-hyperglycemic drugs is to control glycemia in order to prevent stable high-glucose levels in blood that could lead to diabetes-derived complications. The most extended therapies aim to improve both extra-pancreatic insulin resistance and GSIS (Inzucchi et al., 2012). In this section we will focus on how those therapies improve β-cell function and survival under lipotoxic conditions.

5.7 Metformin

Metformin is the first-line option for the treatment of diabetes. Its mechanism of action predominantly involves the inhibition of hepatic glucose production through the inhibition of AMP-activated protein kinase (AMPK) that inactivates the acetyl-CoA carboxylase (ACC) (Zhou et al., 2018). In addition, it directly inhibits the complex I of mitochondrial respiratory chain reducing hepatic glucose production independently of AMPK activation. The effects of metformin in pancreatic β-cells and GSIS are controversial. Some studies reported beneficial effects of metformin on GSIS upon elevated FFA exposure. In contrast, other studies revealed an impaired GSIS and β-cell dysfunction through AMPK activation (Kefas et al., 2004; Lupi et al., 2002a; Patane et al., 2000). Recently, it has been revealed a dose dependent mechanism of metformin that explains the differences shown in previous studies. Bellow 0.5 mM, metformin improves ER stress and oxidative stress by inhibiting NO and ROS production. Moreover, low-dose metformin inhibits Rho Kinase activity that is known to mediate insulin resistance and does not show metformin dependent AMPK action. On the other hand, metformin at high-dose activates AMPK and can trigger β-cell dysfunction (Kim et al., 2019). These data highlight the beneficial effects of low-dose metformin for the prevention of lipotoxic β-cell apoptosis as it is able to prevent ER and oxidative stress derived proapoptotic signaling.

5.8 Glucagon-like peptide 1 based treatment

Glucagon-like peptide 1 (GLP-1) is a gut derived incretin hormone that stimulates insulin secretion, inhibits gastric emptying, suppresses glucagon and reduces appetite. The interaction of GLP-1 with its receptor improves T2D by increasing liver and muscle glucose uptake, enhancing insulin synthesis and GSIS, inducing β-cell proliferation and reducing lipotoxic-β-cell apoptosis (Drucker and Nauck, 2006). Thus, drugs focused on incretin system like GLP-1 receptor agonists have been introduced for the treatment of diabetes. GLP-1 activates adenylate cyclase activity that enhances cyclic adenosine monophosphate (cAMP)-induced protein kinase activation and Ca^{2+} influx that restores the deleterious effect of FFA on islet dynamics (Hodson et al., 2013). Moreover, exendin-4, a GLP-1 analog, has shown beneficial effects in the function of β-cells treated with palmitate as it induces the ER chaperone BiP that lowers ER stress and induces the anti-apoptotic protein JunB. GLP-1 analog exendin-4 is also able to lower the effects of palmitate-induced proapoptotic JNK pathway activation. Thereby the lysosomal function is restored thus allowing a correct autophagic flux upon lipotoxic conditions (Cunha et al., 2009; Zummo et al., 2017) (Fig. 6).

Dipeptidyl peptidase-4 (DPP-4) inhibitors are also established as anti-hyperglycemic agents as they are able to ameliorate the inhibitory effect of the DPP-4 over GLP-1 (Deacon, 2019). Indeed, the DPP-4 inhibitor linagliptin, protected human islets from lipotoxicity and improved β-cell function after being exposed to palmitate (Shah et al., 2013).

5.9 Thiazolidinediones

Thiazolidinediones (TZN) are peroxisome proliferator-activated receptor GAMMA (PPARϒ) activators that primarily improve insulin sensitivity in insulin target tissues (liver, adipose tissue, muscle, etc.) and improve overall hyperglycemic state reducing the metabolic pressure on β-cells (Lu and Zhao, 2018; Shera et al., 2007). In addition, TZNs were found to improve GSIS and the expression of genes involved in β-cell function such as *GLUT2*, insulin receptor substrate 2 (*IRS2*) or *PDX1* (Kim et al., 2013). They are also able to protect β-cell function and viability after the exposure to toxic agents. *In vitro* experiments carried out in MIN6 pancreatic cell line treated with PA showed an increased cell death and reduced GSIS that were partially recovered after TZN treatment. In that study, they showed an increased phosphorylation of NF-*k*B and JNK together with an increased

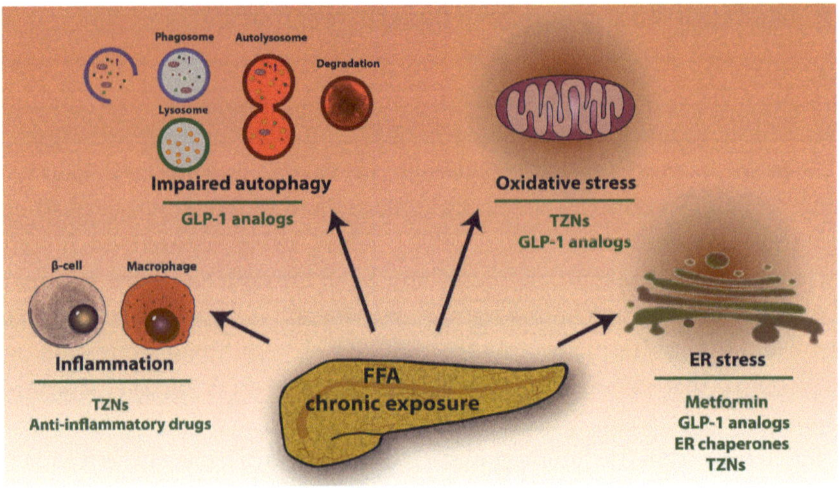

Fig. 6 Molecular mechanism leading to lipotoxic derived β-cell failure and potential therapies. Chronic exposure to FFA and specifically to long-chain FA alone or in combination with high glucose, induces lipotoxic mechanisms in β-cell leading to β-cell death. Those mechanisms include inflammatory crosstalk between β-cells and immune cells, oxidative stress and ROS production that lead to mitochondrial dysfunction, ER stress pathway upregulation and the impairment autophagy and organelle renewal. *In vitro* and *in vivo* studies suggest that different therapies used in the treatment of T2D (shown in green) improve the glucose homeostasis contributing to β-cell survival under lipotoxic conditions.

expression of IL-1β, IL6 and TNF-α that were repressed after TNZ treatment thus improving the overall inflammatory state that contribute to β-cell dysfunction (Hong et al., 2018). TZNs are also able to attenuate lipotoxic-derived oxidative stress as well as ER stress in FFA-treated β-cells and improve ER Ca^{2+} homeostasis (Kono et al., 2012; Saitoh et al., 2008). Thus, the potential of TZN treatment exceed the extra-pancreatic effects, as it is able to improve β-cell function and survival under lipotoxic conditions. However, the use of TZNs for the treatment of diabetic patients is not initially recommended due to their side effects. Indeed, several studies have associated their use to an increase in cardiovascular events or the development of bladder cancer (Colmers et al., 2012) (Fig. 6).

5.10 Anti-inflammatory drugs

Proinflammatory molecules have been shown to contribute to T2D phenotype and β-cell dysfunction as they are able to induce intra-islet macrophage recruitment and if maintained, to promote proapoptotic signaling. In the recent years the development of new anti-inflammatory therapies has allowed studies that tested their efficacy in β-cells exposed to lipotoxic conditions (Lytrivi et al., 2018). IL-1R antagonists have been shown to reverse chemokine secretion in PA exposed islets and to reduce the secretion of several proinflammatory cytokines and the intra-islet macrophage recruitment (Ehses et al., 2009). Other therapeutic agents targeting inflammation in T2D are also under study; however, their effectiveness reducing lipotoxic β-cell dysfunction is still unknown (Lytrivi et al., 2018) (Fig. 6).

5.11 Chemical chaperones

Chemical chaperones are low molecular weight compounds with the ability to influence various components of UPR, stabilize protein conformation and improve ER folding capacity. These include molecules that interact directly with some of the components of UPR signaling pathway; reduce ER stress, inhibit protein degradation; carry antioxidant activity and affect ER calcium signaling (Hetz et al., 2013; Kraskiewicz and FitzGerald, 2012). The two chaperones most extensively studied are 4-phenyl butyric acid (PBA) and the taurine-conjugated derivative of ursodeoxycholic acid TUDCA, which are chemical chaperones that reduce ER stress. It has been demonstrated a good safety profile for both of them (Xiao et al., 2011) and they ameliorate the palmitate-induced ER stress and GSIS in rat β-cells (Choi et al., 2008; Zhu et al., 2013). In obese and overweight non-diabetic individuals treated for 2 weeks with PBA before a lipid infusion over 48 h, PBA prevented a decline in insulin clearance. Being insulin clearance is mediated by similar and partly overlapping pathways affecting insulin sensitivity, including insulin receptor but not insulin receptor substrate (Choice et al., 1998; Poy et al., 2002), it is possible that PBA, may have affected insulin clearance through ER stress reduction or mechanisms independent of ER function. These results point out to a protective effect of PBA against β-cell lipotoxicity (Xiao et al., 2011). Although more studies are required to elucidate if the effects were mediated by alleviation of ER stress or by ER stress–independent mechanisms (Fig. 6).

5.12 Non pharmacological treatments: Caloric restriction and exercise

Lipid metabolism is tightly connected to obesity-related T2D and contributes to β-cell dysfunction. The previously mentioned pharmacological strategies have shown improvements in the maintenance of glucose homeostasis;

however, recent reports underlie the importance of non-pharmacological strategies such as weight loss, throughout caloric-restriction diets or bariatric surgery, in order to recover glucose homeostasis and β-cell functionality (Lean et al., 2018). The DiRECT trial showed that T2D was reversible after 6 months of low-calorie diet and that nearly 60% of the diabetic patients improved insulin response and β-cell capacity after 24 months, reaching similar values than non-diabetic control group (Lean et al., 2019; Taylor et al., 2018; Zhyzhneuskaya et al., 2020). Similarly, patients undergoing bariatric surgery showed an improved glucose homeostasis that was mainly associated to a reduction in pancreatic fat upon weight loss (Lim et al., 2011). Maintained weigh loss improved liver fat content, reduced plasma VDLD1-TG and was also accompanied by a reduction in intrapancreatic fat content. These changes were independent of insulin response as both groups, responders and nonresponders, showed reduced fat and very low density lipoprotein (VLDL) content but only the responders improved glucose homeostasis, which indicates that glucose control appears intrinsic to the β-cell (Zhyzhneuskaya et al., 2020). This underlies the importance of new studies that shed light on the mechanisms underlying β-cell function improvements upon low-calorie diets and weight loss. On the other hand, reduction of pancreatic fat after 2 weeks of exercise was shown in non-diabetic, prediabetic and T2D patients improving β-cell function especially in those with fatty pancreas (Heiskanen et al., 2018).

6. Conclusions

Substantial evidence supports lipotoxicity as a contributing factor to β-cell demise in T2D. It has been determined both in experimental models and metanalyses that lipotoxicity negatively affects many steps of the insulin signaling, production and secretion. Although the underlying mechanisms of lipotoxicity-induced T2M development are still unclear they involve the generation of intracellular lipid metabolites as well as several interrelated and stress-induced pathways.

References

Accili, D., Talchai, S.C., Kim-Muller, J.Y., Cinti, F., Ishida, E., Ordelheide, A.M., Kuo, T., Fan, J., Son, J., 2016. When beta-cells fail: lessons from dedifferentiation. Diabetes Obes. Metab. 18 (Suppl. 1), 117–122.

Acosta-Montano, P., Garcia-Gonzalez, V., 2018. Effects of dietary fatty acids in pancreatic beta cell metabolism, implications in homeostasis. Nutrients 10, 393.

Adams 2nd, J.M., Pratipanawatr, T., Berria, R., Wang, E., Defronzo, R.A., Sullards, M.C., Mandarino, L.J., 2004. Ceramide content is increased in skeletal muscle from obese insulin-resistant humans. Diabetes 53, 25–31.

Akerfeldt, M.C., Laybutt, D.R., 2011. Inhibition of Id1 augments insulin secretion and protects against high-fat diet-induced glucose intolerance. Diabetes 60, 2506–2514.

Alquier, T., Peyot, M.L., Latour, M.G., Kebede, M., Sorensen, C.M., Gesta, S., Ronald Kahn, C., Smith, R.D., Jetton, T.L., Metz, T.O., Prentki, M., Poitout, V., 2009. Deletion of GPR40 impairs glucose-induced insulin secretion in vivo in mice without affecting intracellular fuel metabolism in islets. Diabetes 58, 2607–2615.

Asfari, M., Janjic, D., Meda, P., Li, G., Halban, P.A., Wollheim, C.B., 1992. Establishment of 2-mercaptoethanol-dependent differentiated insulin-secreting cell lines. Endocrinology 130, 167–178.

Ashcroft, F.M., Rorsman, P., 1990. Atp-sensitive K+ channels: a link between B-cell metabolism and insulin secretion. Biochem. Soc. Trans. 18, 109–111.

Bachmann, O.P., Dahl, D.B., Brechtel, K., Machann, J., Haap, M., Maier, T., Loviscach, M., Stumvoll, M., Claussen, C.D., Schick, F., Haring, H.U., Jacob, S., 2001. Effects of intravenous and dietary lipid challenge on intramyocellular lipid content and the relation with insulin sensitivity in humans. Diabetes 50, 2579–2584.

Baynes, H.W., Mideksa, S., Ambachew, S., 2018. The role of polyunsaturated fatty acids (n-3 PUFAs) on the pancreatic beta-cells and insulin action. Adipocytes 7, 81–87.

Belfort, R., Mandarino, L., Kashyap, S., Wirfel, K., Pratipanawatr, T., Berria, R., Defronzo, R.A., Cusi, K., 2005. Dose-response effect of elevated plasma free fatty acid on insulin signaling. Diabetes 54, 1640–1648.

Bensellam, M., Laybutt, D.R., Jonas, J.C., 2012. The molecular mechanisms of pancreatic beta-cell glucotoxicity: recent findings and future research directions. Mol. Cell. Endocrinol. 364, 1–27.

Bensellam, M., Jonas, J.C., Laybutt, D.R., 2018. Mechanisms of beta-cell dedifferentiation in diabetes: recent findings and future research directions. J. Endocrinol. 236, R109–R143.

Biden, T.J., Boslem, E., Chu, K.Y., Sue, N., 2014. Lipotoxic endoplasmic reticulum stress, beta cell failure, and type 2 diabetes mellitus. Trends Endocrinol. Metab. 25, 389–398.

Bjorntorp, P., 1991. Metabolic implications of body fat distribution. Diabetes Care 14, 1132–1143.

Blachnio-Zabielska, A., Baranowski, M., Zabielski, P., Gorski, J., 2010. Effect of high fat diet enriched with unsaturated and diet rich in saturated fatty acids on sphingolipid metabolism in rat skeletal muscle. J. Cell. Physiol. 225, 786–791.

Boni-Schnetzler, M., Boller, S., Debray, S., Bouzakri, K., Meier, D.T., Prazak, R., Kerr-Conte, J., Pattou, F., Ehses, J.A., Schuit, F.C., Donath, M.Y., 2009. Free fatty acids induce a proinflammatory response in islets via the abundantly expressed interleukin-1 receptor I. Endocrinology 150, 5218–5229.

Boslem, E., Meikle, P.J., Biden, T.J., 2012. Roles of ceramide and sphingolipids in pancreatic beta-cell function and dysfunction. Islets 4, 177–187.

Brunzell, J.D., Robertson, R.P., Lerner, R.L., Hazzard, W.R., Ensinck, J.W., Bierman, E.L., Porte Jr., D., 1976. Relationships between fasting plasma glucose levels and insulin secretion during intravenous glucose tolerance tests. J. Clin. Endocrinol. Metab. 42, 222–229.

Busch, A.K., Cordery, D., Denyer, G.S., Biden, T.J., 2002. Expression profiling of palmitate- and oleate-regulated genes provides novel insights into the effects of chronic lipid exposure on pancreatic beta-cell function. Diabetes 51, 977–987.

Bustelo, X.R., 2014. Vav family exchange factors: an integrated regulatory and functional view. Small GTpases 5, 9.

Camastra, S., Manco, M., Mari, A., Baldi, S., Gastaldelli, A., Greco, A.V., Mingrone, G., Ferrannini, E., 2005. Beta-cell function in morbidly obese subjects during free living: long-term effects of weight loss. Diabetes 54, 2382–2389.

Carpentier, A., Mittelman, S.D., Lamarche, B., Bergman, R.N., Giacca, A., Lewis, G.F., 1999. Acute enhancement of insulin secretion by FFA in humans is lost with prolonged FFA elevation. Am. J. Physiol. 276, E1055–E1066.

Carpentier, A., Mittelman, S.D., Bergman, R.N., Giacca, A., Lewis, G.F., 2000. Prolonged elevation of plasma free fatty acids impairs pancreatic beta-cell function in obese nondiabetic humans but not in individuals with type 2 diabetes. Diabetes 49, 399–408.

Carpentier, A.C., Bourbonnais, A., Frisch, F., Giacca, A., Lewis, G.F., 2010. Plasma non-esterified fatty acid intolerance and hyperglycemia are associated with intravenous lipid-induced impairment of insulin sensitivity and disposition index. J. Clin. Endocrinol. Metab. 95, 1256–1264.

Cen, J., Sargsyan, E., Bergsten, P., 2016. Fatty acids stimulate insulin secretion from human pancreatic islets at fasting glucose concentrations via mitochondria-dependent and -independent mechanisms. Nutr. Metab. (Lond.) 13, 59.

Chavez, J.A., Summers, S.A., 2003. Characterizing the effects of saturated fatty acids on insulin signaling and ceramide and diacylglycerol accumulation in 3T3-L1 adipocytes and C2C12 myotubes. Arch. Biochem. Biophys. 419, 101–109.

Chavez, J.A., Knotts, T.A., Wang, L.P., Li, G., Dobrowsky, R.T., Florant, G.L., Summers, S.A., 2003. A role for ceramide, but not diacylglycerol, in the antagonism of insulin signal transduction by saturated fatty acids. J. Biol. Chem. 278, 10297–10303.

Chen, S., Ogawa, A., Ohneda, M., Unger, R.H., Foster, D.W., Mcgarry, J.D., 1994. More direct evidence for a malonyl-CoA-carnitine palmitoyltransferase I interaction as a key event in pancreatic beta-cell signaling. Diabetes 43, 878–883.

Chen, P., Liu, H., Xiang, H., Zhou, J., Zeng, Z., Chen, R., Zhao, S., Xiao, J., Shu, Z., Chen, S., Lu, H., 2019. Palmitic acid-induced autophagy increases reactive oxygen species via the Ca(2+)/PKCalpha/NOX4 pathway and impairs endothelial function in human umbilical vein endothelial cells. Exp. Ther. Med. 17, 2425–2432.

Choi, S.E., Lee, Y.J., Jang, H.J., Lee, K.W., Kim, Y.S., Jun, H.S., Kang, S.S., Chun, J., Kang, Y., 2008. A chemical chaperone 4-PBA ameliorates palmitate-induced inhibition of glucose-stimulated insulin secretion (GSIS). Arch. Biochem. Biophys. 475, 109–114.

Choi, S.E., Lee, S.M., Lee, Y.J., Li, L.J., Lee, S.J., Lee, J.H., Kim, Y., Jun, H.S., Lee, K.W., Kang, Y., 2009. Protective role of autophagy in palmitate-induced INS-1 beta-cell death. Endocrinology 150, 126–134.

Choice, C.V., Howard, M.J., Poy, M.N., Hankin, M.H., Najjar, S.M., 1998. Insulin stimulates pp120 endocytosis in cells co-expressing insulin receptors. J. Biol. Chem. 273, 22194–22200.

Cinti, F., Bouchi, R., Kim-Muller, J.Y., Ohmura, Y., Sandoval, P.R., Masini, M., Marselli, L., Suleiman, M., Ratner, L.E., Marchetti, P., Accili, D., 2016. Evidence of beta-cell dedifferentiation in human type 2 diabetes. J. Clin. Endocrinol. Metab. 101, 1044–1054.

Cline, G.W., Lepine, R.L., Papas, K.K., Kibbey, R.G., Shulman, G.I., 2004. 13C NMR isotopomer analysis of anaplerotic pathways in INS-1 cells. J. Biol. Chem. 279, 44370–44375.

Cnop, M., Hannaert, J.C., Hoorens, A., Eizirik, D.L., Pipeleers, D.G., 2001. Inverse relationship between cytotoxicity of free fatty acids in pancreatic islet cells and cellular triglyceride accumulation. Diabetes 50, 1771–1777.

Cnop, M., Welsh, N., Jonas, J.C., Jorns, A., Lenzen, S., Eizirik, D.L., 2005. Mechanisms of pancreatic beta-cell death in type 1 and type 2 diabetes: many differences, few similarities. Diabetes 54 (Suppl. 2), S97–107.

Cnop, M., Ladriere, L., Hekerman, P., Ortis, F., Cardozo, A.K., Dogusan, Z., Flamez, D., Boyce, M., Yuan, J., Eizirik, D.L., 2007. Selective inhibition of eukaryotic translation

initiation factor 2 alpha dephosphorylation potentiates fatty acid-induced endoplasmic reticulum stress and causes pancreatic beta-cell dysfunction and apoptosis. J. Biol. Chem. 282, 3989–3997.
Colmers, I.N., Bowker, S.L., Majumdar, S.R., Johnson, J.A., 2012. Use of thiazolidinediones and the risk of bladder cancer among people with type 2 diabetes: a meta-analysis. CMAJ 184, E675–E683.
Corkey, B.E., Glennon, M.C., Chen, K.S., Deeney, J.T., Matschinsky, F.M., Prentki, M., 1989. A role for malonyl-CoA in glucose-stimulated insulin secretion from clonal pancreatic beta-cells. J. Biol. Chem. 264, 21608–21612.
Cunha, D.A., Hekerman, P., Ladriere, L., Bazarra-Castro, A., Ortis, F., Wakeham, M.C., Moore, F., Rasschaert, J., Cardozo, A.K., Bellomo, E., Overbergh, L., Mathieu, C., Lupi, R., Hai, T., Herchuelz, A., Marchetti, P., Rutter, G.A., Eizirik, D.L., Cnop, M., 2008. Initiation and execution of lipotoxic ER stress in pancreatic beta-cells. J. Cell Sci. 121, 2308–2318.
Cunha, D.A., Ladriere, L., Ortis, F., Igoillo-Esteve, M., Gurzov, E.N., Lupi, R., Marchetti, P., Eizirik, D.L., Cnop, M., 2009. Glucagon-like peptide-1 agonists protect pancreatic beta-cells from lipotoxic endoplasmic reticulum stress through upregulation of BiP and JunB. Diabetes 58, 2851–2862.
Cunha, D.A., Igoillo-Esteve, M., Gurzov, E.N., Germano, C.M., Naamane, N., Marhfour, I., Fukaya, M., Vanderwinden, J.M., Gysemans, C., Mathieu, C., Marselli, L., Marchetti, P., Harding, H.P., Ron, D., Eizirik, D.L., Cnop, M., 2012. Death protein 5 and p53-upregulated modulator of apoptosis mediate the endoplasmic reticulum stress-mitochondrial dialog triggering lipotoxic rodent and human beta-cell apoptosis. Diabetes 61, 2763–2775.
Deacon, C.F., 2019. Physiology and pharmacology of DPP-4 in glucose homeostasis and the treatment of type 2 diabetes. Front. Endocrinol. 10, 80.
Deeney, J.T., Gromada, J., Hoy, M., Olsen, H.L., Rhodes, C.J., Prentki, M., Berggren, P.O., Corkey, B.E., 2000. Acute stimulation with long chain acyl-CoA enhances exocytosis in insulin-secreting cells (HIT T-15 and NMRI beta-cells). J. Biol. Chem. 275, 9363–9368.
Doliba, N.M., Liu, Q., Li, C., Chen, P., Liu, C., Naji, A., Matschinsky, F.M., 2017. Inhibition of cholinergic potentiation of insulin secretion from pancreatic islets by chronic elevation of glucose and fatty acids: protection by casein kinase 2 inhibitor. Mol. Metab. 6, 1240–1253.
Dranse, H.J., Kelly, M.E., Hudson, B.D., 2013. Drugs or diet?—developing novel therapeutic strategies targeting the free fatty acid family of GPCRs. Br. J. Pharmacol. 170, 696–711.
Drews, G., Krippeit-Drews, P., Dufer, M., 2010. Oxidative stress and beta-cell dysfunction. Pflugers Arch. 460, 703–718.
Drucker, D.J., Nauck, M.A., 2006. The incretin system: glucagon-like peptide-1 receptor agonists and dipeptidyl peptidase-4 inhibitors in type 2 diabetes. Lancet 368, 1696–1705.
Dula, S.B., Jecmenica, M., Wu, R., Jahanshahi, P., Verrilli, G.M., Carter, J.D., Brayman, K.L., Nunemaker, C.S., 2010. Evidence that low-grade systemic inflammation can induce islet dysfunction as measured by impaired calcium handling. Cell Calcium 48, 133–142.
Dziewulska, A., Dobosz, A.M., Dobrzyn, A., Smolinska, A., Kolczynska, K., Ntambi, J.M., Dobrzyn, P., 2020. SCD1 regulates the AMPK/SIRT1 pathway and histone acetylation through changes in adenine nucleotide metabolism in skeletal muscle. J. Cell. Physiol. 235, 1129–1140.

Ebato, C., Uchida, T., Arakawa, M., Komatsu, M., Ueno, T., Komiya, K., Azuma, K., Hirose, T., Tanaka, K., Kominami, E., Kawamori, R., Fujitani, Y., Watada, H., 2008. Autophagy is important in islet homeostasis and compensatory increase of beta cell mass in response to high-fat diet. Cell Metab. 8, 325–332.

Eguchi, K., Manabe, I., Oishi-Tanaka, Y., Ohsugi, M., Kono, N., Ogata, F., Yagi, N., Ohto, U., Kimoto, M., Miyake, K., Tobe, K., Arai, H., Kadowaki, T., Nagai, R., 2012. Saturated fatty acid and TLR signaling link beta cell dysfunction and islet inflammation. Cell Metab. 15, 518–533.

Ehses, J.A., Lacraz, G., Giroix, M.H., Schmidlin, F., Coulaud, J., Kassis, N., Irminger, J.C., Kergoat, M., Portha, B., Homo-Delarche, F., Donath, M.Y., 2009. IL-1 antagonism reduces hyperglycemia and tissue inflammation in the type 2 diabetic GK rat. Proc. Natl. Acad. Sci. U. S. A. 106, 13998–14003.

Eitel, K., Staiger, H., Rieger, J., Mischak, H., Brandhorst, H., Brendel, M.D., Bretzel, R.G., Haring, H.U., Kellerer, M., 2003. Protein kinase C delta activation and translocation to the nucleus are required for fatty acid-induced apoptosis of insulin-secreting cells. Diabetes 52, 991–997.

Ellis, B.A., Poynten, A., Lowy, A.J., Furler, S.M., Chisholm, D.J., Kraegen, E.W., Cooney, G.J., 2000. Long-chain acyl-CoA esters as indicators of lipid metabolism and insulin sensitivity in rat and human muscle. Am. J. Physiol. Endocrinol. Metab. 279, E554–E560.

Elsner, M., Gehrmann, W., Lenzen, S., 2011. Peroxisome-generated hydrogen peroxide as important mediator of lipotoxicity in insulin-producing cells. Diabetes 60, 200–208.

Farese, R.V., Dimarco, P.E., Barnes, D.E., Sabir, M.A., Larson, R.E., Davis, J.S., Morrison, A.D., 1986. Rapid glucose-dependent increases in phosphatidic acid and phosphoinositides in rat pancreatic islets. Endocrinology 118, 1498–1503.

Farfari, S., Schulz, V., Corkey, B., Prentki, M., 2000. Glucose-regulated anaplerosis and cataplerosis in pancreatic beta-cells: possible implication of a pyruvate/citrate shuttle in insulin secretion. Diabetes 49, 718–726.

Ferdaoussi, M., Bergeron, V., Zarrouki, B., Kolic, J., Cantley, J., Fielitz, J., Olson, E.N., Prentki, M., Biden, T., Macdonald, P.E., Poitout, V., 2012. G protein-coupled receptor (GPR)40-dependent potentiation of insulin secretion in mouse islets is mediated by protein kinase D1. Diabetologia 55, 2682–2692.

Flamez, D., Berger, V., Kruhoffer, M., Orntoft, T., Pipeleers, D., Schuit, F.C., 2002. Critical role for cataplerosis via citrate in glucose-regulated insulin release. Diabetes 51, 2018–2024.

Flowers, J.B., Rabaglia, M.E., Schueler, K.L., Flowers, M.T., Lan, H., Keller, M.P., Ntambi, J.M., Attie, A.D., 2007. Loss of stearoyl-CoA desaturase-1 improves insulin sensitivity in lean mice but worsens diabetes in leptin-deficient obese mice. Diabetes 56, 1228–1239.

Fontes, G., Semache, M., Hagman, D.K., Tremblay, C., Shah, R., Rhodes, C.J., Rutter, J., Poitout, V., 2009. Involvement of Per-Arnt-Sim kinase and extracellular-regulated kinases-1/2 in palmitate inhibition of insulin gene expression in pancreatic beta-cells. Diabetes 58, 2048–2058.

Foufelle, F., Ferre, P., 2002. New perspectives in the regulation of hepatic glycolytic and lipogenic genes by insulin and glucose: a role for the transcription factor sterol regulatory element binding protein-1c. Biochem. J. 366, 377–391.

Fribley, A., Zhang, K., Kaufman, R.J., 2009. Regulation of apoptosis by the unfolded protein response. Methods Mol. Biol. 559, 191–204.

Fujiwara, K., Maekawa, F., Yada, T., 2005. Oleic acid interacts with GPR40 to induce Ca2+ signaling in rat islet beta-cells: mediation by PLC and L-type Ca2+ channel and link to insulin release. Am. J. Physiol. Endocrinol. Metab. 289, E670–E677.

Fumagalli, F., Noack, J., Bergmann, T.J., Cebollero, E., Pisoni, G.B., Fasana, E., Fregno, I., Galli, C., Loi, M., Solda, T., D'antuono, R., Raimondi, A., Jung, M., Melnyk, A., Schorr, S., Schreiber, A., Simonelli, L., Varani, L., Wilson-Zbinden, C., Zerbe, O., Hofmann, K., Peter, M., Quadroni, M., Zimmermann, R., Molinari, M., 2016. Corrigendum: translocon component Sec62 acts in endoplasmic reticulum turnover during stress recovery. Nat. Cell Biol. 19, 76.

Galicia-Garcia, U., Benito-Vicente, A., Jebari, S., Larrea-Sebal, A., Siddiqi, H., Uribe, K.B., Ostolaza, H., Martin, C., 2020. Pathophysiology of type 2 diabetes mellitus. Int. J. Mol. Sci. 21, 6275.

Garcia-Ruiz, C., Colell, A., Morales, A., Calvo, M., Enrich, C., Fernandez-Checa, J.C., 2002. Trafficking of ganglioside GD3 to mitochondria by tumor necrosis factor-alpha. J. Biol. Chem. 277, 36443–36448.

Gehrmann, W., Elsner, M., Lenzen, S., 2010. Role of metabolically generated reactive oxygen species for lipotoxicity in pancreatic beta-cells. Diabetes Obes. Metab. 12 (Suppl. 2), 149–158.

Georgiadi, A., Kersten, S., 2012. Mechanisms of gene regulation by fatty acids. Adv. Nutr. 3, 127–134.

Gerber, P.A., Rutter, G.A., 2017. The role of oxidative stress and hypoxia in pancreatic beta-cell dysfunction in diabetes mellitus. Antioxid. Redox Signal. 26, 501–518.

Giacca, A., Xiao, C., Oprescu, A.I., Carpentier, A.C., Lewis, G.F., 2011. Lipid-induced pancreatic beta-cell dysfunction: focus on in vivo studies. Am. J. Physiol. Endocrinol. Metab. 300, E255–E262.

Gray, J.P., Heart, E., 2010. Usurping the mitochondrial supremacy: extramitochondrial sources of reactive oxygen intermediates and their role in beta cell metabolism and insulin secretion. Toxicol. Mech. Methods 20, 167–174.

Grishko, V., Rachek, L., Musiyenko, S., Ledoux, S.P., Wilson, G.L., 2005. Involvement of mtDNA damage in free fatty acid-induced apoptosis. Free Radic. Biol. Med. 38, 755–762.

Grumati, P., Morozzi, G., Holper, S., Mari, M., Harwardt, M.I., Yan, R., Muller, S., Reggiori, F., Heilemann, M., Dikic, I., 2017. Full length RTN3 regulates turnover of tubular endoplasmic reticulum via selective autophagy. Elife 6, e25555.

Guasch-Ferre, M., Becerra-Tomas, N., Ruiz-Canela, M., Corella, D., Schroder, H., Estruch, R., Ros, E., Aros, F., Gomez-Gracia, E., Fiol, M., Serra-Majem, L., Lapetra, J., Basora, J., Martin-Calvo, N., Portoles, O., Fito, M., Hu, F.B., Forga, L., Salas-Salvado, J., 2017. Total and subtypes of dietary fat intake and risk of type 2 diabetes mellitus in the Prevencion con Dieta Mediterranea (PREDIMED) study. Am. J. Clin. Nutr. 105, 723–735.

Gwiazda, K.S., Yang, T.L., Lin, Y., Johnson, J.D., 2009. Effects of palmitate on ER and cytosolic Ca2+ homeostasis in beta-cells. Am. J. Physiol. Endocrinol. Metab. 296, E690–E701.

Hagman, D.K., Hays, L.B., Parazzoli, S.D., Poitout, V., 2005. Palmitate inhibits insulin gene expression by altering PDX-1 nuclear localization and reducing MafA expression in isolated rat islets of langerhans. J. Biol. Chem. 280, 32413–32418.

Haus, J.M., Kashyap, S.R., Kasumov, T., Zhang, R., Kelly, K.R., Defronzo, R.A., Kirwan, J.P., 2009. Plasma ceramides are elevated in obese subjects with type 2 diabetes and correlate with the severity of insulin resistance. Diabetes 58, 337–343.

Heiskanen, M.A., Motiani, K.K., Mari, A., Saunavaara, V., Eskelinen, J.J., Virtanen, K.A., Koivumaki, M., Loyttyniemi, E., Nuutila, P., Kalliokoski, K.K., Hannukainen, J.C., 2018. Exercise training decreases pancreatic fat content and improves beta cell function regardless of baseline glucose tolerance: a randomised controlled trial. Diabetologia 61, 1817–1828.

Hetz, C., Chevet, E., Harding, H.P., 2013. Targeting the unfolded protein response in disease. Nat. Rev. Drug Discov. 12, 703–719.
Hirosumi, J., Tuncman, G., Chang, L., Gorgun, C.Z., Uysal, K.T., Maeda, K., Karin, M., Hotamisligil, G.S., 2002. A central role for JNK in obesity and insulin resistance. Nature 420, 333–336.
Hodson, D.J., Mitchell, R.K., Bellomo, E.A., Sun, G., Vinet, L., Meda, P., Li, D., Li, W.H., Bugliani, M., Marchetti, P., Bosco, D., Piemonti, L., Johnson, P., Hughes, S.J., Rutter, G.A., 2013. Lipotoxicity disrupts incretin-regulated human beta cell connectivity. J. Clin. Invest. 123, 4182–4194.
Hong, S.W., Lee, J., Cho, J.H., Kwon, H., Park, S.E., Rhee, E.J., Park, C.Y., Oh, K.W., Park, S.W., Lee, W.Y., 2018. Pioglitazone attenuates palmitate-induced inflammation and endoplasmic reticulum stress in pancreatic beta-cells. Endocrinol. Metab. 33, 105–113.
Hulver, M.W., Berggren, J.R., Cortright, R.N., Dudek, R.W., Thompson, R.P., Pories, W.J., Macdonald, K.G., Cline, G.W., Shulman, G.I., Dohm, G.L., Houmard, J.A., 2003. Skeletal muscle lipid metabolism with obesity. Am. J. Physiol. Endocrinol. Metab. 284, E741–E747.
Inoue, H., Shirakawa, J., Togashi, Y., Tajima, K., Okuyama, T., Kyohara, M., Tanaka, Y., Orime, K., Saisho, Y., Yamada, T., Shibue, K., Kulkarni, R.N., Terauchi, Y., 2018. Signaling between pancreatic beta cells and macrophages via S100 calcium-binding protein A8 exacerbates beta-cell apoptosis and islet inflammation. J. Biol. Chem. 293, 5934–5946.
Inzucchi, S.E., Bergenstal, R.M., Buse, J.B., Diamant, M., Ferrannini, E., Nauck, M., Peters, A.L., Tsapas, A., Wender, R., Matthews, D.R., American Diabetes Association, European Association for the Study of Diabetes, 2012. Management of hyperglycemia in type 2 diabetes: a patient-centered approach: position statement of the American Diabetes Association (ADA) and the European Association for the Study of Diabetes (EASD). Diabetes Care 35, 1364–1379.
Itoh, Y., Kawamata, Y., Harada, M., Kobayashi, M., Fujii, R., Fukusumi, S., Ogi, K., Hosoya, M., Tanaka, Y., Uejima, H., Tanaka, H., Maruyama, M., Satoh, R., Okubo, S., Kizawa, H., Komatsu, H., Matsumura, F., Noguchi, Y., Shinohara, T., Hinuma, S., Fujisawa, Y., Fujino, M., 2003. Free fatty acids regulate insulin secretion from pancreatic beta cells through GPR40. Nature 422, 173–176.
Jaishy, B., Abel, E.D., 2016. Lipids, lysosomes, and autophagy. J. Lipid Res. 57, 1619–1635.
Janikiewicz, J., Hanzelka, K., Dziewulska, A., Kozinski, K., Dobrzyn, P., Bernas, T., Dobrzyn, A., 2015. Inhibition of SCD1 impairs palmitate-derived autophagy at the step of autophagosome-lysosome fusion in pancreatic beta-cells. J. Lipid Res. 56, 1901–1911.
Johnson, J.D., Han, Z., Otani, K., Ye, H., Zhang, Y., Wu, H., Horikawa, Y., Misler, S., Bell, G.I., Polonsky, K.S., 2004. RyR2 and calpain-10 delineate a novel apoptosis pathway in pancreatic islets. J. Biol. Chem. 279, 24794–24802.
Johnston, L.W., Harris, S.B., Retnakaran, R., Giacca, A., Liu, Z., Bazinet, R.P., Hanley, A.J., 2018. Association of NEFA composition with insulin sensitivity and beta cell function in the prospective metabolism and islet cell evaluation (PROMISE) cohort. Diabetologia 61, 821–830.
Jonas, J.C., Sharma, A., Hasenkamp, W., Ilkova, H., Patane, G., Laybutt, R., Bonner-Weir, S., Weir, G.C., 1999. Chronic hyperglycemia triggers loss of pancreatic beta cell differentiation in an animal model of diabetes. J. Biol. Chem. 274, 14112–14121.
Kaneko, Y.K., Ishikawa, T., 2015. Diacylglycerol signaling pathway in pancreatic beta-cells: an essential role of diacylglycerol kinase in the regulation of insulin secretion. Biol. Pharm. Bull. 38, 669–673.

Kant, S., Standen, C.L., Morel, C., Jung, D.Y., Kim, J.K., Swat, W., Flavell, R.A., Davis, R.J., 2017. A protein scaffold coordinates SRC-mediated JNK activation in response to metabolic stress. Cell Rep. 20, 2775–2783.

Karam, J.H., Grodsky, G.M., Forsham, P.H., 1963. Excessive insulin response to glucose in obese subjects as measured by immunochemical assay. Diabetes 12, 197–204.

Karaskov, E., Scott, C., Zhang, L., Teodoro, T., Ravazzola, M., Volchuk, A., 2006. Chronic palmitate but not oleate exposure induces endoplasmic reticulum stress, which may contribute to INS-1 pancreatic beta-cell apoptosis. Endocrinology 147, 3398–3407.

Karunakaran, U., Moon, J.S., Lee, H.W., Won, K.C., 2015. CD36 initiated signaling mediates ceramide-induced TXNIP expression in pancreatic beta-cells. Biochim. Biophys. Acta 1852, 2414–2422.

Kashyap, S., Belfort, R., Gastaldelli, A., Pratipanawatr, T., Berria, R., Pratipanawatr, W., Bajaj, M., Mandarino, L., Defronzo, R., Cusi, K., 2003. A sustained increase in plasma free fatty acids impairs insulin secretion in nondiabetic subjects genetically predisposed to develop type 2 diabetes. Diabetes 52, 2461–2474.

Keane, D.C., Takahashi, H.K., Dhayal, S., Morgan, N.G., Curi, R., Newsholme, P., 2011. Arachidonic acid actions on functional integrity and attenuation of the negative effects of palmitic acid in a clonal pancreatic beta-cell line. Clin. Sci. (Lond.) 120, 195–206.

Keane, K.N., Cruzat, V.F., Carlessi, R., De Bittencourt Jr., P.I., Newsholme, P., 2015. Molecular events linking oxidative stress and inflammation to insulin resistance and beta-cell dysfunction. Oxid. Med. Cell. Longev. 2015, 181643.

Kefas, B.A., Cai, Y., Kerckhofs, K., Ling, Z., Martens, G., Heimberg, H., Pipeleers, D., Van De Casteele, M., 2004. Metformin-induced stimulation of AMP-activated protein kinase in beta-cells impairs their glucose responsiveness and can lead to apoptosis. Biochem. Pharmacol. 68, 409–416.

Kelpe, C.L., Moore, P.C., Parazzoli, S.D., Wicksteed, B., Rhodes, C.J., Poitout, V., 2003. Palmitate inhibition of insulin gene expression is mediated at the transcriptional level via ceramide synthesis. J. Biol. Chem. 278, 30015–30021.

Kharroubi, I., Ladriere, L., Cardozo, A.K., Dogusan, Z., Cnop, M., Eizirik, D.L., 2004. Free fatty acids and cytokines induce pancreatic beta-cell apoptosis by different mechanisms: role of nuclear factor-kappaB and endoplasmic reticulum stress. Endocrinology 145, 5087–5096.

Kim, J.W., You, Y.H., Ham, D.S., Cho, J.H., Ko, S.H., Song, K.H., Son, H.Y., Suh-Kim, H., Lee, I.K., Yoon, K.H., 2009. Suppression of peroxisome proliferator-activated receptor gamma-coactivator-1alpha normalizes the glucolipotoxicity-induced decreased BETA2/NeuroD gene transcription and improved glucose tolerance in diabetic rats. Endocrinology 150, 4074–4083.

Kim, H.S., Hwang, Y.C., Koo, S.H., Park, K.S., Lee, M.S., Kim, K.W., Lee, M.K., 2013. PPAR-gamma activation increases insulin secretion through the up-regulation of the free fatty acid receptor GPR40 in pancreatic beta-cells. PLoS One 8, E50128.

Kim, H.I., Lee, J.S., Kwak, B.K., Hwang, W.M., Kim, M.J., Kim, Y.B., Chung, S.S., Park, K.S., 2019. Metformin ameliorates lipotoxic beta-cell dysfunction through a concentration-dependent dual mechanism of action. Diabetes Metab. J. 43, 854–866.

Kominato, R., Fujimoto, S., Mukai, E., Nakamura, Y., Nabe, K., Shimodahira, M., Nishi, Y., Funakoshi, S., Seino, Y., Inagaki, N., 2008. Src activation generates reactive oxygen species and impairs metabolism-secretion coupling in diabetic Goto-Kakizaki and ouabain-treated rat pancreatic islets. Diabetologia 51, 1226–1235.

Komiya, K., Uchida, T., Ueno, T., Koike, M., Abe, H., Hirose, T., Kawamori, R., Uchiyama, Y., Kominami, E., Fujitani, Y., Watada, H., 2010. Free fatty acids stimulate autophagy in pancreatic beta-cells via JNK pathway. Biochem. Biophys. Res. Commun. 401, 561–567.

Kono, T., Ahn, G., Moss, D.R., Gann, L., Zarain-Herzberg, A., Nishiki, Y., Fueger, P.T., Ogihara, T., Evans-Molina, C., 2012. PPAR-gamma activation restores pancreatic islet SERCA2 levels and prevents beta-cell dysfunction under conditions of hyperglycemic and cytokine stress. Mol. Endocrinol. 26, 257–271.

Kraskiewicz, H., FitzGerald, U., 2012. Interfering with endoplasmic reticulum stress. Trends Pharmacol. Sci. 33, 53–63.

Kristinsson, H., Smith, D.M., Bergsten, P., Sargsyan, E., 2013. FFAR1 is involved in both the acute and chronic effects of palmitate on insulin secretion. Endocrinology 154, 4078–4088.

Krizhanovskii, C., Kristinsson, H., Elksnis, A., Wang, X., Gavali, H., Bergsten, P., Scharfmann, R., Welsh, N., 2017. EndoC-BetaH1 cells display increased sensitivity to sodium palmitate when cultured in DMEM/F12 medium. Islets 9, E1296995.

Las, G., Serada, S.B., Wikstrom, J.D., Twig, G., Shirihai, O.S., 2011. Fatty acids suppress autophagic turnover in beta-cells. J. Biol. Chem. 286, 42534–42544.

Latour, M.G., Alquier, T., Oseid, E., Tremblay, C., Jetton, T.L., Luo, J., Lin, D.C., Poitout, V., 2007. GPR40 is necessary but not sufficient for fatty acid stimulation of insulin secretion in vivo. Diabetes 56, 1087–1094.

Le Guerroue, F., Eck, F., Jung, J., Starzetz, T., Mittelbronn, M., Kaulich, M., Behrends, C., 2017. Autophagosomal content profiling reveals an LC3C-dependent piecemeal mitophagy pathway. Mol. Cell 68, 786–796 E6.

Lean, M.E., Leslie, W.S., Barnes, A.C., Brosnahan, N., Thom, G., Mccombie, L., Peters, C., Zhyzhneuskaya, S., Al-Mrabeh, A., Hollingsworth, K.G., Rodrigues, A.M., Rehackova, L., Adamson, A.J., Sniehotta, F.F., Mathers, J.C., Ross, H.M., Mcilvenna, Y., Stefanetti, R., Trenell, M., Welsh, P., Kean, S., Ford, I., Mcconnachie, A., Sattar, N., Taylor, R., 2018. Primary care-led weight management for remission of type 2 diabetes (DiRECT): an open-label, cluster-randomised trial. Lancet 391, 541–551.

Lean, M.E.J., Leslie, W.S., Barnes, A.C., Brosnahan, N., Thom, G., Mccombie, L., Peters, C., Zhyzhneuskaya, S., Al-Mrabeh, A., Hollingsworth, K.G., Rodrigues, A.M., Rehackova, L., Adamson, A.J., Sniehotta, F.F., Mathers, J.C., Ross, H.M., Mcilvenna, Y., Welsh, P., Kean, S., Ford, I., Mcconnachie, A., Messow, C.M., Sattar, N., Taylor, R., 2019. Durability of a primary care-led weight-management intervention for remission of type 2 diabetes: 2-year results of the DiRECT open-label, cluster-randomised trial. Lancet Diabetes Endocrinol. 7, 344–355.

Lee, Y., Hirose, H., Ohneda, M., Johnson, J.H., Mcgarry, J.D., Unger, R.H., 1994. Beta-cell lipotoxicity in the pathogenesis of non-insulin-dependent diabetes mellitus of obese rats: impairment in adipocyte-beta-cell relationships. Proc. Natl. Acad. Sci. U. S. A. 91, 10878–10882.

Lee, Y.H., Kim, J., Park, K., Lee, M.S., 2019. Beta-cell autophagy: mechanism and role in beta-cell dysfunction. Mol. Metab. 27s, S92–S103.

Lei, X., Zhang, S., Emani, B., Barbour, S.E., Ramanadham, S., 2010. A link between endoplasmic reticulum stress-induced beta-cell apoptosis and the group via Ca2+-independent phospholipase A2 (iPLA2beta). Diabetes Obes. Metab. 12 (Suppl. 2), 93–98.

Leloup, C., Tourrel-Cuzin, C., Magnan, C., Karaca, M., Castel, J., Carneiro, L., Colombani, A.L., Ktorza, A., Casteilla, L., Penicaud, L., 2009. Mitochondrial reactive oxygen species are obligatory signals for glucose-induced insulin secretion. Diabetes 58, 673–681.

Lemaire, K., Thorrez, L., Schuit, F., 2016. Disallowed and allowed gene expression: two faces of mature islet beta cells. Annu. Rev. Nutr. 36, 45–71.

Lemaire, K., Granvik, M., Schraenen, A., Goyvaerts, L., Van Lommel, L., Gomez-Ruiz, A., In 'T Veld, P., Gilon, P., Schuit, F., 2017. How stable is repression of disallowed genes in pancreatic islets in response to metabolic stress? Plos One 12, E0181651.

Lenaz, G., 2001. The mitochondrial production of reactive oxygen species: mechanisms and implications in human pathology. IUBMB Life 52, 159–164.

Leung, N., Sakaue, T., Carpentier, A., Uffelman, K., Giacca, A., Lewis, G.F., 2004. Prolonged increase of plasma non-esterified fatty acids fully abolishes the stimulatory effect of 24 hours of moderate hyperglycaemia on insulin sensitivity and pancreatic beta-cell function in obese men. Diabetologia 47, 204–213.

Lewis, G.F., Carpentier, A., Adeli, K., Giacca, A., 2002. Disordered fat storage and mobilization in the pathogenesis of insulin resistance and type 2 diabetes. Endocr. Rev. 23, 201–229.

Li, W.W., Alexandre, S., Cao, X., Lee, A.S., 1993. Transactivation of the grp78 promoter by Ca2+ depletion. A comparative analysis with A23187 and the endoplasmic reticulum Ca(2+)-ATPase inhibitor thapsigargin. J. Biol. Chem. 268, 12003–12009.

Li, N., Frigerio, F., Maechler, P., 2008. The sensitivity of pancreatic beta-cells to mitochondrial injuries triggered by lipotoxicity and oxidative stress. Biochem. Soc. Trans. 36, 930–934.

Lim, E.L., Hollingsworth, K.G., Aribisala, B.S., Chen, M.J., Mathers, J.C., Taylor, R., 2011. Reversal of type 2 diabetes: normalisation of beta cell function in association with decreased pancreas and liver triacylglycerol. Diabetologia 54, 2506–2514.

Lin, J.H., Walter, P., Yen, T.S., 2008. Endoplasmic reticulum stress in disease pathogenesis. Annu. Rev. Pathol. 3, 399–425.

Liu, Y.Q., Jetton, T.L., Leahy, J.L., 2002. Beta-cell adaptation to insulin resistance. Increased pyruvate carboxylase and malate-pyruvate shuttle activity in islets of nondiabetic Zucker fatty rats. J. Biol. Chem. 277, 39163–39168.

Lu, P., Zhao, Z., 2018. Advances on PPARgamma research in the emerging era of precision medicine. Curr. Drug Targets 19, 663–673.

Lu, D., Mulder, H., Zhao, P., Burgess, S.C., Jensen, M.V., Kamzolova, S., Newgard, C.B., Sherry, A.D., 2002. 13C NMR isotopomer analysis reveals a connection between pyruvate cycling and glucose-stimulated insulin secretion (GSIS). Proc. Natl. Acad. Sci. U. S. A. 99, 2708–2713.

Lupi, R., Del Guerra, S., Fierabracci, V., Marselli, L., Novelli, M., Patane, G., Boggi, U., Mosca, F., Piro, S., Del Prato, S., Marchetti, P., 2002a. Lipotoxicity in human pancreatic islets and the protective effect of metformin. Diabetes 51 (Suppl. 1), S134–S137.

Lupi, R., Dotta, F., Marselli, L., Del Guerra, S., Masini, M., Santangelo, C., Patane, G., Boggi, U., Piro, S., Anello, M., Bergamini, E., Mosca, F., Di Mario, U., Del Prato, S., Marchetti, P., 2002b. Prolonged exposure to free fatty acids has cytostatic and pro-apoptotic effects on human pancreatic islets: evidence that beta-cell death is caspase mediated, partially dependent on ceramide pathway, and Bcl-2 regulated. Diabetes 51, 1437–1442.

Ly, L.D., Xu, S., Choi, S.K., Ha, C.M., Thoudam, T., Cha, S.K., Wiederkehr, A., Wollheim, C.B., Lee, I.K., Park, K.S., 2017. Oxidative stress and calcium dysregulation by palmitate in type 2 diabetes. Exp. Mol. Med. 49, E291.

Lytrivi, M., Igoillo-Esteve, M., Cnop, M., 2018. Inflammatory stress in islet beta-cells: therapeutic implications for type 2 diabetes? Curr. Opin. Pharmacol. 43, 40–45.

MacDonald, M.J., 1995. Feasibility of a mitochondrial pyruvate malate shuttle in pancreatic islets. Further implication of cytosolic NADPH in insulin secretion. J. Biol. Chem. 270, 20051–20058.

Maechler, P., Wollheim, C.B., 1999. Mitochondrial glutamate acts as a messenger in glucose-induced insulin exocytosis. Nature 402, 685–689.

Maestre, I., Jordan, J., Calvo, S., Reig, J.A., Cena, V., Soria, B., Prentki, M., Roche, E., 2003. Mitochondrial dysfunction is involved in apoptosis induced by serum withdrawal and fatty acids in the beta-cell line INS-1. Endocrinology 144, 335–345.
Malodobra-Mazur, M., Dziewulska, A., Kozinski, K., Dobrzyn, P., Kolczynska, K., Janikiewicz, J., Dobrzyn, A., 2014. Stearoyl-CoA desaturase regulates inflammatory gene expression by changing DNA methylation level in 3T3 adipocytes. Int. J. Biochem. Cell Biol. 55, 40–50.
Manukyan, L., Ubhayasekera, S.J., Bergquist, J., Sargsyan, E., Bergsten, P., 2015. Palmitate-induced impairments of beta-cell function are linked with generation of specific ceramide species via acylation of sphingosine. Endocrinology 156, 802–812.
Marselli, L., Thorne, J., Dahiya, S., Sgroi, D.C., Sharma, A., Bonner-Weir, S., Marchetti, P., Weir, G.C., 2010. Gene expression profiles of beta-cell enriched tissue obtained by laser capture microdissection from subjects with type 2 diabetes. PLoS One 5, E11499.
Martinez, S.C., Tanabe, K., Cras-Meneur, C., Abumrad, N.A., Bernal-Mizrachi, E., Permutt, M.A., 2008. Inhibition of Foxo1 protects pancreatic islet beta-cells against fatty acid and endoplasmic reticulum stress-induced apoptosis. Diabetes 57, 846–859.
Masini, M., Bugliani, M., Lupi, R., Del Guerra, S., Boggi, U., Filipponi, F., Marselli, L., Masiello, P., Marchetti, P., 2009. Autophagy in human type 2 diabetes pancreatic beta cells. Diabetologia 52, 1083–1086.
Maycox, P.R., Deckwerth, T., Hell, J.W., Jahn, R., 1988. Glutamate uptake by brain synaptic vesicles. Energy dependence of transport and functional reconstitution in proteoliposomes. J. Biol. Chem. 263, 15423–15428.
Meglasson, M.D., Matschinsky, F.M., 1986. Pancreatic islet glucose metabolism and regulation of insulin secretion. Diabetes Metab. Rev. 2, 163–214.
Merksamer, P.I., Trusina, A., Papa, F.R., 2008. Real-time redox measurements during endoplasmic reticulum stress reveal interlinked protein folding functions. Cell 135, 933–947.
Miyauchi, S., Hirasawa, A., Ichimura, A., Hara, T., Tsujimoto, G., 2010. New frontiers in gut nutrient sensor research: free fatty acid sensing in the gastrointestinal tract. J. Pharmacol. Sci. 112, 19–24.
Mollazadeh, H., Atkin, S.L., Butler, A.E., Ruscica, M., Sirtori, C.R., Sahebkar, A., 2018. The effect of statin therapy on endoplasmic reticulum stress. Pharmacol. Res. 137, 150–158.
Moon, J.S., Karunakaran, U., Suma, E., Chung, S.M., Won, K.C., 2020. The role of CD36 in type 2 diabetes mellitus: beta-cell dysfunction and beyond. Diabetes Metab. J. 44, 222–233.
Morgan, N.G., 2009. Fatty acids and beta-cell toxicity. Curr. Opin. Clin. Nutr. Metab. Care 12, 117–122.
Morgan, M.J., Liu, Z.G., 2011. Crosstalk of reactive oxygen species and NF-kappaB signaling. Cell Res. 21, 103–115.
Moulle, V.S., Ghislain, J., Poitout, V., 2017. Nutrient regulation of pancreatic beta-cell proliferation. Biochimie 143, 10–17.
Mozaffarian, D., 2016. Dietary and policy priorities for cardiovascular disease, diabetes, and obesity: a comprehensive review. Circulation 133, 187–225.
Nakamura, M.T., Yudell, B.E., Loor, J.J., 2014. Regulation of energy metabolism by long-chain fatty acids. Prog. Lipid Res. 53, 124–144.
Newgard, C.B., McGarry, J.D., 1995. Metabolic coupling factors in pancreatic beta-cell signal transduction. Annu. Rev. Biochem. 64, 689–719.
Newsholme, P., Keane, D., Welters, H.J., Morgan, N.G., 2007. Life and death decisions of the pancreatic beta-cell: the role of fatty acids. Clin. Sci. (Lond.) 112, 27–42.
Newsholme, P., Rebelato, E., Abdulkader, F., Krause, M., Carpinelli, A., Curi, R., 2012. Reactive oxygen and nitrogen species generation, antioxidant defenses, and beta-cell function: a critical role for amino acids. J. Endocrinol. 214, 11–20.

Newsholme, P., Cruzat, V.F., Keane, K.N., Carlessi, R., De Bittencourt Jr., P.I., 2016. Molecular mechanisms of ROS production and oxidative stress in diabetes. Biochem. J. 473, 4527–4550.

Ng, M. C., Shriner, D., Chen, B. H., Li, J., Chen, W. M., Guo, X., Liu, J., Bielinski, S. J., Yanek, L. R., Nalls, M. A., Comeau, M. E., Rasmussen-Torvik, L. J., Jensen, R. A., Evans, D. S., Sun, Y. V., An, P., Patel, S. R., Lu, Y., Long, J., Armstrong, L. L., Wagenknecht, L., Yang, L., Snively, B. M., Palmer, N. D., Mudgal, P., Langefeld, C. D., Keene, K. L., Freedman, B. I., Mychaleckyj, J. C., Nayak, U., Raffel, L. J., Goodarzi, M. O., Chen, Y. D., Taylor, H. A., Jr., Correa, A., Sims, M., Couper, D., Pankow, J. S., Boerwinkle, E., Adeyemo, A., Doumatey, A., Chen, G., Mathias, R. A., Vaidya, D., Singleton, A. B., Zonderman, A. B., Igo, R. P., Jr., Sedor, J. R., FIND Consortium, Kabagambe, E. K., Siscovick, D. S., Mcknight, B., Rice, K., Liu, Y., Hsueh, W. C., Zhao, W., Bielak, L. F., Kraja, A., Province, M. A., Bottinger, E. P., Gottesman, O., Cai, Q., Zheng, W., Blot, W. J., Lowe, W. L., Pacheco, J. A., Crawford, D. C., eMERGE Consortium, DIAGRAM Consortium, Grundberg, E., MuTHER Consortium, Rich, S. S., Hayes, M. G., Shu, X. O., Loos, R. J., Borecki, I. B., Peyser, P. A., Cummings, S. R., Psaty, B. M., Fornage, M., Iyengar, S. K., Evans, M. K., Becker, D. M., Kao, W. H., Wilson, J. G., Rotter, J. I., Sale, M. M., Liu, S., Rotimi, C. N., Bowden, D. W. & MEta-analysis of type 2 DIabetes in African Americans Consortium 2014. Meta-analysis of genome-wide association studies in African Americans provides insights into the genetic architecture of type 2 diabetes. PLoS Genet., 10, E1004517.

Nolan, C.J., Madiraju, M.S., Delghingaro-Augusto, V., Peyot, M.L., Prentki, M., 2006. Fatty acid signaling in the beta-cell and insulin secretion. Diabetes 55 (Suppl. 2), S16–S23.

Ntambi, J.M., Miyazaki, M., 2004. Regulation of stearoyl-CoA desaturases and role in metabolism. Prog. Lipid Res. 43, 91–104.

Oh, Y.S., 2015. Mechanistic insights into pancreatic beta-cell mass regulation by glucose and free fatty acids. Anat. Cell Biol. 48, 16–24.

Oh, Y.S., Bae, G.D., Baek, D.J., Park, E.Y., Jun, H.S., 2018. Fatty acid-induced lipotoxicity in pancreatic beta-cells during development of type 2 diabetes. Front. Endocrinol. 9, 384.

Okado-Matsumoto, A., Fridovich, I., 2001. Subcellular distribution of superoxide dismutases (SOD) in rat liver: Cu,Zn-SOD in mitochondria. J. Biol. Chem. 276, 38388–38393.

O'Neill, C.M., Lu, C., Corbin, K.L., Sharma, P.R., Dula, S.B., Carter, J.D., Ramadan, J.W., Xin, W., Lee, J.K., Nunemaker, C.S., 2013. Circulating levels of IL-1B+IL-6 cause ER stress and dysfunction in islets from prediabetic male mice. Endocrinology 154, 3077–3088.

Oshima, M., Pechberty, S., Bellini, L., Gopel, S.O., Campana, M., Rouch, C., Dairou, J., Cosentino, C., Fantuzzi, F., Toivonen, S., Marchetti, P., Magnan, C., Cnop, M., Le Stunff, H., Scharfmann, R., 2020. Stearoyl CoA desaturase is a gatekeeper that protects human beta cells against lipotoxicity and maintains their identity. Diabetologia 63, 395–409.

Oyadomari, S., Takeda, K., Takiguchi, M., Gotoh, T., Matsumoto, M., Wada, I., Akira, S., Araki, E., Mori, M., 2001. Nitric oxide-induced apoptosis in pancreatic beta cells is mediated by the endoplasmic reticulum stress pathway. Proc. Natl. Acad. Sci. U. S. A. 98, 10845–10850.

Pacher, P., Beckman, J.S., Liaudet, L., 2007. Nitric oxide and peroxynitrite in health and disease. Physiol. Rev. 87, 315–424.

Palomer, X., Pizarro-Delgado, J., Barroso, E., Vazquez-Carrera, M., 2018. Palmitic and oleic acid: the Yin and Yang of fatty acids in type 2 diabetes mellitus. Trends Endocrinol. Metab. 29, 178–190.

Paolisso, G., Gambardella, A., Amato, L., Tortoriello, R., D'amore, A., Varricchio, M., D'onofrio, F., 1995. Opposite effects of short- and long-term fatty acid infusion on insulin secretion in healthy subjects. Diabetologia 38, 1295–1299.

Patane, G., Piro, S., Rabuazzo, A.M., Anello, M., Vigneri, R., Purrello, F., 2000. Metformin restores insulin secretion altered by chronic exposure to free fatty acids or high glucose: a direct metformin effect on pancreatic beta-cells. Diabetes 49, 735–740.

Pickrell, A.M., Youle, R.J., 2015. The roles of PINK1, parkin, and mitochondrial fidelity in Parkinson's disease. Neuron 85, 257–273.

Pinto, B.A.S., Franca, L.M., Laurindo, F.R.M., Paes, A.M.A., 2019. Unfolded protein response: cause or consequence of lipid and lipoprotein metabolism disturbances? Adv. Exp. Med. Biol. 1127, 67–82.

Plaisance, V., Perret, V., Favre, D., Abderrahmani, A., Yang, J.Y., Widmann, C., Regazzi, R., 2009. Role of the transcriptional factor C/EBPbeta in free fatty acid-elicited beta-cell failure. Mol. Cell. Endocrinol. 305, 47–55.

Poitout, V., Robertson, R.P., 2002. Minireview: secondary beta-cell failure in type 2 diabetes—a convergence of glucotoxicity and lipotoxicity. Endocrinology 143, 339–342.

Poitout, V., Robertson, R.P., 2008. Glucolipotoxicity: fuel excess and beta-cell dysfunction. Endocr. Rev. 29, 351–366.

Poudyal, H., Panchal, S.K., Ward, L.C., Brown, L., 2013. Effects of ALA, EPA and DHA in high-carbohydrate, high-fat diet-induced metabolic syndrome in rats. J. Nutr. Biochem. 24, 1041–1052.

Poy, M.N., Yang, Y., Rezaei, K., Fernstrom, M.A., Lee, A.D., Kido, Y., Erickson, S.K., Najjar, S.M., 2002. CEACAM1 regulates insulin clearance in liver. Nat. Genet. 30, 270–276.

Prentki, M., 1996. New insights into pancreatic beta-cell metabolic signaling in insulin secretion. Eur. J. Endocrinol. 134, 272–286.

Prentki, M., Corkey, B.E., 1996. Are the beta-cell signaling molecules malonyl-CoA and cystolic long-chain Acyl-CoA implicated in multiple tissue defects of obesity and NIDDM? Diabetes 45, 273–283.

Prentki, M., Madiraju, S.R., 2012. Glycerolipid/free fatty acid cycle and islet beta-cell function in health, obesity and diabetes. Mol. Cell. Endocrinol. 353, 88–100.

Prentki, M., Vischer, S., Glennon, M.C., Regazzi, R., Deeney, J.T., Corkey, B.E., 1992. Malonyl-CoA and long chain Acyl-CoA esters as metabolic coupling factors in nutrient-induced insulin secretion. J. Biol. Chem. 267, 5802–5810.

Prentki, M., Joly, E., El-Assaad, W., Roduit, R., 2002. Malonyl-CoA signaling, lipid partitioning, and glucolipotoxicity: role in beta-cell adaptation and failure in the etiology of diabetes. Diabetes 51 (Suppl. 3), S405–S413.

Prentki, M., Matschinsky, F.M., Madiraju, S.R., 2013. Metabolic signaling in fuel-induced insulin secretion. Cell Metab. 18, 162–185.

Quan, W., Hur, K.Y., Lim, Y., Oh, S.H., Lee, J.C., Kim, K.H., Kim, G.H., Kim, S.W., Kim, H.L., Lee, M.K., Kim, K.W., Kim, J., Komatsu, M., Lee, M.S., 2012. Autophagy deficiency in beta cells leads to compromised unfolded protein response and progression from obesity to diabetes in mice. Diabetologia 55, 392–403.

Rhee, J.S., Betz, A., Pyott, S., Reim, K., Varoqueaux, F., Augustin, I., Hesse, D., Sudhof, T.C., Takahashi, M., Rosenmund, C., Brose, N., 2002. Beta phorbol ester- and diacylglycerol-induced augmentation of transmitter release is mediated by Munc13s and not by PKCs. Cell 108, 121–133.

Rice Bradley, B.H., 2018. Dietary fat and risk for type 2 diabetes: a review of recent research. Curr. Nutr. Rep. 7, 214–226.

Richardson, S.J., Willcox, A., Bone, A.J., Foulis, A.K., Morgan, N.G., 2009. Islet-associated macrophages in type 2 diabetes. Diabetologia 52, 1686–1688.

Riserus, U., Willett, W.C., Hu, F.B., 2009. Dietary fats and prevention of type 2 diabetes. Prog. Lipid Res. 48, 44–51.
Robson-Doucette, C.A., Sultan, S., Allister, E.M., Wikstrom, J.D., Koshkin, V., Bhattacharjee, A., Prentice, K.J., Sereda, S.B., Shirihai, O.S., Wheeler, M.B., 2011. Beta-cell uncoupling protein 2 regulates reactive oxygen species production, which influences both insulin and glucagon secretion. Diabetes 60, 2710–2719.
Roche, E., Farfari, S., Witters, L.A., Assimacopoulos-Jeannet, F., Thumelin, S., Brun, T., Corkey, B.E., Saha, A.K., Prentki, M., 1998. Long-term exposure of beta-INS cells to high glucose concentrations increases anaplerosis, lipogenesis, and lipogenic gene expression. Diabetes 47, 1086–1094.
Roduit, R., Nolan, C., Alarcon, C., Moore, P., Barbeau, A., Delghingaro-Augusto, V., Przybykowski, E., Morin, J., Masse, F., Massie, B., Ruderman, N., Rhodes, C., Poitout, V., Prentki, M., 2004. A role for the malonyl-CoA/long-chain Acyl-CoA pathway of lipid signaling in the regulation of insulin secretion in response to both fuel and nonfuel stimuli. Diabetes 53, 1007–1019.
Ruderman, N., Prentki, M., 2004. Amp kinase and malonyl-CoA: targets for therapy of the metabolic syndrome. Nat. Rev. Drug Discov. 3, 340–351.
Saitoh, Y., Chun-Ping, C., Noma, K., Ueno, H., Mizuta, M., Nakazato, M., 2008. Pioglitazone attenuates fatty acid-induced oxidative stress and apoptosis in pancreatic beta-cells. Diabetes Obes. Metab. 10, 564–573.
Sakuma, K., Yabuki, C., Maruyama, M., Abiru, A., Komatsu, H., Negoro, N., Tsujihata, Y., Takeuchi, K., Habata, Y., Mori, M., 2016. Fasiglifam (TAK-875) has dual potentiating mechanisms via Galphaq-GPR40/FFAR1 signaling branches on glucose-dependent insulin secretion. Pharmacol. Res. Perspect. 4, E00237.
Salehi, A., Flodgren, E., Nilsson, N.E., Jimenez-Feltstrom, J., Miyazaki, J., Owman, C., Olde, B., 2005. Free fatty acid receptor 1 (FFA(1)R/GPR40) and its involvement in fatty-acid-stimulated insulin secretion. Cell Tissue Res. 322, 207–215.
Salt, I.P., Johnson, G., Ashcroft, S.J., Hardie, D.G., 1998. AMP-activated protein kinase is activated by low glucose in cell lines derived from pancreatic beta cells, and may regulate insulin release. Biochem. J. 335 (Pt. 3), 533–539.
Samad, F., Hester, K.D., Yang, G., Hannun, Y.A., Bielawski, J., 2006. Altered adipose and plasma sphingolipid metabolism in obesity: a potential mechanism for cardiovascular and metabolic risk. Diabetes 55, 2579–2587.
Scharfmann, R., Pechberty, S., Hazhouz, Y., Von Bulow, M., Bricout-Neveu, E., Grenier-Godard, M., Guez, F., Rachdi, L., Lohmann, M., Czernichow, P., Ravassard, P., 2014. Development of a conditionally immortalized human pancreatic beta cell line. J. Clin. Invest. 124, 2087–2098.
Scheenen, W.J., Wollheim, C.B., Pozzan, T., Fasolato, C., 1998. Ca2+ depletion from granules inhibits exocytosis. A study with insulin-secreting cells. J. Biol. Chem. 273, 19002–19008.
Schmitz-Peiffer, C., 2002. Protein kinase C and lipid-induced insulin resistance in skeletal muscle. Ann. N. Y. Acad. Sci. 967, 146–157.
Shah, P., Ardestani, A., Dharmadhikari, G., Laue, S., Schumann, D.M., Kerr-Conte, J., Pattou, F., Klein, T., Maedler, K., 2013. The DPP-4 inhibitor linagliptin restores beta-cell function and survival in human isolated islets through GLP-1 stabilization. J. Clin. Endocrinol. Metab. 98, E1163–E1172.
Shapiro, H., Shachar, S., Sekler, I., Hershfinkel, M., Walker, M.D., 2005. Role of GPR40 in fatty acid action on the beta cell line INS-1E. Biochem. Biophys. Res. Commun. 335, 97–104.
Shera, A.S., Jawad, F., Maqsood, A., 2007. Prevalence of diabetes in Pakistan. Diabetes Res. Clin. Pract. 76, 219–222.

Shimabukuro, M., Higa, M., Zhou, Y.T., Wang, M.Y., Newgard, C.B., Unger, R.H., 1998a. Lipoapoptosis in beta-cells of obese prediabetic fa/fa rats. Role of serine palmitoyltransferase overexpression. J. Biol. Chem. 273, 32487–32490.

Shimabukuro, M., Zhou, Y.T., Levi, M., Unger, R.H., 1998b. Fatty acid-induced beta cell apoptosis: a link between obesity and diabetes. Proc. Natl. Acad. Sci. U. S. A. 95, 2498–2502.

Sicari, D., Delaunay-Moisan, A., Combettes, L., Chevet, E., Igbaria, A., 2020. A guide to assessing endoplasmic reticulum homeostasis and stress in mammalian systems. FEBS J. 287, 27–42.

Silverstein, R.L., Li, W., Park, Y.M., Rahaman, S.O., 2010. Mechanisms of cell signaling by the scavenger receptor CD36: implications in atherosclerosis and thrombosis. Trans. Am. Clin. Climatol. Assoc. 121, 206–220.

Solinas, G., Naugler, W., Galimi, F., Lee, M.S., Karin, M., 2006. Saturated fatty acids inhibit induction of insulin gene transcription by JNK-mediated phosphorylation of insulin-receptor substrates. Proc. Natl. Acad. Sci. U. S. A. 103, 16454–16459.

Spijker, H.S., Ravelli, R.B., Mommaas-Kienhuis, A.M., Van Apeldoorn, A.A., Engelse, M.A., Zaldumbide, A., Bonner-Weir, S., Rabelink, T.J., Hoeben, R.C., Clevers, H., Mummery, C.L., Carlotti, F., De Koning, E.J., 2013. Conversion of mature human beta-cells into glucagon-producing alpha-cells. Diabetes 62, 2471–2480.

Stancill, J.S., Cartailler, J.P., Clayton, H.W., O'connor, J.T., Dickerson, M.T., Dadi, P.K., Osipovich, A.B., Jacobson, D.A., Magnuson, M.A., 2017. Chronic beta-cell depolarization impairs beta-cell identity by disrupting a network of Ca(2+)-regulated genes. Diabetes 66, 2175–2187.

Steer, S.A., Scarim, A.L., Chambers, K.T., Corbett, J.A., 2006. Interleukin-1 stimulates beta-cell necrosis and release of the immunological adjuvant HMGB1. PLoS Med. 3, E17.

Stein, D.T., Esser, V., Stevenson, B.E., Lane, K.E., Whiteside, J.H., Daniels, M.B., Chen, S., Mcgarry, J.D., 1996. Essentiality of circulating fatty acids for glucose-stimulated insulin secretion in the fasted rat. J. Clin. Invest. 97, 2728–2735.

Stein, D.T., Stevenson, B.E., Chester, M.W., Basit, M., Daniels, M.B., Turley, S.D., Mcgarry, J.D., 1997. The insulinotropic potency of fatty acids is influenced profoundly by their chain length and degree of saturation. J. Clin. Invest. 100, 398–403.

Steneberg, P., Rubins, N., Bartoov-Shifman, R., Walker, M.D., Edlund, H., 2005. The FFA receptor GPR40 links hyperinsulinemia, hepatic steatosis, and impaired glucose homeostasis in mouse. Cell Metab. 1, 245–258.

Storgaard, H., Jensen, C.B., Vaag, A.A., Volund, A., Madsbad, S., 2003. Insulin secretion after short- and long-term low-grade free fatty acid infusion in men with increased risk of developing type 2 diabetes. Metabolism 52, 885–894.

Straczkowski, M., Kowalska, I., Nikolajuk, A., Dzienis-Straczkowska, S., Kinalska, I., Baranowski, M., Zendzian-Piotrowska, M., Brzezinska, Z., Gorski, J., 2004. Relationship between insulin sensitivity and sphingomyelin signaling pathway in human skeletal muscle. Diabetes 53, 1215–1221.

Summers, S.A., 2006. Ceramides in insulin resistance and lipotoxicity. Prog. Lipid Res. 45, 42–72.

Supale, S., Li, N., Brun, T., Maechler, P., 2012. Mitochondrial dysfunction in pancreatic beta cells. Trends Endocrinol. Metab. 23, 477–487.

Talchai, C., Xuan, S., Lin, H.V., Sussel, L., Accili, D., 2012. Pancreatic beta cell dedifferentiation as a mechanism of diabetic beta cell failure. Cell 150, 1223–1234.

Tang, S., Luo, F., Feng, Y.M., Wei, X., Miao, H., Lu, Y.B., Tang, Y., Ding, D.F., Jin, J.F., Zhu, Q., 2017. Neutral ceramidase secreted via exosome protects against palmitate-induced apoptosis in INS-1 cells. Exp. Clin. Endocrinol. Diabetes 125, 130–135.

Taylor, R., Al-Mrabeh, A., Zhyzhneuskaya, S., Peters, C., Barnes, A.C., Aribisala, B.S., Hollingsworth, K.G., Mathers, J.C., Sattar, N., Lean, M.E.J., 2018. Remission of human type 2 diabetes requires decrease in liver and pancreas fat content but is dependent upon capacity for beta cell recovery. Cell Metab. 28, 547–556 E3.

Tersey, S.A., Levasseur, E.M., Syed, F., Farb, T.B., Orr, K.S., Nelson, J.B., Shaw, J.L., Bokvist, K., Mather, K.J., Mirmira, R.G., 2018. Episodic beta-cell death and dedifferentiation during diet-induced obesity and dysglycemia in male mice. FASEB J. 32, Fj201800150rr.

Thomas, T.P., Pek, S.B., 1992. Diacylglycerol inhibits potassium-induced calcium influx and insulin release by a protein kinase-C-independent mechanism in HIT T-15 islet cells. Endocrinology 131, 1985–1992.

Thomas, H.E., Mckenzie, M.D., Angstetra, E., Campbell, P.D., Kay, T.W., 2009. Beta cell apoptosis in diabetes. Apoptosis 14, 1389–1404.

Tiedge, M., Lortz, S., Drinkgern, J., Lenzen, S., 1997. Relation between antioxidant enzyme gene expression and antioxidative defense status of insulin-producing cells. Diabetes 46, 1733–1742.

Unger, R.H., 1995. Lipotoxicity in the pathogenesis of obesity-dependent NIDDM. Genetic and clinical implications. Diabetes 44, 863–870.

Veret, J., Coant, N., Berdyshev, E.V., Skobeleva, A., Therville, N., Bailbe, D., Gorshkova, I., Natarajan, V., Portha, B., Le Stunff, H., 2011. Ceramide synthase 4 and de novo production of ceramides with specific N-acyl chain lengths are involved in glucolipotoxicity-induced apoptosis of INS-1 beta-cells. Biochem. J. 438, 177–189.

Victor, V.M., Rocha, M., Herance, R., Hernandez-Mijares, A., 2011. Oxidative stress and mitochondrial dysfunction in type 2 diabetes. Curr. Pharm. Des. 17, 3947–3958.

Wallin, T., Ma, Z., Ogata, H., Jorgensen, I.H., Iezzi, M., Wang, H., Wollheim, C.B., Bjorklund, A., 2010. Facilitation of fatty acid uptake by CD36 in insulin-producing cells reduces fatty-acid-induced insulin secretion and glucose regulation of fatty acid oxidation. Biochim. Biophys. Acta 1801, 191–197.

Wang, X., Zhou, L., Li, G., Luo, T., Gu, Y., Qian, L., Fu, X., Li, F., Li, J., Luo, M., 2007. Palmitate activates AMP-activated protein kinase and regulates insulin secretion from beta cells. Biochem. Biophys. Res. Commun. 352, 463–468.

Weir, G.C., 2020. Glucolipotoxicity, beta-cells, and diabetes: the emperor has no clothes. Diabetes 69, 273–278.

Wigger, L., Cruciani-Guglielmacci, C., Nicolas, A., Denom, J., Fernandez, N., Fumeron, F., Marques-Vidal, P., Ktorza, A., Kramer, W., Schulte, A., Le Stunff, H., Liechti, R., Xenarios, I., Vollenweider, P., Waeber, G., Uphues, I., Roussel, R., Magnan, C., Ibberson, M., Thorens, B., 2017. Plasma dihydroceramides are diabetes susceptibility biomarker candidates in mice and humans. Cell Rep. 18, 2269–2279.

Wu, J.J., Quijano, C., Chen, E., Liu, H., Cao, L., Fergusson, M.M., Rovira, I.I., Gutkind, S., Daniels, M.P., Komatsu, M., Finkel, T., 2009. Mitochondrial dysfunction and oxidative stress mediate the physiological impairment induced by the disruption of autophagy. Aging (Albany NY) 1, 425–437.

Xiao, C., Giacca, A., Lewis, G.F., 2011. Sodium phenylbutyrate, a drug with known capacity to reduce endoplasmic reticulum stress, partially alleviates lipid-induced insulin resistance and beta-cell dysfunction in humans. Diabetes 60, 918–924.

Yajima, H., Komatsu, M., Yamada, S., Straub, S.G., Kaneko, T., Sato, Y., Yamauchi, K., Hashizume, K., Sharp, G.W., Aizawa, T., 2000. Cerulenin, an inhibitor of protein acylation, selectively attenuates nutrient stimulation of insulin release: a study in rat pancreatic islets. Diabetes 49, 712–717.

Yamamoto, S., Kuramoto, K., Wang, N., Situ, X., Priyadarshini, M., Zhang, W., Cordoba-Chacon, J., Layden, B.T., He, C., 2018. Autophagy differentially regulates insulin production and insulin sensitivity. Cell Rep. 23, 3286–3299.

Yaney, G.C., Korchak, H.M., Corkey, B.E., 2000. Long-chain acyl CoA regulation of protein kinase C and fatty acid potentiation of glucose-stimulated insulin secretion in clonal beta-cells. Endocrinology 141, 1989–1998.

Yang, W., Tanaka, Y., Bundo, M., Hirokawa, N., 2014. Antioxidant signaling involving the microtubule motor KIF12 is an intracellular target of nutrition excess in beta cells. Dev. Cell 31, 202–214.

Yaribeygi, H., Bo, S., Ruscica, M., Sahebkar, A., 2020. Ceramides and diabetes mellitus: an update on the potential molecular relationships. Diabet. Med. 37, 11–19.

Yu, C., Chen, Y., Cline, G.W., Zhang, D., Zong, H., Wang, Y., Bergeron, R., Kim, J.K., Cushman, S.W., Cooney, G.J., Atcheson, B., White, M.F., Kraegen, E.W., Shulman, G.I., 2002. Mechanism by which fatty acids inhibit insulin activation of insulin receptor substrate-1 (IRS-1)-associated phosphatidylinositol 3-kinase activity in muscle. J. Biol. Chem. 277, 50230–50236.

Yu, L., Mcphee, C.K., Zheng, L., Mardones, G.A., Rong, Y., Peng, J., Mi, N., Zhao, Y., Liu, Z., Wan, F., Hailey, D.W., Oorschot, V., Klumperman, J., Baehrecke, E.H., Lenardo, M.J., 2010. Termination of autophagy and reformation of lysosomes regulated by mTOR. Nature 465, 942–946.

Yuzefovych, L., Wilson, G., Rachek, L., 2010. Different effects of oleate vs. palmitate on mitochondrial function, apoptosis, and insulin signaling in L6 skeletal muscle cells: role of oxidative stress. Am. J. Physiol. Endocrinol. Metab. 299, E1096–E1105.

Zhou, J., Massey, S., Story, D., Li, L., 2018. Metformin: an old drug with new applications. Int. J. Mol. Sci. 19, 2863.

Zhu, Q., Zhong, J.J., Jin, J.F., Yin, X.M., Miao, H., 2013. Tauroursodeoxycholate, a chemical chaperone, prevents palmitate-induced apoptosis in pancreatic beta-cells by reducing ER stress. Exp. Clin. Endocrinol. Diabetes 121, 43–47.

Zhyzhneuskaya, S.V., Al-Mrabeh, A., Peters, C., Barnes, A., Aribisala, B., Hollingsworth, K.G., Mcconnachie, A., Sattar, N., Lean, M.E.J., Taylor, R., 2020. Time course of normalization of functional beta-cell capacity in the diabetes remission clinical trial after weight loss in type 2 diabetes. Diabetes Care 43, 813–820.

Zummo, F.P., Cullen, K.S., Honkanen-Scott, M., Shaw, J.A.M., Lovat, P.E., Arden, C., 2017. Glucagon-like peptide 1 protects pancreatic beta-cells from death by increasing autophagic flux and restoring lysosomal function. Diabetes 66, 1272–1285.

Further reading

World Health Organization, 2020. Obesity and Overweight. World Health Organization, Geneva, Switzerland.

FSC
www.fsc.org
MIX
Papier | Fördert gute Waldnutzung
FSC® C083411

Kolibri 360 GmbH
Ferdinand-Jühlke-Straße 7
99095 Erfurt, Deutschland
produktsicherheit@kolibri360.de